Lecture Notes in Mathematics

Editors:
A. Dold, Heidelberg
F. Takens, Groningen
B. Teissier, Paris

Springer
Berlin
Heidelberg
New York
Barcelona
Hong Kong
London
Milan
Paris
Singapore
Tokyo

Vassili N. Kolokoltsov

Semiclassical Analysis for Diffusions and Stochastic Processes

Springer

Author

Vassili N. Kolokoltsov
Department of Mathematics,
Statistics & Operational Research
The Nottingham Trent University
Burton Street
Nottingham NG1 4BU, United Kingdom

E-mail: vk@maths.ntu.ac.uk

Cataloging-in-Publication Data applied for

Die Deutsche Bibliothek - CIP-Einheitsaufnahme

Kolokoltsov, Vassili N.:
Semiclassical analysis for diffusions and stochastic processes / Vassili N. Kolokoltsov. -
Berlin ; Heidelberg ; New York ; Barcelona ; Hong Kong ; London ;
Milan ; Paris ; Singapore ; Tokyo : Springer, 2000
(Lecture notes in mathematics ; 1724)
ISBN 3-540-66972-8

Mathematics Subject Classification (2000): Primary: 35K05, 60E07, 60F10,
60J35, 60J60, 60J75, 81Q20, 81S40 Secondary: 34B15, 49J55, 60G15, 60G17,
70H20, 81P15

ISSN 0075-8434
ISBN 3-540-66972-8 Springer-Verlag Berlin Heidelberg New York

Springer-Verlag is a company in the BertelsmannSpringer publishing group.
© Springer-Verlag Berlin Heidelberg 2000
Printed in Germany

Typesetting: Camera-ready TeX output by the author
Printed on acid-free paper   SPIN: 10700351   41/3143/du   543210

To my teachers V.P. Maslov and A.M. Stepin

Preface

This monograph is devoted mainly to the analytical study of the differential, pseudo-differential and stochastic evolution equations describing the transition probabilities of various Markov random processes. They include (i) diffusions (in particular, degenerate diffusions), (ii) more general jump-diffusions, especially stable jump-diffusions driven by stable noise or stable Lévy processes, which are becoming more and more popular in modelling various phenomena in science, engineering and economics, (iii) complex stochastic Schrödinger equations which correspond to models of quantum open systems which have been extensively studied recently. The main results of the book concern the existence, qualitative properties, two-sided estimates, path integral representation, and small time and semiclassical asymptotics for the Green function (or fundamental solution) of these equations, which represent the transition probability densities of the corresponding random processes. Applications to the theory of large deviations and to the sample path properties of random trajectories are presented. The proofs of the main theorem require some auxiliary results from other areas, which seem to be of independent interest. For example, a special chapter is devoted to the study of the boundary value problem for Hamiltonian systems which constitute the "classical part" of the semiclassical approximation. Some relevant topics in spectral asymptotics are also discussed. Most of the results of the book are new.

The background necessary for reading the book has been reduced to a minimum and consists of an elementary knowledge of probability, complex and functional analysis, and calculus. The parts dealing with stochastic equations can be omitted by those not acquainted with stochastic analysis.

As a short guide to the content of the book let me indicate that it can be conditionally divided into the following parts (which are strongly related idealogically but are very weakly dependent formally): (i) asymptotics for diffusion Sect. 1.1-1.3, 2.1-2.4, Ch. 3,4 (ii) asymptotics for stable jump-diffusions Sect. 2.1, 2.5, Ch. 5,6, Ap. B-F, (iii) asymptotics for complex stochastic Schrödinger and diffusion equations Sect. 1.4,1.5,2.1, 2.6,2.7, Ch. 7, (iv) miscellaneous topics including spectral asymptotics and path integral representation for the Schrödinger and diffusion equation Ch. 8, 9, Ap. G,H.

I am grateful to many people for fruitful discussions, especially to S. Albeverio, D.B. Applebaum, V.P. Belavkin, A.M. Chebotarev, Z. Coelho, S. Dobrokhotov, K.D. Elworthy, V. Gavrilenko, A. Hilbert, R.L. Hudson, N. Jacob, V.P. Maslov, R.L. Schilling, O.G. Smolyanov, A. Truman and A. Tyukov. Let me mention with special gratitude R.L. Hudson and R.L. Schilling for reading the manuscript and making lots of useful comments. I am thankful for support to all members of my big family.

V.N. Kolokoltsov, September 1999

CONTENTS

Introduction

We present first the context and motivation for the present study, and then discuss in more detail the content of the book and main results. Let us recall the main connections between the theory of stochastic processes, evolutionary pseudo-differential equations (ΨDE), and positivity preserving semigroups; for a fuller discussion see Appendix C,D and the references given there, for instance, [Ja]. By definition, a Feller semigroup is a strongly continuous semigroup T_t, $t \geq 0$, of linear contractions on the Banach space $C_0(\mathcal{R}^d)$ of continuous function on $\mathcal{R}^d$ vanishing at infinity such that, for $u \in C_0(\mathcal{R}^d)$, $0 \leq u \leq 1$ implies $0 \leq T_t u \leq 1$ for all t. It follows that T_t can be extended by continuity and monotonicity to all bounded continuous functions and this extension does not increase constants. Conversely, one readily sees that if a strongly continuous semigroup of linear operators on the space of bounded continuous functions preserves positivity (i.e. $u \geq 0$ implies $T_t u \geq 0$), does not increase constants and takes $C_0(\mathcal{R}^d)$ to itself, then T_t is a Feller semigroup. An important result of probability theory states that to each Feller semigroup there corresponds a Markov process $X(t,x)$ (here t is the time and x is the initial point), taking values in $\mathcal{R}^d$ and defined on some probability space, such that $(T_t u)(x)$ is the expectation of the random function $u(X(t,x))$ at any time t. Processes corresponding in this way to Feller semigroups are called Feller processes. It follows that if L is the generator of the semigroup T_t (i.e. L is a linear operator defined on a dense subspace of $C_0(\mathcal{R}^d)$ such that $Lu = \lim_{t \to 0}((T_t u - u)/t)$ on this subspace), then the Green function, or, in an alternative terminology, the fundamental solution, $u_G(t, x, x_0)$, of the Cauchy problem for the evolutionary equation

$$\frac{\partial u}{\partial t} = Lu, \quad t \geq 0 \tag{0.1}$$

(i.e. the solution to this equation with the Dirac initial data $u|_{t=0} = \delta(x - x_0)$), which can in general be defined only in the sense of distributions, gives the transition probability (from x to x_0 in time t) of the process X. Thus, if the Green function exists as a continuous (or even only measurable) function, then it coincides with the transition probability density of the corresponding process, which implies in particular that the transition probabilities of this process are absolutely continuous with respect to Lebesgue measure.

A Feller process is called space-homogeneous if its transition probabilities from x to x_0 in any time t depend only on the difference $x - x_0$. Such processes are also called processes with independent identically distributed increments. Then the corresponding semigroup has the property that all T_t commute with space translations, i.e. $T_t \theta_a = \theta_a T_t$ for all t and a, where $\theta_a u(x) = u(x+a)$. The famous Lévy-Khintchine theorem states that the generator of a general space-homogeneous Feller semigroup is an integro-differential operator of the form

$$(Lu)(x) = \left(A, \frac{\partial u}{\partial x} \right)$$

$$+\frac{1}{2}\, tr\,\left(G\frac{\partial^2 u}{\partial x^2}\right) + \int_{\mathcal{R}^d\setminus\{0\}}\left(u(x+\xi)-u(x)-\frac{(\frac{\partial u}{\partial x},\xi)}{1+|\xi|^2}\right)\nu(d\xi), \qquad (0.2)$$

where $\nu(d\xi)$ is a Lévy measure on $\mathcal{R}^d\setminus\{0\}$, i.e. a sigma-finite Borel measure such that the integral $\int \min(1,|\xi|^2)\,\nu(d\xi)$ is finite, $A = \{A^j\}_{j=1}^d$ is a vector (called the drift) and $G = \{G_{ij}\}$ is a non-negative (non-negative definite) matrix (called the matrix of diffusion coefficients). In less concise notations, we may write

$$\left(A,\frac{\partial u}{\partial x}\right) = A^j\frac{\partial u}{\partial x^j}, \quad tr\,\left(G\frac{\partial^2 u}{\partial x^2}\right) = G_{ij}\frac{\partial^2 u}{\partial x^i \partial x^j}$$

in (0.2) and in all formulas that follow. Here (and everywhere in the book) the summation convention over repeated suffices is in force. Moreover, the general form of the (pseudo-differential) generators of positivity preserving space-homogeneous semigroups is very similar, namely, they have the form $L-a$, where L is as above and a is a non-negative constant. The most important classes of space-homogeneous Feller process are Gaussian diffusions (when $\nu = 0$ in (0.2)), compound Poisson processes (when $G = 0$, $A = 0$, and the measure ν is finite), and stable non-Gaussian Lévy motions of index $\alpha \in (0,2)$ (when $G = 0$ and $\nu(d\xi) = |\xi|^{-(1+\alpha)}d|\xi|\mu(ds)$ where $s = \xi/|\xi|$ and μ is an arbitrary finite measure on the sphere S^{d-1}).

The fundamental theorem of Courrège states that under the natural assumption that all infinitely differentiable functions with compact support belong to the domain of L, and for any such function f the function Lf is continuous, the generator L of a general not necessarily space-homogeneous Feller semigroup has form (0.2) but with variable coefficients, i.e.

$$(Lu)(x) = \left(A(x),\frac{\partial u}{\partial x}\right)$$

$$+\frac{1}{2}\, tr\,\left(G(x)\frac{\partial^2 u}{\partial x^2}\right) + \int_{\mathcal{R}^d\setminus\{0\}}\left(u(x+\xi)-u(x)-\frac{(\frac{\partial u}{\partial x},\xi)}{1+|\xi|^2}\right)\nu(x,d\xi), \qquad (0.3)$$

where A, G and ν depend measurably on x. However that this is only a necessary condition, and the problem of when operators of form (0.3) do in fact define a Feller semigroup is in general quite non-trivial.

The connection with the theory of ΨDO comes from a simple observation that operator (0.3) is a pseudo-differential operator that can be written in the pseudo-differential form as follows:

$$(Lu)(x) = \Psi(x,-i\Delta)u, \qquad (0.3')$$

where

$$\Psi(x,p) = i(A(x),p) - \frac{1}{2}(G(x)p,p) + \int_{\mathcal{R}^d\setminus\{0\}}\left(e^{i(p,\xi)}-1-\frac{i(p,\xi)}{1+\xi^2}\right)\nu(x,d\xi).$$

We shall call processes (and the corresponding semigroups) with generators of form (0.3) Feller-Courrège processes (resp. semigroups) or jump-diffusions, because the trajectories of such processes need not be continuous, as in the case of a diffusion, but generally have jumps. The theory of processes with generators of form (0.3) is currently rapidly developing, and different authors use different names for them. Sometimes, these processes are called Feller processes with pseudo-differential generators, sometimes they are called the Lévy-type processes, or diffusions with jumps, see e.g. [Ja], [Ho], [Schi], [JS], [RY], and references therein. The generator of the general positivity preserving semigroup has the form $L - a(x)$ with L of form (0.3) and $a(x)$ a non-negative function of x.

As was noted above, the solution to the Cauchy problem for equation (0.1), (0.3) with the initial function $u(x)$ is given by the formula $E(u(X(t,x)))$, where E denotes the expectation with respect to the measure of the corresponding Feller process X. The solution to the Cauchy problem for the corresponding more general positivity preserving equation

$$\frac{\partial u}{\partial t} = (L - a(x))u, \quad t \geq 0, \tag{0.4}$$

can be then expressed by the Feynmann-Kac formula

$$E\left(\exp\{-\int_0^t a(X(s,x))\,ds\}u(X(t,x))\right)$$

(see Appendix G, where a more general complex version of this formula is discussed). The exploitation of such probabilistic formulae for the solution to the Cauchy problem of equation (0.1), (0.3) constitute the probabilistic approach to the study of positivity preserving pseudo-differential equations. However, in this book we shall study such equations analytically assuming some smoothness assumptions on G, A, ν. We will mainly consider the following classes of such equations:

(i) diffusions (when $\nu = 0$) including the case of degenerate diffusions, (i.e. G is degenerate); (ii) stable jump-diffusions (i.e. $G = 0$ and $\nu(x, d\xi) = |\xi|^{-(1+\alpha)} d|\xi| \mu(x, ds)$) and their natural generalisations, stable-like diffusions that differ in that the index α is not a constant but also depends on x, and (iii) the combinations of these processes and their perturbations by compound Poisson processes, especially the truncated (or localised) stable jump- diffusions with the Lévy measure $\nu(d\xi) = \Theta_a(|\xi|)|\xi|^{-(1+\alpha)} d|\xi| \mu(x, ds)$. The behaviour of the latter processes is similar to stable jump-diffusions locally, but differs essentially at large distances or times.

The study of stable processes and their generalisations is motivated by the ever-increasing use of these processes in modelling many processes in engineering, natural sciences and economics (see e.g [ST], [Zo], [KSZ]). In particular, they are widely used in plasma physics and astronomy (see e.g. [Lis] or [Cha]).

The discussion above concerned the concept of pointwise positivity. In quantum physics and non-commutative analysis, a more general notion of positivity is developed (see e.g. [ApB], [Da3], [Li], [LP], [AH3], [AHO], [Be5] for the description and discussion of quantum positivity preserving mappings). The simplest examples of the corresponding equations give the second order partial differential equations with complex coefficients, which we shall call complex diffusion equations or complex Schrödinger equations, because they include standard diffusion and Schrödinger equations as particular cases and behave in many ways like these particular cases. Generally, the corresponding positivity preserving semigroups can be described by means of quantum stochastic equations. We shall also study an important class of these more general models, which we call complex stochastic diffusions. In physics, the corresponding stochastic equations are often called stochastic Schrödinger equations (SSE) and appear now to be the central objects in the study of quantum open systems (see e.g. [BHH], [QO], and Appendix A). The simplest example of a SSE has the form

$$d\psi = (\frac{1}{2}G\Delta\psi - V(x)\psi - \frac{1}{2}|c(x)|^2\psi)\,dt + c(x)\psi\,dB, \qquad (0.5)$$

where dB denotes the Ito differential of the Brownian motion B, G is a nondegenerate complex matrix with non-negative real part, and $c(x), V(x)$ are complex functions, the real part of V being non-negative. This equation can be formally written in form (0.4), (0.3) with $\nu = 0$, $A = 0$, G complex, and a time dependent complex random $a(x)$ (which makes sense in terms of distributions). Even without a stochastic term complex diffusions or complex Schrödinger equations of type (0.5) (with complex G and/or V) have many applications, see e.g. [Berr] or [BD].

We shall be mainly concerned with the properties of the Green functions (or transition probability densities) for the equations described above, in particular with their small time asymptotics and two-sided estimates. Our main purpose is to develop the method of semiclassical approximation (WKB method) for these Green functions. Semiclassical approximation, which in quantum mechanics means asymptotics with respect to small Planck constant, in probability means the asymptotics with respect to small amplitude of jumps, called also small diffusion asymptotics. Formally, it means (see also Appendix D) that instead of equation (0.1), (0.3)-(0.3') one considers the equation

$$h\frac{\partial u}{\partial t} = \Psi(x, -ih\Delta)u,$$

or, in integrao-differential form,

$$\frac{\partial u}{\partial t} = \left(A(x), \frac{\partial u}{\partial x}\right)$$

$$+\frac{h}{2}\,tr\left(G(x)\frac{\partial^2 u}{\partial x^2}\right) + \int_{\mathcal{R}^d\backslash\{0\}}\left(\frac{u(x+h\xi) - u(x)}{h} - \frac{(\frac{\partial u}{\partial x}, \xi)}{1+|\xi|^2}\right)\nu(x, d\xi), \qquad (0.6)$$

and looks for the asymptotics of its solutions as $h \to 0$. Since the solutions of (0.6) can be expressed in terms of an infinite dimensional integral (by the

Feynman-Kac formula), a search for the small time or small h asymptotics for these solutions can be considered as the study of an infinite dimensional Laplace method, in other words, as the study of the asymptotic expansions of certain infinite dimensional integrals (over a suitable path space) of Laplace type.

The construction and investigation of the Green functions (or fundamental solutions) for equation (0.1) with a pseudo-differential operator (ΨDO) L constitutes one of the central problem in the theory of linear evolutionary differential or pseudo-differential equations (ΨDE). This is because, by linearity, the solution of the Cauchy problem for equation (0.1) with arbitrary initial function can be expressed by the convolution of this initial function with the Green function. In particular, the Green function of the ΨDE (0.1) defines completely the corresponding semigroup $T_t = e^{tL}$. Moreover, in the case of a positivity preserving semigroup, the Green function defines the finite dimensional distributions for the corresponding Markov process, which, by the celebrated Kolmogorov reconstruction theorem, define this Markov process uniquely up to a natural equivalence. The study of the Green functions of evolutionary, and particularly parabolic, ΨDE utilises different techniques. It has a long history and remains a field of intensive mathematical investigations. Let us now review the principle relevant results on the Green function for equations (0.5), (0.6) and its asymptotics.

The simplest, and the most studied equation of type (0.6) is the second-order parabolic differential equation

$$\frac{\partial u}{\partial t} = \frac{1}{2}htr\left(G(x)\frac{\partial^2 u}{\partial x^2}\right) + \left(A(x), \frac{\partial u}{\partial x}\right), \qquad (0.7)$$

called also the diffusion equation, or the heat conduction equation. In the latter terminology, its fundamental solution is often referred to as to the heat kernel. Equation (0.7) is called uniformly parabolic if

$$\Lambda^{-1}|\xi|^2 \le G_{ij}(x)\xi^i\xi^j \le \Lambda|\xi|^2 \iff \Lambda^{-1} \le G(x) \le \Lambda \qquad (0.8)$$

uniformly for all ξ and x. The existence of the Green function for a uniformly parabolic equation (0.7) with bounded and uniformly Hölder continuous coefficients $G(x), A(x)$ is extensively presented in the literature, see e.g., [IKO], [LSU]. This result is obtained by the classical Lévy method, which reduces (using Duhamel's principle) the construction of the Green function to the solution of a certain integral equation, which in turn can be solved by the regular perturbation theory. This method also provides the estimate

$$0 \le u_G(t, x, x_0, h) \le C u_2(t, x, x_0, h; a) \qquad (0.9)$$

for the Green function in the domain $\{t \in (0, T], x, x_0 \in \mathcal{R}^d\}$, where C and a are constants depending on T, d, h, Λ and the maximum of the amplitudes and Hölder constants of all coefficients, and where

$$u_2(t, x, x_0, h; a) = (2\pi aht)^{-d/2}\exp\{-\frac{|x - x_0|^2}{2aht}\} \qquad (0.10)$$

is the solution of equation (0.7) with the matrix $G = a1_d$ constant multiple of the identity, and with vanishing A (notice that the index 2 in our notation u_2 stands for the second order). This result implies, in particular, the well-posedness of the Cauchy problem for the uniformly parabolic equation (0.7). For the equations of divergence form

$$\frac{\partial u}{\partial t} = \frac{1}{2} h \frac{\partial}{\partial x^i} \left(G_{ij}(x) \frac{\partial u}{\partial x^j} \right) \tag{0.11}$$

more powerful estimates were obtained by D. Aronson. Namely, it was proved in [Aro1], [Aro2] that under condition (0.8),

$$C^{-1} u_2(t, x, x_0, h; a_1) \le u_G(t, x, x_0, h) \le C u_2(t, x, x_0, h; a_2) \tag{0.12}$$

uniformly for all $t > 0, x, x_0$ with C, a_1, and a_2 depending only on d, Λ and h but not on the Hölder constants, and are valid for all times. Various proofs of this result are available now, see e.g. [PE],[Da1] or [NS] and references therein. The estimate (0.12) is closely connected with (in a sense, it is equivalent to) the famous Harnack inequality for the positive solutions of (0.11) obtained first by Moser in [Mos1], [Mos2], and it generalises the previously obtained Nash inequalities [Nas]. Estimates of the form (0.12) can only partially be extended to general uniformly parabolic equations (0.7), namely under some additional assumptions on the coefficients, and only for finite times.

If the non-negative matrix G is degenerate at some (or all) points, equations of the form (0.7) are called degenerate parabolic equations. To analyse these equations, it is sometimes convenient to rewrite them in a slightly different form. Recall that vector fields in $\mathcal{R}^d$ are just first-order differential operators of the form $f(x) \frac{\partial}{\partial x}$ with some function f. Clearly, for given G and A one can find vector fields $X_0, X_1, ... X_k$ with some k such that equation (0.7) can be written in the form

$$\frac{\partial u}{\partial t} = h \cdot (X_1^2 + X_2^2 + ... + X_k^2) u + X_0 u, \tag{0.13}$$

which is called the Hörmander form (or sum of squares of vector fields form) of a second-order parabolic equation. For example, one can take $k = d$ and $X_j = a_{jk}(x) \frac{\partial}{\partial x^k}$ for $j = 1, .., d$, where the matrix $\{a_{ij}\}$ is the non-negative symmetric square root of the matrix G. In this book, we shall not use this form of the degenerate equation (0.7), but it is important to mention it, because of the large number of investigations carried out for equation (0.13). The convenience of this form for the use of probabilistic methods lies in the fact that the evolution of the diffusion process defined by equation (0.13) can be described directly by the stochastic differential equation

$$dx = X_0(x) \, dt + \sqrt{h} \sum_{i=1}^{k} X_i(x) \, d_S W_i(t),$$

where $d_S W_i$ denote the Stratonovich differential of the standard Wiener process. To formulate the main results on the existence of the Green function for degenerate equations (0.13), one needs some notions from differential geometry. For a point $x \in \mathcal{R}^d$, denote by T_x the tangent space to $\mathcal{R}^d$ at the point x, and for any linear space V of vector fields in $\mathcal{R}^d$, denote by V_x the subspace of T_x consisting of the evaluations at x of the vector fields in V. Clearly, if the condition of uniform parabolicity (0.8) is fulfilled, the vectors $X_1, ..., X_k$ in the representation (0.13) of equation (0.7) generate a linear space of maximal dimension d at each point. The simplest generalisation of this condition is the following. Suppose that the vector fields X_j, $j = 0, ..., k$, are infinitely smooth. Let $LA(X_1, ..., X_k)$ be the Lie algebra of the vector fields generated by $X_1, ..., X_k$. One says that equation (0.13) satisfies the strong Hörmander condition, if at each point x, the vector space $LA(X_1, ..., X_k)_x$ has the maximal dimension d. It was shown in [Hor] that if equation (0.13) satisfies the strong Hörmander condition, then it has an infinitely smooth Green functions. This result was generalised in [IK], where a complete description was given of infinitely smooth equations (0.13) having smooth Green function. Let $Id = Id(X_0; X_1, ..., X_k)$ be the ideal in the Lie algebra $LA(X_0, ..., X_k)$ generated by the vector fields $X_1, ..., X_k$. It was proved in [IK] that if the space Id_x has maximal dimension d at any point x, then equation (0.13) has an infinitely smooth Green function. The attempt to obtain a probabilistic proof of Hörmander's result gave rise to a new powerful tool of probability theory: the Malliavin calculus [Mal] (for a full exposition see [Bel] or [Nu]).

Notice that form (0.13) of equation (0.7) is invariant, and therefore the results of [Hor] and [IK] can be formulated naturally for heat conduction equations of form (0.13) on manifolds. For a Riemannian manifold the most natural second-order operator is the Laplace-Beltrami operator. One can show that for any strongly parabolic equation (0.13) on a compact manifold M, there exists a Riemannian metric on M such that the operator $X_1^2 + ... + X_k^2$ is the Laplace-Beltrami operator for the corresponding Riemannian structure. It turns out (see e.g. [Bis], [SC]) that equations of form (0.13) on a smooth manifold, which are not strongly parabolic, but satisfy the strong Hörmander condition, similarly define a certain generalisation of Riemannian geometry, which is called hypoelliptic geometry, with the corresponding distance on M being called semi-Riemannian. Some estimates of Aronson type (0.12) for the diffusions on manifolds satisfying the strong Hörmander condition can be found in [JS1], [KS], [Le3], where powerful probabilistic techniques are used.

Turning to the discussion of the existence of the Green function for more general ΨDE (0.6) we note that, although the well-posedness of the Cauchy problem is now proved under rather general assumptions (see e.g. [Ja],[Ho] and references therein), very little is known concerning the existence and the properties of the Green functions for these ΨDE. In [Koch], the existence of the Green function (and an upper bound for it in terms of some rational expression) is obtained for ΨDE (0.1) with L being a finite sum of ΨDO with homogeneous symbols of orders $\alpha_1 > ... > \alpha_k$, $\alpha_1 > 1$, and with Hölder continuous coefficients. This includes, in particular, the case of stable jump-diffusions with index of stability $\alpha > 1$. In [Neg], [KN], the existence of a Green function was obtained for the case of stable-like processes with varying index stability and with infinitely smooth coefficients.

It is surely the case that to study the properties of Feller semigroups and the corresponding random processes, not only the existence of the Green function is important, but also the description of its behaviour in the different domains of its arguments and parameters. That is why, in the mathematical literature, much attention was given to the study of its asymptotics for large or small times and distances, as well as its small diffusion (or small h) asymptotics. From the point of view of the theory of stochastic processes, the small time asymptotics of the Green functions are connected with the local properties of the trajectories of the corresponding Markov processes, and the large time asymptotics concern the ergodic properties of these processes. In this monograph, we give little attention (apart from Chapter 1, see discussion below) to the large time asymptotics, and therefore we will not discuss this topic. Notice only that the semiclassical approximation (i.e. $h \to 0$ asymptotics) considered when $t \to \infty$, or more precisely, rather non-trivial and unusual asymptotics for $t = h^{-\kappa}$, $\kappa > 0$, $t \to \infty$, can be used (see [DKM1], [DKM2] and Chapter 8) for the study of the semiclassical approximations of stationary problems.

The study of small time and small h asymptotics of the Green functions of parabolic equations began with the famous papers [Var1], [Var2], [Var3]. In particular, it was proved in [Var2] that for the Green function (or heat kernel) $u_G(t, x, x_0, h)$ of a uniformly parabolic equation (0.7), (0.8) with vanishing drift, i.e. of the equation

$$\frac{\partial u}{\partial t} = \frac{1}{2} h G_{ij}(x) \frac{\partial^2 u}{\partial x_i \partial x_j} \tag{0.14}$$

one has

$$\lim_{t \to 0} [-t \log u_G(t, x, x_0, h)] = d^2(x, x_0)/2h, \tag{0.15}$$

where the function $d(x, x_0)$ is the distance in $\mathcal{R}^d$ corresponding to the Riemannian structure given by the quadratic form $(G(x))_{ij}^{-1} dx^i dx^j$ in $\mathcal{R}^d$. This result is closely connected with a similar result from [Var2] on the boundary value problem, which states that if D is a region with the boundary ∂D, and if $\phi(x, \lambda)$ is the solution to the stationary equation

$$\frac{1}{2} G_{ij}(x) \frac{\partial^2 \phi}{\partial x_i \partial x_j} = \lambda \phi$$

in D with $\lambda > 0$ and the boundary condition $\phi = 1$ on ∂D, then

$$\lim_{\lambda \to \infty} \left[-(2\lambda)^{-1/2} \log \phi(x, \lambda) \right] = d(x, \partial D).$$

We shall not in this book exploit this connection of the asymptotics of the Cauchy problem with boundary-value problems (see e.g. [DF1], [DF2] and references therein for some recent results in this direction). As was noted in [Var3], it easily follows from (0.15) that the Green function $u_G(t, x, x_0)$ of uniformly parabolic equation (0.14) also satisfies the following "large deviation principle" (locally, i.e. for small t):

$$\lim_{h \to 0}[-h \log u_G(t, x, x_0, h)] = d^2(x, x_0)/2t. \qquad (0.16)$$

A technique for proving formulas of type (0.15), (0.16) was also developed in [Bo], [FW]. One can find in these papers, and in [Var1]-[Var3], some discussion on the significance of asymptotic formulas (0.15), (0.16) for large deviations and other problems of probability theory.

Formulas (0.15), (0.16) give only logarithmic asymptotics of the Green function for small t and h (and only in the case of vanishing drift). Later the complete small time asymptotic expansions for the heat kernel of general uniformly parabolic equations (0.7) were intensively studied, especially locally, i.e. for small $|x - x_0|$. Different expositions can be found in many papers, see e.g. [Mol], [Kana], [Az2], [Ue] (resp. [Ki], [MC1]) for small t (resp. small h) asymptotics. The outcome of these investigations can be stated as follows. For $t \le t_0$ with some $t_0 > 0$ and bounded positive h, the Green function $u_G(t, x, x_0, h)$ of equation (0.7), (0.8) (or even the more general equation (0.22) below) whose coefficients together with their derivatives are uniformly bounded (in fact, this condition can be weakened) can be written in the form

$$(2\pi h t)^{-d/2}\chi(|x - x_0|)\phi(t, x, x_0, h) \exp\{-S(t, x, x_0)/h\} + O(\exp\{-\frac{\Omega}{th}\}), \qquad (0.17)$$

where $\chi(v)$ is a smooth mollifier (or cut-off function), which equals one for small v and vanishes outside a neighbourhood of the origin, the remainder $O(\exp\{-\frac{\Omega}{th}\})$ is a uniformly bounded function decreasing exponentially in x when $x \to \infty$. Here the phase S is the solution to a certain problem in the calculus of variations, has the form

$$S(t, x, x_0) = \frac{d^2(x, x_0)}{2t} + O(|x - x_0|) + O(t),$$

and satisfies a certain first order partial differential equation (Hamilton-Jacobi equation). The amplitude ϕ has the form

$$\phi(t, x, x_0, h) = \phi(t, x)(1 + O(th))$$

or more precisely

$$\phi(t, x, x_0, h) = \phi(t, x)(1 + \sum_{j=1}^{k} \phi_j(t, x, x_0)h^j + O(h^{k+1}t^{k+1})), \qquad (0.18)$$

for any $k \in \mathcal{N}$, with ϕ and all ϕ_j having regular asymptotic expansions (in positive integer powers of t) for small times t. In fact, though all the ingredients of formula (0.17) are well known (see previously cited papers and [M1]-[M3], [DKM1]) its full proof (including the estimates for possibly focal large x) seems not to exist in the literature and will be given in Chapter 3 of this book.

Notice that asymptotics of type (0.17), (0.18) are called multiplicative, because the remainder enters this formulas multiplicatively, unlike the case of the WKB asymptotics (see e.g. [MF1]) for the Schrödinger equation

$$ih\frac{\partial \psi}{\partial t} = -\frac{h^2}{2}\Delta \psi + V(x)\psi.$$

Namely, the WKB asymptotics (or semiclassical approximations) of the solutions to this equation are usually written in the form

$$\psi(t, x) = \phi(t, x)\exp\{iS(t, x)/h\} + O(h)$$

with real S, ϕ, i.e. they have the form of a rapidly oscillating function with a remainder $O(h)$ that enters this formula additively. Moreover, the estimates for this remainder are given in terms of its L^2-norm, they are not pointwise estimates, as in the case of asymptotics for parabolic equations discussed above.

As an important corollary of the small time asymptotics of the Green functions of parabolic equations, one obtains the small time asymptotic expansion of these Green functions on the diagonal $x = x_0$, which, for $h = 1$ say, has the form

$$u_G(t, x, x, 1) = (2\pi t)^{-d/2}(\sum_{j=0}^{k} a_j t^j + O(t^{k+1})) \tag{0.19}$$

for any $k \in \mathcal{N}$. If one considers a diffusion equation on a compact manifold one can integrate this expansion over x to obtain the asymptotic expansion of the trace of the Green function. The coefficients of this expansion have important geometrical interpretations. These results can be partially generalised to parabolic partial differential equations of higher order, in particular the expansion of the Green function on the diagonal $x = x_0$ (see e.g. [Gr]). The literature on the asymptotics of the trace of the Green function is very extensive (see e.g. [AJPS] and its bibliography for a discussion of the applications of these expansions in theoretical physics). In Chapter 4, we shall consider the expansions of the trace for the case of degenerate invariant diffusions on manifolds.

The asymptotic formula (0.15) was first generalised to the case of degenerate diffusions of form (0.13) with the strong Hörmander condition in [Le1], [Le2]. In that case $d(x, x_0)$ becomes the semi-Riemannian distance, mentioned above, see [Bi]. These arguments of papers [Le1], [Le2] are based on various probabilistic techniques such as stochastic differential equations, Malliavin calculus, and the theory of large deviations, for the latter in this context see e.g. [Az1]. For different results on the full small time expansions of the heat kernel of equation (0.13) with the strong Hörmander condition we refer the reader to papers [BA1], [BA2], [BAG], [BAL], [Le4] and to the references therein, see also [CMG] for the applications of these results to the study of the fine properties of hypoelliptic diffusions. In view of the large literature on the degenerate diffusion with the strong Hörmander condition, we shall not consider this type of degeneracy in this book, but will confine our attention to a different class of degenerate diffusion, which are, in a sense, uniformly degenerate.

Notice that formula (0.16) for equation (0.14) follows from (0.15) by means of a simple change of the time variable t. Moreover, one can encode formulas (0.15), (0.16) in the one formula

$$\lim_{h\to 0,t\to 0} ht\log u_G(t,x,x_0,h) = -d^2(x,x_0)/2. \qquad (0.20)$$

As follows from (0.17), similar result holds for general uniformly parabolic equations (0.7). Moreover, for these equations, the full asymptotic expansions in small h and small time coincide: expanding ϕ_j from (0.18) as a power series in t one obtains small time asymptotics, because the small in h remainder is also small in t. It turns out that this property of semiclassical asymptotics of uniformly parabolic equations is shared by a large class of degenerate diffusions which are considered in detail in this book. But this is not the case for either general degenerate diffusion or for equation (0.6) with non-vanishing Lévy measure ν. For these more general equations, the small h and small time asymptotics are different, which makes the study of the small h asymptotics essentially more difficult, even locally.

One of the main goals of this book is to generalise small time and small h asymptotics given in (0.17) for uniformly parabolic diffusion equations to more general equations of form (0.6) or (0.5).

Problems of different kind arise when one looks for global semiclassical approximations, i.e. for the asymptotics as $h \to 0$ for all (not necessarily small) t and x, x_0. In the case of Schrödinger equation, this globalisation of local semiclassical asymptotics was first carried out rigorously by V.P. Maslov (see [M1], [M4]), by means of his canonical operator method. To globalise the asymptotics of diffusion equations, and more general equations of tunnel type (see Chapter 6), Maslov proposed in [M2], [M3] a modification of his canonical operator, namely the tunnel canonical operator. However, a complete rigorous justification of the global asymptotics constructed by this method was carried out only in the particular case of the equation

$$h\frac{\partial u}{\partial t} = \frac{h^2}{2}\Delta u - V(x)u, \qquad (0.21)$$

where Δ is the standard Laplacian in $\mathcal{R}^d$ (see [DKM1], [DKM2], [KM2]). In fact, the problem with this justification lies not in globalisation, but in the justification of local (as discussed above) multiplicative small h asymptotics. Maslov's procedure [M2] leads automatically (and rigorously) to global asymptotics of tunnel equations, whenever the local asymptotics are justified. In this book, after obtaining rigorous local asymptotics of some classes of equations (0.6), we are going to develop a simplified version of Maslov's approach to obtain global semiclassical asymptotics by a simple integral formula, thus avoiding a beautiful and rather general but very sophisticated construction of the tunnel canonical operator; the latter can be found in [M3] or [DKM1].

Let us now describe in more detail the contents of the book.

The preliminary Chapter 1 deals with Gaussian diffusions, i.e. with equations of form (0.1), (0.2) with vanishing ν, and its generalisations of type (0.5).

We start with a (slightly different) exposition of the results of [CME] classifying the small time asymptotics of the Green functions of the Gaussian diffusions in terms of the matrix G. Then we present some results on the large time behaviour of Gaussian processes, in particular, estimating the escape rate of some classical Gaussian processes such as the integral of the Brownian motion or Ornstein-Uhlenbeck process, and then developing scattering theory for small perturbations of these processes on the classical (Newton systems driven by white noise) and quantum (unitary stochastic Schrödinger equation) levels. In section 1.4 we derive the exact Green function for equation (0.5) with quadratic $V(x)$ and $c(x) = x$, and then describe a remarkable property of this equation, namely the effect of Gaussianisation and localisation of its solutions, which has an important physical interpretation: it describes the spontaneous collapse of the wave-state of a quantum particle under continuous observation of its position (called also the watch-dog effect). The Gaussian case allows a direct and explicit description because the Green function in that case can be written explicitly. It can then be used to check the results of any asymptotic method developed to deal with the general case.

Since the logarithmic asymptotics $\lim_{h \to 0}(h \log u_G)$ of the Green function is expected to be expressed in terms of the solutions of of a certain Hamilton-Jacobi equation, which in its turn is expressed by means of the solutions to a boundary value problem of a corresponding Hamiltonian system, we devote Chapter 2 to the study of the boundary value problem (existence, uniqueness, and asymptotics of the solutions) for the relevant Hamiltonian systems. First, for completeness, the connection between Hamiltonian formalism and the calculus of variations is described in a compact and self-contained form in the introductory Sect. 2.1, with emphasis on the facts that are most relevant to asymptotic methods, and on the approaches that can be used for the study of degenerate Hamiltonians and singular Lagrangians. Sections 2.2-2.4 are devoted to a class of Hamiltonians depending quadratically on momenta, which are called regular, and which include degenerate as well as non-degenerate Hamiltonians. The main definition of general regular Hamiltonians is given in Sect. 2.4 and in previous sections the most important examples are investigated in detail. The classification of regular degenerate Hamiltonians corresponds (in a sense, explained in Chapter 3) to the classification of the small time asymptotics of Gaussian diffusions, and is described using Young schemes well known in the theory of representations, because they classify the irreducible representations of the semi-simple Lie groups. Regular degenerate Hamiltonians appear naturally when considering the problems of the minimisation of functionals with Lagrangians depending on higher derivatives. Section 2.5 is devoted to more general Hamiltonians having exponential growth in momenta and relating to the equations of type (0.6) with non-vanishing ν. The methods developed in this section can be applied also to some other classes of Hamiltonians. In Sections 2.6, 2.7 the cases of stochastic and complex Hamiltonians are studied, and the corresponding problems of the complex and stochastic calculus of variations are discussed. The results of this chapter give the main ingredient in the construction of the asymptotic solutions of equation (0.5), (0.6) developed further, they give "the classical part" of the semiclassical approximation. However, this chapter is also of independent interest and can be read independently of the rest of the book.

Chapter 3 is devoted to the construction of the small time $(t \to 0)$ and/or small diffusion $(h \to 0)$ asymptotics of the solution of a large class of second order parabolic equations

$$h\frac{\partial u}{\partial t} = \frac{1}{2}h^2 tr\left(G(x)\frac{\partial^2 u}{\partial x^2}\right) + h\left(A(x), \frac{\partial u}{\partial x}\right) - V(x), \qquad (0.22)$$

which we call regular and which include the cases of non-degenerate and also some classes of degenerate matrices $G(x)$. To describe this class notice that equation (0.22) can be written in the "pseudo-differential form" as

$$h\frac{\partial u}{\partial t} = H\left(x, -h\frac{\partial}{\partial x}\right)u \qquad (0.23)$$

with the Hamiltonian function

$$H(x,p) = \frac{1}{2}(G(x)p, p) - (A(x), p) - V(x). \qquad (0.24)$$

The class of regular diffusions corresponds to Hamiltonians (0.24) belonging to the class of regular Hamiltonians as defined in Sections 2.2-2.4. Therefore, these regular diffusions are also classified in terms of the Young schemes. It turns out that this class of diffusions is characterised by the property that the main term of the small time asymptotics of their Green functions, i.e. of the solutions of (0.23), (0.24) with initial data $\delta(x - x_0)$, in a neighbourhood of x_0 is the same as this main term for their Gaussian diffusion approximations

$$\frac{\partial u}{\partial t} = \frac{1}{2}h\,tr\left(G(x_0)\frac{\partial^2 u}{\partial x^2}\right) + \left(A(x_0) + \frac{\partial A}{\partial x}(x_0)(x - x_0), \frac{\partial u}{\partial x}\right), \qquad (0.25)$$

around x_0. Equivalently, these diffusions are characterised by the fact that the asymptotics of their Green functions for small t and/or small h in a neighbourhood of x_0 can be expressed in the form

$$C(h)(1 + O(h))\phi(t, x)\exp\{-\frac{1}{h}S(t, x)\}, \qquad (0.26)$$

where $C(h)$ is some normalising constant, and both the "amplitude" ϕ and the phase S has the form of a regular (i.e. in series of non-negative powers) asymptotic expansion in $(t, x - x_0)$, multiplied by t^{-M} with some positive M (different for ϕ and S). Small time asymptotics for this class of diffusions are effectively constructed in Sections 3.1-3.3, and the justifications are given in Section 3.4. In Section 3.5, the global (i.e. for any finite t or x, including focal points) small diffusion asymptotics $(h \to 0)$ are given for regular diffusions, presenting in particular the solution of the large deviation problem. In Section

3.6, the problem of constructing the small diffusion asymptotics for more general, non-regular, diffusions is discussed. For such diffusions, in the representation (0.26) for its small time (and/or small diffusion) asymptotics, if it exists, the amplitude and the phase have more complicated singularities at $(0, x_0)$.

Chapter 4 is devoted to the invariant regular degenerate diffusions on cotangent bundles associated with compact Riemannian manifolds. These invariant diffusions turn out to be the curvilinear analogues of the Ornstein-Uhlenbeck process, and in the simplest case of the absence of "friction", they can naturally be called stochastic geodesic flows. In the latter case these processes are completely defined by the geometry, i.e. by the Riemannian structure. One can generalise the expansion (0.19) of the Green function on the diagonal to this case of degenerate diffusion. Integrating this expansion yields (as in the well known case of non-degenerate invariant diffusions on manifolds) the asymptotic expansion of the trace of the Green function for small times t with geometric invariants as coefficients in this series.

Chapter 5 is devoted to the study of the Green function (transition probability density) for stable jump-diffusions, i.e. for the Feller semigroups defined by equation (0.6) with $G = 0$ and the Lévy measure ν of the form $|\xi|^{-(1+\alpha)} d|\xi| \mu(x, ds)$, $s = \xi/|\xi|$, and also for some natural generalisations of these processes, for instance, for the stable-like jump-diffusions and for the corresponding truncated processes. Two-sided estimates and small time asymptotics are obtained for these Green functions generalising the well known results for the case of standard non-degenerate diffusions. Some preliminary work on the asymptotic properties of finite-dimensional stable laws is also carried out, and the application of the results obtained to sample path properties (lim sup behaviour and the distribution of the maximum) of the corresponding stochastic processes are given. This chapter presents, on the one hand, the necessary preliminary results for the asymptotic theory of the next chapter, and on the other hand, presents an introduction to the analytical study of stable jump-diffusions, which is of independent interest, and which can be read essentially independently of other parts of the book.

The technically most difficult result of the book is given in Chapter 6, which is devoted to the generalisation of Varadhan's large deviation principle (0.16) to the case of localised stable jump-diffusions, and to the corresponding more precise local and global small h asymptotics for equations of type (0.6). The proof uses various techniques from Chapters 2,3,5 and Appendices B-F, and is based on the construction of a rather complicated uniform small time and small h asymptotics of the Green function, which differ from the form (0.26), and which is obtained by glueing together (in a nontrivial way) the asymptotics of form (0.26) with the small time asymptotics obtained in the previous chapter. The form of this uniform asymptotics was actually proposed as early as in [M2] for an even more general class of equations, called in [M1] tunnel equations, but without rigorous proof.

Chapter 7 is devoted to the asymptotics of the Cauchy problem for the equations of complex stochastic diffusion of type (0.5). A specific feature of the semiclassical asymptotics for these equations is the necessity to deal with complex and/or stochastic characteristics, which complicates essentially both the formal construction and the justification of these asymptotics. Together with the semiclassical asymptotics constructed and justified in Sections 7.1, 7.2, we obtain in Section 7.3 a complex analogue of the large deviation principle and two-sided estimates for the Green function (or the complex heat kernel) for these equations. In Section 7.4 we discuss shortly another approach to the construction of the solutions of these equations based on the theory of inifinite dimensional Fresnel integrals. It is well known that the semi-classical asymptotics of the diffusion or Schrödinger equations can be recovered from the path integral representation by means of (sometimes only heuristically defined) infinite dimensional stationary phase or Laplace methods. In the case of the complex stochastic diffusion considered here, this procedure requires the use of the general infinite dimensional saddle-point method. It is worth noting that the results on the semiclassical asymptotics of complex Schrödinger equation obtained in Chapter 7, are used in Chapter 9 for the purpose of regularisation in the construction of the path integral for the standard Schrödinger equation.

Some of the results of this chapter can be extended to more general equations, which are intensively studied now in the quantum theory of open systems and in non-commutative probability (see [BK] for the first steps in this direction), namely to quantum stochastic equations of the type

$$d\psi + \left(\frac{i}{h}H + K\right)\psi\,dt = (L^- dA^- + L^+ dA^+ + LdN)\psi, \qquad (0.27)$$

where dA^-, dA^+, dN are quantum stochastic differentials (quantum noise) acting in the appropriate Fock space (see e.g. [HP], [ApB], [Be4], [Par]) and H, K, L^-, L^+, L are pseudo-differential operators in $L^2(\mathcal{R}^d)$.

This book is mainly devoted to the semi-classical theory of the Green function (heat kernel) of second order parabolic equations and more general pseudo-differential equations describing the evolution of stochastic processes. Asymptotic spectral analysis of second order differential operators, though being linked in many senses with the theory of evolutionary equations, has its own methods and problems. Many books are devoted to this subject (see e.g. [He], [M5],[M9], [Rob], [SV]), and we shall not present here a systematic study of this field. Instead, in Chapter 8, we shall introduce briefly (referring to the original papers for the proofs and developments) three topics in asymptotic spectral analysis closely connected with probability and with the semi-classical methods developed for the corresponding evolutionary equations. Namely, in section 1 we discuss the problem of the asymptotic calculations of the low lying eigenvalues of the Schrödinger operator, describing some recently obtained asymptotic formula for the splitting of low lying eigenvalues corresponding to symmetric potential wells and demonstrating how this formula works on a concrete physical example of the discrete ϕ^4-model on tori. In section 2, we present some results on the low lying eigenvalues of a diffusion operator, giving in particular a theorem on the probabilistic interpretation of these eigenvalues (in terms of the life-times of the diffusions in fundamental domains). Finally, in Section 3, we present a construction of the quasi-modes of the diffusion operator around a closed stable trajectory of the corresponding classical system.

In Chapter 9 we develop an approach to the rigorous construction of path integrals. This approach incorporates several known approaches but is extended in a way to be applied to the heat, Schrödinger and complex stochastic diffusion equations in a unified way covering a rather large class of potentials. Moreover, this approach leads to the construction of a measure on the Cameron-Martin space of curves with L^2-integrable derivatives such that the solutions to the Schrödinger or heat equations can be written as the expectations of the exponentials of the classical actions on the paths of a certain stochastic process. This measure has a natural representation in Fock space, which gives rise to other stochastic representations of the solutions to the Schrödinger equation, for example, in terms of the Wiener or another general Lévy process. The Fock representation also puts our path integral in the framework of noncommutative stochastic calculus.

Appendices are devoted to various topics and results which are used in the main text. Appendix A introduces the main equation of the theory of continuous quantum measurement (called the stochastic Schrödinger equation, state diffusion model, or Belavkin filtering equation), which is the main motivation for the study of stochastic complex diffusions in Chapter 7. Appendix B is devoted to the proof of asymptotic formulas for Laplace integrals with complex phase, and to estimates for the remainder in these formulas (the latter are usually neglected in expositions of Laplace asymptotic formulas, but are of vital importance for our purposes). Appendices C and D present (for completeness) essentially well-known material on the characteristic functions of stable laws and the corresponding pseudo-differential operators, as well as some simple estimates which follow (more or less) directly from the definitions, and which are used in Chapter 6. In Appendix E, the global smooth equivalence of smooth convex functions with non-degenerate minima is proved. This is a very natural result, which however the author did not find in the literature. Appendix F presents in a compact form the main results on the unimodality of finite dimensional symmetric distributions, which are used in Chapter 5. In Appendix G, a general scheme for the construction of the complex measures on path space and for the representation of the solutions of evolutionary equations in terms of the path integral is presented. An application of this scheme leads, in particular, to an important representation of the Schrödinger equation, with a potential which is the Fourier transform of a finite Borel measure, in terms of the expectation of a certain functional with respect to a compound Poisson process. This representation is one of the main sources for the theory developed in Chapter 9. In Appendix H we sketch the main approaches to the definitions of path integral, where this integral is defined as a certain generalised functional and not as a genuine integral (in a sense of Riemann or Lebesgue) with respect to a bona fide σ-additive measure. Some possible developments of the results displayed in the book are discussed in Appendix J together with related open problems.

CHAPTER 1. GAUSSIAN DIFFUSIONS

1. Gaussian diffusions. Probabilistic and analytic approaches

The simplest case of a parabolic second order equation is the Gaussian diffusion, whose Green function can be written explicitly as the exponential of a quadratic form. This chapter is devoted to this simplest kind of diffusion equation. In the first section we collect some well-known general facts about Gaussian diffusions pointing out the connection between probabilistic and analytic approaches to its investigation. In the next section the complete classification of its small time asymptotics is given, which is due essentially to Chaleyat-Maurel [CME]. We give a slightly different exposition stressing also the connection with the Young schemes. Sections 1.3-1.5 are devoted to the long time behaviour of Gaussian and complex stochastic Gaussian diffusions, and their (deterministic) perturbations.

A *Gaussian diffusion operator* is a second order differential operator of the form

$$L = \left(Ax, \frac{\partial}{\partial x} \right) + \frac{1}{2} tr \left(G \frac{\partial^2}{\partial x^2} \right), \tag{1.1}$$

where $x \in \mathcal{R}^m$, A and G are $m \times m$-matrices, the matrix G being symmetric and non-negative-definite. The corresponding parabolic equation $\partial u / \partial t = Lu$ can be written more explicitly as

$$\frac{\partial u}{\partial t} = A_{ij} x_i \frac{\partial u}{\partial x_j} + \frac{1}{2} G_{ij} \frac{\partial^2 u}{\partial x_i \partial x_j}. \tag{1.2}$$

The *Green function* of the corresponding Cauchy problem is by definition the solution $u_G(t, x, x_0)$ of (1.2) with initial condition

$$u_G(0, x, x_0) = \delta(x - x_0),$$

where δ is Dirac's δ-function.

We shall say that a square matrix is degenerate or singular, if its determinant vanishes.

The following fact is well-known.

Proposition 1.1 *If the matrix*

$$E = E(t) = \int_0^t e^{A\tau} G e^{A'\tau} \, d\tau \tag{1.3}$$

is non-singular (at least for small $t > 0$), then

$$u_G(t, x, x_0) = (2\pi)^{-m/2} (\det E(t))^{-1/2} \exp\{-\frac{1}{2} \left(E^{-1}(x_0 - e^{At}x), x_0 - e^{At}x \right)\}. \tag{1.4}$$

Non-singularity of E is a necessary and sufficient condition for the Green function of (1.2) to be smooth in x, t for $t > 0$.

We shall sketch several proofs of this simple but important result.

First proof. Let $f(t,p)$ be the Fourier transform of $u_G(t,x,x_0)$ with respect to the variable x. Then $f(t,p)$ satisfies the equation

$$\frac{\partial f}{\partial t} = -\left(A'p, \frac{\partial f}{\partial p}\right) - \left(\frac{1}{2}(Gp,p) + \operatorname{tr} A\right) f$$

and the initial condition $f(0,p) = (2\pi)^{-m/2} \exp\{-ipx_0\}$. Solving this linear first order partial differential equation by means of the standard method of characteristics, yields

$$f = (2\pi)^{-m/2} \exp\{-i(x_0, e^{-A't}p) - \frac{1}{2}(E(t)e^{-A't}p, e^{-A't}p) - t \operatorname{tr} A\}.$$

Taking the inverse Fourier transform of f and changing the variable of integration $p \mapsto q = e^{-A't}p$ one gets

$$u_G = (2\pi)^{-m} \int \exp\{i(e^{At}x - x_0, q) - \frac{1}{2}(E(t)q, q)\}\, dq,$$

which is equal to (1.4).

Second proof. This proof uses the theory of Gaussian stochastic processes. Namely, we associate with the operator (1.2) the stochastic process defined by the stochastic differential equation

$$dX = AX\, dt + \sqrt{G}\, dW, \tag{1.5}$$

where $\sqrt{G}$ is the symmetric non-negative-definite square root of G and W is the standard m-dimensional Brownian motion (or Wiener process). Its solution with initial data $X(0) = X_0$ is given by

$$X(t) = e^{At}X_0 + \int_0^t e^{A(t-\tau)}\sqrt{G}\, dW(\tau).$$

Direct calculations show that the correlation matrix of the process $X(t)$ is given by formula (1.4). Therefore, the probability density of the transition $x \to x_0$ in time t is given by (1.3) and by the general theory of diffusion processes (see, e.g. [Kal]), this transition probability is just the Green function for the Cauchy problem of equation (1.2).

Other proofs. Firstly, one can check by direct calculations that the function given by (1.4) satisfies equation (1.2) and the required initial condition. Secondly, one can deduce (1.4) using the WKB method, as shown at the end of Section 3.1. Lastly, one can also get (1.4) by the method of "Gaussian substitution", which will be described in Section 1.4, where the Green function for stochastic complex Gaussian diffusions will be constructed in this way.

We discuss now the connection between the non-singularity property of E and a general analytic criterion for the existence of a smooth Green function for second order parabolic equations. It is convenient to work in coordinates, where the matrix G is diagonal. It is clear that the change of the variables $x \to Cx$, for some non-singular matrix C, changes the coefficients of the operator L by the law

$$A \to CAC^{-1}, \quad G \to CGC'.$$

Therefore, one can always choose coordinates such that

$$L = (Ax, \frac{\partial}{\partial x}) + \frac{1}{2} \sum_{j=n+1}^{m} \frac{\partial^2}{\partial x_j^2}, \tag{1.6}$$

where $m - n = rank\, G$. It is convenient to introduce a special notation for the coordinates involving the second derivatives. From now we shall denote by $x = (x_1, ..., x_n)$ the first n coordinates and by $y = (y_1, ..., y_k)$, where $k = m - n$, the remaining ones. In other words, the coordinate space is considered to be decomposed into the direct sum

$$\mathcal{R}^m = \mathcal{R}^{n+k} = \mathcal{R}^n \oplus \mathcal{R}^k = X \oplus Y, \tag{1.7}$$

and L can be written in the form

$$L = L_0 + \frac{1}{2}\Delta_y, \tag{1.6'}$$

where Δ_y is the Laplace operator in the variables y, and

$$L_0 = (A^{xx}x, \frac{\partial}{\partial x}) + (A^{xy}y, \frac{\partial}{\partial x}) + (A^{yx}x, \frac{\partial}{\partial y}) + (A^{yy}y, \frac{\partial}{\partial y})$$

with

$$A = \begin{pmatrix} A^{xx} & A^{xy} \\ A^{yx} & A^{yy} \end{pmatrix} \tag{1.8}$$

according to the decomposition (1.7). The operator (1.6') has the so called Hörmander form. Application of the general theory of such operators (see, e.g. [IK]) to the case of the operator (1.6') gives the following result.

Proposition 1.2. *Let Id be the ideal generated by $\frac{\partial}{\partial y_1}, ..., \frac{\partial}{\partial y_k}$ in the Lie algebra of linear vector fields in $\mathcal{R}^{n+k}$ generated by L_0 and $\frac{\partial}{\partial y_1}, ..., \frac{\partial}{\partial y_k}$. The equation*

$$\frac{\partial u}{\partial t} = (L_0 + \frac{1}{2}\Delta_y)u$$

in $\mathcal{R}^{n+k}$ has a smooth Green function for $t > 0$ if and only if the dimension of the ideal Id is maximal, i.e. it is equal to $n + k$.

Due to Proposition 1.1, the condition of Proposition 1.2 is equivalent to the non-singularity of the matrix E. We shall prove this fact directly in the next section by giving as well the classification of the main terms of the small time asymptotics of the matrix E^{-1} for the processes satisfying the conditions of Proposition 1.2. We conclude this section by a simple description of the ideal Id.

Lemma 1.1 *The ideal* Id *is generated as a linear space by the vector fields*

$$\frac{\partial}{\partial y_j}, [\frac{\partial}{\partial y_j}, L_0], [[\frac{\partial}{\partial y_j}, L_0], L_0], ..., \quad j = 1, ..., k,$$

or, more explicitly, by $\frac{\partial}{\partial y_j}$, $j = 1, ...k$, *and the vector fields, whose coordinates in the basis* $\{\frac{\partial}{\partial x_i}\}$, $i = 1, ..., n$, *are given by the columns of the matrices* $(A^{xx})^l A^{xy}$, $l = 0, 1, ...$

Proof.

$$[\frac{\partial}{\partial y_j}, L_0] = (A^{yy})_{ij}\frac{\partial}{\partial y_i} + (A^{xy})_{ij}\frac{\partial}{\partial x_i}.$$

Therefore, taking the first order commutators we obtain the vector fields $v_j^1 = (A^{xy})_{ij}\frac{\partial}{\partial x_i}$, whose coordinates in the basis $\{\frac{\partial}{\partial x_i}\}$ are given by the columns of the matrix A^{xy}. The commutators $[v_i^1, \frac{\partial}{\partial y_j}]$ do not produce new independent vectors. Therefore, new vector fields can be obtained only by the second order commutators

$$[v_j^1, L_0] = (A^{xy})_{ij}(A^{xx})_{li}\frac{\partial}{\partial x_l} + (A^{xy})_{ij}(A^{yx})_{li}\frac{\partial}{\partial y_l},$$

which produce a new set of vector fields $v_j^2 = (A^{xx}A^{xy})_{lj}\frac{\partial}{\partial x_l}$ whose coordinates in the basis $\{\frac{\partial}{\partial x_i}\}$ are the columns of the matrix $A^{xx}A^{xy}$. The proof is completed by induction.

2. Classification of Gaussian diffusions by Young schemes

Let $H_0 = A^{xy}Y$ and let H_m, $m = 1, 2, ...$, be defined recurrently by the equation

$$H_m = A^{xx}H_{m-1} + H_{m-1}.$$

In coordinate description, H_m is the subspace of $X = \mathcal{R}^n$ generated by the columns of the matrices $A^{xy}, A^{xx}A^{xy}, ..., (A^{xx})^m A^{xy}$. Let M be the minimal natural number such that $H_M = H_{M+1}$. Clearly M is well defined and $0 \le M \le n - dim\, H_0$. Moreover, $dim\, Id = n + k$, iff $H_M = X$, or equivalently iff $dim\, H_m = n$.

Lemma 2.1. *If the correlation matrix* E *given by (1.3) corresponding to the operator (1.6') is non-singular, then* $dim\, Id = n + k$.

Proof. Suppose $dim Id < n + k$ and consequently $dim\, H_M < n$. Then one can choose a basis in X whose first $n - dim\, H_M$ vectors belong to the orthogonal complement $H_M^\perp$ of H_M. In this basis, the first $n - dim\, H_M$ rows of the matrix A given by (1.8) vanish and therefore the matrix $e^{AT}Ge^{A'T}$ has the same property for all t, and therefore so does the matrix E. Thus E is singular.

These arguments show in fact that if $dim\, Id < n + k$, one can reduce the Gaussian process defined by the operator L to a process living in a Eucleadian space of lower dimension. From now on we suppose that $dim\, Id = n + k$ and thus $H_M = X$. The natural number $M + 1$ will be called further the degree of singularity of the Gaussian diffusion. A finite non-increasing sequence of natural numbers is called a *Young scheme*. Young schemes play an important role in the representation theory of classical groups. The (clearly non-increasing) sequence $\mathcal{M}$ of $M + 2$ numbers $m_{M+1} = dim\, Y = k$, $m_M = dim\, H_0$, $m_{M-1} = dim\, H_1 - dim\, H_0, ..., m_0 = dim\, H_M - dim\, H_{M-1}$ will be called the Young scheme of the operator (1.6'). As we shall show these schemes completely define the main term of the small time asymptotics of the inverse matrix E^{-1} and therefore of the transition probability or the Green function of the corresponding Gaussian diffusion. To this end , let us decompose $X = H_M$ in the orthogonal sum:

$$X = X_0 \oplus ... \oplus X_{M-1} \oplus X_M, \tag{2.1}$$

where X_J, $J = 0, ..., M$ are defined by the equation $X_{M-J} \oplus H_{J-1} = H_J$, i.e. each X_{M-J} is the orthogonal complement of H_{J-1} in H_J. To simplify the notation we shall sometimes denote Y by X_{M+1}. The coordinates are therefore decomposed in series $(x, y) = (x^0, ..., x^{M+1})$ with the dimension of each series $x^J = (x_1^J, ..., x_{m_J}^J)$, $J = 0, ..., M + 1$, being defined by the entry m_J of the Young scheme $\mathcal{M}$. Evidently in these coordinates the blocks $A_{J,J+I}$ of the matrix A vanish whenever $I > 1$ for all $J = 0, ..., M$. Let A_J, $J = 0, ..., M$, denote the blocks $A_{J,J+1}$ of A, which are $(m_J \times m_{J+1})$-matrices (with m_J rows and m_{J+1} columns) of rank m_J, and let

$$\alpha_J = A_J A_{J+1}...A_M, \quad J = 0, ..., M.$$

Let us find the main term $E_{IJ}^0(t)$ of the expansion in small t of the blocks $E_{IJ}(t)$ of the correlation matrix (1.3).

Lemma 2.2. *In the chosen coordinates, the blocks $E_{IJ}(t)$ of the matrix (1.3) are given by*

$$E_{IJ}(t) = E_{IJ}^0(t)(1 + O(t)) = \frac{t^{2M+3-(I+J)}\alpha_I \alpha_J'(1 + O(t))}{(2M + 3 - (I + J))(M + 1 - I)!(M + 1 - J)!}. \tag{2.2}$$

Proof. Let us calculate first the main term of the expansion of the blocks of the matrix $\Omega(t) = e^{At}Ge^{A't}$ in the integral in (1.3), taking into account that according to our assumptions the block $G_{M+1,M+1}$ of G is the unit matrix and all other blocks of G vanish. Writing

$$\Omega(t) = \sum_{p=0}^{\infty} t^p \Omega(p),$$

one has

$$\Omega(p) = \sum_{q=0}^{p} \frac{1}{q!(p - q)!} A^q G(A^{p-q})'. \tag{2.3}$$

It is easy to see that for $p < (2M + 2) - (I + J)$ the blocks $\Omega(p)_{IJ}$ vanish and for $p = 2M + 2 - (I + J)$ only one term in sum (2.3) survives, namely that with $q = M + 1 - I$, $p - q = M + 1 - J$. Consequently, for this value of p

$$\Omega(p)_{IJ} = \frac{1}{(M - I + 1)!} \frac{1}{(M - J + 1)!} \alpha_I \alpha'_J,$$

which implies (2.2).

It turns out that the matrix $E^0(t) = \{E^0_{IJ}(t)\}$ is invertible and its inverse in appropriate coordinates depends only on the Young scheme $\mathcal{M}$. The following result is crucial.

Lemma 2.3. *There exist orthonormal coordinates in Y and (not necessarily orthonormal) coordinates in X_J, $J = 0, ..., M$, such that in these coordinates, each matrix A_J has the form*

$$A_J = (0, 1_{m_J}), \quad J = 0, ..., M,$$

where 1_{m_J} denote the $(m_J \times m_J)$ unit matrix.

Remark. We need the coordinates in Y to be orthonormal in order not to destroy the simple second order part of the operator (1.6').

Proof. Consider the chain

$$X_{M+1} \to X_M = H_0 \to H_1 = X_{M-1} \oplus H_0$$

$$\to H_2 = X_{M-2} \oplus H_1 \to ... \to H_M = X_0 \oplus H_{M-1},$$

where the first arrow stands for the linear map A^{xy} and all other stand for A^{xx}. Taking the projection on the first term in each $X_{M-J} \oplus H_{J-1}$ yields the chain

$$X_{M+1} \to X_M \to X_{M-1} \to X_{M-2} \to ... \to X_0, \tag{2.4}$$

where each arrow stands for the composition $\tilde{A} = Pr \circ A$ of the map A and the corresponding projection. Since

$$X_{M+1} \supset Ker\tilde{A}^{M+1} \supset ... \supset Ker\tilde{A}^2 \supset Ker\tilde{A},$$

one can expand $Y = X_{M+1}$ as the orthogonal sum

$$X_{M+1} = Ker\tilde{A}^1 \oplus (Ker\tilde{A}^2 \ominus Ker\tilde{A}^1) \oplus ...$$

$$\oplus (Ker\tilde{A}^{M+1} \ominus Ker\tilde{A}^M) \oplus (X_{M+1} \ominus Ker\tilde{A}^{M+1}),$$

where $Ker\tilde{A}^j \ominus Ker\tilde{A}^{j-1}$ means the orthogonal complement of $Ker\tilde{A}^{j-1}$ in $Ker\tilde{A}^j$. Now choose an orthonormal basis in X_{M+1} which respects this decomposition, i.e. the first $(m_{M+1} - m_M)$ elements of this basis belong to the first term of this decomposition, the next $(m_M - m_{M-1})$ elements belong to the second term and so on. The images of the basis vectors under the action of $\tilde{A}^j$ in the chain (2.4) define the basis in X_{M-j+1} (not necessarily orthonormal).

The coordinates in $X \oplus Y$ defined in such a way satisfy the requirements of the Lemma.

Corollary. *In the coordinates of Lemma 2.3, if $J \geq I$, then*

$$\alpha_I = (0, 1_{m_I}), \quad \alpha_I \alpha_J' = (0, 1_{m_I}), \tag{2.5}$$

where the matrix 0 in the first and second formulas are vanishing matrices with m_I rows and $m_M - m_I$ (resp. $m_J - m_I$) columns.

Lemma 2.4. *If $E_{IJ}^0(t)$ is defined by (2.2) and (2.5), then by changing the order of the basis vectors one can transform the matrix $E^0(t)$ to the block-diagonal matrix with m_0 square blocks Λ_{M+2}, $(m_0 - m_1)$ square blocks $\Lambda_{M+1},...,$ $(m_M - m_{M-1})$ blocks Λ_2, and $m_{M+1} - m_M$ one-dimensional blocks $\Lambda_1 = t$, where $\Lambda_p(t)$ denotes the $(p \times p)$-matrix with entries*

$$\Lambda_p(t)_{ij} = \frac{t^{2p+1-(i+j)}}{(2p+1-(i+j))(p-i)!(p-j)!}, \quad i,j = 1,...,p. \tag{2.6}$$

Proof. It is straightforward. Let us only point out that the block representation of E^0 in the blocks E_{IJ}^0 corresponds to the partition of the coordinates in the parts corresponding to the rows of the Young scheme $\mathcal{M}$, and the representation of E^0 in the block-diagonal form with blocks (2.6) stands for the partition of coordinates in parts corresponding to the columns of the Young scheme $\mathcal{M}$.

Example. If the Young scheme $\mathcal{M} = (3, 2, 1)$, i.e. if $M = 1$, $dim Y = dim X = 3$ and $X = X_0 \oplus X_1$ with $\dim X_0 = 1$, $\dim X_1 = 2$, the matrices A and $E^0(t)$ in the coordinates of Lemma 2.3 have the forms respectively

$$\begin{pmatrix} \star & 0 & 1 & 0 & 0 & 0 \\ \star & \star & \star & 0 & 1 & 0 \\ \star & \star & \star & 0 & 0 & 1 \\ \star & \star & \star & \star & \star & \star \\ \star & \star & \star & \star & \star & \star \\ \star & \star & \star & \star & \star & \star \end{pmatrix}, \quad \text{and} \quad \begin{pmatrix} \frac{t^5}{5\cdot2\cdot2} & 0 & \frac{t^4}{4\cdot2} & 0 & 0 & \frac{t^3}{3\cdot2} \\ 0 & \frac{t^3}{3} & 0 & 0 & \frac{t^2}{2} & 0 \\ \frac{t^4}{4\cdot2} & 0 & \frac{t^3}{3} & 0 & 0 & \frac{t^2}{2} \\ 0 & 0 & 0 & t & 0 & 0 \\ 0 & \frac{t^2}{2} & 0 & 0 & t & 0 \\ \frac{t^3}{3\cdot2} & 0 & \frac{t^2}{2} & 0 & 0 & t \end{pmatrix},$$

where the entries denoted by $\star$ are irrelevant. Clearly by change of order of the basis, E^0 can be transformed to

$$\begin{pmatrix} \frac{t^5}{5\cdot2\cdot2} & \frac{t^4}{4\cdot2} & \frac{t^3}{3\cdot2} & 0 & 0 & 0 \\ \frac{t^4}{4\cdot2} & \frac{t^3}{3} & \frac{t^2}{2} & 0 & 0 & 0 \\ \frac{t^3}{3\cdot2} & \frac{t^2}{2} & t & 0 & 0 & 0 \\ 0 & 0 & 0 & \frac{t^3}{3} & \frac{t^2}{2} & 0 \\ 0 & 0 & 0 & \frac{t^2}{2} & t & 0 \\ 0 & 0 & 0 & 0 & 0 & t \end{pmatrix} = \begin{pmatrix} \Lambda_3(t) & 0 & 0 \\ 0 & \Lambda_2(t) & 0 \\ 0 & 0 & \Lambda_1(t) \end{pmatrix}.$$

Lemma 2.5.

$$(i) \qquad \det \Lambda_p(t) = t^{p^2} \frac{2! \cdot ... \cdot (p-1)!}{p!(p+1)!...(2p-1)!},$$

(ii) *the matrix $E^0(t)$ is non-singular and in the coordinates of Lemma 2.3 its determinant is equal to*

$$\det E^0(t) = (\det \Lambda_1(t))^{m_{M+1}-m_M}(\det \Lambda_2(t))^{m_M-m_{M-1}}...(\det \Lambda_{M+2}(t))^{m_0-m_1},$$
(2.7)

(iii) the maximal negative power of t in the small time asymptopics of the entries of $(E^0(t))^{-1}$ is $-(2M+3)$ and there are exactly m_0 entries that have this maximal power.

Proof. (ii) and (iii) follow directly from Lemma 2.4. To prove (i) notice that

$$\det \Lambda_p(t) = [2!...(p-1)!]^{-2} \det \lambda_p(t),$$

where $\lambda_p(t)$ is the matrix with entries

$$\lambda_p(t)_{ij} = \frac{t^{2p+1-(i+j)}}{2p+1-(i+j)}.$$
(2.8)

In order to see clearly the structure of these matrices, let us write down explicitly the first three representatives:

$$\lambda_1(t) = t, \quad \lambda_2(t) = \begin{pmatrix} \frac{t^3}{3} & \frac{t^2}{2} \\ \frac{t^2}{2} & t \end{pmatrix}, \quad \lambda_3(t) = \begin{pmatrix} \frac{t^5}{5} & \frac{t^4}{4} & \frac{t^3}{3} \\ \frac{t^4}{4} & \frac{t^3}{3} & \frac{t^2}{2} \\ \frac{t^3}{3} & \frac{t^2}{2} & t \end{pmatrix}.$$

The determinant of these matrices is well known (see, e.g. [Ak]) and can be easily found to be:

$$\det \lambda_p(t) = \frac{[2! \cdot 3! \cdot ... \cdot (p-1)!]^3}{p!(p+1)!...(2p-1)!} t^{p^2},$$
(2.9)

which implies (i).

The entries of the matrices $\Lambda_p(t)^{-1}$ and therefore of $(E^0(t))^{-1}$ can be calculated by rather long recurrent formulas. It turns out that these entries are always integer multipliers of negative powers of t, for example, $(\Lambda_1(t))^{-1} = \frac{1}{t}$,

$$(\Lambda_2(t))^{-1} = \begin{pmatrix} \frac{12}{t^3} & -\frac{6}{t^2} \\ -\frac{6}{t^2} & \frac{4}{t} \end{pmatrix}, \quad (\Lambda_3(t))^{-1} = \begin{pmatrix} \frac{6!}{t^5} & -\frac{6!}{2t^4} & \frac{60}{t^3} \\ -\frac{6!}{2t^4} & \frac{192}{t^3} & -\frac{36}{t^2} \\ \frac{60}{t^3} & -\frac{36}{t^2} & \frac{9}{t} \end{pmatrix}.$$

Therefore, we have proved the following result.

Theorem 2.1. *For an arbitrary Gaussian diffusion in the Eucleadian space $\mathcal{R}^m$ whose correlation matrix is non-singular (or equivalently, whose transition probability has smooth density), there exists a scheme $\mathcal{M} = (m_{M+1}, m_M, ..., m_0)$ such that $\sum_{J=0}^{M+1} m_J = m$, and a coordinate system z in $\mathcal{R}^m$ such that in these coordinates the inverse $E(t)^{-1}$ of the correlation matrix (1.3) has the entries*

$$E_{ij}^{-1}(t) = (E^0)_{ij}^{-1}(t)(1 + (t)),$$

where $E^0(t)$ is an invertible matrix that depends only on Y_L, and which is described in Lemmas 2.4, 2.5. The Green function for small t has the form

$$u_G(t,z;z_0) = (2\pi)^{-n/2}(1+O(t))\exp\{-\frac{1}{2}(E^{-1}(t)(z_0 - e^{At}z), z_0 - e^{At}z)\}$$

$$\times[(\det\Lambda_1(t))^{k-m_M}(\det\Lambda_2(t))^{m_M-m_{M-1}}...(\det\Lambda_{M+2}(t))^{m_0-m_1}]^{-1/2}. \quad (2.10)$$

In particular, the coefficient of the exponential in (2.8) has the form of a constant multiple of $t^{-\alpha}$ with

$$\alpha = \frac{1}{2}[(m_{M+1} - m_M) + 2^2(m_M - m_{M-1} + ... + (M+2)^2 m_0]. \quad (2.11)$$

Conversely we have

Theorem 2.2. *For any Young scheme $\mathcal{M}$ satisfying the conditions of Theorem 2.1, there exists a Gaussian diffusion, for which the small time asymptotics of its Green function is (2.8). Moreover, there exists a Gaussian diffusion for which the matrix $E^0(t)$ of the principle term of the asymptotic expansion of E is the exact correlation matrix.*

For example, if in the example with $\mathcal{M} = (3,2,1)$ considered above, one places zero instead of all the entries denoted by "stars" in the expression for A, one gets a diffusion for which $E^0(t)$ is the exact correlation matrix.

Notice that the case of a Young scheme consisting of only one element (i.e. the case $M+1=0$) corresponds to the case of non-singular diffusion.

3. Long time behaviour of the Green function for Gaussian diffusions

This section lies somewhat apart from the main line of the exposition. Here the large time asymptotics is discussed for some classes of Gaussian diffusions including the most commonly used Ornstein-Uhlenbeck and oscillator processes. One aim of this section is to demonstrate that the small time asymptotics classification given in the previous section has little to do with the large time behaviour. Even the property of non-singularity of the matrix G of second derivatives in the expression for the corresponding operator L has little relevance. The crucial role in the long time behaviour description belongs to the eigenvectors of the matrix A together with a "general position" property of these eigenvectors with respect to the matrix G. We consider two particular cases of the operator (1.1) with the matrix A being antisymmetric and with A having only real eigenvalues.

First let A be antisymmetric so that the evolution e^{At} is orthogonal and there exists a unitary matrix U such that $U^{-1}AU$ is diagonal. Let the rank of A be $2n \leq m$. Then one can write down the spectrum of A as

$$i\lambda_1, ..., i\lambda_k, -i\lambda_1, ..., -i\lambda_k, 0, ..., 0$$

with $\lambda_1 > \lambda_2 > ... > \lambda_k > 0$, and to order the unit eigenvectors as follows

$$v_1^1, ..., v_{j_1}^1, v_1^2, ..., v_{j_2}^2, ..., v_1^k, ..., v_{j_k}^k,$$

$$v_1^{k+1}, ..., v_{j_1}^{k+1}, v_1^{k+2}, ..., v_{j_2}^{k+2}, ..., v_1^{2k}, ..., v_{j_k}^{2k}, v_1^{2k+1}, ..., v_{m-2n}^{2k+1}, \qquad (3.1)$$

where $j_1 + ... + j_k = n$, the vectors $v_1^l, ..., v_{j_l}^l$, $l = 1, ..., k$, and their complex conjugates $v_1^{k+l} = \bar{v}_1^l, ..., v_{j_l}^{k+l} = \bar{v}_{j_l}^l$ correspond to the eigenvalues $i\lambda_l$ and $-i\lambda_l$ respectively, and the vectors $v_1^{2k+1}, ..., v_{m-2n}^{2k+1}$ belong to the kernel of A. With these notation, the columns of U are the components of the vectors (3.1) and an arbitrary operator $B : \mathcal{R}^m \mapsto \mathcal{R}^m$ is represented in the basis (3.1) by a matrix $\beta = U^* B U$ given by rectangular blocks β_{IJ}, $I, J = 1, ..., 2k + 1$. The correlation matrix (1.3) becomes $U\tilde{E}(t)U^*$ with

$$\tilde{E}(t) = \int_0^t D(s) U^* G U D^*(s) \, ds = \int_0^t D(s) \Gamma D^*(s) \, ds,$$

where $D(s)$ is diagonal with diagonal elements $e^{\pm i\lambda_j}$ and 1, and the matrix $\Gamma = U^* G U$ consists of the blocks

$$(\Gamma_{IJ})_{lp} = (\bar{v}_l^I, G v_p^J), \quad l = 1, ..., I, \ p = 1, ... J.$$

For $I = J$ these blocks are clearly nonnegative-definite selfadjoint $(j_I \times j_I)$-matrices.

Proposition 3.1. *If for all $I = 1, ..., 2k + 1$ the square blocks Γ_{II} are non-singular, then*

$$\det E(t) = t^m \prod_{I=1}^{2k+1} \det \Gamma_{II}(1 + O(t)),$$

as $t \to \infty$, moreover

$$(\tilde{E}^{-1})_{IJ} = \begin{cases} (t\Gamma_{II})^{-1}(1 + O(\frac{1}{t})), & J = I, \\ O(t^{-2}), & J \neq I. \end{cases}$$

Proof. There are algebraic manipulations, which we omit.

Notice that the non-singularity assumption in Proposition 3.1 is quite different from the non-singularuty assumption of the matrix G that defines the second order part of the diffusion operator. In order to meet the hypothesis of Proposition 3.1 it is enough that the rank of G be equal to the maximal multiplicity of the eigenvalues of A. For example, if the eigenvalues of A are different, then the hypothesis of Proposition 3.1 means just that $(\bar{v}_j, G v_j) \neq 0$ for all eigenvalues v_j of A, and it can be satisfied by the one-dimensional projection. From Proposition 3.1 it follows that the large time asymptotics of the Green function (1.4) in this situation is similar to the standard diffusion with the unit matrix G and vanishing drift.

Corollary. *Let the hypothesis of Proposition 3.1 hold and let all (necessarily positive) eigenvalues of all blocks $(\Gamma^{II})^{-1}$, $I = 1, ..., 2k+1$, lie inside the interval*

$[\beta_1, \beta_2]$. *Then for arbitrary $\epsilon > 0$ and sufficiently large t, the Green function (1.4) satisfies the two-sided estimates*

$$(2\pi t)^{-m/2} \left(\prod_{I=1}^{2k+1} \det \Gamma^{II} \right)^{-1/2} (1 - \epsilon) \exp\{-\frac{1}{2}\beta_2 \|x_0 - e^{At}x\|^2\} \le u_G(t, x; x_0)$$

$$\le (2\pi t)^{-m/2} \left(\prod_{I=1}^{2k+1} \det \Gamma^{II} \right)^{-1/2} (1 + \epsilon) \exp\{-\frac{1}{2}\beta_1 \|x_0 - e^{At}x\|^2\}.$$

An example of this situation is given by the stochastic process defined by the motion of the classical oscillator perturbed by a force given by white noise.

Now let us assume that A has only real eigenvalues λ_j. For simplicity assume A is diagonalisable with k positive eigenvalues $\mu_1 \ge \dots \ge \mu_k > 0$, l vanishing eigenvalues, and $m - k - l$ negative eigenvalues $0 > -\nu_{k+l+1} \ge \dots \ge -\nu_m$. The matrix (1.3) has the entries $E_{ij} = (e^{(\lambda_i + \lambda_j)t} - 1)(\lambda_i + \lambda_j)^{-1}G_{ij}$ if $\lambda_i + \lambda_j \ne 0$, and $E_{ij} = tG_{ij}$, if $\lambda_i + \lambda_j = 0$. One easily derives the following:

Proposition 3.2. *Let the quadratic matrices B_1, B_2, B_3 be non-degenerate, where $(B_1)_{ij} = (\mu_i + \mu_j)^{-1}G_{ij}$ with $i, j = 1, ..., k$, $(B_2)_{ij} = tG_{ij}$ with $i, j = k + 1, ..., k + l$, and $(B_3)_{ij} = (\nu_i + \nu_j)^{-1}G_{ij}$ with $i, j = k + l + 1, ..., n$. Then, as $t \to \infty$,*

$$\det E(t) = \exp\{2 \sum_{j=1}^{k} \mu_j t\} t^l \det B_1 \det B_2 \det B_3 (1 + O(t)),$$

moreover, $(E(t)^{-1})_{ij}$ is exponentially small whenever i or j does not exceed k, and $(E(t)^{-1})_{ij}$ have a finite limit whenever both $i, j > k + l$.

This result implies the corresponding asymptotics for the Green function (1.4). An example of this situation is given by the diffusion operator of the Orstein-Uhlenbeck process. The cases where A has nontrivial Jordan blocks can be considered similarly. Let us point out finally that two-dimensional diffusions described by

$$A = \begin{pmatrix} 0 & 1 \\ -\beta & 0 \end{pmatrix}, G = \begin{pmatrix} 0 & 0 \\ 0 & 1 \end{pmatrix}, \quad \text{and} \quad A = \begin{pmatrix} 0 & 1 \\ 0 & -\beta \end{pmatrix}, G = \begin{pmatrix} 0 & 0 \\ 0 & 1 \end{pmatrix},$$

(the oscillator process and Orstein-Uhlenbeck process respectively) belong to the two different classes described in Propositions 3.1 and 3.2 respectively but from the point of view of the small time asymptotics classification of the previous section they belong to the same class given by the Young scheme (1,1).

The results of this section can be used to estimate the escape rate of transient Gaussian processes defined by equation (1.5) and also of perturbations of them; see [AHK1],[AK],[K2], and section 1.5.

4. Complex stochastic Gaussian diffusion

It is well-known that the Green function of the Cauchy problem for partial differential equations depending quadratically on position and derivatives, i.e. on x and $\partial/\partial x$, has Gaussian form, see e.g. Proposition 1.1 for the case of diffusion. It was realised recently that stochastic generalisations of such equations are of importance for many applications. We present here a simple method for effective calculation of the corresponding Green functions. However, in order not to get lost in complexities we shall not consider the most general case but reduce the exposition to a class of such equations which contains the most important examples for the theory of stochastic filtering, quantum stochastic analysis and continuous quantum measurements. Namely, let us consider the equation

$$d\psi = \frac{1}{2}(G\Delta\psi - \beta x^2\psi)\,dt + \alpha x\psi\,dB, \quad \psi = \psi(t,x,[B]), \tag{4.1}$$

where $x \in \mathcal{R}^m$, $dB = (dB^1, ..., dB^m)$ is the stochastic differential of the standard Brownian motion in $\mathcal{R}^m$, and G, β, α are complex constants such that $|G| > 0$, $ReG \geq 0$, $Re\beta \geq |\alpha|^2$. The last two conditions ensure the conservativity of the system, namely that the expectation of $\|\psi\|^2$ is not increasing in time, which one checks by the formal application of the Ito formula. To justify these calculations one needs actually the well-posedness of the Cauchy problem for equation (4.1) that follows for instance from the explicit expression for the Green function given below. We suppose also for simplicity that ImG and $Im\beta$ are nonnegative.

Let us discuss the main examples. If G, α, β are real, (4.1) is the so called stochastic heat equation, which is also the simplest example of Zakai's equation [Za] of stochastic filtering theory. Its exact Green function was apparently first given in [TZ1] for $\alpha = G = \beta = 1$ and in [TZ2] for $\alpha = G = 1$, $\beta > 1$. In the latter case it was called in [TZ2] the stochastic Mehler formula, because in the deterministic case $\alpha = 0$, G, β positive this formula describes the evolution of the quantum oscillator in imaginary time and is sometimes called Mehler's formula. If α, G are purely imaginary and $Re\beta = |\alpha|^2$, (4.1) represents a unitary stochastic Schrödinger equation, which appeared recently in stochastic models of unitary evolution, see [HP], [TZ1], [K2]. It can be obtained by the formal quantisation of a classical system describing a Newtonian free particle or an oscillator perturbed by a white noise force (see next Section). The explicit Green function for that equation was given in [TZ1]. If G is imaginary, α real (more generally, α complex) and $Re\beta = |\alpha|^2$, equation (4.1) represents the simplest example of the Belavkin quantum filtering equation describing the evolution of the state of a free quantum particle or a quantum oscillator (when $Im\beta = 0$ or $Im\beta > 0$ respectively) under continuous (indirect but non-demolition) observation of its position, see [Be1],[Be2] and Appendix A. When $\alpha = \beta = 1$, $G = i$, the Green function was constructed in [BK], [K4] and a generalisation was given in [K1]. It is worth mentioning that in this case, it is physically more relevant (see Appendix A) to consider the Brownian motion B (which is interpreted as the output process) to have nonzero mean, and to be connected with the standard (zero-mean) Wiener process W (called in the filtering theory the innovation process) by the equation $dB = dW + 2\langle x \rangle_\psi \, dt$, where $\langle x \rangle_\psi = \int x|\psi|^2(x) \, dx \|\psi\|^{-1}$ is the mean position of the normalised state ψ. Since in quantum mechanics one is interested in normalised solutions, it is convenient to rewrite equation (4.1) (using Ito's formula) in terms of the normalised state $\phi = \psi\|\psi\|^{-1}$ and the "free" Wiener process W to obtain, in the simplest case $\alpha = \beta = 1$, $G = i$, the following norm preserving but nonlinear equation

$$d\phi = \frac{1}{2}\left(i\Delta\phi - (x - \langle x \rangle_\phi)^2\phi\right) dt + (x - \langle x \rangle_\phi)\phi \, dW. \qquad (4.2)$$

This equation and its generalisations are extensively discussed in current physical literature on open systems, see e.g. [BHH], [QO], and also Appendix A. It is worth while noting that the equation

$$d\phi = \frac{1}{2}(\beta\Delta\phi - Gx^2\phi) \, dt + \alpha i\frac{\partial}{\partial p}\phi \, dB, \quad \phi = \phi(t, p, [B]), \qquad (4.1')$$

which can be obtained from (4.1) by Fourier transformation, decribes (under the appropriate choice of the parameters) the evolution of a quantum particle or an oscillator under the continuous observation of its momentum (see [Bel]).

In view of so many examples it seems reasonable to give a unified deduction of a formula describing all these situations, which is done below in Theorem 4.1.

To obtain the Green function for equation (4.1) we calculate first the dynamics of the Gaussian functions

$$\psi(x) = \exp\{-\frac{\omega}{2}x^2 + zx - \gamma\}, \qquad (4.3)$$

where ω, γ and the coordinates of the vector z are complex constants and $\mathrm{Re}\,\omega > 0$. It turns out that the Gaussian form of a function is preserved by the dynamics defined by (4.1).

Proposition 4.1. *For an initial function of Gaussian form (4.3) with arbitrary initial ω_0, z_0, β_0, solution to the Cauchy problem for (4.1) exists and has Gaussian form (4.3) for all $t > 0$. Moreover, the dynamics of the coefficients ω, z, γ is given by the differential equations*

$$\begin{cases} \dot{\omega} = -G\omega^2 + (\beta + \alpha^2) \\ dz = -\omega Gz \, dt + \alpha \, dB \\ \dot{\gamma} = \frac{1}{2}G(\omega - z^2) \end{cases} \qquad (4.4)$$

Proof. Let us write down the dynamics of z with undetermined coefficients $dz = z_t \, dt + z_B \, dB$ and let us assume the dynamics of ω to be non-stochastic: $d\omega = \dot{\omega} \, dt$, $d\gamma = \dot{\gamma} \, dt$. (This is justified because inserting the differentials of ω or γ in (4.1) with non-vanishing stochastic terms, yields a contradiction.) Inserting ψ of form (4.3) in (4.1) and using Ito's formula, yields

$$-\frac{1}{2}\dot{\omega}x^2 \, dt + x \, dz - \dot{\gamma} \, dt + \frac{1}{2}x^2 z_B^2 \, dt$$

$$= \frac{1}{2}[G((-\omega x + z)^2 - \omega) - \beta x^2] \, dt + \alpha x \, dB.$$

Comparing coefficients of dB, $x^2 dt$, $x dt$, dt yields (4.4).

Remark. For the purposes of quantum mechanics it is often convenient to express the Gaussian function (4.2) in the equivalent form

$$g_{q,p}^{\omega}(x) = c \exp\{-\frac{\omega}{2}(x - q)^2 + ipx\}, \tag{4.5}$$

where real q and p are respectively the mean position and the mean momentum of the Gaussian function. One deduces from (4.4) that the dynamics of these means under the evolution (4.1) is given by the equations

$$\begin{cases} dq = \frac{1}{Re\omega}[Im(G\omega)p - Re(\beta + \alpha^2)q] \, dt + \frac{Re\alpha}{Re\omega} \, dB \\ dp = -\frac{1}{Re\omega}[Im(\bar{\omega}(\beta + \alpha^2))q + |\omega|^2 p Re G] \, dt + \frac{Im(\bar{\omega}\alpha)}{Re\omega} \, dB. \end{cases}$$

The solution of equation (4.4) can be found explicitly. Namely, let

$$\sigma = \sqrt{\frac{\beta + \alpha^2}{G}} \equiv \sqrt{\left|\frac{\beta + \alpha^2}{G}\right|} \exp\{\frac{i}{2} \arg \frac{\beta + \alpha^2}{G}\}. \tag{4.6}$$

Since $Im\, G \geq 0$ and $Re\, \beta \geq |\alpha|^2$ one sees that $-\frac{\pi}{2} \leq \arg \sigma \leq \frac{\pi}{2}$. The solution to the first equation in (4.4) is

$$\omega(t) = \begin{cases} \sigma \frac{\omega_0 \coth(\sigma Gt) + \sigma}{\omega_0 + \sigma \coth(\sigma Gt)}, & \sigma \neq 0 \\ \omega_0(1 + tG\omega_0)^{-1}, & \sigma = 0. \end{cases} \tag{4.7}$$

In the case $\omega_0 \neq \sigma$ the first formula in (4.7) can be also written in the form

$$\omega = \sigma \coth(\sigma Gt + \Omega(\omega_0)), \quad \Omega(\omega_0) = \frac{1}{2} \log \frac{\omega_0 + \sigma}{\omega_0 - \sigma}. \tag{4.8}$$

Thia implies the following result.

Proposition 4.2. *For an arbitrary solution of form (4.3) of equation (4.1)*

$$\lim_{t \to \infty} \omega(t) = \sigma.$$

From the second equation of (4.4) one gets

$$z(t) = \exp\{-G \int_0^t \omega(s) \, ds\}(z_0 + \int_0^t \exp\{G \int_0^\tau \omega(s) \, ds\} \, dB(\tau)). \tag{4.9}$$

Furthermore, since

$$\int_0^t \coth(\sigma G\tau + \Omega(\omega_0)) \, d\tau = \frac{1}{G\sigma} \left[\log \sinh(\sigma Gt + \Omega(\omega_0)) - \log \frac{\sigma}{\sqrt{\omega_0^2 - \sigma^2}}\right],$$

inserting (4.7) in (4.9) yields

$$z(t) = (\sinh(\sigma Gt + \Omega(\omega_0)))^{-1} \left[\frac{\sigma}{\sqrt{\omega_0^2 - \sigma^2}} z_0 + \alpha \int_0^t \sinh(\sigma G\tau + \Omega(\omega_0)) \, dB(\tau) \right]$$

(4.10)

for $\sigma \neq 0$, and similarly for $\sigma = 0$

$$z(t) = (1 + tG\omega_0)^{-1} \left(z_0 + \alpha \int_0^t (1 + \tau G\omega_0) \, dB(\tau) \right).$$

(4.11)

From the last equation of (4.4) one gets

$$\gamma(t) = \gamma_0 + \frac{G}{2} \int_0^t (\omega(\tau) - z^2(\tau)) \, d\tau,$$

(4.12)

and thus the following result is proved.

Proposition 4.3. *The coefficients of the Gaussian solution (4.3) of equation (4.1) are given by (4.7), (4.8), (4.10)-(4.12).*

We can prove now the main result of the Section.

Theorem 4.1. *The Green function $u_G(t, x; x_0)$ of equation (4.1) exists and has the Gaussian form*

$$u_G(t, x; x_0) = C_G^m \exp\{-\frac{\omega_G}{2}(x^2 + x_0^2) + \beta_G x x_0 - a_G x - b_G x_0 - \gamma_G\}, \quad (4.13)$$

where the coefficients C_G, ω_G, β_G are deterministic (they do not depend on the Brownian trajectory $B(t)$) and are given by

$$\omega_G = \sigma \coth(\sigma Gt), \quad \beta_G = \sigma(\sinh(\sigma Gt))^{-1}, \quad C_G = \left(\frac{2\pi}{\sigma} \sinh(\sigma Gt) \right)^{-1/2},$$

(4.14)

and

$$\omega_G = \beta_G = \frac{1}{tG}, \quad C_G = (2\pi tG)^{-1/2}$$

(4.15)

for the cases $\sigma \neq 0$ and $\sigma = 0$ respectively; the other coefficients are given by

$$a_G = \alpha(\sinh(\sigma Gt))^{-1} \int_0^t \sinh(\sigma G\tau) \, dB(\tau),$$

(4.16)

$$b_G = \sigma G \int_0^t \frac{a(\tau)}{\sinh(\sigma G\tau)} \, d\tau, \quad \gamma_G = \frac{G}{2} \int_0^t a^2(\tau) \, d\tau$$

(4.17)

for $\sigma \neq 0$ and

$$a_G = \frac{\alpha}{t} \int_0^t \tau \, dB(\tau), \quad b_G = \int_0^t \frac{a(\tau)}{\tau} \, d\tau, \quad \gamma_G = \frac{1}{2} \int_0^t a^2(\tau) \, d\tau$$

(4.18)

for $\sigma = 0$.

Remark. It follows in particular that the Green function (4.13) is continuous everywhere except for the case when σG is purely imaginary, in which case (4.14) has periodical singularities.

Proof. Since the Dirac δ-function is the weak limit of Gaussian functions

$$\psi_0^\epsilon = (2\pi\epsilon)^{-1/2} \exp\{-(x - \xi)/2\epsilon\},$$

as $\epsilon \to 0$, we can calculate $u_G(t, x, x_0)$ as a limit of solutions ψ^ϵ of form (4.3) with initial data

$$w_0^\epsilon = \frac{1}{\epsilon}, \quad z_0^\epsilon = \frac{\xi}{\epsilon}, \quad \gamma_0^\epsilon = \frac{\xi^2}{2\epsilon} + \frac{1}{2}\log 2\pi\epsilon. \tag{4.19}$$

Since clearly

$$\Omega(w_0^\epsilon) = \epsilon\sigma + O(\epsilon^2), \quad (w_0^\epsilon - \sigma^2)^{-1/2} z_0^\epsilon \to x_0,$$

as $\epsilon \to 0$, substituting (4.19) in (4.7), (4.10), (4.11) yields

$$\lim_{\epsilon \to 0} w^\epsilon = \sigma \coth(\sigma G t),$$

$$\lim z^\epsilon = (\sinh(\sigma G t))^{-1}\left(\sigma x_0 + \alpha \int_0^t \sinh(\sigma G t)\, dB(\tau)\right)$$

for $\sigma \neq 0$ and

$$\lim_{\epsilon \to 0} w^\epsilon = \frac{1}{tG}, \quad \lim_{\epsilon \to 0} z^\epsilon = \frac{1}{tG}\left(x_0 + \alpha \int_0^t \tau G\, dB(\tau)\right)$$

for $\sigma = 0$, which implies (4.16),(4.18) and the first two formulas in (4.14), (4.15). Let us calculate γ. If $\sigma \neq 0$,

$$\int_0^t w^\epsilon(\tau)\, d\tau = \sigma \int_0^t \coth(\sigma G \tau + \Omega(w_0^\epsilon))\, d\tau$$

$$= \frac{\log \sinh(\sigma G t + \Omega(w_0^\epsilon)) - \log \frac{\sigma}{\sqrt{w_0^\epsilon - \sigma^2}}}{G} = \frac{\log \sinh(\sigma G t) - \log(\epsilon\sigma) + o(1)}{G},$$

and

$$\int_0^t z^2(\tau)\, d\tau = -\frac{\sigma}{G}\xi^2 \coth(\sigma G t) + \frac{\xi^2}{G\epsilon} + o(1)$$

$$+ \int_0^t \frac{2x_0\sigma\alpha \int_0^\tau \sinh(\sigma G s)\, dB(s) + \left(\alpha \int_0^\tau \sinh(\sigma G s)\, dB(s)\right)^2}{\sinh^2(\sigma G t)}\, d\tau.$$

Substituting these formulas in (4.12) and taking the limit as $\epsilon \to 0$ yields the remaining formulas. The simpler case $\sigma = 0$ is dealt with similarly.

It is easy now to write down the Green function $\tilde{u}_G(t, p, p_0)$ of equation (4.1'). Since (4.1') is obtained from (4.1) by the Fourier transformation,

$$\tilde{u}_G(t, p, p_0) = \frac{1}{(2\pi)^m} \int \int u_G(t, x, \xi) \exp\{i\xi p_0 - ixp\} \, d\xi dx. \qquad (4.20)$$

To evaluate this integral it is convenient to change the variables (x, ξ) to $y = x + \xi$, $\eta = x - \xi$. Then (4.20) takes the form

$$\tilde{u}_G = \left(\frac{C_G}{(4\pi)}\right)^m e^{-\gamma_G} \int \exp\{-\frac{\omega_G - \beta_G}{4}y^2 - \frac{1}{2}(a_G + b_G - ip_0 + ip)y\} \, dy$$

$$\times \int \exp\{-\frac{\omega_G + \beta_G}{4}\eta^2 - \frac{1}{2}(a_G - b_G + ip_0 + ip)\eta\} \, d\eta.$$

It is easy to evaluate these Gaussian integrals using the fact that $\omega_G^2 - \beta_G^2 = \sigma^2$. This yields the following result.

Proposition 4.4. *The Green function of equation (4.1') has the form*

$$\tilde{u}_G(t, p; p_0) = \left(\frac{C_G}{\sigma}\right)^m \exp\{-\frac{\omega_G}{2\sigma^2}(p^2 + p_0^2) + \frac{\beta_G}{\sigma^2}pp_0 - \tilde{a}_G p - \tilde{b}_G p_0 - \tilde{\gamma}_G\}, \quad (4.21)$$

where

$$\tilde{a}_G = -\frac{i}{\sigma^2}(\omega_G a_G + \beta_G b_G), \quad \tilde{b}_G = \frac{i}{\sigma^2}(\omega_G b_G + \beta_G a_G),$$

$$\tilde{\gamma}_G = \gamma_G - \frac{\omega_G(a_G^2 + b_G^2)}{2\sigma^2} - \frac{\beta_G a_G b_G}{\sigma^2}.$$

The explicit formulas for the Green functions of equations (4.1), (4.1') can be used in estimating the norms of the corresponding integral operators (giving the solution to the Cauchy problem of these equations) in different L_p spaces, as well as in some spaces of analytic functions, see [K7].

The Gaussian solutions constructed here can serve as convenient examples to test various asymptotic methods. Moreover, they often present the principle term of an asymptotic expansion with respect to a small time or a small diffusion coefficient for more complicated models. This will be explained in detail in the following chapters. Furthermore, since the Gaussian solutions are globally defined for all times, they can be used to study the behaviour of the solutions as time goes to infinity and to provide a basis for scattering theory. In the next section we give some results in this direction. To conclude this section let us mention another interesting property of equation 4.1 and the corresponding nonlinear equation 4.2. It was shown that the Gaussian form is preserved by the evolution defined by these equations. However, they certainly have non-Gaussian solutions as well. An interesting fact concerning these solutions is the effect of Gaussianisation, which means that every solution is asymptotically Gaussian as $t \to \infty$. Moreover, unlike the case of the unitary Schrödinger equation of a free quantum particle, where all solutions are asymptotically free waves e^{ipx} (that is,

Gaussian packets with the infinite dispersion), the solutions of (4.1), (4.2) tend to a Gaussian function (4.3) with a fixed finite non-vanishing ω. This fact has an important physical interpretation (it is called the watchdog effect for continuous measurement, or the continuous collapse of the quantum state). For Gaussian solutions it is obvious (see Proposition 4.2) and was observed first in [Di2], [Be2]. For general initial data it was proved by the author in [K4] (with improvements in [K7]). We shall now state (without proof) the precise result.

Theorem 4.2 [K4],[K7]. *Let ϕ be the solution of the Cauchy problem for equation (4.2) with an arbitrary initial function $\phi_0 \in L_2$, $\|\phi_0\| = 1$. Then for a.a. trajectories of the innovating Wiener process W,*

$$\|\phi - \pi^{1/4} g^{1-i}_{q(t),p(t)}\| = O(e^{-\gamma t})$$

as $t \to \infty$, for arbitrary $\gamma \in (0,1)$, where

$$\begin{cases} q(t) = q_W + p_W t + W(t) + \int_0^t W(s)\,ds + O(e^{-\gamma t}) \\ p(t) = p_W + W + O(e^{-\gamma t}) \end{cases}$$

for some random constants q_W, p_W.

It follows in particular that the mean position of the solution behaves like the integral of the Brownian motion, which provides one of the motivations for the study of this process in the next section.

We note also that finite dimensional analogues of the localisation under continuous measurement and its applications are discussed in [K8], [Ju], [K14], where the notion of the *coefficient of the quality of measurement* was introduced to estimate this localisation quantitatively.

5. The escape rate for Gaussian diffusions and scattering theory for its perturbations

In this Section we show how the results of the two previous Sections can be used to estimate the escape rate for Gaussian diffusions and to develop the scattering theory for small perturbations. The results of this section will not be used in other parts of the book. We shall show first how one estimates the escape rate in the simplest nontrivial example, namely for the integral of the Brownian motion. Then more general models will be discussed including the stochastic Schrödinger equation.

Let $W(t)$ be the standard Wiener process in $\mathcal{R}^n$. Consider the system

$$\dot{x} = p, \quad dp = -\epsilon V'(x)\,dt + dW \qquad (5.1)$$

with some positive smooth function $V(x)$ and some $\epsilon > 0$. This system describes a Newtonian particle in a potential field V disturbed by a white noise force. The global existence theorem for this system was proved in [AHZ]. Firstly, if V vanishes, the solution $x(t)$ of (3.1) is simply the integral of the Brownian motion $Y(t) = \int_0^t W(s)\,ds$.

Theorem 5.1 [K6],[K2]. *If $n > 1$, then, almost surely, $Y(t) \to \infty$, as $t \to \infty$, moreover,*

$$\liminf_{t \to \infty}(|Y(t)|/g(t)) = \infty \tag{5.2}$$

for any positive increasing function $g(t)$ on $\mathcal{R}_+$ such that $\int_0^\infty (g(t)t^{-3/2})^n \, dt$ is a convergent integral.

Remark. For example, for any $\delta > 0$ the function

$$g(t) = t^{\frac{3}{2}-\frac{1}{n}-\delta}$$

satisfies the conditions of the Theorem.

Proof. For an event B in the Wiener space we shall denote by $P(B)$ the probability of B with respect to the standard Wiener measure. The theorem is a consequence of the following assertion. Let A be a fixed positive constant and let $B_{A,g}^t$ be the event in Wiener space which consists of all trajectories W such that the set $\{Y(s) : s \in [t, t+1]\}$ has nonempty intersection with the cube $[-Ag(t), Ag(t)]^n$. Then

$$P(B_{A,g}^t) = (O(g(t)t^{-3/2}) + O(t^{-1}))^n. \tag{5.3}$$

In fact, (5.3) implies that $\sum_{m=1}^\infty P(B_{A,g}^m) < \infty$, if the conditions of the Theorem hold. Then by the first Borell-Cantelli lemma, only a finite number of the events $B_{A,\beta}^m$ can hold. Hencs there exists a constant T such that $Y(t) \notin [-Ag([t]), Ag([t])]^n$ for $t > T$, where $[t]$ denotes the integer part of t. This implies the result.

Let us now prove (5.3). Clearly, it is enough to consider the case $n = 1$. The density $p_t(x,y)$ of the joint distribution of $W(t)$ and $Y(t)$ is well known to be

$$p_t(x, y) = \frac{\sqrt{3}}{\pi t^2} \exp\left\{ -\frac{2}{t}x^2 + \frac{6}{t^2}xy - \frac{6}{t^3}y^2 \right\}.$$

In particular,

$$p_t(x, y) \le \frac{\sqrt{3}}{\pi t^2} \exp\left\{ -\frac{x^2}{2t} \right\}. \tag{5.4}$$

It is clear that

$$P(B_{A,g}^t) = P(Y(t) \in [-Ag(t), Ag(t)])$$

$$+2 \int_{Ag(t)}^\infty dy \int_{-\infty}^{+\infty} p_t(x, y) P\left(\min_{0 \le \tau \le 1} (y + \tau x + \int_0^\tau W(s)\, ds) < Ag(t) \right) dx \tag{5.5}$$

The first term of (5.5) is given by

$$\frac{1}{\sqrt{2\pi t^3}} \int_{-Ag(t)}^{Ag(t)} \exp\left\{ -\frac{y^2}{2t^3} \right\} dy$$

and is of order $O(g(t)t^{-3/2})$. The second term can be estimated from above by the integral

$$2 \int_{Ag(t)}^{\infty} dy \int_{-\infty}^{+\infty} p_t(x,y) P\left(\min_{0 \leq \tau \leq 1} \tau x + \min_{0 \leq \tau \leq 1} W(\tau) < Ag(t) - y \right) dx.$$

We decompose this integral in the sum $I_1 + I_2 + I_3$ of three integrals, whose domain of integration in the variable x are $\{x \geq 0\}$, $\{Ag(t) - y \leq x \leq 0\}$, and $\{x < Ag(t) - y\}$ respectively. We shall show that the integrals I_1 and I_2 are of order $O(t^{-3/2})$ and the integral I_3 is of order $O(t^{-1})$, which will complete the proof of (5.3).

It is clear that

$$I_1 = 2 \int_{Ag(t)}^{\infty} dy \int_0^{\infty} p_t(x,y) P\left(\min_{0 \leq \tau \leq 1} W(\tau) < Ag(t) - y \right) dx.$$

Enlarging the domain of integration in x to the whole line, integrating over x, and using the well known distribution for the minimum of the Brownian motion we obtain

$$I_1 \leq \frac{2}{\pi \sqrt{t^3}} \int_{Ag(t)}^{\infty} \exp\left\{ -\frac{y^2}{2t^3} \right\} dy \int_{y-Ag(t)}^{\infty} \exp\left\{ -\frac{z^2}{2} \right\} dz.$$

Changing the order of integration we can rewrite the last expression in the form

$$\frac{2}{\pi \sqrt{t^3}} \int_0^{\infty} \exp\left\{ -\frac{z^2}{2} \right\} dz \int_{Ag(t)}^{Ag(t)+z} \exp\left\{ -\frac{y^2}{2t^3} \right\} dy.$$

Consequently,

$$I_1 \leq \frac{2}{\pi \sqrt{t^3}} \int_0^{\infty} z \exp\left\{ -\frac{z^2}{2} \right\} dz = O(t^{-3/2}).$$

We continue with I_2. Making the change of variable $x \mapsto -x$ we obtain

$$I_2 = 2 \int_{Ag(t)}^{\infty} dy \int_0^{y-Ag(t)} p_t(-x,y) P\left(\min_{0 \leq \tau \leq 1} < Ag(t) - y + x \right) dx.$$

Making the change of the variable $s = y - Ag(t)$ and using the distribution of the minimum of the Brownian motion we obtain that

$$I_2 = 2 \int_0^{\infty} ds \int_0^s p_t(-x, s + Ag(t)) dx \sqrt{\frac{2}{\pi}} \int_{s-x}^{\infty} \exp\left\{ -\frac{z^2}{2} \right\} dz.$$

Estimating $p_t(x,y)$ by (5.4) and changing the order of integration we get

$$I_2 \leq \frac{4\sqrt{3}}{\sqrt{2\pi} \pi t^2} \int_0^{\infty} dz \exp\left\{ -\frac{z^2}{2} \right\} \int_0^{\infty} dx \exp\left\{ -\frac{x^2}{2t} \right\} \int_x^{x+z} ds.$$

The last integral is clearly of order $O(t^{-3/2})$. It remains to estimate the integral I_3. We have

$$I_3 = 2 \int_{Ag(t)}^{\infty} dy \int_{y-Ag(t)}^{\infty} p_t(-x,y)\,dx = 2 \int_0^{\infty} p_t(-x,y)\,dx \int_{Ag(t)}^{Ag(t)+x} dy$$

$$\leq \frac{2\sqrt{3}}{\pi t^2} \int_0^{\infty} x \exp\{-\frac{x^2}{2t}\}\,dx = O(t^{-1}).$$

The proof is complete.

It is evident that the method of the proof is rather general and can be applied to other processes whenever a reasonable estimate for the transition probability at large times is available. For example, one easily obtains the following generalisation of the previous result.

Theorem 5.2. *Let Y_k be the family of processes defined recurrently by the formulas $Y_k = \int_0^t Y_{k-1}(s)\,ds$, $k = 1, 2, ...$, with $Y_0 = W$ being the standard n-dimensional Brownian motion. If $n > 1$ and $f(t)$ is an increasing positive function for which the integral $\int_1^{\infty} (f(t)t^{-(k+1/2)})^n\,dt$ is convergent, then*

$$\liminf_{t \to \infty} (|Y_k(t)|/f(t)) = +\infty$$

with probability one.

The same method can be used to estimate the rate of escape for the processes discussed in Section 3. For example, for the Ornstein-Uhlenbeck process defined by the stochastic differential system

$$\begin{cases} dX = v\,dt, \\ dv = -\beta v\,dt + dW \end{cases} \tag{5.6}$$

with constant $\beta > 0$, the application of this method together with the estimate of Proposition 3.2 leads to the following result.

Theorem 5.3 [AK]. *Let $n \geq 3$ and let $f(t)$ be an increasing positive function such that $\int_1^{\infty} (f(t)/\sqrt{t})^n\,dt < \infty$. Let $X(t,[W])$, $v(t,[W])$ denote a solution of (5.6). Then almost surely*

$$\liminf_{t \to \infty} (|X(t,[W])|/f(t)) = \infty. \tag{5.7}$$

Similar results were obtained in [AK] for the processes described in Proposition 3.1. Infinite-dimensional generalisations of these results are also given in [AK]. Theorems 5.1-5.3 allow to develop the scattering theory for small perturbations of the corresponding Gaussian diffusions. For example, the following result is a simple corollary of Theorem 5.1 and standard arguments of deterministic scattering theory.

Theorem 5.4 [AHK1]. *Let $n > 2$ and let the vector valued function $F(x) = V'(x)$ is uniformly bounded, locally Lipschitz continuous and suppose furthermore that there exist constants $C > 0$ and $\alpha > 4n/(3n-2)$ such that*

$$|K(x)| \leq C|x|^{-\alpha} \; \forall \, x \in \mathcal{R}^n, \tag{5.8}$$

$$|K(x) - K(y)| \leq C r^{-\alpha} |x - y| \ \forall \ x, y : |x|, |y| > r. \tag{5.9}$$

Then for any pair $(x_\infty, p_\infty) \in \mathcal{R}^{2n}$ *and for almost all* W *there exists a unique pair* (x_0, p_0) *(depending on* W*) such that the solution* $(\tilde{x}, \tilde{p})$ *to the Cauchy problem for system (5.1) with initial data* (x_0, p_0) *has the following limit behaviour:*

$$\lim_{t \to \infty} (\tilde{p}(t) - W(t) - p_\infty) = 0, \tag{5.10}$$

$$\lim_{t \to \infty} \left(\tilde{x}(t) - \int_0^t W(s) \, ds - x_\infty - t p_\infty \right) = 0. \tag{5.11}$$

Moreover, the mapping $\Omega_+([W]) : (x_\infty, p_\infty) \mapsto (x_0, p_0)$, *which can naturally be called the random wave operator, is an injective measure preserving mapping* $\mathcal{R}^{2n} \mapsto \mathcal{R}^{2n}$.

It is worth mentioning that the assumptions on the force F in the theorem are weaker than those usually adopted to prove the existence of wave operators for deterministic Newtonian systems. In particular, the long range Coulomb potential satisfies the assumption of Theorem 5.4. The reason for this lies in Theorem 5.1 which states that a particle driven by white noise force tends to infinity faster than linearly in time. The question whether Ω_+ is surjective or not can be considered as the question of asymptotic completeness of the wave operator Ω_+. The following weak result was obtained by means of Theorem 5.1 and certain estimates for the probability density for the processes defined by the system (5.1).

Theorem 5.5 [AHK2]. *Let* $F(x) = V'(x)$ *be bounded locally Lipschitz continuous function from* $L_2(\mathcal{R}^n)$ *and* $n > 2$. *Then there exists* $\epsilon_0 > 0$ *such that for arbitrary* $\epsilon \in (0, \epsilon_0]$ *and any* (x_0, p_0) *there exists with probability one a pair* (x_∞, p_∞) *such that (5.10),(5.11) hold for the solution of the Cauchy problem for (5.1) with initial data* (x_0, p_0).

Hence, if the conditions of Theorems 5.4 and 5.5 are satisfied, then for small $\epsilon > 0$ the random wave operator for the scattering defined by system (5.1) exists and is a measure preserving bijection (i.e. it is complete).

Similarly one can obtain the existence of the random wave operator for small perturbations of the Ornstein-Uhlenbeck process (5.6) (see details in [AK]).

The stochastic Newtonian system (5.1) formally describes the dynamics of particle in the (formal) potential field $V(x) - x \dot{W}$. The formal Schrödinger equation for the corresponding quantised system would have the form

$$ih\dot{\psi} = (-\frac{h^2}{2}\Delta + V(x))\psi - x\psi\dot{W}. \tag{5.12}$$

To write this equation in a rigorous way, one should use stochastic differentials and thus one obtains

$$ih \, d\psi = (-\frac{h^2}{2}\Delta + V(x))\psi \, dt - x\psi \, d_S W. \tag{5.13}$$

Using the transformation rule for going from the Stratonovich differential to the Ito one $\psi \, d_S W = \psi dW + \frac{1}{2} d\psi dW$ one gets the Ito form of stochastic Schrödinger equation

$$ih \, d\psi = (-\frac{h^2}{2}\Delta + V(x))\psi \, dt - \frac{i}{2h}x^2\psi \, dt - x\psi \, dW. \qquad (5.14)$$

This equation is one of the simplest (and also most important) examples of a Hudson-Parthasarathy quantum stochastic evolution (with unbounded coefficients) [HP] describing in general the coupling of a given quantum system with boson reservoir (the latter being, in particular, the simplest model of a measuring apparatus). Formally, one can easily verify (using Ito calculus) that the evolution defined by (5.14) is almost surely unitary. To make these calculations rigorous one should use the well-posedness theorem for the Cauchy problem of equation (5.14) obtained in [K1] for measurable bounded potentials V. The idea of the proof is to develop a perturbation theory, starting from equation (5.14) with vanishing potential, i.e. from the equation

$$ih \, d\phi = -\frac{h^2}{2}\Delta\phi \, dt - \frac{i}{2h}x^2\phi \, dt - x\phi \, dW. \qquad (5.15)$$

This equation has the form (4.1) with purely imaginary α, G and real β and was considered in detail in the previous Section. The properties of equation (5.15) obtained there can be used also for the development of the scattering theory for equation (5.14). Namely, using Theorem 5.1 and the Gaussian solutions (4.2) of equation (5.15) as the test solutions for the Cook method [Coo] one obtains (see details of the proof in [K2]) the existence of the wave operator for the scattering defined by the stochastic Schrödinger equation (5.14), namely, the following result.

Theorem 5.6 [K2]. *Let the potential V in (5.14) belong to the class $L_r(\mathcal{R}^n)$ for some $r \in [2, n)$ and let the dimension n be greater than 2. Then for each solution of (4.3) (defined by an initial function $\psi_0 \in L_2(\mathcal{R}^n)$) there exists with probability one a solution ϕ of (5.15) such that, in $L_2(\mathcal{R}^n)$,*

$$\lim_{t\to\infty} (\psi(t) - \phi(t)) = 0.$$

This result is a more or less straightforward generalisation of the corresponding deterministic result. Apparently a deeper theory is required for the consideration of the perturbations of the general equation (4.1), because already the "free" dynamics for this case is much more complicated, as Theorem 4.2 states.

CHAPTER 2. BOUNDARY VALUE PROBLEM FOR HAMILTONIAN SYSTEMS

1. Rapid course in calculus of variations

In this preliminary section we present in a compact form the basic facts of the calculus of variations which are relevant to the asymptotical methods developed further. Unlike most standard courses in calculus of variations, see e.g. [Ak], [ATF], [GH], we develop primarily the Hamiltonian formalism in order to include in the theory the case of degenerate Hamiltonians, whose Lagrangians are singular (everywhere discontinuous) and for which in consequence the usual method of obtaining the formulas for the first and second variations (which lead to the basic Euler-Lagrange equations) makes no sense. Moreover, we draw more attention to the absolute minimum, (and not only to local minima), which is usually discussed in the framework of the so called direct methods of the calculus of variations.

A) Hamiltonian formalism and the Weierstrass condition. Let $H = H(x, p)$ be a smooth real-valued function on $\mathcal{R}^{2n}$. By "smooth" we shall always mean existence of as many continuous derivatives as appears in formulas and conditions of theorems. For the main results of this section it is enough to consider H to be twice continuously differentiable. Let $X(t, x_0, p_0), P(t, x_0, p_0)$ denote the solution of the Hamiltonian system

$$\begin{cases} \dot{x} = \frac{\partial H}{\partial p}(x, p) \\ \dot{p} = -\frac{\partial H}{\partial x}(x, p) \end{cases} \tag{1.1}$$

with initial conditions (x_0, p_0) at time zero. The projections on the x-space of the solutions of (1.1) are called *characteristics of the Hamiltonian H*, or *extremals*. Suppose for some x_0 and $t_0 > 0$, and all $t \in (0, t_0]$, there exists a neighbourhood of the origin in the p-space $\Omega_t(x_0) \in \mathcal{R}^n$ such that the mapping $p_0 \mapsto X(t, x_0, p_0)$ is a diffeomorphism from $\Omega_t(x_0)$ onto its image and, moreover, this image contains a fixed neighbourhood $D(x_0)$ of x_0 (not depending on t). Then the family $\Gamma(x_0)$ of solutions of (1.1) with initial data (x_0, p_0), $p_0 \in \Omega_t(x_0)$, will be called *the family (or field) of characteristics starting from x_0 and covering $D(x_0)$ in times $t \leq t_0$*. The discussion of the existence of this family $\Gamma(x_0)$ for different Hamiltonians is one of the main topics of this chapter and will be given in the following sections. Here, we shall suppose that the family exists, and therefore there exists a smooth function

$$p_0(t, x, x_0) : (0, t_0] \times D(x_0) \mapsto \Omega_t(x_0)$$

such that

$$X(t, x_0, p_0(t, x, x_0)) = x. \tag{1.2}$$

The family $\Gamma(x_0)$ defines two natural vector fields in $(0, t_0] \times D(x_0)$, namely, with each point of this set are associated the momentum and velocity vectors

$$p(t, x) = P(t, x_0, p_0(t, x, x_0)), \quad v(t, x) = \frac{\partial H}{\partial p}(x, p(t, x)) \tag{1.3}$$

of the solution of (1.1) joining x_0 and x in time t.

Furthermore, to each solution $X(t, x_0, p_0), P(t, x_0, p_0)$ of (1.1) corresponds the action function defined by the formula

$$\sigma(t, x_0, p_0) = \int_0^t \left(P(\tau, x_0, p_0) \dot{X}(\tau, x_0, p_0) - H(X(\tau, x_0, p_0), P(\tau, x_0, p_0)) \right) d\tau.$$
(1.4)

Due to the properties of the field of characteristics $\Gamma(x_0)$, one can define locally the *two-point function* $S(t, x, x_0)$ as the action along the trajectory from $\Gamma(x_0)$ joining x_0 and x in time t, i.e.

$$S(t, x, x_0) = \sigma(t, x_0, p_0(t, x, x_0)).$$
(1.5)

Using the vector field $p(t, x)$ one can rewrite it in the equivalent form

$$S(t, x, x_0) = \int_0^t \left(p(\tau, x) \, dx - H(x, p(\tau, x)) \, d\tau \right),$$
(1.6)

the curvilinear integral being taken along the characteristic $X(\tau, x_0, p_0(t, x; x_0))$.

The following statement is a central result of the classical calculus of variations.

Proposition 1.1. *As a function of (t, x) the function $S(t, x, x_0)$ satisfies the Hamilton-Jacobi equation*

$$\frac{\partial S}{\partial t} + H(x, \frac{\partial S}{\partial x}) = 0$$
(1.7)

in the domain $(0, t_0] \times D(x_0)$, *and moreover*

$$\frac{\partial S}{\partial x}(t, x) = p(t, x).$$
(1.8)

Proof. First we prove (1.8). This equation can be rewritten as

$$P(t, x_0, p_0) = \frac{\partial S}{\partial x}(t, X(t, x_0, p_0))$$

or equivalently as

$$P(t, x_0, p_0) = \frac{\partial \sigma}{\partial p_0}(t, x_0, p_0(t, x, x_0)) \frac{\partial p_0}{\partial x}(t, x, x_0).$$

Due to (1.2) the inverse matrix to $\frac{\partial p_0}{\partial x}(t, x, x_0)$ is $\frac{\partial X}{\partial p_0}(t, x_0, p_0(t, x, x_0))$. It follows that equation (1.8) written in terms of the variables (t, p_0) has the form

$$P(t, x_0, p_0) \frac{\partial X}{\partial p_0}(t, x_0, p_0) = \frac{\partial \sigma}{\partial p_0}(t, x_0, p_0).$$
(1.9)

This equality clearly holds at $t = 0$ (both parts vanish). Moreover, differentiating (1.9) with respect to t one gets using (1.1) (and omitting some arguments for brevity) that

$$-\frac{\partial H}{\partial x}\frac{\partial X}{\partial p_0} + P\frac{\partial^2 X}{\partial t\partial p_0} = \frac{\partial P}{\partial p_0}\frac{\partial H}{\partial p} + P\frac{\partial^2 X}{\partial t\partial p_0} - \frac{\partial H}{\partial p}\frac{\partial P}{\partial p_0} - \frac{\partial H}{\partial x}\frac{\partial X}{\partial p_0},$$

which clearly holds. Therefore, (1.9) holds for all t, which proves (1.8).

To prove (1.7), let us first rewrite it as

$$\frac{\partial \sigma}{\partial t} + \frac{\partial \sigma}{\partial p_0}\frac{\partial p_0}{\partial t}(t, x) + H(x, p(t, p_0(t, x))) = 0.$$

Substituting for $\frac{\partial \sigma}{\partial t}$ from (1.4) and for $\frac{\partial \sigma}{\partial p_0}$ from (1.9) yields

$$P(t, x_0, p_0)\dot{X}(t, x_0, p_0) + P(t, x_0, p_0)\frac{\partial X}{\partial p_0}(t, x_0, p_0)\frac{\partial p_0}{\partial t} = 0. \tag{1.10}$$

On the other hand, differentiating (1.2) with respect to t yields

$$\frac{\partial X}{\partial p_0}(t, x_0, p_0)\frac{\partial p_0}{\partial t} + \dot{X}(t, x_0, p_0) = 0,$$

which proves (1.10).

We now derive some consequences of Proposition 1.1 showing in particular what it yields for the theory of optimisation.

Corollary 1. *The integral in the r.h.s. of (1.6) does not depend on the path of integration, i.e. it has the same value for all smooth curves $x(\tau)$ joining x_0 and x in time t and lying completely in the domain $D(x_0)$.*

Proof. This is clear, because, by (1.7) and (1.8), this is the integral of a complete differential.

In the calculus of variations, the integral on the r.h.s. of (1.6) is called the invariant Hilbert integral and it plays the crucial role in this theory.

Let the Lagrange function $L(x, v)$ be defined as the Legendre transform of $H(x, p)$ in the variable p, i.e.

$$L(x, v) = \max_p(pv - H(x, p)), \tag{1.11}$$

and let us define the functional

$$I_t(y(.)) = \int_0^t L(y(\tau), \dot{y}(\tau)) \, d\tau \tag{1.12}$$

on all piecewise-smooth curves (i.e. these curves are continuous and have continuous derivatives everywhere except for a finite number of points, where the left and right derivatives exist) joining x_0 and x in time t, i.e. such that $y(0) = x_0$

and $y(t) = x$. Together with the invariant Hilbert integral, an important role in the calculus of variations belongs to the so called Weierstrass function $W(x, q, p)$ defined (in the Hamiltonian picture) as

$$W(x, q, p) = H(x, q) - H(x, p) - (q - p, \frac{\partial H}{\partial p}(x, p)). \qquad (1.13)$$

One says that the Weierstrass condition holds for a solution $(x(\tau), p(\tau))$ of system (1.1), if $W(x(\tau), q, p(\tau)) \geq 0$ for all τ and all $q \in \mathcal{R}^n$. Note that if the Hamiltonian H is convex (even non-strictly) in the variable p, then the Weierstrass function is non-negative for any choice of its arguments, thus in this case the Weierstrass condition holds trivially for all curves.

Corollary 2. (Weierstrass sufficient condition for a relative minimum). *If the Weierstrass condition holds on a trajectory $X(\tau, x_0, p_0)$, $P(\tau, x_0, p_0)$ of the field $\Gamma(x_0)$ joining x_0 and x in time t (i.e. such that $X(t, x_0, p_0) = x$), then the characteristic $X(\tau, x_0, p_0)$ provides a minimum for the functional (1.12) over all curves lying completely in $D(x_0)$. Furthermore $S(t, x, x_0)$ is the corresponding minimal value.*

Proof. For any curve $y(\tau)$ joining x_0 and x in time t and lying in $D(x_0)$ one has (from (1.11)):

$$I_t(y(.)) = \int_0^t L(y(\tau), \dot{y}(\tau)) \, d\tau \geq \int_0^t (p(t, y(\tau))\dot{y}(\tau) - H(y(\tau), p(\tau, y(\tau)))) \, d\tau.$$
$$(1.14)$$

By Corollary 1, the r.h.s. here is just $S(t, x, x_0)$. It remains to prove that $S(t, x, x_0)$ gives the value of I_t on the characteristic $X(\tau, x_0, p_0(t, x.x_0))$. It is enough to show that

$$P(\tau, x_0, p_0)\dot{X}(\tau, x_0, p_0) - H(X(\tau, x_0, p_0), P(\tau, x_0, p_0))$$

equals $L(X(\tau, x_0, p_0), \dot{X}(\tau, x_0, p_0))$, where $p_0 = p_0(t, x, x_0)$, i.e. that

$$P(\tau, x_0, p_0)\frac{\partial H}{\partial p}(X(\tau, x_0, p_0), P(\tau, x_0, p_0)) - H(X(\tau, x_0, p_0), P(\tau, x_0, p_0))$$

$$\geq q\frac{\partial H}{\partial p}(X(\tau, x_0, p_0), P(\tau, x_0, p_0)) - H(X(\tau, x_0, p_0), q)$$

for all q. But this inequality is just the Weierstrass condition, which completes the proof.

Remark. In the more usual Lagrangian picture, i.e. in terms of the variables x, v connected with the canonical variables x, p by the formula $v(x, p) = \frac{\partial H}{\partial p}(x, p)$, the Weierstrass function (1.13) takes its original form

$$W(x, v_0, v) = L(x, v) - L(x, v_0) - (v - v_0, \frac{\partial L}{\partial v}(x, v_0))$$

and the invariant Hilbert integral (1.6) in terms of the field of velocities (or slopes) $v(t,x)$ (see (1.3)) takes the form

$$\int \frac{\partial L}{\partial v}(x,v)\,dx - \left((v,\frac{\partial L}{\partial v}(x,v)) - L(x,v)\right)\,dt.$$

Before formulating the next result let us recall a fact from convex analysis: if H is convex (but not necessarily strictly) and smooth, and L is its Legendre transform (1.11), then H is in its turn the Legendre transform of L, i.e.

$$H(x,p) = \max_v(vp - L(x,v)), \qquad (1.15)$$

moreover, the value of v furnishing maximum in this expression is unique and is given by $v = \frac{\partial H}{\partial p}$. The proof of this fact can be found e.g. in [Roc]. In fact, we use it either for strictly convex H (with $\frac{\partial^2 H}{\partial p^2} > 0$ everywhere), or for quadratic Hamiltonians, and for both these cases the proof is quite straightforward.

Corollary 3. *If H is (possibly non-strictly) convex, then the characteristic of the family Γ joining x_0 and x in time t is the unique curve minimising the functional I_t (again in the class of curves lying in $D(x_0)$).*

Proof. From the fact of the convex analysis mentioned above, the inequality in (1.14) will be strict whenever $\dot y(\tau) \neq v(\tau,y)$ (the field of velocities v was defined in (1.3)), which proves the uniqueness of the minimum.

B) Conjugate points and Jacobi's theory. The system in variations corresponding to a solution $x(\tau), p(\tau)$ of (1.1) is by definition the linear (non-homogeneous) system

$$\begin{cases} \dot v = \frac{\partial^2 H}{\partial p \partial x}(x(\tau),p(\tau))v + \frac{\partial^2 H}{\partial p^2}(x(\tau),p(\tau))w, \\ \dot w = -\frac{\partial^2 H}{\partial x^2}(x(\tau),p(\tau))v - \frac{\partial^2 H}{\partial x \partial p}(x(\tau),p(\tau))w. \end{cases} \qquad (1.16)$$

This equation holds clearly for the derivatives of the solution with respect to any parameter, for instance, for the characteristics from the family $\Gamma(x_0)$ it is satisfied by the matrices

$$v = \frac{\partial X}{\partial p_0}(\tau,x_0,p_0), \quad w = \frac{\partial P}{\partial p_0}(\tau,x_0,p_0).$$

The system (1.16) is called the Jacobi equation (in Hamiltonian form). One sees directly that (1.16) is itself a Hamiltonian system corresponding to the quadratic inhomogeneous Hamiltonian

$$\frac{1}{2}\left(\frac{\partial^2 H}{\partial x^2}(x(\tau),p(\tau))v,v\right) + \left(\frac{\partial^2 H}{\partial p \partial x}(x(\tau),p(\tau))v,w\right) + \frac{1}{2}\left(\frac{\partial^2 H}{\partial p^2}(x(\tau),p(\tau))w,w\right)$$
$$(1.17)$$

Two points $x(t_1), x(t_2)$ on a characteristic are called conjugate, if there exists a solution of (1.16) on the interval $[t_1,t_2]$ such that $v(t_1) = v(t_2) = 0$ and v does not vanish identically on $[t_1,t_2]$.

Proposition 1.2 (Jacobi condition in Hamiltonian form). *Suppose the Hamiltonian H is strictly convex and smooth. If a characteristic $x(\tau)$ contains two conjugate points $x(t_1), x(t_2)$, then for any $\delta > 0$, its interval $[x(t_1), x(t_2 + \delta)]$ does not yield even a local minimum for the functional (1.12) among the curves joining $x(t_1)$ and $x(t_2 + \delta)$ in time $t_2 - t_1 + \delta$.*

The standard proof of this statement (see any textbook in the calculus of variations, for example [Ak]) uses the Lagrangian formalism and will be sketched at the end of subsection D) in a more general situation. In Sect. 3, we shall present a Hamiltonian version of this proof, which can be used for various classes of degenerate Hamiltonians.

C) Connections between the field of extremals and the two-point function. These connections are systematically used in the construction of WKB-type asymptotics for pseudo-differential equations.

Let us first write the derivatives of $S(t, x, x_0)$ with respect to x_0.

Proposition 1.3. *Let the assumptions of Proposition 1.1 hold for all x_0 in a certain domain. Then*

$$\frac{\partial S}{\partial x_0}(t, x, x_0) = -p_0(t, x, x_0). \tag{1.18}$$

Moreover, as a function of t, x_0, the function $S(t, x, x_0)$ satisfies the Hamilton-Jacobi equation corresponding to the Hamiltonian $\tilde{H}(x, p) = H(x, -p)$.

Proof. If the curve $(x(\tau), p(\tau))$ is a solution of (1.1) joining x_0 and x in time t, then the curve $(\tilde{x}(\tau) = x(t - \tau),\ \tilde{p}(\tau) = -p(t - \tau))$ is the solution of the Hamiltonian system with Hamiltonian $\tilde{H}$ joining the points x and x_0 in time t. Both statements of the Proposition follow directly from this observation and Proposition 1.1.

Corollary 1. *If $\frac{\partial X}{\partial p_0}(t, x_0, p_0)$ is a non-degenerate matrix, then*

$$\frac{\partial^2 S}{\partial x^2}(t, x, x_0) = \frac{\partial P}{\partial p_0}(t, x_0, p_0) \left(\frac{\partial X}{\partial p_0}(t, x_0, p_0) \right)^{-1}, \tag{1.19}$$

$$\frac{\partial^2 S}{\partial x_0^2}(t, x, x_0) = \left(\frac{\partial X}{\partial p_0}(t, x_0, p_0) \right)^{-1} \frac{\partial X}{\partial x_0}(t, x_0, p_0), \tag{1.20}$$

$$\frac{\partial^2 S}{\partial x_0 \partial x}(t, x, x_0) = -\left(\frac{\partial X}{\partial p_0}(t, x_0, p_0) \right)^{-1}. \tag{1.21}$$

Proof. This follows from (1.2), (1.8), and (1.18) by differentiating.

Formula (1.19) combined with the following result, which is a consequence of (1.8) and Taylor's formula, can be used for the asymptotic calculations of S.

Corollary 2. *Let $\tilde{x}(t, x_0), \tilde{p}(t, x_0)$ denote the solution of (1.1) with initial data $(x_0, 0)$. Then*

$$S(t, x, x_0) = S(t, \tilde{x}, x_0) + (\tilde{p}(t, x_0), x - \tilde{x})$$

$$+ \int_0^1 (1 - \theta) \left(\frac{\partial^2 S}{\partial x^2}(t, \tilde{x} + \theta(x - \tilde{x}), x_0)(x - \tilde{x}), x - \tilde{x} \right) d\theta. \qquad (1.22)$$

Finally let us mention here the formula for the "composition of Jacobians" after splitting of a characteristic. The function $J(t, x, x_0) = \det \frac{\partial X}{\partial p_0}(t, x_0, p_0)$ is called the Jacobian (corresponding to the family $\Gamma(x_0)$).

Proposition 1.4. *Under the assumptions of Proposition 1.3 let $t_1 + t_2 \le t_0$ and define*

$$f(\eta) = S(t_1, x, \eta) + S(t_2, \eta, x_0).$$

Denote $q = p_0(t_1 + t_2, x; x_0)$ and

$$\tilde{\eta} = X(t_2, x_0, q), \quad \tilde{p} = P(t_2, x_0, q).$$

Then

$$\frac{\partial X}{\partial p_0}(t_1 + t_2, x_0, q) = \frac{\partial X}{\partial p_0}(t_1, \tilde{\eta}, \tilde{p}) \frac{\partial^2 f}{\partial \eta^2}(\tilde{\eta}) \frac{\partial X}{\partial p_0}(t_2, x_0, q).$$

In particular,

$$\det \frac{\partial^2 f}{\partial \eta^2}(\tilde{\eta}) = J(t_1 + t_2, x, x_0) J^{-1}(t_1, x, \tilde{\eta}) J^{-1}(t_2, \tilde{\eta}, x_0). \qquad (1.23)$$

Proof. Let us represent the map $X(t_1 + t_2, x_0, p_0)$ as the composition of two maps

$$(x_0, p_0) \mapsto (\eta = X(t_2, x_0, p_0), p_\eta = P(t_2, x_0, p_0))$$

and $(\eta, p_\eta) \mapsto X(t_1, \eta, p_\eta)$. Then

$$\frac{\partial X}{\partial p_0}(t_1 + t_2, x_0, p_0) = \frac{\partial X}{\partial x_0}(t_1, \eta, p_\eta) \frac{\partial X}{\partial p_0}(t_2, x_0, p_0) + \frac{\partial X}{\partial p_0}(t_1, \eta, p_\eta) \frac{\partial P}{\partial p_0}(t_2, x_0, p_0)$$

For $p_0 = q$ we have $\eta = \tilde{\eta}, p_\eta = \tilde{p}_\eta$. Substituting in the last formula for the derivatives $\frac{\partial X}{\partial p_0}$ and $\frac{\partial X}{\partial x_0}$ in terms of the second derivatives of the two-point function by means of (1.19)-(1.21), yields the Proposition.

D) Lagrangian formalism. We discuss now the Lagrangian approach to the calculus of variations studying directly the minimisation problem for functionals depending on the derivatives of any order. Let the function L on $\mathcal{R}^{n(\nu+2)}$, the Lagrangian, be given, and assume that it has continuous derivatives of order up to and including $\nu + 2$ in all its arguments. Consider the integral functional

$$I_t(y(.)) = \int_0^t L(y(\tau), \dot{y}(\tau), ..., y^{(\nu+1)}(\tau) \, d\tau) \qquad (1.24)$$

on the class of functions y on $[0, t]$ having continuous derivatives up to and including order ν, having a piecewise-continuous derivative of order $\nu + 1$ and satisfying the boundary conditions

$$\begin{cases} y(0) = a_0, & \dot{y}(0) = a_1, & ..., & y^{(\nu)}(0) = a_\nu, \\ y(t) = b_0, & \dot{y}(t) = b_1, & ..., & y^{(\nu)}(t) = b_\nu. \end{cases} \qquad (1.25)$$

Such functions will be called admissible. Suppose now that there exists an admissible function $\bar{y}(\tau)$ minimising the functional (1.24). It follows that for any function $\eta(\tau)$ on $[0, t]$ having continuous derivatives of order up to and including ν, having a piecewise-continuous derivative of order $\nu+1$, and vanishing together with all its derivatives up to and including order ν at the end points $0, t$, the function

$$f(\epsilon) = I_t(y(.) + \epsilon\eta(.))$$

has a minimum at $\epsilon = 0$ and therefore its derivative at this point vanishes:

$$f'(0) = \int_0^t \left(\frac{\partial \bar{L}}{\partial y}\eta + \frac{\partial \bar{L}}{\partial \dot{y}}\dot{\eta} + ... + \frac{\partial \bar{L}}{\partial y^{(\nu+1)}}\eta^{(\nu+1)} \right) = 0,$$

where we denote $\frac{\partial \bar{L}}{\partial y^{(j)}} = \frac{\partial L}{\partial y^{(j)}}(\bar{y}(\tau), \dot{\bar{y}}(\tau), ..., \bar{y}^{(\nu+1)}(\tau))$. Integrating by parts and using boundary conditions for η, yields

$$\int_0^t \left[\left(\frac{\partial \bar{L}}{\partial \dot{y}} - \int_0^\tau \frac{\partial \bar{L}}{\partial y}\, ds \right) \dot{\eta}(\tau) + \frac{\partial \bar{L}}{\partial \ddot{y}}\ddot{\eta}(\tau) + ... + \frac{\partial \bar{L}}{\partial y^{(\nu+1)}}\eta^{(\nu+1)}(\tau) \right] d\tau = 0.$$

Continuing this process of integrating by parts one obtains

$$\int_0^t g(\tau)\eta^{(\nu+1)}(\tau)\, d\tau = 0, \tag{1.26}$$

where

$$g(\tau) = \frac{\partial \bar{L}}{\partial y^{(\nu+1)}}(\tau) - \int_0^\tau \frac{\partial \bar{L}}{\partial y^{(\nu)}}(\tau_1)\, d\tau_1 + \int_0^\tau \left(\int_0^{\tau_1} \frac{\partial \bar{L}}{\partial y^{(\nu-1)}}(\tau_2)\, d\tau_2 \right) d\tau_1 + ...$$

$$+(-1)^{\nu+1} \int_0^\tau ... \left(\int_0^{\tau_\nu} \frac{\partial \bar{L}}{\partial y}(\tau_{\nu+1})\, d\tau_{\nu+1} \right) ...d\tau_1. \tag{1.27}$$

Proposition 1.5 (Second lemma of the calculus of variations). *If a continuous function g on $[0, t]$ satisfies (1.26) for all η in the class described above, then g is a polynomial of order ν, i.e. there exist constants $c_0, c_1, ..., c_\nu$ such that*

$$g(\tau) = c_0 + c_1\tau + ... + c_\nu\tau^\nu. \tag{1.28}$$

Proof. There exist constants $c_0, c_1, ..., c_\nu$ such that

$$\int_0^t (g(\tau) - c_0 - c_1\tau - ... - c_\nu\tau^\nu)\tau^j\, d\tau = 0 \tag{1.29}$$

for all $j = 0, 1, ..., \nu$. In fact, (1.29) can be written in the form

$$c_0\frac{t^{j+1}}{j+1} + c_1\frac{t^{j+2}}{j+2} + ... + c_\nu\frac{t^{j+\nu+1}}{j+\nu+1} = \int_0^t g(\tau)\tau^j\, d\tau, \quad j = 0, ..., \nu.$$

The system of linear equations for $c_0, ..., c_\nu$ has a unique solution, because its matrix is of form (1.2.8) and has non-vanishing determinant given by (1.2.9). Let us set

$$\bar{\eta}(\tau) = \int_0^\tau \frac{(\tau - s)^\nu}{\nu!} [g(s) - c_0 - c_1 s - ... - c_\nu s^\nu] \, ds.$$

It follows that

$$\bar{\eta}^{(j)}(\tau) = \int_0^\tau \frac{(\tau - s)^{\nu-j}}{(\nu - j)!} [g(s) - c_0 - c_1 s - ... - c_\nu s^\nu] \, ds, \quad j = 0, ..., \nu,$$

and therefore, by (1.28), $\bar{\eta}$ satisfies the required conditions and one can take $\eta = \bar{\eta}$ in (1.26), which gives

$$\int_0^t (g(\tau) - c_0 - c_1\tau - ... - c_\nu\tau^\nu) g(\tau) \, d\tau = 0.$$

Using (1.29) one can rewrite this equation in the form

$$\int_0^t (g(\tau) - c_0 - c_1\tau - ... - c_\nu\tau^\nu)^2 \, d\tau = 0,$$

which implies (1.28).

Equation (1.28), with $g(\tau)$ given by (1.27), is called the Euler equation in the integral form. Solutions of this equation are called extremals of functional (1.24). The following simple but important result was first stated by Hilbert in the case of the standard problem, i.e. when $\nu = 0$.

Proposition 1.6 (Hilbert's theorem on the regularity of extremals). Let $\bar{y}(\tau)$ be an admissible curve for problem (1.24) satisfying equation (1.27), (1.28), and let the matrix

$$\frac{\partial^2 \bar{L}}{(\partial y^{(\nu+1)})^2}(\tau) = \frac{\partial^2 L}{(\partial y^{(\nu+1)})^2}(\bar{y}(\tau), \dot{\bar{y}}(\tau), ..., \bar{y}^{(\nu+1)}(\tau))$$

be positive-definite for all $\tau \in [0, t]$. Then $\bar{y}(\tau)$ has continuous derivatives of order up to and including $2(\nu + 1)$; moreover, it satisfies the Euler differential equation

$$\frac{\partial L}{\partial y} - \frac{d}{d\tau}\frac{\partial L}{\partial \dot{y}} + \frac{d^2}{d^2\tau}\frac{\partial L}{\partial \ddot{y}} - ... + (-1)^{\nu+1}\frac{d^{\nu+1}}{d\tau^{\nu+1}}\frac{\partial L}{\partial y^{(\nu+1)}} = 0. \tag{1.30}$$

Proof. By the assumption of the Proposition and the implicit function theorem, one can solve equation (1.27),(1.28) (at least locally, in a neighbourhood of any point) for the last derivative $y^{(\nu+1)}$. This implies that $\bar{y}^{(\nu+1)}$ is a continuously differentiable function, moreover, differentiating (with respect to τ) the formula for $\bar{y}^{(\nu+1)}$ thus obtaines, one gets by induction the existence of the required number of derivatives. (In fact, if L is infinitely differentiable, one gets by this method that $\bar{y}$ is also infinitely differentiable.) Once the regularity of $\bar{y}$ is proved, one obtains (1.30) differentiating (1.27),(1.28) $\nu + 1$ times.

Remark. The standard Euler-Lagrange equations correspond to (1.27), (1.28) and (1.30) with $\nu = 0$. It is worth mentioning also that if one assumes from the beginning that a minimising curve has $2(\nu + 1)$ derivatives one can easily obtain differential equations (1.30) directly, without using the (intuitively not much appealing) integral equations (1.27), (1.28).

We now present the Hamiltonian form of the Euler-Lagrange equation (1.30) thus giving the connection between the Lagrangian formalism of the calculus of variations and the Hamiltonian theory developed above. For this purpose, let us introduce the canonical variables $x = (x_0, ..., x_\nu)$ by $x_0 = y, x_1 = \dot{y}, ..., x_\nu = y^{(\nu)}$ and $p = (p_0, ..., p_\nu)$ by the equations

$$\begin{cases} p_\nu = \frac{\partial L}{\partial y^{(\nu+1)}}(y, \dot{y}, ..., y^{(\nu)}, y^{(\nu+1)}), \\ p_{\nu-1} = \frac{\partial L}{\partial y^{(\nu)}} - \frac{d}{d\tau}\left(\frac{\partial L}{\partial y^{(\nu+1)}}\right), \\ \cdots \\ p_0 = \frac{\partial L}{\partial y} - \frac{d}{d\tau}\left(\frac{\partial L}{\partial \dot{y}}\right) + ... + (-1)^\nu \frac{d^\nu}{d\tau^\nu}\left(\frac{\partial L}{\partial u^{(\nu+1)}}\right). \end{cases} \quad (1.31)$$

Proposition 1.7. Let $\frac{\partial^2 L}{(\partial y^{(\nu+1)})^2} \geq \delta$ everywhere for some $\delta > 0$. Then the equations (1.31) can be solved for $y^{(\nu+1)}, ..., y^{(2\nu+1)}$. Moreover, $y^{(\nu+l+1)}$ does not depend on $p_0, ..., p_{\nu-l-1}$, i.e. for all l:

$$y^{(\nu+l+1)} = f_l(x, p_\nu, p_{\nu-1}, ..., p_{\nu-l}). \quad (1.32)$$

Proof. Due to the assumptions of the Proposition, the first equation in (1.31) can be solved for $y^{(\nu+1)}$:

$$y^{(\nu+1)} = f_0(x_0, ..., x_\nu, p_\nu). \quad (1.33)$$

The second equation in (1.31) takes the form

$$p_{\nu-1} = \frac{\partial L}{\partial y^{(\nu)}}(x, y^{(\nu+1)}) - \frac{\partial^2 L}{\partial y^{(\nu+1)}\partial y}x_1 - ... - \frac{\partial^2 L}{\partial y^{(\nu+1)}\partial y^{(\nu-1)}}x_\nu$$
$$- \frac{\partial^2 L}{\partial y^{(\nu+1)}\partial y^{(\nu)}}y^{(\nu+1)} - \frac{\partial^2 L}{(\partial y^{(\nu+1)})^2}y^{(\nu+2)}.$$

One solves this equation with respect to $y^{(\nu+2)}$ and proceeding in the same way one obtains (1.32) for all l by induction.

The following fundamental result can be checked by direct calculations that we omit (see e.g. [DNF]).

Proposition 1.8 (Ostrogradski's theorem). *Under the assumptions of Proposition 1.7, the Lagrangian equations (1.30) are equivalent to the Hamiltonian system (1.1) for the canonical variables $x = (x_0, ..., x_\nu), p = (p_0, ..., p_\nu)$, with the Hamiltonian*

$$H = x_1 p_0 + ... + x_\nu p_{\nu-1} + f_0(x, p_\nu)p_\nu - L(x_0, ..., x_\nu, f_0(x, p_\nu)), \quad (1.34)$$

where f_0 is defined by (1.33).

The most important example of a Lagrangian satisfying the assumptions of Proposition 1.7 is when L is a quadratic form with respect to the last argument, i.e.

$$L(x_0, ..., x_\nu, z) = \frac{1}{2}(g(x)(z + \alpha(x)), z + \alpha(x)) + V(x). \qquad (1.35)$$

The corresponding Hamiltonian (1.34) has the form

$$H = x_1 p_0 + ... + x_\nu p_{\nu-1} + \frac{1}{2}(g^{-1}(x)p_\nu, p_\nu) - (\alpha(x), p_\nu) - V(x). \qquad (1.36)$$

To conclude, let us sketch the proof of Jacobi's condition, Prop. 1.2, for Hamiltonians of form (1.34) (which are degenerate for $\nu > 0$) corresponding to functionals depending on higher derivatives. One verifies similarly to Proposition 1.8 that the Jacobi equations (1.16), being the Hamiltonian equations for the quadratic approximation of the Hamiltonian (1.34), are equivalent to the Lagrangian equation for the quadratic approximation $\tilde{L}$ of the Lagrangian L around the characteristic $x(.)$. Moreover, this (explicitly time- dependent) Lagrangian $\tilde{L}(\eta, \eta', ..., \eta^{(\nu+1)})$ turns out to be the Lagrangian of the second variation of (1.24), i.e. of the functional

$$\frac{d^2}{d\epsilon^2}|_{\epsilon=0}(I_t(x(.) + \epsilon\eta(.)).$$

For this functional $\eta = 0$ clearly furnishes a minimum. However, if the point $x(s), s \in (0, t)$ is conjugate to $x(0)$, then a continuous curve $\bar{\eta}$ equal to a nontrivial solution of Jacobi's equation on the interval $[0, s]$ and vanishing on $[s, t]$ provides a broken minimum (with derivative discontinuous at s) to this functional, which is impossible by Proposition 1.6. Notice that we have only developed the theory for time-independent Lagrangians, but one sees that including an explicit dependence on time t does not affect the theory.

2. Boundary value problem for non-degenerate Hamiltonians

This section is devoted to the boundary value problem for the system (1.1) with the Hamiltonian

$$H(x, p) = \frac{1}{2}(G(x)p, p) - (A(x), p) - V(x), \qquad (2.1)$$

where $G(x)$ is a uniformly strictly positive matrix, i.e. $G(x)^{-1}$ exists for all x and is uniformly bounded. The arguments used for this simple model (where the main results are known) are given in a form convenient for generalisations to more complex models discussed later. We first prove the existence of the field of characteristics $\Gamma(x_0)$, i.e. the uniqueness and existence for the local boundary value problem for system (1.1), and then the existence of the global minimum for functional (1.12), which also gives the global existence for the boundary value problem. Finally the asymptotic formulas for solutions are given. Before proceeding with the boundary value problem, one needs some estimates on the solutions of the Cauchy problem for the Hamiltonian system (1.1). For the case of the Hamiltonian (2.1), (1.1) takes the form

$$\begin{cases} \dot{x} = G(x)p - A(x) \\ \dot{p}_i = -\frac{1}{2}(\frac{\partial G}{\partial x_i}p, p) + (\frac{\partial A}{\partial x_i}, p) + \frac{\partial V}{\partial x_i}, \quad i = 1, ..., m, \end{cases} \tag{2.2}$$

where we have written the second equation for each coordinate separately for clarity.

Lemma 2.1. *For an arbitrary $x_0 \in \mathcal{R}^m$ and an arbitrary open bounded neighbourhood $U(x_0)$ of x_0, there exist positive constants t_0, c_0, C such that if $t \in (0, t_0]$, $c \in (0, c_0]$ and $p_0 \in B_{c/t}$, then the solution $X(s, x_0, p_0), P(s, x_0, p_0)$ of (2.2) with initial data (x_0, p_0) exists on the interval $[0, t]$, and for all $s \in [0, t]$,*

$$X(s, x_0, p_0) \in U(x_0), \quad \|P(s, x_0, p_0)\| < C(\|p_0\| + t). \tag{2.3}$$

Proof. Let $T(t)$ be the time of exit of the solution from the domain $U(x_0)$, namely

$$T(t) = \min\left(t, \sup\{s : X(s, x_0, p_0) \in U(x_0), P(s, x_0, p_0) < \infty\}\right).$$

Since G, A and their derivatives in x are continuous, it follows that for $s \leq T(t)$ the growth of $\|X(s, x_0, p_0) - x_0\|$, $\|P(s, x_0, p_0)\|$ is bounded by the solution of the system

$$\begin{cases} \dot{x} = K(p + 1) \\ \dot{p} = K(p^2 + 1) \end{cases}$$

with the initial conditions $x(0) = 0, p(0) = \|p_0\|$ and some constant K. The solution of the second equation is

$$p(s) = \tan(Ks + \arctan p(0)) = \frac{p(0) + \tan Ks}{1 - p(0)\tan Ks}.$$

Therefore, if $\|p_0\| \leq c/t$ with $c \leq c_0 < 1/\tilde{K}$, where $\tilde{K}$ is chosen in such a way that $\tan Ks \leq \tilde{K}s$ for $s \leq t_0$, then

$$1 - \|p_0\|\tan Ks > 1 - \|p_0\|\tilde{K}s \geq 1 - c_0\tilde{K}$$

for all $s \leq T(t)$. Consequently, for such s,

$$\|P(s, x_0, p_0)\| \leq \frac{\|p_0\| + \tan Ks}{1 - c_0\tilde{K}}, \quad \|X(s, x_0, p_0) - x_0\| \leq Ks + K\frac{c + s\tan Ks}{1 - c_0\tilde{K}}.$$

We have proved the required estimate but only for $T(t)$ instead of t. However, if one chooses t_0, c_0 in such a way that the last inequality implies $X(s, x_0, p_0) \in U(x_0)$, it will follow that $T(t) = t$. Indeed, if $T(t) < t$, then either $X(T(t), x_0, p_0)$

belongs to the boundary of $U(x_0)$ or $P(T(t), x_0, p_0) = \infty$, which contradicts the last inequalities. The lemma is proved.

Lemma 2.2 *There exist $t_0 > 0$ and $c_0 > 0$ such that if $t \in (0, t_0]$, $c \in (0, c_0]$, $p_0 \in B_{c/t}$, then*

$$\frac{1}{s}\frac{\partial X}{\partial p_0}(s, x_0, p_0) = G(x_0) + O(c + t), \qquad \frac{\partial P}{\partial p_0}(s, x_0, p_0) = 1 + O(c + t) \qquad (2.4)$$

uniformly for all $s \in (0, t]$.

Proof. Differentiating the first equation in (2.2) yields

$$\ddot{x}_i = \frac{\partial G_{ik}}{\partial x_l}(x)\dot{x}_l p_k + G_{ik}(x)\dot{p}_k - \frac{\partial A_i}{\partial x_l}\dot{x}_l$$

$$= \left(\frac{\partial G_{ik}}{\partial x_l}p_k - \frac{\partial A_i}{\partial x_l}\right)(G_{lj}p_j - A_l) + G_{ik}\left(-\frac{1}{2}(\frac{\partial G}{\partial x_k}p, p) + (\frac{\partial A}{\partial x_k}, p) + \frac{\partial V}{\partial x_k}\right).$$
$$(2.5)$$

Consequently, differentiating the Taylor expansion

$$x(s) = x_0 + \dot{x}(0)s + \int_0^s (s - \tau)\ddot{x}(\tau)\, d\tau \qquad (2.6)$$

with respect to the initial momentum p_0 and using (2.3) one gets

$$\frac{\partial X}{\partial p_0}(s, x_0, p_0) = G(x_0)s$$

$$+ \int_0^s \left(O(1 + \|p_0\|^2)\frac{\partial X}{\partial p_0}(\tau, x_0, p_0) + O(1 + \|p_0\|)\frac{\partial P}{\partial p_0}(\tau, x_0, p_0)\right)(s - \tau)\, d\tau.$$
$$(2.7)$$

Similarly differentiating $p(s) = p_0 + \int_0^s \dot{p}(\tau)\, d\tau$ one gets

$$\frac{\partial P}{\partial p_0}(s, x_0, p_0)$$

$$= 1 + \int_0^s \left(O(1 + \|p_0\|^2)\frac{\partial X}{\partial p_0}(\tau, x_0, p_0) + O(1 + \|p_0\|)\frac{\partial P}{\partial p_0}(\tau, x_0, p_0)\right)\, d\tau. \quad (2.8)$$

Let us now regard the matrices $v(s) = \frac{1}{s}\frac{\partial X}{\partial p_0}(s, x_0, p_0)$ and $u(s) = \frac{\partial P}{\partial p_0}(s, x_0, p_0)$ as elements of the Banach space $M_m[0, t]$ of continuous $m \times m$-matrix-valued functions $M(s)$ on $[0, t]$ with norm $\sup\{\|M(s)\| : s \in [0, t]\}$. Then one can write equations (2.7), (2.8) in abstract form

$$v = G(x_0) + L_1 v + \tilde{L}_1 u, \qquad u = 1 + L_2 v + \tilde{L}_2 u,$$

where $L_1, L_2, \tilde{L}_1, \tilde{L}_2$ are linear operators in $M_m[0, t]$ with norms $\|L_i\| = O(c^2 + t^2)$ and $\|\tilde{L}_i\| = O(c + t)$. This implies (2.4) for c and t small enough. In fact, from the second equation we get $u = 1 + O(c+t) + O(c^2 + t^2)v$, substituting this equality in the first equation yields $v = G(x_0) + O(c+t) + O(c^2 + t^2)v$, and solving this equation with respect to v we obtain the first equation in (1.4).

Now we are ready to prove the main result of this section, namely the existence of the family $\Gamma(x_0)$ of the characteristics of system (2.2) starting from x_0 and covering a neighbourhood of x_0 in times $t \leq t_0$.

Theorem 2.1. *(i) For each $x_0 \in \mathcal{R}^m$ there exist c and t_0 such that for all $t \leq t_0$ the mapping $p_0 \mapsto X(t, x_0, p_0)$ defined on the ball $B_{c/t}$ is a diffeomorphism onto its image.*

(ii) For an arbitrary small enough c there exist positive $r = O(c)$ and $t_0 = O(c)$ such that the image of this diffeomorphism contains the ball $B_r(x_0)$ for all $t \leq t_0$.

Proof. (i) Note first that, by Lemma 2.2, the mapping $p_0 \mapsto X(t, x_0, p_0)$ is a local diffeomorphism for all $t \leq t_0$. Furthermore, if $p_0, q_0 \in B_{c/t}$, then

$$X(t, x_0, p_0) - X(t, x_0, q_0) = \int_0^1 \frac{\partial X}{\partial p_0}(t, x_0, q_0 + s(p_0 - q_0))\, ds\, (p_0 - q_0)$$

$$= t(G(x_0) + O(c+t))(p_0 - q_0) \tag{2.9}$$

Therefore, for c and t sufficiently small, the r.h.s. of (2.9) cannot vanish if $p_0 - q_0 \neq 0$.

(ii), (iii) We must prove that for $x \in B_r(x_0)$ there exists $p_0 \in B_{c/t}$ such that $x = X(t, x_0, p_0)$, or equivalently, that

$$p_0 = p_0 + \frac{1}{t}G(x_0)^{-1}(x - X(t, x_0, p_0)).$$

In other words the mapping

$$F_x : p_0 \mapsto p_0 + \frac{1}{t}G(x_0)^{-1}(x - X(t, x_0, p_0)) \tag{2.10}$$

has a fixed point in the ball $B_{c/t}$. Since every continuous mapping from a ball to itself has a fixed point, it is enough to prove that F_x takes the ball $B_{c/t}$ in itself, i.e. that

$$\|F_x(p_0)\| \leq c/t \tag{2.11}$$

whenever $x \in B_r(x_0)$ and $\|p_0\| \leq c/t$. By (2.3), (2.5) and (2.6)

$$X(t, x_0, p_0) = x_0 + t(G(x_0)p_0 - A(x_0)) + O(c^2 + t^2),$$

and therefore it follows from (2.10) that (2.11) is equivalent to

$$\|G(x_0)^{-1}(x - x_0) + O(t + c^2 + t^2)\| \leq c,$$

which certainly holds for $t \leq t_0$, $|x - x_0| \leq r$ and sufficiently small r, t_0 whenever c is chosen small enough.

Corollary. *If (i) either A, V, G and their derivatives are uniformly bounded, or (ii) if G is a constant and A and V'' are uniformly bounded together with their derivatives, then there exist positive r, c, t_0 such that for any $t \in (0, t_0]$ and any x_1, x_2 such that $|x_1 - x_2| \leq r$ there exists a solution of system (2.2) with the boundary conditions*

$$x(0) = x_1, \quad x(t) = x_2.$$

Moreover, this solution is unique under the additional assumption that $\|p(0)\| \leq c/t$.

Proof. The case (i) follows directly from Theorem (2.1). Under assumptions (ii), to get the analog of Lemma 2.1 one should take in its proof the system

$$\begin{cases} \dot{x} = K(p+1) \\ \dot{p} = K(1+p+x) \end{cases}$$

as a bound for the solution of the Hamiltonian system. This system is linear (here the asumption that G is constant plays the role) and its solutions can be easily estimated. the rest of the proof remains the same.

The proof of the existence of the boundary value problem given above is not constructive. However, when the well-posedness is given, it is easy to construct approximate solutions up to any order in small t for smooth enough Hamiltonians. Again one begins with the construction of the asymptotic solution for the Cauchy problem.

Proposition 2.1. *If the functions G, A, V in (2.1) have $k+1$ continuous bounded derivatives, then for the solution of the Cauchy problem for equation (2.2) with initial data $x(0) = x_0$, $p(0) = p_0$ one has the asymptotic formulas*

$$X(t, x_0, p_0) = x_0 + tG(x_0)p_0 - A(x_0)t + \sum_{j=2}^{k} Q_j(t, tp_0) + O(c+t)^{k+1}, \quad (2.12)$$

$$P(t, x_0, p_0) = p_0 + \frac{1}{t}\left[\sum_{j=2}^{k} P_j(t, tp_0) + O(c+t)^{k+1}\right], \quad (2.13)$$

where $Q_j(t, q) = Q_j(t, q^1, ..., q^m)$, $P_j(t, q) = P_j(t, q^1, ..., q^m)$ are homogeneous polynomials of degree j with respect to all their arguments with coefficients depending on the values of G, A, V and their derivatives up to order j at the point x_0. Moreover, one has the following expansion for the derivatives with respect to initial momentum

$$\frac{1}{t}\frac{\partial X}{\partial p_0} = G(x_0) + \sum_{j=1}^{k} \tilde{Q}_j(t, tp_0) + O(c+t)^{k+1}, \quad (2.14)$$

$$\frac{\partial P}{\partial p_0} = 1 + \sum_{j=1}^{k} \tilde{P}_j(t, tp_0) + O(c+t)^{k+1}, \quad (2.15)$$

where $\tilde{Q}_j, \tilde{P}_j$ are again homogeneous polynomials of degree j, but now they are matrix-valued.

Proof. This follows directly by differentiating equations (2.2), then using the Taylor expansion for its solution up to k-th order and estimating the remainder using Lemma 2.1.

Proposition 2.2. *Under the hypotheses of Proposition 2.1, the function $p_0(t, x, x_0)$ (defined by (1.2)), which, by Theorem 2.1, is well-defined and smooth in $B_R(x_0)$, can be expended in the form*

$$p_0(t, x, x_0) = \frac{1}{t}G(x_0)^{-1}\left[(x - x_0) + A(x_0)t + \sum_{j=2}^{k} P_j(t, x - x_0) + O(c + t)^{k+1}\right],$$
(2.16)

where $P_j(t, x - x_0)$ are certain homogeneous polynomials of degree j in all their arguments.

Proof. It follows from (2.12) that $x - x_0$ can be expressed as an asymptotic power series in the variable $(p_0 t)$ with coefficients that have asymptotic expansions in powers of t. This implies the existence and uniqueness of the formal power series of form (2.16) solving equation (2.12) with respect to p_0. The well-posedness of this equation (which follows from Theorem 2.1) completes the proof.

Proposition 2.3. *Under the assumptions of Proposition 2.1, the two-point function $S(t, x, x_0)$ defined in (1.5), can be expended in the form*

$$S(t, x, x_0) = \frac{1}{2t}(x - x_0 + A(x_0)t, G(x_0)^{-1}(x - x_0 + A(x_0)t))$$

$$+ \frac{1}{t}(V(x_0)t^2 + \sum_{j=3}^{k} P_j(t, x - x_0) + O(c + t)^{k+1}),$$
(2.17)

where the P_j are again polynomials in t and $x - x_0$ of degree j (and the term quadratic in $x - x_0$ is written explicitly).

Proof. One first finds the asymptotic expansion for the action $\sigma(t, x_0, p_0)$ defined in (1.4). For Hamiltonian (2.1) one gets that $\sigma(t, x_0, p_0)$ equals

$$\int_0^t \left[\frac{1}{2}(G(X(\tau, x_0, p_0))P(\tau, x_0, p_0), P(\tau, x_0, p_0)) + V(X(\tau, x_0, p_0))\right] d\tau,$$

and using (2.12), (2.13) one obtains

$$\sigma(t, x_0, p_0) = \frac{1}{t}\left[\frac{1}{2}(p_0 t, G(x_0)p_0 t) + V(x_0)t^2 + \sum_{j=3}^{k} P_j(t, tp_0) + O(c + t)^{k+1}\right],$$

where P_j are polynomials of degree $\leq j$ in p_0. Inserting the asymptotic expansion (2.16) for $p_0(t, x, x_0)$ in this formula yields (2.17).

Remark. One can calculate the coefficients of the expansion (2.17) directly from the Hamilton-Jacobi equation without solving the boundary value problem for (2.2) (as we shall do in the next chapter). The theory presented above explains why the asymptotic expansion has such a form and justifies the formal calculation of its coefficients by means of, for example, the method of undetermined coefficients.

By Theorem 2.1, all the assumptions of Proposition 1.1 hold for the Hamiltonian (2.1); moreover, for all x_0, the domain $D(x_0)$ can be chosen as the ball $B_r(x_0)$. It was proved in Corollary 2 of Proposition 1.1 that the two-point function $S(t, x, x_0)$, defined by (1.5),(1.6) in a neighbourhood of x_0, is equal to the minimum of the functional (1.12) over all curves lying in the domain $B_r(x_0)$. We shall show first that (at least for r sufficiently small) it is in fact the global minimum (i.e. among all curves, not only those lying in $B_r(x_0)$).

Proposition 2.4. *Let the potential V be uniformly bounded from below. Then there exists $r_1 \leq r$ such that for $x \in B_{r_1}(x_0)$ the function $S(t, x, x_0)$ defined by (1.5) gives the global minimum of the functional (1.12).*

Proof. It follows from asymptotic representation (2.17) (in fact, from only its first two terms) that there exist $r_1 \leq r$ and $t_1 \leq t_0$ such that for a $\delta > -t_0 \inf V$

$$\max_{\|x-x_0\|=r_1} S(t_1, x, x_0) \leq \min_{t \leq t_1} \min_{\|y-x_0\|=r} S(t, y, x_0) - \delta. \qquad (2.18)$$

Because $L(x, \dot{x}) - V(x)$ is positive, the result of Proposition 2.4 follows from (2.18) using the fact that, on the one hand, the functional (1.12) depends additively on the curve, and on the other hand, if a continuous curve $y(\tau)$ joining x_0 and $x \in B_{r_1}(x_0)$ in time t_1 is not completely contained in $B_r(x_0)$, there exists $t_2 < t_1$ such that $|y(t_2) - x_0| = r$ and $y(\tau) \in B_r(x_0)$ for $t \leq t_2$.

The following result is a consequence of the Hilbert regularity theorem (Proposition 1.6). However we shall give another proof, which is independent of the Lagrangian formalism and which will be convenient for some other situations.

Proposition 2.5. *Suppose that the absolute minimum of the functional (1.12) for the Hamiltonian (2.1) is attained by a piecewise smooth curve. Then this curve is in fact a smooth characteristic, i.e. a solution of (2.2).*

Proof. Suppose that $y(\tau)$ gives a minimum for I_t and at $\tau = s$ its derivative is discontinuous (or it is not a solution of (2.2) in a neighbourhood of $\tau = s$). Then for δ sufficiently small, $|y(s + \delta) - y(s - \delta)| < r_1$, where r_1 is defined in Proposition 2.4, and replacing the segment $[y(s - \delta), y(s + \delta)]$ of the curve $y(\tau)$ by the characteristic joining $y(s-\delta)$ and $y(s+\delta)$ in time 2δ one gets a trajectory, whose action is be less then that of $y(\tau)$. This contradiction completes the proof.

Proposition 2.6. (Tonelli's theorem). *Under the assumptions of Proposition 2.4, for arbitrary $t > 0$ and x_0, x, there exists a solution $(x(\tau), p(\tau))$ of (2.2) with the boundary conditions $x(0) = x_0, x(t) = x$ such that the characteristic $x(\tau)$ attains the global minimum for the corresponding functional (1.12).*

Proof. Suppose first that the functions G, V, A are uniformly bounded together with all their derivatives. Let $t \leq t_0$ and suppose that $k r_1 < |x - $

$x_0| \leq (k+1)r_1$ for some natural number k. Then there exists a piecewise-smooth curve $y(s)$ joining x_0 and x in time t such that it has not more than k points $y(s_1), ..., y(s_k)$, where the derivatives $\dot{y}(s)$ is discontinuous, and for all j, $|y(s_j) - y(s_{j-1})| \leq r_1$ and which is an extremal on the interval $[s_{j-1}, s_j]$. The existence of such a curve implies in particular that $J = \inf I_t$ is finite. Now let $y_n(s)$ be a minimising sequence for (1.12), i.e. a sequence of piecewise- smooth curves joining x_0 and x in time t such that $\lim_{n \to \infty} I_t(y_n(.)) = J$. Comparing the action along $y_n(s)$ with the action along $y(s)$ using (2.18) one concludes that, for $n > n_0$ with some n_0, all curves $y_n(s)$ lie entirely in $B_{kr}(x) \cup B_r(x_0)$. Consequently, one can define a finite sequence of points $y_n(t_j(n)), j \leq k$, on the curve y_n recursively by

$$t_j(n) = \sup\{t > t_{j-1} : |y_n(t) - y_n(t_{j-1}(n))| < r\}.$$

Since all $y_n(t_j(n))$ belong to a compact set, it follows that there exists a subsequence, which will again be denoted by y_n such that the number of the $\{t_j\}$ does not depend on n and the limits $t_j = \lim_{n \to \infty} t_j(n)$ and $y_j = \lim_{n \to \infty} y_n(t_j(n))$ exist for all j. Consider now the sequence of curves $\tilde{y}_n(s)$ constructed by the following rule: on each interval $[t_{j-1}(n), t_j(n)]$ the curve $\tilde{y}_n$ is the extremal joining $y_n(t_{j-1}(n))$ and $y_n(t_j(n))$ in time $t_j(n) - t_{j-1}(n)$. By corollary 2 of Proposition 1.1, the limit of the actions along $\tilde{y}_n$ is also J. But, clearly, the sequence $\tilde{y}_n$ tends to a broken extremal $\tilde{y}$ (whose derivatives may be discontinuous only in points t_j) with the action J, i.e. $\tilde{y}(s)$ gives a minimum for (1.12). By Proposition 2.5, this broken extremal is in fact everywhere smooth. Finally, one proves the result of Proposition 2.6 for all $t > 0$ by a similar procedure using the splitting of the curves of a minimising sequence into parts with the time length less than t_0 and replacing these parts of curves by extremals. The case when G, A, V are not uniformly bounded, is proved by a localisation argument. Namely, any two points can be placed in a large ball, where everything is bounded.

We shall describe now the set of regular points for a variational problem with fixed ends. As we shall see in the next chapter, these are the points for which the WKB-type asymptotics of the corresponding diffusion takes the simplest form.

Let us fix a point x_0. We say that the pair (t, x) is a regular point (with respect to x_0), if there exists a unique characteristic of (2.2) joining x_0 and x in time t and furnishing a minimum to the corresponding functional (1.12), which is not degenerate in the sense that the end point $x(t)$ is not conjugate to x_0 along this characteristic, which implies in particular that $\frac{\partial X}{\partial p_0}(t, x_0, p_0)$ is not degenerate.

Proposition 2.7. *For arbitrary x_0, the set $Reg(x_0)$ of regular points is an open connected and everywhere dense set in $\mathcal{R}_+ \times \mathcal{R}^m$. Moreover, for arbitrary (t,x), all pairs $(\tau, x(\tau))$ with $\tau < t$ lying on any minimising characteristic joining x_0 and x in time t are regular. For any fixed t, the set $\{x : (t, x) \in Reg(x_0)\}$ is open and everywhere dense in $\mathcal{R}^m$.*

Proof. The second statement is a direct consequence of Proposition 2.5. In its turn, this statement, together with Proposition 2.6, implies immediately that the set $Reg(x_0)$ is everywhere dense and connected. In order to prove that this set is open, suppose that (t, x) is regular, and therefore $\frac{\partial X}{\partial p_0}(t, x_0, p_0)$ is non-degenerate. By the inverse function theorem this implies the existence of a continuous family of characteristics emanating of x_0 and coming to any point in a neighbourhood of (t, x). Then by the argument of the proof of Corollary 2 to Proposition 1.1 one proves that each such characteristic furnishes a local minimum to (1.12). Since at (t, x) this local minimum is in fact global, one easily gets that the same holds for neighbouring points. The last statement of the Proposition is a consequence of the others.

At the beginning of section 1.1 we defined the two-point function $S(t, x, x_0)$ locally as the action along the unique characteristic joining x_0 and x in time t. Then we proved that it gives a local minimum, and then that it gives even a global minimum for the functional (1.12), when the distance between x_0 and x is small enough. As a consequence of the last propositions, one can claim this connection to be global.

Proposition 2.8. *Let us define the two-point function $S(t, x, x_0)$ for all $t > 0, x, x_0$ as the global minimum of the functional (1.12). Then, in the case of the Hamiltonian (2.1), $S(t, x, x_0)$ is an everywhere finite and continuous function, which for all all (t, x, x_0) is equal to the action along a minimising characteristic joining x_0 and x in time t. Moreover, on the set $Reg(x_0)$ of regular points, $S(t, x, x_0)$ is smooth and satisfies the Hamilton-Jacobi equation (1.7).*

Remark 1. As stated in Proposition 2.8, the two-point function $S(t, x, x_0)$ is almost everywhere smooth and almost everywhere satisfies the Hamilton-Jacobi equation. In the theory of generalised solutions of Hamilton-Jacobi-Bellman equation (see e.g. [KM1], [KM2]) one proves that $S(t, x, x_0)$ is in fact the generalised solution of the Cauchy problem for equation (1.7) with discontinuous initial data: $S(0, x_0, x_0) = 0$ and $S(0, x, x_0) = +\infty$ for $x \neq x_0$.

Remark 2. An important particular case of the situation considered in this section is the case of a purely quadratic Hamiltonian, namely when $H = (G(x)p, p)$. The solutions of the corresponding system (1.1) (or more precisely, their projections on x-space) are called geodesics defined by the Riemanian metric given by the matrix $g(x) = G^{-1}(x)$. For this case, theorem 2.1 reduces to the well known existence and uniqueness of minimising geodesics joining points with sufficiently small distance between them. The proofs for this special case are essentially simpler, because geodesics enjoy the following homogeneity property. If $(x(\tau), p(\tau))$ is a solution of the corresponding Hamiltonian system, then the pair $(x(\epsilon t), \epsilon p(\epsilon t))$ for any $\epsilon > 0$ is a solution as well. Therefore, having the local diffeomorphism for some $t_0 > 0$ one automatically gets the results for all $t \leq t_0$.

Remark 3. There seems to be no reasonable general criterion for uniqueness of the solution of the boundary value problem, as is shown even by the case of geodesic flows (where uniqueness holds only under the assumption of negative curvature). Bernstein's theorem (see, e.g. [Ak]) is one of the examples of (very restrictive) conditions that assure global uniqueness. Another example is provided by the case of constant diffusion and constant drift, which we shall now discuss.

Proposition 2.9 [M6, DKM1]. *Suppose that in the Hamiltonian (2.1)* $G(x) = G$ *and* $A(x) = A$ *are constant and the matrix of second derivatives of* V *is uniformly bounded. Then the estimates (2.4) for the derivatives are global. More precisely,*

$$\frac{1}{t}\frac{\partial X}{\partial p_0}(t, x_0, p_0) = G(x_0) + O(t^2), \qquad \frac{\partial P}{\partial p_0}(s, x_0, p_0) = 1 + O(t^2), \quad (2.19)$$

$$\frac{\partial^2 X}{\partial p_0^2} = O(t^4), \qquad \frac{\partial^3 X}{\partial p_0^3} = O(t^4) \tag{2.20}$$

and therefore

$$\frac{\partial^2 S}{\partial x^2} = \frac{1}{tG}(1 + O(t^2)), \qquad \frac{\partial^2 S}{\partial x_0^2} = \frac{1}{tG}(1 + O(t^2)), \qquad \frac{\partial^2 S}{\partial x \partial x_0} = -\frac{1}{tG}(1 + O(t^2))$$

$$\tag{2.21}$$

uniformly for all $t \leq t_0$ *and all* p_0. *Moreover, for some* $t_0 > 0$ *the mapping* $p_0 \mapsto X(t, x_0, p_0)$ *is a global diffeomorphism* $\mathcal{R}^n \mapsto \mathcal{R}^n$ *for all* $t \leq t_0$, *and thus the boundary value problem for the corresponding Hamiltonian system has a unique solution for small times and arbitrary end points* x_0, x *such that this solution provides an absolute minimum in the corresponding problem of the calculus of variations.*

Sketch of the proof. It follows from the assumptions that the Jacobi equation (1.16) has uniformly bounded coefficients. This implies the required estimates for the derivatives of the solutions to the Hamiltonian system with respect to p_0. This, in turn, implies the uniqueness of the boundary value problem for small times and arbitrary end points x_0, x. The corresponding arguments are given in detail in a more general (stochastic) situation in Section 7.

Similar global uniqueness holds for the model of the next section, namely for the Hamiltonian (3.4), if g is a constant matrix, $\alpha(x, y) = y$ and $b(x, y) = b(x), V(x, y) = V(x)$ do not depend on y and are uniformly bounded together with their derivatives of first and second order; see details in [KM2].

3. Regular degenerate Hamiltonians of the first rank

We are turning now to the main topic of this chapter, to the investigation of the boundary value problem for degenerate Hamiltonians. As in the first chapter, we shall suppose that the coordinates (previously denoted by x) are divided into two parts, $x \in \mathcal{R}^n$ and $y \in \mathcal{R}^k$ with corresponding momenta $p \in \mathcal{R}^n$ and $q \in \mathcal{R}^k$ respectively, and that H is non-degenerate with respect to q. More precisely, H has the form

$$H(x, y, p, q) = \frac{1}{2}(g(x, y)q, q) - (a(x, y), p) - (b(x, y), q) - V(x, y), \tag{3.1}$$

where $g(x, y)$ is a non-singular positive-definite $(k \times k)$-matrix such that

$$\Lambda^{-1} \leq g(x, y) \leq \Lambda \tag{3.2}$$

for all x, y and some positive Λ. It is natural to try to classify the Hamiltonians of form (3.1) in a neighbourhood of a point (x_0, y_0) by their quadratic approximations

$$\tilde{H}_{x_0, y_0}(x, y, p, q) = -\left(a(x_0, y_0) + \frac{\partial a}{\partial x}(x_0, y_0)(x - x_0) + \frac{\partial a}{\partial y}(x_0, y_0)(y - y_0), p\right)$$

$$+ \frac{1}{2}(g(x_0, y_0)q, q) - \tilde{V}_{x_0, y_0}(x, y)$$

$$- \left[b(x_0, y_0) + \frac{\partial b}{\partial x}(x_0, y_0)(x - x_0) + \frac{\partial b}{\partial y}(x_0, y_0)(y - y_0), q\right], \tag{3.3}$$

where $\tilde{V}_{x_0, y_0}$ is the quadratic term in the Taylor expansion of $V(x, y)$ near x_0, y_0. For the Hamiltonian (3.3), the Hamiltonian system (1.1) is linear and its solutions can be investigated by means of linear algebra. However, it turns out that the qualitative properties of the solutions of the boundary value problem for H are similar to those of its approximation (3.3) only for rather restrictive class of Hamiltonians, which will be called regular Hamiltonians. In the next section we give a complete description of this class and further we shall present an example showing that for non-regular Hamiltonians the solution of the boundary value problem may not exist even locally, even though it does exist for its quadratic approximation. In this section we investigate in detail the simplest and the most important examples of regular Hamiltonians, which correspond in the quadratic approximation to the case, when $k \geq n$ and the matrix $\frac{\partial a}{\partial y}(x_0, y_0)$ has maximal rank. For this type of Hamiltonian we shall make a special notational convention, namely we shall label the coordinates of the variables x and q by upper indices, and those of the variables y and p by low indices. The sense of this convention will be clear in the next chapter when considering the invariant diffusions corresponding to these Hamiltonians.

Definition. *The Hamiltonian of form (3.1) is called regular of the first rank of degeneracy, if $k \geq n$, g does not depend on y, the functions a, b, V are polynomials in y of degrees not exceeding 1,2 and 4 respectively with uniformly bounded coefficients depending on x, the polynomial V is bounded from below, and the rank of $\frac{\partial a}{\partial y}(x, y)$ everywhere equals to its maximal possible value n.*

Such a Hamiltonian can be written in the form

$$H = \frac{1}{2}(g(x)q, q) - (a(x) + \alpha(x)y, p) - (b(x) + \beta(x)y + \frac{1}{2}(\gamma(x)y, y), q) - V(x, y), \tag{3.4}$$

or more precisely as

$$H = \frac{1}{2}g_{ij}(x)q^i q^j - (a^i(x) + \alpha^{ij}y_j)p_i - (b_i(x) + \beta_i^j(x)y_j + \frac{1}{2}\gamma_i^{jl}y_j y_l)q^i - V(x, y), \tag{3.4'}$$

where $V(x, y)$ is a polynomial in y of degree ≤ 4, bounded from below, and $rank\ \alpha(x) = n$. The Hamiltonian system (1.1) for this Hamiltonian has the form

$$\begin{cases} \dot{x}^i = -(a^i(x) + \alpha^{ij}(x)y_j) \\ \dot{y}_i = -(b_i(x) + \beta^j_i(x)y_j + \frac{1}{2}\gamma^{jl}_i(x)y_jy_m) + g_{ij}(x)q^j \\ \dot{q}^i = \alpha^{ji}(x)p_j + (\beta^i_j(x) + \gamma^{il}_j(x)y_l)q^j + \frac{\partial V}{\partial y_i}(x,y) \\ \dot{p}_i = (\frac{\partial a^j}{\partial x^i} + \frac{\partial \alpha^{jl}}{\partial x^i}y_l)p_j + \left(\frac{\partial b_j}{\partial x^i} + \frac{\partial \beta^l_j}{\partial x^i}y_l + \frac{1}{2}\frac{\partial \gamma^{lm}_j}{\partial x^i}y_ly_m \right)q^j - \frac{1}{2}\frac{\partial g_{jl}}{\partial x^i}q^jq^l + \frac{\partial V}{\partial x^i}, \end{cases}$$

$$(3.5)$$

where for brevity we have omitted the dependence on x of the coefficients in the last equation.

Proposition 3.1. *There exist constants* K, t_0, *and* c_0 *such that for all* $c \in (0, c_0]$ *and* $t \in (0, t_0]$, *the solution of the system (3.5) with initial data* $(x(0), y(0), p(0), q(0))$ *exists on the interval* $[0, t]$ *whenever*

$$|y(0)| \leq \frac{c}{t}, \quad |q(0)| \leq \frac{c^2}{t^2}, \quad |p(0)| \leq \frac{c^3}{t^3}, \tag{3.6}$$

and on this interval

$$|x - x(0)| \leq Kt(1 + \frac{c}{t}), \quad |y - y(0)| \leq Kt(1 + \frac{c^2}{t^2}), \tag{3.7}$$

$$|q - q(0)| \leq Kt(1 + \frac{c^3}{t^3}), \quad |p - p(0)| \leq Kt(1 + \frac{c^4}{t^4}). \tag{3.8}$$

Proof. Estimating the derivatives of the magnitudes $|y|, |q|, |p|$ from (3.5), one sees that their growths do not exceed the growths of the solutions of the system

$$\begin{cases} \dot{y} = \sigma(1 + q + y^2) \\ \dot{q} = \sigma(1 + p + yq + y^3) \\ \dot{p} = \sigma(1 + yp + q^2 + y^4) \end{cases} \tag{3.9}$$

for some constant σ, and with initial values $y_0 = |y(0)|, q_0 = |q(0)|, p_0 = |p(0)|$. Suppose (3.6) holds. We claim that the solution of (3.9) exists on the interval $[0, t]$ as a convergent series in $\tau \in [0, t]$. For example, let us estimate the terms of the series

$$p(t) = p_0 + t\dot{p}_0 + \frac{1}{2}t^2\ddot{p}_0 + ...,$$

where the $p_0^{(j)}$ are calculated from (3.9). The main observation is the following. If one allocates the degrees 1,2 and 3 respectively to the variables y, q, and p, then the right hand sides of (3.9) have degrees 2, 3 and 4 respectively. Moreover, $p_0^{(j)}$ is a polynomial of degree $j + 3$. Therefore, one can estimate

$$p_0^{(j)} \leq \nu_j \sigma^j \left(1 + \left(\frac{c}{t}\right)^{j+3}\right)$$

where ν_j are natural numbers depending only on combinatorics. A rough estimate for ν_j is

$$\nu_j \le 4 \cdot (4 \cdot 4) \cdot (4 \cdot 5)...(4 \cdot (k+2)) = \frac{(j+2)!}{6} 4^j,$$

because each monomial of degree $\le d$ (with coefficient 1) can produce after differentiation in t not more than $4d$ new monomials of degree $\le (d+1)$ (again with coefficient 1). Consequently, the series for $p(t)$ is dominated by

$$p_0 + \sum_{j=1}^{\infty} \frac{t^j}{j!} \frac{(j+2)!}{6} (4\sigma)^j \left(1 + \left(\frac{c}{t}\right)^{j+3}\right) \le K \left(1 + \left(\frac{c}{t}\right)^3\right),$$

for some K, if $4\sigma c_0 < 1$ and $4\sigma t_0 < 1$. Using similar estimates for q and y we prove the existence of the solution of (3.5) on the interval $[0, t]$ with the estimates

$$|y| \le K \left(1 + \frac{c}{t}\right), \quad |q| \le K \left(1 + \left(\frac{c}{t}\right)^2\right), \quad |p| \le K \left(1 + \left(\frac{c}{t}\right)^3\right), \quad (3.10)$$

which directly imply (3.7).(3.8).

Using Proposition 3.1 we shall obtain now more precise formulae for the solution of (3.5) for small times. We shall need the development of the solutions of the Cauchy problem for (3.5) in Taylor series up to orders 1,2,3 and 4 respectively for p, q, y, and x. The initial values of the variables will be denoted by x_0, y^0, p^0, q_0. In order to simplify the forms of rather long expressions that appear after differentiating the equations of (3.5), it is convenient to use the following pithy notation: μ will denote an arbitrary (uniformly) bounded function in x and expressions such as μy^j will denote polynomials in $y = (y_1,...y_k)$ of degree j with coefficients of the type $O(\mu)$. Therefore, writing $p(t) = p(0) + \int_0^t \dot{p}(\tau) d\tau$ and using the last equation from (3.5) one gets

$$p = p(0) + \int_0^t [(\mu + \mu y)p + (\mu + \mu y + \mu y^2)q + \mu q^2 + \mu + \mu y + \mu y^2 + \mu y^3 + \mu y^4] d\tau. \quad (3.11)$$

Differentiating the third equation in (3.5) yields

$$\ddot{q}^i = \frac{\partial \alpha^{ji}}{\partial x^m} \dot{x}^m p_j + \alpha^{ji} \dot{p}_j + \left(\frac{\partial \beta_j^i}{\partial x^m} \dot{x}^m + \frac{1}{2} \frac{\partial \gamma_j^{im}}{\partial x^l} \dot{x}^l y_m + \gamma_j^{im} \dot{y}_m\right) q^j$$

$$+ (\beta_j^i + \gamma_j^{im} y_m)\dot{q}^j + \frac{\partial^2 V}{\partial y_i \partial x^m} \dot{x}^m + \frac{\partial^2 V}{\partial y_i \partial y_m} \dot{y}_m, \quad (3.12)$$

and from the representation $q(t) = q(0) + t\dot{q}(0) + \int_0^t (t - \tau)\ddot{q}(\tau) d\tau$ one gets

$$q^i = q_0^i + \left[\alpha^{ji}(x_0)p_j^0 + (\beta_j^i(x_0) + \gamma_j^{im}(x_0)y_m^0)q_0^j + \frac{\partial V}{\partial y_i}(x_0, y^0)\right] t$$

$$+ \int_0^t (t-\tau)[(\mu+\mu y)p+(\mu+\mu y+\mu y^2)q+\mu q^2+\mu+\mu y+\mu y^2+\mu y^3+\mu y^4]\,d\tau. \quad (3.13)$$

Furthermore,

$$\ddot{y}_i = \left(\frac{\partial b_i}{\partial x^m} + \frac{\partial \beta_i^j}{\partial x^m} y_j + \frac{1}{2} \frac{\partial \gamma_i^{jl}}{\partial x^m} y_j y_l \right) (a^m + a^{mk} y_k)$$

$$\dot{-} (\beta_i^j + \gamma_i^{jl} y_l) \left(g_{jm} q^m - (b_j + \beta_j^m y_m + \frac{1}{2} \gamma_j^{mk} y_m y_k) \right)$$

$$- \frac{\partial g_{ij}}{\partial x^m}(a^m + a^{ml} y_l)q^j + \left(a^{lj} p_l + (\beta_l^j + \gamma_l^{jm} y_m)q^l + \frac{\partial V}{\partial y_j} \right), \quad (3.14)$$

or, in concise notation,

$$\ddot{y} = \mu + \mu y + \mu y^2 + \mu y^3 + (\mu + \mu y)q + \mu p.$$

It follows that

$$y^{(3)} = (\mu + \mu y + \mu y^2 + \mu y^3 + \mu y^4) + (\mu + \mu y + \mu y^2)q + \mu q^2 + (\mu + \mu y)p. \quad (3.15)$$

Let $(\tilde{x}, \tilde{y}, \tilde{p}, \tilde{q}) = (\tilde{x}, \tilde{y}, \tilde{p}, \tilde{q})(t, x_0, y^0, 0, 0)$ denote the solution of (3.5) with initial condition $(x_0, y^0, 0, 0)$. From (3.14), (3.15) one obtains

$$y = \tilde{y} + t g(x_0)q_0 + \frac{1}{2}t^2 \left[g(x_0)\alpha(x_0)p^0 + (\omega + (\Omega y^0))q_0 \right]$$

$$+ \int_0^t (t-\tau)^2[(\mu+\mu y+\mu y^2+\mu y^3+\mu y^4)+(\mu+\mu y+\mu y^2)q+\mu q^2+(\mu+\mu y)p]\,d\tau,$$

$$(3.16)$$

where the matrices ω and Ωy^0 are given by

$$\omega_{il} = g_{ij}(x_0)\beta_l^j(x_0) - \beta_i^j(x_0)g_{jl}(x_0) - \frac{\partial g_{il}}{\partial x^j}(x_0)a_j(x_0), \quad (3.17)$$

$$(\Omega y^0)_{il} = \left(g_{ij}(x_0)\gamma_l^{jm}(x_0)g_{jl}(x_0) - \frac{\partial g_{il}}{\partial x^j}(x_0)\alpha_{jm}(x_0) \right) y_m^0. \quad (3.18)$$

Differentiating the first equation in (3.5) yields

$$\ddot{x}^i = \left(\frac{\partial a^i}{\partial x^m} + \frac{\partial a^{ij}}{\partial x^m} y_j \right) (a^m + a^{ml} y_l) - \alpha_{ij}\dot{y}_j,$$

and therefore

$$(x^{(3)})^i = \left[\left(2\frac{\partial a^{ij}}{\partial x^m}a^{ml} + \frac{\partial a^{il}}{\partial x^m}a^{mj} \right) y_l + 2\frac{\partial a^{ij}}{\partial x^m}a^m + \frac{\partial a^i}{\partial x^m}a^{mj} \right] \dot{y}_j$$

$$-\alpha^{ij}\ddot{y}_j + \mu + \mu y + \mu y^2 + \mu y^3.$$

In particular, the consice formula for $x^{(3)}$ is the same as for $\ddot{y}$ and for $\dot{q}$. Let us write now the formula for x which one gets by keeping three terms of the Taylor series:

$$x = \tilde{x} - \frac{1}{2}t^2\alpha(x_0)g(x_0)q_0 - \frac{1}{6}t^3[(\alpha g\alpha')(x_0)p^0 + (\omega' + (\Omega'y^0))q_0]$$

$$+ \int_0^t (t-\tau)^3[(\mu + \mu y + \mu y^2 + \mu y^3 + \mu y^4) + (\mu + \mu y + \mu y^2)q + \mu q^2 + (\mu + \mu y)p]\, d\tau,$$

$$(3.19)$$

where the entries of the matrices ω' and $\Omega'y^0$ are given by

$$(\omega')_k^i = \alpha^{ij}(x_0)\omega_{jk} - \left(2\frac{\partial\alpha^{ij}}{\partial x^m}a^m + \frac{\partial a^i}{\partial x^m}\alpha^{mj}\right)(x_0)g_{jk}(x_0), \qquad (3.20)$$

$$(\Omega'y^0)_k^i = \alpha^{ij}(x_0)(\Omega y^0)_{jk} - \left(2\frac{\partial\alpha^{ij}}{\partial x^m}a^{ml} + \frac{\partial a^{il}}{\partial x^m}\alpha^{mj}\right)(x_0)g_{jk}(x_0)y_l^0. \quad (3.21)$$

One can now obtain the asymptotic representation for the solutions of (3.5) and their derivatives with respect to the initial momenta in the form needed to prove the main result on the well-posedness of the boundary value problem. In the following formulas, δ will denote an arbitrary function of order $O(t+c)$, and α_0 will denote the matrix $\alpha(x_0)$, with similar notation for other matrices at initial point x_0. Expanding the notations of Sect. 2.1 we shall denote by $(X, Y, P, Q)(t, x_0, y^0, p^0, q_0)$ the solution of the Cauchy problem for (3.5).

Proposition 3.2. *Under the assumptions of Proposition 3.1, for the solutions of (3.5) one has*

$$X = \tilde{x} - \frac{1}{2}t^2\alpha_0 g_0 q_0 - \frac{1}{6}t^3[\alpha_0 g_0\alpha_0'p^0 + (\omega' + (\Omega'y^0))q_0] + \delta^4, \qquad (3.22)$$

$$Y = \tilde{y} + tg_0 q_0 + \frac{1}{2}t^2[g_0\alpha_0'p^0 + (\omega + (\Omega y^0))q_0] + \frac{1}{t}\delta^4, \qquad (3.23)$$

where the matrices $\omega, \omega', (\Omega y^0), (\Omega'y^0)$ are defined by (3.17),(3.18),(3.20),(3.21), and the solution $(\tilde{x}, \tilde{y}, \tilde{p}, \tilde{q})$ of (3.5) with initial conditions $(x_0, y^0, 0, 0)$ is given by

$$\tilde{x} = x_0 - (a_0 + \alpha_0 y^0)t + O(t^2), \quad \tilde{y} = y^0 + O(t), \qquad (3.24)$$

$$\tilde{p} = O(t), \quad \tilde{q} = O(t). \qquad (3.25)$$

Moreover,

$$\frac{\partial X}{\partial p^0} = -\frac{1}{6}t^3(\alpha_0 g_0\alpha_0' + \delta), \quad \frac{\partial X}{\partial q_0} = -\frac{1}{2}t^2(\alpha_0 g_0 + \frac{t}{3}(\omega' + \Omega'y^0) + \delta^2), \quad (3.26)$$

$$\frac{\partial Y}{\partial p^0} = \frac{1}{2}t^2(g_0\alpha_0' + \delta), \quad \frac{\partial Y}{\partial q_0} = t(g_0 + \frac{t}{2}(\omega + \Omega y^0) + \delta^2), \qquad (3.27)$$

$$\frac{\partial P}{\partial p^0} = 1 + \delta, \quad \frac{\partial P}{\partial q_0} = \frac{\delta^2}{t}, \tag{3.28}$$

$$\frac{\partial Q}{\partial p^0} = t(\alpha_0' + \delta), \quad \frac{\partial Q}{\partial q_0} = 1 + t(\beta_0' + \gamma' y^0) + \delta^2, \tag{3.29}$$

where $(\beta_0' + \gamma_0' y^0)_j^i = \beta_j^i(x_0) + \gamma_j^{im}(x_0) y_m^0$.

Proof. (3.22)-(3.25) follow directly from (3.16), (3.19) and the estimates (3.10). They imply also that the matrices

$$v_1 = \frac{1}{t^3}\frac{\partial X}{\partial p^0}, \quad u_1 = \frac{1}{t^2}\frac{\partial Y}{\partial p^0}, \quad v_2 = \frac{1}{t^2}\frac{\partial X}{\partial q_0}, \quad u_2 = \frac{1}{t}\frac{\partial Y}{\partial q_0}, \tag{3.30}$$

are bounded (on the time interval defined by Proposition 3.1). Let us consider them as elements of the Banach space of continuous matrix-valued functions on $[0, t]$. Differentiating (3.11) with respect to p^0, q^0 and using (3.10) yields

$$\frac{\partial P}{\partial p^0} = 1 + v_1 O(tc^3 + c^4 + t^2 c^2) + u_1 O(c^3 + tc^2) + \frac{\partial P}{\partial p^0}O(t + c) + \frac{\partial Q}{\partial p^0}O(t + c + \frac{c^2}{t}), \tag{3.31}$$

$$\frac{\partial P}{\partial q_0} = v_2 O(c^3 + \frac{c^4}{t} + tc^2) + u_2 O(\frac{c^3}{t} + c^2) + \frac{\partial P}{\partial q_0}O(t + c) + \frac{\partial Q}{\partial q_0}O(t + c + \frac{c^2}{t}). \tag{3.32}$$

Similarly, from (3.13)

$$\frac{\partial Q}{\partial p^0} = t\alpha_0 + v_1 O(t^2 c^3 + tc^4 + t^3 c^2) + u_1 O(tc^3 + t^2 c^2) + \frac{\partial P}{\partial p^0}O(t^2 + ct) + \frac{\partial Q}{\partial p^0}O(t^2 + c^2), \tag{3.33}$$

$$\frac{\partial Q}{\partial q_0} = 1 + t(\beta_0' + \gamma_0' y^0) + v_2 O(tc^3 + c^4 + t^2 c^2) + u_2 O(c^3 + tc^2)$$

$$+ \frac{\partial P}{\partial q_0}O(t^2 + tc) + \frac{\partial Q}{\partial q^0}O(t^2 + tc + c^2). \tag{3.34}$$

From (3.33) one has

$$\frac{\partial Q}{\partial p^0} = (1 + \delta^2)\left[t\delta\frac{\partial P}{\partial p^0} + v_1 t\delta^4 + u_1 t\delta^3 + t\alpha_0\right]. \tag{3.35}$$

Inserting this in (3.31) yields

$$\frac{\partial P}{\partial p^0} = 1 + \delta\frac{\partial P}{\partial p^0} + v_1 \delta^4 + u_1 \delta^3,$$

and therefore

$$\frac{\partial P}{\partial p^0} = 1 + \delta + v_1 \delta^4 + u_1 \delta^3, \tag{3.36}$$

$$\frac{\partial Q}{\partial p^0} = t(\alpha_0 + \delta + v_1 \delta^4 + u_1 \delta^3). \tag{3.37}$$

Similarly, from (3.32),(3.34) one gets

$$\frac{\partial P}{\partial q_0} = \frac{1}{t}(\delta^2 + v_2\delta^4 + u_2\delta^3), \tag{3.38}$$

$$\frac{\partial Q}{\partial q_0} = 1 + t(\beta_0' + \gamma_0'y^0) + \delta^2 + v_2\delta^4 + u_2\delta^3. \tag{3.39}$$

Furthermore, differentiating (3.16) with respect to p_0, q_0 yields

$$u_1 = \frac{1}{2}g_0\alpha_0' + v_1\delta^4 + u_1\delta^3 + \frac{\partial P}{\partial p_0}\delta + \frac{\partial Q}{\partial p_0}\frac{\delta^2}{t},$$

$$u_2 = g_0 + \frac{t}{2}(\omega + \Omega y^0) + v_2\delta^4 + u_2\delta^3 + \frac{\partial P}{\partial q_0}\delta t + \frac{\partial Q}{\partial q_0}\delta^2.$$

By (3.36)-(3.39), this implies

$$u_1 = \frac{1}{2}g_0\alpha_0' + \delta + v_1\delta^4 + u_1\delta^3,$$

$$u_2 = g_0 + \frac{t}{2}(\omega + \Omega y^0) + \delta^2 + v_2\delta^4 + u_2\delta^3.$$

Similarly, differentiating (3.19) yields

$$v_1 = -\frac{1}{6}\alpha_0 g_0\alpha_0' + \delta + v_1\delta^4 + u_1\delta^3,$$

$$v_2 = -\frac{1}{2}\alpha_0 g_0 - \frac{t}{6}(\omega' + \Omega'y^0) + \delta^2 + v_2\delta^4 + u_2\delta^3.$$

From the last 4 equations one easily obtains

$$u_1 = \frac{1}{2}g_0\alpha_0' + \delta, \quad u_2 = g_0 + \frac{t}{2}(\omega + \Omega y^0) + \delta^2,$$

$$v_1 = -\frac{1}{6}\alpha_0 g_0\alpha_0' + \delta, \quad v_2 = -\frac{1}{2}\alpha_0 g_0 - \frac{t}{6}(\omega' + \Omega'y^0) + \delta^2,$$

which is equivalent to (3.26)-(3.29). Formulas (3.30), (3.31) then follow from (3.36)-(3.39).

We shall prove now the main result of this section.

Theorem 3.1. *(i) There exist positive real numbers c and t_0 (depending only on x_0) such that for all $t \le t_0$ and $\|y\| \le c/t$, the mapping $(p^0, q_0) \mapsto (X, Y)(t, x_0, y^0, p^0, q_0)$ defined on the polydisc $B_{c^3/t^3} \times B_{c^2/t^2}$ is a diffeomorphism onto its image.*

(ii) There exists $r > 0$ such that by reducing c and t_0 if necessary, one can assume that the image of this diffeomorphism contains the polydisc $B_{r/t}(\tilde{y}) \times B_r(\tilde{x})$. These c, t_0, r can be chosen smaller than an arbitrary positive number δ.

(iii) Assume the matrix $(\alpha(x)g(x)\alpha'(x))^{-1}$ is uniformly bounded. Then c, t_0, r do not depend on x_0.

Proof. (i) From Proposition 3.2, one gets

$$\begin{pmatrix} X \\ Y \end{pmatrix}(t, x_0, y^0, p^0, q_0) - \begin{pmatrix} X \\ Y \end{pmatrix}(t, x_0, y^0, \pi^0, \xi_0) = tG_\delta(t, x_0)\begin{pmatrix} p^0 - \pi^0 \\ q_0 - \xi_0 \end{pmatrix},$$

where

$$G_\delta(t, x_0) = \begin{pmatrix} -\frac{1}{6}t^2(\alpha_0 g_0 \alpha_0' + \delta) & -\frac{1}{2}t(\alpha_0 g_0 + \delta) \\ \frac{1}{2}t(g_0 \alpha_0' + \delta) & g_0 + \delta \end{pmatrix}. \tag{3.40}$$

To prove (i), it is therefore enough to prove that the matrix (3.40) is invertible. We shall show that its determinant does not vanish. Using the formula

$$\det \begin{pmatrix} a & b \\ c & d \end{pmatrix} = \det d \cdot \det(a - bd^{-1}c) \tag{3.41}$$

(which holds for arbitrary matrices a, b, c, d with invertible d) one gets

$$\det G_\delta(t, x_0) = \det(g_0 + \delta)$$

$$\times \det\left[-\frac{1}{6}t^2(\alpha_0 g_0 \alpha_0' + \delta) + \frac{1}{4}t^2(\alpha_0 g_0 + \delta)(g_0^{-1} + \delta)(g_0 \alpha_0' + \delta)\right].$$

The second factor is proportional to $\det(\alpha_0 g_0 \alpha_0' + \delta)$ and therefore, neither factor vanishes for small δ.

(ii) As in the proof of Theorem 2.1, one notes that the existence of (x, y) such that $(x, y) = (X, Y)(t, x_0, y^0, p^0, q_0)$ is equivalent to the existence of a fixed point of the map

$$F_{x,y}\begin{pmatrix} p^0 \\ q_0 \end{pmatrix} = \begin{pmatrix} p^0 \\ q_0 \end{pmatrix} + t^{-1}G(t, x_0)^{-1}\begin{pmatrix} x - X(t, x_0, y^0, p^0, q_0) \\ y - Y(t, x_0, y^0, p^0, q_0) \end{pmatrix},$$

where

$$G(t, x_0) = \begin{pmatrix} -\frac{1}{6}t^2\alpha_0 g_0 \alpha_0' & -\frac{1}{2}t(\alpha_0 g_0 + \frac{1}{3}t(\omega' + \Omega'y^0)) \\ \frac{1}{2}tg_0\alpha_0' & g_0 + \frac{t}{2}(\omega + \Omega y^0) \end{pmatrix}.$$

By (3.22), (3.23),

$$F_{x,y}\begin{pmatrix} p^0 \\ q_0 \end{pmatrix} = t^{-1}G(t, x_0)^{-1}\begin{pmatrix} x - \tilde{x} + \delta^4 \\ y - \tilde{y} + \delta^4/t \end{pmatrix}.$$

To estimate $G(t, x_0)^{-1}$ notice first that $G(t, x_0)$ is of form (3.40) and therefore it is invertible. Moreover, by the formula

$$\begin{pmatrix} a & b \\ c & d \end{pmatrix}^{-1} = \begin{pmatrix} (a - bd^{-1}c)^{-1} & -a^{-1}b(d - ca^{-1}b)^{-1} \\ -d^{-1}c(a - bd^{-1}c)^{-1} & (d - ca^{-1}b)^{-1} \end{pmatrix} \tag{3.42}$$

(which holds for an invertible matrix $\begin{pmatrix} a & b \\ c & d \end{pmatrix}$ with invertible blocks a, d), one gets that $G(t, x_0)$ has the form

$$\begin{pmatrix} O(t^{-2}) & O(t^{-1}) \\ O(t^{-1}) & O(1) \end{pmatrix}.$$

Therefore, to prove that $F_{x,y}(p^0, q_0) \in B_{c^3/t^3} \times B_{c^2/t^2}$ one must show that

$$O(\|x - \tilde{x}\| + t\|y - \tilde{y}\| + \delta^4) \leq c^3,$$

which is certainly satisfied for small c, t_0 and r.

(iii) Follows directly from the above proof, because all parameters depend on the estimates for the inverse of the matrix (3.40).

Since $\tilde{x} - x_0 = O(t)$ for any fixed y^0, it follows that for arbitrary (x_0, y^0) and sufficiently small t, the polydisc $B_{r/2}(x_0) \times B_{r/2t}(y_0)$ belongs to the polydisc $B_r(\tilde{x}) \times B_{r/t}(\tilde{y})$. Therefore, due to Theorem 3.1, all the assumptions of Proposition 1.1 are satisfied for Hamiltonians (3.4), and consequently, all the results of Proposition 1 and its corollaries hold for these Hamiltonians. We shall prove now, following the same line of arguments as in the previous section, that the two-point function $S(t, x, y; x_0, y^0)$ in fact gives the global minimum for the corresponding functional (1.12). To get the necessary estimates of S, we need a more precise formula for the inverse of the matrix (3.40).

Let us express $G_\delta(t, x_0)$ in the form

$$G_\delta(t, x_0) = \begin{pmatrix} 1 & 0 \\ 0 & \sqrt{g_0} \end{pmatrix} \begin{pmatrix} -\frac{1}{6}t^2(A_0 A_0' + \delta) & -\frac{1}{2}(A_0 + \delta) \\ \frac{1}{2}t(A_0' + \delta) & 1 + \delta \end{pmatrix} \begin{pmatrix} 1 & 0 \\ 0 & \sqrt{g_0} \end{pmatrix},$$

where $A_0 = \alpha_0\sqrt{g_0}$. Denoting the matrix $g_0^{-1/2}$ by J and using (3.42) yields that $G_\delta(t, x_0)^{-1}$ equals

$$\begin{pmatrix} 1 & 0 \\ 0 & J \end{pmatrix} \begin{pmatrix} \frac{12}{t^2}(A_0 A_0')^{-1}(1 + \delta) & \frac{3}{t}(A_0 A_0')^{-1} A_0 \beta^{-1}(1 + \delta) \\ -\frac{6}{t}A_0'(A_0 A_0')^{-1}(1 + \delta) & -\beta^{-1} + \delta \end{pmatrix} \begin{pmatrix} 1 & 0 \\ 0 & J \end{pmatrix},$$

$$(3.43)$$

where $-\beta = 1 - \frac{3}{2}A_0'(A_0 A_0')^{-1} A_0$. In the simplest case, when $k = n$, i.e. when the matrix $\alpha(x)$ is square non-degenerate, (3.43) reduces to

$$G_\delta(t, x_0)^{-1} = \begin{pmatrix} \frac{12}{t^2}(\alpha_0 g_0 \alpha_0')^{-1}(1 + \delta) & \frac{6}{t}(g_0 \alpha_0')^{-1}(1 + \delta) \\ -\frac{6}{t}(\alpha_0 g_0)^{-1}(1 + \delta) & -2g_0^{-1} + \delta \end{pmatrix}. \qquad (3.44)$$

To write down the formula for the general case, let us decompose $Y = \mathcal{R}^k$ as the orthogonal sum $Y = Y_1 \oplus Y_2$, where $Y_2 = Ker\, A_0$. Let us denote again by A_0 the restriction of A_0 to Y_1. Then A_0^{-1} exists and $\beta|_{Y_1} = -\frac{1}{2}, \beta|_{Y_2} = 1$. With respect to this decomposition,

$$G_\delta(t, x)^{-1} = \begin{pmatrix} 1 & 0 & 0 \\ 0 & J_{11} & J_{12} \\ 0 & J_{21} & J_{22} \end{pmatrix}$$

$$\times \begin{pmatrix} \frac{12}{t^2}(A_0 A_0')^{-1}(1+\delta) & \frac{6}{t}(A_0')^{-1}(1+\delta) & \frac{\delta}{t} \\ -\frac{6}{t}A_0^{-1}(1+\delta) & -2+\delta & \delta \\ \frac{\delta}{t} & \delta & 1+\delta \end{pmatrix} \begin{pmatrix} 1 & 0 & 0 \\ 0 & J_{11} & J_{12} \\ 0 & J_{21} & J_{22} \end{pmatrix}. \qquad (3.45)$$

From (1.19), (3.30), (3.31) one gets

$$\frac{\partial^2 S}{\partial(x,y)^2}(t,x,y,x_0,y^0) = \frac{1}{t}\begin{pmatrix} 1+\delta & \delta^2/t \\ t(\alpha_0'+\delta) & 1+\delta \end{pmatrix} G_\delta(t,x_0)^{-1}. \qquad (3.46)$$

In the case $n = k$, this takes the form

$$\frac{\partial^2 S}{\partial(x,y)^2}(t,x,y,x_0,y^0) = \begin{pmatrix} \frac{12}{t^3}(\alpha_0 g_0 \alpha_0')^{-1}(1+\delta) & \frac{6}{t^2}(g_0 \alpha_0')^{-1}(1+\delta) \\ \frac{6}{t^2}(\alpha_0 g_0)^{-1}(1+\delta) & \frac{4}{t}g_0^{-1}+\delta \end{pmatrix}, \qquad (3.47)$$

and in the general case, from (3.45)-(3.46) one gets (omitting some arguments for brevity) that $\partial^2 S/\partial(x,y)^2$ equals

$$= \begin{pmatrix} \frac{12}{t^3}((A_0 A_0')^{-1}+\delta) & \frac{6}{t^2}((A_0')^{-1}J_{11}+\delta) & \frac{6}{t^2}((A_0')^{-1}J_{12}+\delta) \\ \frac{6}{t^2}(J_{11}A_0^{-1}+\delta) & \frac{1}{t}(4J_{11}^2+J_{12}J_{21}+\delta) & \frac{1}{t}(4J_{11}J_{12}+J_{12}J_{22}+\delta) \\ \frac{6}{t^2}(J_{21}A_0^{-1}+\delta) & \frac{1}{t}(4J_{21}J_{11}+J_{22}J_{21}+\delta) & \frac{1}{t}(4J_{21}J_{12}+J_{22}^2+\delta) \end{pmatrix}.$$

$$(3.48)$$

Therefore, from (1.22) one gets the following.

Proposition 3.3. *For small t and (x,y) in a neighbourhood of (x_0, y^0) the two-point function $S(t,x,y;x_0,y^0)$ is equal to*

$$\frac{6}{t^3}((A_0^{-1}+\delta)(x-\tilde{x}), A_0^{-1}(x-\tilde{x})) + \frac{6}{t^2}((J_{11}+\delta)(y_1-\tilde{y}_1),(A_0^{-1}+\delta)(x-\tilde{x}))$$

$$+\frac{6}{t^2}((J_{12}+\delta)(y_2-\tilde{y}_2),(A_0^{-1}+\delta)(x-\tilde{x}))$$

$$+\frac{1}{2t}(y_1-\tilde{y}_1,(4J_{11}^2+J_{12}J_{21}+\delta)(y_1-\tilde{y}_1))$$

$$+\frac{1}{2t}(y_2-\tilde{y}_2,(4J_{21}J_{12}+J_{22}^2+\delta)(y_2-\tilde{y}_2))$$

$$+\frac{1}{t}(y_1-\tilde{y}_1,(4J_{11}J_{12}+J_{21}J_{22}+\delta)(y_2-\tilde{y}_2)). \qquad (3.49)$$

Expanding $\tilde{x},\tilde{y}$ in t we can present this formula in terms of x_0, y^0. Let us write down the main term (when $\delta = 0$), for brevity only in the case $k = n$:

$$S_0(t,x,y,x_0,y^0) = \frac{6}{t^3}(g_0^{-1}\alpha_0^{-1}(x-x_0),\alpha_0^{-1}(x-x_0))$$

$$+\frac{6}{t^2}(g_0^{-1}\alpha_0^{-1}(x-x_0),y+y_0+2\alpha_0^{-1}a_0)$$

$$+\frac{2}{t}[(y,g_0^{-1}y)+(y,g_0^{-1}y_0)+(y_0,g_0^{-1}y_0)]$$

$$+\frac{6}{t}\left[(y + y_0, g_0^{-1}\alpha_0^{-1}a_0) + (g_0^{-1}\alpha_0^{-1}a_0, \alpha_0^{-1}a_0)\right].$$

Equivalently, this can be expressed in a manifestly positive form

$$S_0(t, x, y, x_0, y^0) = \frac{1}{2t}(y - y_0, g_0^{-1}(y - y_0)) + \frac{6}{t^3}(g_0^{-1}z, z) \qquad (3.50)$$

with

$$z = \alpha_0^{-1}(x - x_0) + \frac{t}{2}(y + y_0 + 2\alpha_0^{-1}a_0).$$

Using these expressions one proves the following property of extremals quite similarly to the proof of Proposition 2.4.

Proposition 3.4. *There exists $r' \leq r$ such that for $(x, y) \in B_{r'/2}(x_0) \times B_{r'/2t}(y_0)$, the solution of the boundary value problem $(x, y)(0) = (x_0, y^0)$ and $(x, y)(t) = (x, y)$ for system (3.5), which exists and is unique under the additional assumption $(p^0, q_0) \in B_{c^3/t^3} \times B_{c^2/t^2}$, furnishes the absolute minimum for the functional (1.12) corresponding to the Hamiltonian (3.4).*

Proposition 3.5. *Let the absolute minimum of the functional (1.12) for Hamiltonian (3.4) be given by a piecewise-smooth curve. Then this curve is in fact a smooth characteristic and it contains no conjugate points.*

Proof. The first part is proved as in Proposition 2.5. Let us prove the Jacobi condition for the degenerate Hamiltonian (3.4), i.e. that a minimising characteristic does not contain conjugate points. First of all we claim that if a solution $(x(\tau), p(\tau))$ of the Hamiltonian system (1.1) corresponding to an arbitrary Hamiltonian of form (2.1) with non-negative-definite (but possibly singular) matrix G furnishes a minimum for the functional (1.12), then the curve $\eta(\tau) = 0$ gives a minimum for the functional (1.12) defined on curves with fixed endpoints $\eta(0) = \eta(t) = 0$ and corresponding to the quadratic time dependent Hamiltonian $\tilde{H}_t$ of form (1.17). In fact, let the extremal $(x(\tau), p(\tau))$ furnish a minimum for (1.12) corresponding to the Hamiltonian (2.1). This implies that for arbitrary $\epsilon > 0$ and smooth $\eta(\tau)$ such that $\eta(0) = \eta(t) = 0$:

$$\int_0^t \max_w [(p(\tau) + \epsilon w)(\dot{x}(\tau) + \epsilon \dot{\eta}(\tau))$$

$$-H(x + \epsilon\eta, p(\tau) + \epsilon w) - p(\tau)\dot{x}(\tau) + H(x(\tau), p(\tau))]\, d\tau \geq 0.$$

Since H is a polynomial in p of second degree, one has

$$H(x + \epsilon\eta, p + \epsilon w) = H(x, p) + \epsilon\frac{\partial H}{\partial x}(x, p)\eta + \epsilon\frac{\partial H}{\partial p}(x, p)w + O(\epsilon^3)$$

$$+\frac{1}{2}\epsilon^2\left[\left(\frac{\partial^2 H}{\partial x^2}(x, p)\eta, \eta\right) + 2\left(\frac{\partial^2 H}{\partial x \partial p}(x, p)(\eta + O(\epsilon)), w\right)\right.$$

$$\left. + \left(\left(\frac{\partial^2 H}{\partial p^2}(x, p) + O(\epsilon)\right)w, w\right)\right]$$

with $O(\epsilon)$ independent of w. Substituting this expression in the previous formula, integrating by parts the term $p(\tau)\dot{\eta}$, then using equations (1.1) and dividing by ϵ^2 yields

$$\int_0^t \max_w[w\dot{\eta} - \frac{1}{2}\left(\frac{\partial^2 H}{\partial x^2}(x,p)\eta, \eta\right) - 2\left(\frac{\partial H}{\partial x \partial p}(x,p)(\eta + O(\epsilon)), w\right)$$

$$-\frac{1}{2}\left(\left(\frac{\partial^2 H}{\partial p^2}(x,p) + O(\epsilon)\right)w, w\right)]\,d\tau + O(\epsilon) \geq 0.$$

Taking the limit as $\epsilon \to 0$ one gets

$$\int_0^t \max_w(w\dot{\eta} - \tilde{H}_t(\eta, w))\,d\tau \geq 0,$$

as claimed.

Now let H have the form (3.4). Then $\tilde{H}_t$ has this form as well, only it is time dependent. One sees easily that with this additional generalisation the analogue of Theorem 3.1 is still valid and therefore, the first part of Proposition 3.5 as well. Suppose now that on a characteristic $x(\tau)$ of (3.5) that furnishes minimum for the corresponding functional (1.12), the points $x(s_1), x(s_2)$ are conjugate, where $0 \leq s_1 < s_2 < t$. Then there exists a solution v, w of the Jacobi equation (1.16) on the interval $[s_1, s_2]$ such that $v(s_1) = v(s_2) = 0$ and v is not identically zero. Then the curve $\tilde{v}(s)$ on $[0, t]$, which is equal to v on $[s_1, s_2]$ and vanishes outside this interval, gives a minimum for the functional (1.12) corresponding to Hamiltonian $\tilde{H}_t$. But this curve is not smooth at $\tau = s_2$, which contradicts to the first statement of Proposition 3.5.

We can now prove the analogue of Tonelli's theorem for the Hamiltonian (3.4), namely the global existence of the boundary value problem.

Proposition 3.6. *For arbitrary $t > 0$ and arbitrary x_0, y^0, x, y, there exists a solution of the Hamiltonian system (3.5) with boundary conditions $(x, y)(0) = (x_0, y^0)$, $(x, y)(t) = (x, y)$, which furnish global minimum for the corresponding functional (1.12).*

Proof. The only difference from the proof of Proposition 3.5 is that the radius of balls can depend on y^0, but this is not of importance, because the proof is given by means of exhausting $\mathcal{R}^m$ by compact sets.

As a consequence, we have

Proposition 3.7. *Propositions 2.7 and 2.8 hold also for Hamiltonians of the form (3.4).*

To conclude we give some estimates on the derivatives of the two-point function

Proposition 3.8. *For arbitrary j and $l \leq j$*

$$\frac{\partial^j S}{\partial x^l \partial y^{j-l}}(t, \tilde{x}, \tilde{y}, x_0, y^0) = t^{-(l+1)} R(t, y_0),$$

where $R(t, y_0)$ has a regular asymptotic expansion in powers of t and y^0.

   Proof. This is proved by induction on j using (1.19).

   This Proposition, together with (1.19), suggests that the function $tS(t, \tilde{x} + x, \tilde{y} + y; x_0, y^0)$ can be expressed as a regular asymptotic expansion in the variables x/t and y. This important consequence will be used in the next chapter for effective calculations of the two-point function. We shall also need there estimates for the higher derivatives of the solutions of the Cauchy problem for (3.5) with respect to initial momenta, which one easily gets from Theorem 3.1 together with the Taylor expansion of the solutions up to any order.

   Proposition 3.9. *Let x^0, x^1, p^0, p^1 denote respectively x, y, p, q. The following estimates hold*

$$\frac{\partial^2 X^I}{\partial p_0^J \partial p_0^K} = O\left(t^{6-I-J-K}\right).$$

More generally, if H has sufficiently many bounded derivatives, then

$$\frac{\partial^K X^I}{\partial p_0^{I_1} \dots \partial p_0^{I_K}} = O\left(t^{3K-I-I_1-\dots-I_K}\right).$$

4. General regular Hamiltonians depending quadratically on momenta

   We now consider here general regular Hamiltonians (RH). These are the Hamiltonians for which, roughly speaking, the boundary-value problem enjoys the same properties as for their quadratic (or Gaussian) approximation. As we shall see in the next chapter, the main term of the small time asymptotics for the corresponding diffusion is then also the same as for the Gaussian diffusion approximation. In fact, the motivation for the following definition will be better seen when we consider formal power series solutions of the corresponding Hamilton-Jacobi equation in the next chapter, but rigorous proofs seem to be simpler to carry out for boundary value-problem for Hamiltonian systems.

   Since the Gaussian diffusions were classified in the previous chapter by means of the Young schemes, it is clear that RH should also be classified by these schemes.

   Definition. *Let $\mathcal{M} = \{m_{M+1} \geq m_M \geq \dots \geq m_0 > 0\}$ be a non-degenerate sequence of positive integers (Young scheme). Let X^I denote Eucleadian space $\mathcal{R}^{m_I}$ of dimension m_I with coordinates x^I, $I = 0, \dots, M$, and $Y = X^{M+1} = \mathcal{R}^{M+1}$. Let $p_I, I = 0, \dots, M$, and $q = p^{M+1}$ be the momenta corresponding to x^I and y respectively. The $\mathcal{M}$-degree, $\deg_{\mathcal{M}} P$, of a polynomial P in the variables $x^1, \dots, x^M, y = x^{M+1}$ is by definition the degree, which one gets prescribing the degree I to the variable x^I, $I = 0, \dots, M + 1$. A RH corresponding to a given Young scheme is by definition a function of the form*

$$H(x, y, p, q) = \frac{1}{2}(g(x^0)q, q) - R_1(x, y)p_0 - \dots$$

$$-R_{M+1}(x,y)p_M - R_{M+2}(x,y)q - R_{2(M+2)}(x,y), \tag{4.1}$$

where the $R_I(x,y)$ are (vector-valued) polynomials in the variables $x^1,...x^M, y = x^{M+1}$ of the $\mathcal{M}$-degree $\deg_{\mathcal{M}} R_I = I$ with smooth coefficients depending on x^0, and $g(x^0)$ depends only on the variable x^0 and is nondegenerate everywhere. Moreover, the matrices $\frac{\partial R_I}{\partial x^I}$ (which, due to the condition on $\deg_{\mathcal{M}} R_I$, depend only on x^0) have everywhere maximal rank, equal to m_{I-1}, and the polynomial $R_{2(M+1)}$ is bounded from below. When the coefficients of the polynomials R_I are uniformly bounded in x_0, we shall say that the RH has bounded coefficients.

All results of the previous section hold for this more general class of Hamiltonians with clear modifications. The proofs are similar, but with notationally heavier. We omit the details and give only the main estimates for the derivatives of the solution of the corresponding Hamilton system with respect to the initial momenta. These estimates play a central role in all proofs. To obtain these estimates, one should choose the convenient coordinates in a neighbourhood of initial point, which were described in the previous chapter, in Theorem 1.2.1. Let us note also that the assumption of the boundedness of the coefficients of the polynomials in (4.1) insures the uniformity of all estimates with respect to the initial value x_0^0, and is similar to the assumptions of boundedness of the functions A, V, G defining the non-degenerate Hamiltonians of Section 2.

Theorem 4.1. *There exist positive constants K, t_0, c_0 such that for all $c \in (0, c_0], t \in (0, t_0]$ the solution of the Hamiltonian system (1.1) corresponding to the regular Hamiltonian (4.1) exists on the interval $[0, t]$ whenever the initial values of the variables satisfy the estimates*

$$|x_0^1| \le \frac{c}{t}, ..., |x_0^{M+1}| \le \left(\frac{c}{t}\right)^{M+1},$$

$$|p_0^{M+1}| \le \left(\frac{c}{t}\right)^{M+2}, ..., |p_0^1| \le \left(\frac{c}{t}\right)^{2M+2}, \quad |p_0^0| \le \left(\frac{c}{t}\right)^{2M+3}.$$

On the interval $0 < t < t_0$ the growth of the solution is governed by the estimates

$$|X(t)^I| \le K\left(1 + \left(\frac{c}{t}\right)^I\right), \quad |P(t)^I| \le K\left(1 + \left(\frac{c}{t}\right)^{2M+3-I}\right), \quad I = 0, ..., M+1, \tag{4.2}$$

the derivatives with respect to initial momenta have the form

$$\left[\frac{\partial(X^0, ..., X^{M+1})}{\partial(P_0^0, ..., P_0^{M+1})}\right]_{IJ} = t^{2M+3-I-J}\beta_{IJ}(1 + O(t)), \tag{4.3}$$

$$\left[\frac{\partial(X^0, ..., X^{M+1})}{\partial(P_0^0, ..., P_0^{M+1})}\right]_{IJ}^{-1} = t^{-(2M+3-I-J)}\gamma_{IJ}(1 + O(t)), \tag{4.4}$$

where β_{IJ}, γ_{IJ} are matrices of the maximal rank $\min(m_I, m_J)$, and for higher derivatives one has the estimates

$$\frac{\partial^K X^I}{\partial p_0^{I_1}...\partial p_0^{I_K}} = O\left(t^{(3+2M)K-I-I_1-...-I_K}\right). \tag{4.5}$$

Clearly, the Lagrangians corresponding to degenerate Hamiltonians are singular. However, it turns out that the natural optimisation problems corresponding to degenerate regular Hamiltonians are problems of the calculus of variations for functionals depending on higher derivatives. To see this, consider first a Hamiltonian (3.1) such that $n = k$ and the map $y \mapsto a(x,y)$ is a diffeomorphism for each x. Then the change of variables $(x,y) \mapsto (x,z)$: $z = -a(x,y)$ implies the change of the momenta $p_{old} = p_{new} - (\partial a/\partial x)^t q_{new}$, $q_{old} = -(\partial a/\partial y)^t p_{new}$ and the Hamiltonian (3.1) takes the form

$$\frac{1}{2}\left(\frac{\partial a}{\partial y} g(x, y(x,z))(\frac{\partial a}{\partial y})^t q, q\right) + (z, p)$$

$$+ \left(\frac{\partial a}{\partial y} b(x, y(x,z)) - \frac{\partial a}{\partial x} z, q\right) - V(x, y(x,z)).$$

In the case of Hamiltonian (3.4), the new Hamiltonian takes the form

$$\frac{1}{2}\left(\alpha(x)g(x)\alpha^t(x)q, q\right) + (z, p)$$

$$+ \left(\alpha(x)(b(x) + \beta(x)y + \frac{1}{2}(\gamma(x)y, y)) - (\frac{\partial a}{\partial x} + \frac{\partial \alpha}{\partial x} y, z), q\right) - V(x, y)$$

with $y = -\alpha(x)^{-1}(z + a(x))$. This Hamiltonian is still regular of form (3.4), but at the same time it has the form (1.36) of a Hamiltonian corresponding to the problem of the calculus of variations with Lagrangian depending on first and second derivatives. Therefore, all results of the previous section correspond to the solution of the problems of that kind. In general, not all regular Hamiltonians (4.1) can be transformed to the form (1.36) but only a subclass of them. General results on regular Hamiltonians give the existence and the estimates for the solutions of problems with Lagrangian (1.35). For example, one has the following result.

Theorem 4.2. *Let $x_0, ..., x_n \in \mathcal{R}^n$ and let a smooth function L be given by the formula*

$$L(x_0, ..., x_m, z) = \frac{1}{2}\left(g(x_0)(z + \alpha(x_0, ..., x_m), z + \alpha(x_0, ..., x_m)) + V(x_0, ..., x_m)\right),$$
(4.6)

with $g(x_0)$ being strictly positive-definite matrix, α and V being polynomials of m-degree $m + 1$ and $2(m + 1)$ respectively, where $\deg_m x_j = j$, and V being positive. Then there exists a solution of equation (1.30) with boundary conditions (1.25), which provides the absolute minimum for functional (1.24).

One also can specify the additional conditions under which this solution is unique for small times.

We have noticed that the Hamiltonians with the Young scheme whose entries are equal may correspond to the Lagrangians depending on $(M + 2)$ derivatives. The general RH with the Young scheme $(m_{M+1}, m_M, ..., m_0)$ may correspond to the variational problems with Lagrangians depending on $M + 2$ derivatives of

m_0 variables, $M + 1$ derivatives of $(m_1 - m_0)$ variables and so on. Theorem 4.2 can be generalised to cover these cases as well.

To conclude this section let us give a simple example of non-regular Hamiltonian, whose quadratic approximation is regular of the first rabk (at least in a neighbourhood of almost every point) but for which the solution of the boundary value problem does not exist even locally. In this simplest case x and y are one-dimensional and H even does not depend on x. Let

$$H(x, y, p, q) = -f(y)p + \frac{1}{2}q^2 \qquad (4.7)$$

with an everywhere positive function f, which therefore can not be linear (consequently H is not regular). The corresponding Hamiltonian system has the form

$$\dot{x} = -f(y), \quad \dot{y} = q, \quad \dot{p} = 0, \quad \dot{q} = f'(y)p.$$

Therefore $\dot{x}$ is always negative and there is no solution of the Hamiltonian system joining (x_0, y_0) and (x, y) whenever $x > x_0$, even for small positive t. On the other hand, if $f(y)$ is a nonlinear diffeomorphism, say $f(y) = y^3 + y$, then it is not difficult to prove the global existence of the solutions to the boundary value problem for the corresponding Hamiltonian (4.7), though H is still non-regular. In fact, regularity ensures not only the existence of the solutions but also some "nice" asymptotics for them.

5. Hamiltonians of exponential growth in momenta

In this section we generalise partially the results of Section 2 to some non-degenerate Hamiltonians, which are not quadratic in momentum. First we present a theorem of existence and local uniqueness for a rather general class of Hamiltonians and then give some asymptotic formulas for the case mainly of interest, when the Hamiltonians increase exponentially in momenta.

Definition 1. *We say that a smooth function $H(x, p)$ on $\mathcal{R}^{2m}$ is a Hamiltonian of uniform growth, if there exist continuous positive functions $C(x)$, $\kappa(x)$ on $\mathcal{R}^n$ such that*

(i) $\frac{\partial^2 H}{\partial p^2}(x, p) \geq C^{-1}(x)$ for all x, p;

(ii) if $|p| \geq \kappa(x)$, the norms of all derivatives of H up to and including the third order do not exceed $C(x)H(x, p)$ and moreover,

$$C^{-1}(x)H(x, p) \leq \left\| \frac{\partial H}{\partial p}(x, p) \right\| \leq C(x)H(x, p), \qquad (5.1)$$

$$\left\| \frac{\partial g}{\partial p}(x, p) \right\| \leq C(x)H(x, p)\frac{\partial^2 H}{\partial p^2}(x, p), \qquad (5.2)$$

where

$$g(x, p) = \frac{\partial^2 H}{\partial p \partial x}\frac{\partial H}{\partial p} - \frac{\partial^2 H}{\partial p^2}\frac{\partial H}{\partial x}; \qquad (5.3)$$

(iii) for some positive continuous function $\delta(x)$ one has

$$\left| \left(\frac{\partial^2 H}{\partial p^2}(x,p) \right)^{-1} \right| \le C(x) \left| \left(\frac{\partial^2 H}{\partial p^2}(x+y,p+q) \right)^{-1} \right| \qquad (5.4)$$

whenever $|y| \le \delta$, $|q| \le \delta$, $|p| \ge \kappa(x)$.

The main properties of the boundary-value problem for such Hamiltonians are given in Theorems 5.1, 5.2 below. The function $H(x,p) = \alpha(x)\cosh p$ with $\alpha(x) > 0$ is a simple example. In fact in this book we are interested in the finite-dimensional generalisations of this example, which are described in Theorem 5.3 below.

Let a Hamiltonian of uniform growth be given. Following the same plan of investigation as for quadratic Hamiltonians we study first the Cauchy problem for the corresponding Hamiltonian system.

Proposition 5.1. *For an arbitrary neighbourhood $U(x_0)$ of x_0 there exist positive K, c_0, t_0 such that if $H(x_0, p_0) \le c/t$ with $c \le c_0$, $t \le t_0$, then the solution $X(s) = X(s, x_0, p_0)$, $P(s) = P(s, x_0, p_0)$ of the Hamiltonian system exists on $[0,t]$; moreover, on this interval $X(s) \in U(x_0)$, $\|P(s) - p_0\| \le K(t+c)$ and*

$$\frac{\partial X(s)}{\partial p_0} = s \frac{\partial^2 H}{\partial p^2}(x_0, p_0)(1 + O(c)), \qquad \frac{\partial P(s)}{\partial p_0} = 1 + O(c). \qquad (5.5)$$

If in addition, the norms of the derivatives of H of order up to and including k do not exceed $C(x)H(x,p)$ for large p, then $\frac{\partial^l X(s)}{\partial p_0^l} = O(c)$ for $l \le k - 2$. In particular,

$$\frac{\partial^2 X(s)}{\partial p_0^2}(s) = s \frac{\partial^3 H}{\partial p^3}(x_0, p_0) + O(c^2).$$

Proof. We can suppose that $\|p_0\| > 2 \max_{x \in U(x_0)} \kappa(x)$, because the case of p_0 from any fixed compact is considered trivially). Let

$$T(t) = \min(t, \sup\{s > 0 : X(s) \in U(x_0), \quad \|P(s)\| > \kappa\}).$$

Using Definition 5.1 and the conservation of H along the trajectories of the Hamiltonian flow one obtains

$$\|\dot{P}(s)\| \le \left| \frac{\partial H}{\partial x} \right| \le C(X(s))H(X(s), P(s)) = C(X(s))H(x_0, p_0) \le C(X(s))\frac{c}{t};$$

hence $|P(s) - p_0| = O(c)$ and similarly $|X(s) - x_0| = O(c)$ for $s \le T(t)$. If one chooses small c_0 in such a way that the last inequalities would imply $X(s) \in U(x_0)$ and $|P(s)| > \kappa(X(s))$, the assumption $T(t) < t$ would lead to a contradiction. Consequently, $T(t) = t$ for such c_0. It remains to prove (5.5). For p_0 (or, equivalently, c/t) from any bounded neighbourhood of the origin the result is trivial. Let us suppose therefore again that $|p_0|$ is large enough. Following the lines of the proof of Lemma 2.2 let us differentiate the integral form of the Hamiltonian equations

$$\begin{cases} X(s) = x_0 + s\frac{\partial H}{\partial p}(x_0, p_0) + \int_0^s (s - \tau)g(X, P)(\tau)\, d\tau \\ P(s) = p_0 - \int_0^s \frac{\partial H}{\partial x}(X, P)(\tau)\, d\tau \end{cases}, \qquad (5.6)$$

to obtain

$$\begin{cases} \frac{\partial X(s)}{\partial p_0} = s\frac{\partial^2 H}{\partial p^2}(x_0, p_0) + \int_0^s (s-\tau)\left(\frac{\partial g}{\partial x}\frac{\partial X}{\partial p_0} + \frac{\partial g}{\partial p}\frac{\partial P}{\partial p_0}\right)(\tau)\,d\tau \\ \frac{\partial P(s)}{\partial p_0} = 1 - \int_0^s \left(\frac{\partial^2 H}{\partial x^2}\frac{\partial X}{\partial p_0} + \frac{\partial^2 H}{\partial x \partial p}\frac{\partial P}{\partial p_0}\right)d\tau \end{cases}$$

Considering now the matrices $v(s) = \frac{1}{s}\frac{\partial X(s)}{\partial p_0}$ and $u(s) = \frac{\partial P(s)}{\partial p_0}$ as vectors of the Banach space of continuous $m \times m$-matrix-valued functions $M(s)$ on $[0, t]$ with the norm $\sup\{\|M(s)\| : s \in [0, t]\}$ one deduces from the previous equations that

$$\begin{cases} v = \frac{\partial^2 H}{\partial p^2}(x_0, p_0) + O(t^2)H^2(x_0, p_0)v + O(t)\max_s |\frac{\partial g}{\partial p}(X(s), P(s))|u \\ u = 1 + O(t^2)H(x_0, p_0)v + O(t)H(x_0, p_0)u \end{cases}$$

Due to (5.2),(5.4),

$$\left|\frac{\partial g}{\partial p}(X(s), P(s))\right|\left|\left(\frac{\partial^2 H}{\partial p^2}(x_0, p_0)\right)^{-1}\right|$$

$$\leq C(x_0)\left|\frac{\partial g}{\partial p}(X(s), P(s))\right|\left|\left(\frac{\partial^2 H}{\partial p^2}(x(s), P(s))\right)^{-1}\right|$$

$$\leq H(X(s), P(s))C(X(s))C(x_0) = C(X(s))C(x_0)H(x_0, p_0),$$

and thus the previous system of equations can be written in the form

$$\begin{cases} v = \frac{\partial^2 H}{\partial p^2}(x_0, p_0) + O(c^2)v + O(c)\frac{\partial^2 H}{\partial p^2}(x_0, p_0)u \\ u = 1 + O(tc)v + O(c)u \end{cases}$$

From the second equation one gets

$$u = (1 + O(c))(1 + (tc)v), \tag{5.7}$$

and inserting this in the first one yields

$$v = \frac{\partial^2 H}{\partial p^2}(x_0, p_0)(1 + O(c)) + O(c^2)v,$$

and consequently

$$v = (1 + O(c))\frac{\partial^2 H}{\partial p^2}(x_0, p_0),$$

which yields the first equation in (5.5). Inserting it in (5.7) yields the second equation in (5.5). Higher derivatives with respect to p_0 can be estimated similarly by differentiating (5.6) sufficient number of times and using induction. Proposition is proved.

Theorem 5.1. *For any $c \leq c_0$ with small enough c_0 there exists t_0 such that for all $t \leq t_0$ the map $p_0 \mapsto X(t, x_0, p_0)$ defined on the domain $D_c =$*

$\{p_0 : H(x_0, p_0) \le c/t\}$ *is a diffeomorphism on its image, which contains the ball* $B_{cr}(x_0)$ *and belongs to the ball* $B_{cr^{-1}}(x_0)$ *for some* r *(that can be chosen arbitrary close to* $C^{-1}(x_0)c$ *whenever* c *is small enough).*

Proof. From Proposition 5.1 one concludes that $\frac{1}{t}\frac{\partial X(t)}{\partial p_0}$ is bounded from below in D_c by some positive constant. This implies that the map under consideration is a diffeomorphism on its image, which one shows by the same arguments as in the proof of Theorem 2.1. To estimate this image, let us estimate $X(t, x_0, p_0) - x_0$ on the boundary of the domain D_c, namely when $H(x_0, p_0) = c/t$. From (5.6) it follows that

$$\|X(t, x_0, p_0) - x_0\| = \|t\frac{\partial H}{\partial p}(x_0, p_0) + \int_0^t (t-s)g(x, p)(s)\, ds\|$$

$$\ge tC^{-1}(x_0)H(x_0, p_0) - t^2 C(x)^2 H(x_0, p_0)^2 = tc^{-1}(x_0)H(x_0, p_0)(1 + O(c)).$$

Since the image of the boundary of D_c is homeomorphic to S^{m-1} and therefore divides the space into two open connected components, it follows from the last estimate that the ball with the centre at x_0 and the radius rt belongs to the image of D_c, where r can be chosen arbitrary close to $C^{-1}(x_0)c$, if c is sufficiently small. Similarly one proves that

$$\|X(t, x_0, p_0) - x_0\| \le tC(x)H(x_0, p_0)(1 + O(c)),$$

which implies the required upper bound for the image of D_c.

Proposition 5.2. *There exist* t_0, r, c *such that if* $|x - x_0| \le rc$, *the solution to the boundary value problem with the condition* $x(0) = x_0$, $x(t) = x$, *for the Hamiltonian system with the Hamiltonian* H *exists for all* $t \le t_0$ *and is unique under additional condition that* $H(x_0, p_0) \le c/t$. *If* $|x - x_0|/t$ *be outside a fixed neighbourhood of the origin, then the initial momentum* p_0 *and the initial velocity* $\dot{x}_0$ *on this solution satisfy the estimates*

$$r\frac{|x - x_0|}{t} \le H(x_0, p_0) \le \frac{|x - x_0|}{rt}, \quad \dot{x}_0 = \frac{x - x_0}{t}(1 + O(|x - x_0|)) \qquad (5.8)$$

and the two-point function $S(t, x, x_0)$ *on this solution has the form*

$$S(t, x, x_0) = tL(x_0, \frac{x - x_0}{t}(1 + O(|x - x_0|))) + O(|x - x_0|^2). \qquad (5.9)$$

If $C(x), C^{-1}(x), \kappa(x), \kappa^{-1}(x), \delta(x)$ *from Definition 5.1 are bounded (for all* x *uniformly), the constants* t_0, r, c *can be chosen independently of* x_0.

Proof. Everything, except for the formula for S, follows directly from the previous theorem and the estimates used in its proof. To prove (5.9), we write

$$S(t, x, x_0) = \int_0^t L(x_0 + O(|x - x_0|), \dot{x}_0 + O(|x - x_0|)H^2(x_0, p_0))\, d\tau$$

$$= tL(x_0, \frac{x - x_0}{t}(1 + O(|x - x_0|))) + O(t|x - x_0|)\frac{\partial L}{\partial x}(x_0 + O(c), \frac{x - x_0}{t}(1 + O(c))),$$

which implies (5.9), since

$$\frac{\partial L}{\partial x}(x, v) = -\frac{\partial H}{\partial x}\left(x, \frac{\partial L}{\partial v}(x, v)\right). \tag{5.10}$$

Proposition 5.3. *For $\dot{x}_0$ (or equivalently $(x - x_0)/t$) from a bounded neighbourhood of $\frac{\partial H}{\partial p}(x, 0)$*

$$\frac{1}{t}S(t, x; x_0) = L(x_0, \frac{x - x_0}{t}) + O(t); \tag{5.11}$$

moreover, if $H(x, 0) = 0$ for all x, then

$$L(x_0, \dot{x}_0) = \frac{1}{2}\left(\left[\frac{\partial^2 H}{\partial p^2}(x_0, 0)\right]^{-1}\left(\dot{x}_0 - \frac{\partial H}{\partial p}(x_0, 0)\right), \dot{x}_0 - \frac{\partial H}{\partial p}(x_0, 0)\right) + ...,$$
$$\tag{5.12}$$

$$\frac{1}{t}S(t, x, x_0) = \frac{1}{2}(1 + O(t))\left(\left[\frac{\partial^2 H}{\partial p^2}(x_0, 0)\right]^{-1}\frac{x - \tilde{x}(t, x_0)}{t}, \frac{x - \tilde{x}(t, x_0)}{t}\right) + ...,$$
$$\tag{5.13}$$

where ... in (5.12),(5.13) denote the higher terms of the expansion with respect to $\dot{x}_0 - \frac{\partial H}{\partial p}(x_0, 0)$ and

$$\frac{x - \tilde{x}(t, x_0)}{t} = \frac{x - x_0}{t} - \frac{\partial H}{\partial p}(x_0, 0) + O(t)$$

respectively, and each coefficient in series (5.12) differs from the corresponding coefficient of (5.13) by the value of the order $O(t)$.

Remark. The number of available terms in asymptotic series (5.12) or (5.13) depends of course on the number of existing derivatives of H.

Proof. Formula (5.11) is proved similarly to (5.9). Next, since $H(x, 0) = 0$, the Lagrangian $L(x_0, v)$ (resp. the function $S(t, x; x_0)$) has its minimum equal to zero at the point $v = \frac{\partial H}{\partial p}(x_0, 0)$ (resp. at $x = \tilde{x}(t, x_0)$, where $\tilde{x}$ is as usual the solution of the Hamiltonian system with initial data $x(0) = x_0, p(0) = 0$). At last, one compares the coefficients in series (5.12),(5.13) using (1.19) and the obvious relations

$$\frac{\partial^k X}{\partial p_0^k} = t\frac{\partial^{k+1} H}{\partial p_0^{k+1}}(x_0, p_0) + O(c^2), \qquad \frac{\partial^k P}{\partial p_0^k} = O(t), \quad k > 1, \tag{5.14}$$

which hold for p_0 from any bounded domain.

Now we can prove the smooth equivalence of the two-point functions of the Hamiltonian $H(x, p)$ and the corresponding Hamiltonian $H(x_0, p)$ with a fixed x_0. This result plays a key role in the construction of the semi-classical asymptotics for the Feller processes given in Chapter 6.

Theorem 5.2. *Let the assumptions of Theorem 5.1 hold and $H(x, 0) = 0$ for all x. Then there exists a smooth map $z(t, v, x_0)$ defined for v from the ball of the radius rc/t such that*

(i) for fixed t, x_0 the map $v \mapsto z(t, v, x_0)$ is a diffeomorphism on its image,

(ii) for v from a bounded domain

$$\|z(t, x, x_0) - v\| = O(t)\|v\| + O(t), \tag{5.15}$$

(ii) if v is outside a neighbourhood of the origin, then

$$z(t, v, x_0) = (1 + \omega(t, v))D_t v + O(t), \tag{5.16}$$

where $\omega(t, v) = O(|x - x_0|)$ is a scalar function and D_t is a linear diffeomorphism of $\mathcal{R}^d$ of the form $1 + O(t)$ with a uniformly bounded derivative in t;

(iii) z takes $S(t, x, x_0)$ into $tL(x, (x - x_0)/t)$, i.e.

$$L(x_0, z(t, \frac{x - x_0}{t}, x_0), x_0) = \frac{1}{t}S(t, x, x_0). \tag{5.17}$$

Proof. It follows from Propositions 5.2, 5.3 and E2. More precisely, one repeats the proof of Propositions E1, E2 of Appendix E to obtain a diffeomorphism that takes the function $L(x_0, v)$ in the function $S(t, x, x_0)/t$ considered both as the functions of $v = (x - x_0)/t$ and depending on t, x_0 as parameters. Due to Proposition 5.3, the linear and the local parts D_3, D_2 of the required diffeomorphism have the form $1 + O(t)$. Due to (5.9), the dilatation coefficient ω from (E1) has the order $O(|x - x_0|)$. To get (5.16) one needs then only take in account the necessary shift on the difference between minimum points of L and S which is of the order $O(t)$ due to Proposition 5.3.

We shall concentrate now on a more concrete class of Hamiltonians and shall obtain for these Hamiltonians more exact estimates of the objects introduced above. For any vector p we shall denote by $\bar{p}$ a unit vector in the direction p, i.e. $\bar{p} = p/\|p\|$.

Definition 5.2. *We say that a smooth function $H(x, p)$ on $\mathcal{R}^{2m}$ is a Hamiltonian of exponential growth, if there exist two positive continuous functions $a(x, \bar{p}) \geq b(x, \bar{p})$, on $\mathcal{R}^m \times S^{m-1}$ and a positive continuous function $C(x)$ on $\mathcal{R}^m$ such that*

(i) for p outside a neighbourhood of the origin, the norms of all derivatives of H up to and including the third order do not exceed $C(x)H(x, p)$;

(ii) $H(x, p) \leq C(x) \exp\{a(x, \bar{p})|p|\}$ for all x, p;

(iii) $\frac{\partial^2 H}{\partial p^2}(x, p) \geq C^{-1}(x) \exp\{b(x, \bar{p})|p|\}$ for all x, p.

Notice that the condition on the growth of the Hamiltonian implies that the matrix of the second derivatives of the corresponding Lagrangian $L(x, \dot{x})$ tends to zero, as $\dot{x} \to \infty$ (see Remark after Theorem 5.3), which means that the corresponding problem of the calculus of variations has certain degeneracy at infinity. Nevertheless, similarly to the case of Hamiltonians from Definition 5.1, one can show the existence of the solution to the boundary-value problem (with the condition $x(0) = x_0, x(t) = x$) for the Hamiltonian systems with Hamiltonians of exponential growth and the uniqueness of such solution for $|x - x_0| < t^\Delta$, where $\Delta \in (0, 1)$ depends on the difference $a(x, \bar{p}) - b(x, \bar{p})$ and can be chosen arbitrary small whenever this difference can be chosen arbitrary small. We are not going into detail, because actually we are interested in a more restrictive class of Hamiltonians that satisfy both Definitions 5.1 and 5.2. This class of Hamiltonians is provided by the Lévy-Khintchine formula with the Lévy measure having finite support, namely, the Hamiltonians of this class are given by the formula

$$H(x, p) = \frac{1}{2}(G(x)p, p) - (A(x), p) + \int_{\mathcal{R}^m \setminus \{0\}} \left(e^{-(p,\xi)} - 1 + \frac{(p, \xi)}{1 + \xi^2} \right) d\nu_x(\xi),$$

(5.18)

where $G(x)$ is a nonnegative matrix, ν_x is a so called Lévy measure on $\mathcal{R}^m \setminus \{0\}$, which means that

$$\int_{\mathcal{R}^m \setminus \{0\}} \min(\xi^2, 1) \, d\nu_x(\xi) < \infty$$

for all x, and the support of the Lévy measure ν is supposed to be a bounded set in $\mathcal{R}^m$. The last assumption insures that function (5.18) is well defined for all complex p and is an entire function with respect to p. We suppose that all $G(x), A(x), \nu_x$ are continuously differentiable (at least thrice). Notice that

$$\left(\frac{\partial^2 H}{\partial p^2}(x, p)v, v \right) = (G(x)v, v) + \int_{\mathcal{R}^m \setminus \{0\}} (\xi, v)^2 e^{-(p,\xi)} \, d\nu_x(\xi),$$

and therefore $\frac{\partial^2 H}{\partial x^2}$ is always nonnegative. Moreover, one sees directly that for Hamiltonian (5.18), properties (i),(ii) from Definition 5.2 hold with the function a being the support function of the set $-supp\,\nu_x$, i.e.

$$a(x, \bar{p}) = \max\{(\bar{p}, -\xi) : -\xi \in supp\,\nu_x\}.$$

(5.19)

The following results give simple sufficient conditions on ν_x that ensure that corresponding function (5.18) is a Hamiltonian of exponential growth. We omit rather simple proofs.

Proposition 5.4. *(i) Let the function β on $\mathcal{R}^m \times S^{m-1}$ be defined by the formula*

$$\beta(x, \bar{v}) = \sup\{r : r\bar{v} \in supp\,\nu_x\}.$$

If β is continuous and everywhere positive, then Hamiltonian (5.18) is of exponential growth with the function a defined in (5.19) and any continuous $b(x, p) < a(x, p)$.

(ii) If there exists $\epsilon > 0$ such that for any $\bar{v} \in S^{m-1}$ the convex hull of $supp\,\nu_x \cap \{\xi : (\xi, \bar{v}) \geq \epsilon\}$ depends continuously on x and has always nonempty

interior, then function (5.18) is of exponential growth with the function a defined as above and

$$b(x, \bar{v}) = \min_{w \in S^{m-1}} \max\{|(w, \xi)| : \xi \in supp\, \nu_x \cap \{\xi : (-\xi, \bar{v}) \geq \epsilon\}\}. \qquad (5.20)$$

Examples. Let $m = 2$, G and A vanish in (5.18), and let $\nu_x = \nu$ does not depend on x. If the support of ν consists of only three points, then H of form (5.18) is not of exponential growth. Actually in this case $\frac{\partial^2 H}{\partial p^2}$ tends to zero, as p tends to infinity along some directions, and one can show that the boundary-value problem have no solution for some pairs of points x, x_0. On the other hand, if the support of ν consists of four vertices of a square with the centre at the origin, then H of form (5.18) again is not of exponential growth. However, it satisfies the condition of Proposition 5.4 (ii) with $\epsilon = 0$, and one can prove that for this Hamiltonian the boundary-value problem is always solvable.

In order that function (5.18) would satisfy all conditions of Definition 5.1, it seems necessarily to make some assumptions on the behaviour of ν near the boundary of its support. We are not going to describe the most general assumptions of that kind. In the next statement we give only the simplest sufficient conditions.

Theorem 5.3. *Let ν_x have a convex support, containing the origin as an inner point, with a smooth boundary, $\partial\, supp\, \nu_x$, depending smoothly on x and having nowhere vanishing curvature, and moreover, let $\nu_x(d\xi) = f(x, \xi)\, d\xi$ in a neighbourhood of $\partial\, supp\, \nu_x$ with a continuous f not vanishing on $\partial\, supp\, \nu_x$. Then H of form (5.18) satisfies the requirements of both Definitions 5.1 and 5.2 with $a(x, \bar{p})$ given by (5.19), and moreover for large p and some continuous $C(x)$*

$$C^{-1}(x)|p|^{-(m+1)/2} \exp\{a(x, \bar{p})|p|\} \leq H(x, p) \leq C(x)|p|^{-(m+1)/2} \exp\{a(x, \bar{p})|p|\}, \qquad (5.21)$$

$$\frac{\partial^2 H}{\partial p^2}(x, p) \geq C^{-1}(x)\frac{H(x, p)}{|p|},$$

$$C^{-1}(x)\frac{H^d(x, p)}{|p|^{d-1}} \leq \det \frac{\partial^2 H}{\partial p^2}(x, p) \leq C(x)\frac{H^d(x, p)}{|p|^{d-1}}, \qquad (5.22)$$

and

$$\max(|g(x, p)|, |\frac{\partial g}{\partial p}(x, p)|) \leq C(x)\frac{H(x, p)}{|p|},$$

Remark. If the above $f(x, \xi)$ vanishes at $\partial\, supp\, \nu_x$, but has a non-vanishing normal derivative there, then the same holds but with $m + 2$ instead of $m + 1$ in (5.21).

Proof. Clear that for p from any compact set H is bounded together with all its derivatives and the matrix of the second derivatives is bounded from below by a positive constant. It is also obvious that (5.22) implies (5.2). In order to obtain the precise asymptotics of H as $p \to \infty$ notice first that for large p

the behaviour of H and its derivatives is the same (asymptotically) as by the function

$$\tilde{H}(x,p) = \int_{U(x,\nu)} \exp\{|p|(\bar{p},\xi)\} f(x,-\xi) \, d\xi,$$

where $U(x,\nu)$ is an arbitrary small neighbourhood of $\partial \, supp \, \nu_x$. To estimate this integral we consider it as the Laplace integral with the large parameter $|p|$ (depending on the additional bounded parameter $\bar{p}$). The phase of this integral $S(\xi,\bar{p}) = (\xi,\bar{p})$ takes its maximum at the unique point $\xi_0 = \xi_0(\bar{p},x)$ on the boundary $\partial \, supp \, \nu_x$ of the domain of integration, the unit vector $\bar{p}$ provides an outer normal vector to this boundary at ξ_0, and the value of this maximum is given by the support function (5.19). Moreover, this maximum is not degenerate (due to the condition of not vanishing curvature) in the sense that the normal derivative of S at ξ_0 (the derivative with respect to ξ in the direction $\bar{p}$) does not vanish (because it equals $\bar{p}$) and the $(m-1) \times (m-1)$-matrix $A(x,\xi_0)$ of the second derivatives of S restricted to $\partial \, supp \, \nu_x$ at ξ_0 is not degenerate. Thus by the Laplace method (see e.g. Proposition B5) one finds for large $|p|$

$$H(x,p) = |p|^{-(d+1)/2} \exp\{a(x,\bar{p})|p|\} f(x,-\xi_0)$$

$$\times (2\pi)^{(d-1)/2} (\det A(x,\xi_0))^{-1/2} (1 + O(|p|^{-1})). \qquad (5.23)$$

Similarly one finds that $\frac{\partial H}{\partial p}(x,p)$ and $\left(\frac{\partial^2 H}{\partial p^2}(x,p)v,v\right)$ equal respectively to

$$|p|^{-(d+1)/2} \exp\{a(x,\bar{p})|p|\} f(x,-\xi_0)\xi_0$$

$$\times (2\pi)^{(d-1)/2} (\det A(x,\xi_0))^{-1/2} (1 + O(|p|^{-1})). \qquad (5.24)$$

and

$$|p|^{-(d+1)/2} \exp\{a(x,\bar{p})|p|\} f(x,-\xi_0)(\xi_0,v)^2$$

$$\times (2\pi)^{(d-1)/2} (\det A(x,\xi_0))^{-1/2} (1 + O(|p|^{-1})). \qquad (5.25)$$

Similarly one finds the asymptotic representations for other derivatives of H, which implies (5.1),(5.21) and the required upper bounds for all derivatives of H. To get a lower bound for the eigenvalues of the matrix of the second derivatives of H notice that due to the above formulas $\left(\frac{\partial^2 H}{\partial p^2}(x,p)v,v\right)$ is of the same order as $H(x,p)$ whenever v is not orthogonal to ξ_0. If $v = v_0$ is such that $(v_0,\xi_0) = 0$, then the major term of the corresponding asymptotic expansion vanishes, which means the drop in at least one power of $|p|$. To get (5.22) one must show that the second term in this expansion does not vanish for any such v_0. This follows from the general explicit formula for this term (see e.g. in Proposition B5) and the fact that the amplitude in the corresponding Laplace integral has zero of exactly second order at ξ_0. To complete the proof of the Proposition, it remains to note that writing down the major terms of the expansions of $\frac{\partial^2 H}{\partial p^2}, \frac{\partial H}{\partial x}, \frac{\partial^2 H}{\partial p \partial x}, \frac{\partial H}{\partial p}$ one sees that the terms proportional to $|p|^{-(d+1)}$ cancel in the expansions for g or its derivative in p, which implies the required estimates for g.

Remark. Notice that from the formulas

$$v = \frac{\partial H}{\partial p}\left(x, \frac{\partial L}{\partial v}(x, v)\right), \quad \frac{\partial^2 L}{\partial v^2}(x, v) = \left(\frac{\partial^2 H}{\partial p^2}\right)^{-1}\left(x, \frac{\partial L}{\partial v}(x, v)\right), \quad (5.26)$$

connecting the derivatives of H and its Legendre transform L, it follows that if H is a Hamiltonian from Theorem 5.3, the first (resp. the second) derivative of the corresponding Lagrangian $\frac{\partial L}{\partial v}(x, v)$ (resp. $\frac{\partial^2 L}{\partial v^2}(x, v)$ increases like $\log|v|$ (resp. decreases like $|v|^{-1}$) as $|v| \to \infty$.

Proposition 5.5. *For a Hamiltonian from Theorem 3.1, if $|x - x_0| \le rc$, $t \le t_0$ and $(x-x_0)/t$ does not approach the origin, one has the following estimates for the initial momentum $p_0 = p_0(t, x, x_0)$ and the two-point function $S(t, x, x_0)$ of the solution to the boundary-value problem with conditions $x(0) = x_0$, $x(t) = x$ (recall that the existence and uniqueness of this solution is proved in Theorem 5.1):*

$$|p_0|\left(1 + \frac{O(\log(1 + |p_0|))}{|p_0|}\right) = \frac{1}{a(x_0, \bar{p}_0)}\log(1 + \frac{|x - x_0|}{t}) + O(1), \quad (5.27)$$

$$-\sigma t + C|x - x_0| \le S(t, x; x_0) \le \sigma t + C|x - x_0|. \quad (5.28)$$

with some constants σ, C.

Proof. Estimate (5.27) follow directly from (5.8) and (5.24). Next, from (5.22) one has

$$|\log t|^{-1}\frac{|x - x_0|}{Ct} \le \frac{\partial^2 H}{\partial p^2}(x_0, p_0) \le C\frac{|x - x_0|}{t} \quad (5.29)$$

for small t and some constant C, which implies, due to (5.5), that for θ not approaching zero

$$C^{-1}|x - x_0|^{-1} \le \left(\frac{\partial X}{\partial p_0}\right)^{-1}(t, x, x_0 + \theta(x - x_0)) \le |\log t|C|x - x_0|^{-1}. \quad (5.30)$$

Hence, from (1.22) and (5.13) one obtains

$$-\sigma t + C|x - x_0| \le S(t, x; x_0) \le \sigma t + C|x - x_0||\log t|.$$

In order to get rig of $\log t$ on the r.h.s. (which is not very important for our purposes) one needs to argue similarly to the proof of (5.31) below. We omit the details.

Now one can use the same arguments as in Section 2 to get the following

Proposition 5.6. *The statements (and the proofs) of Proposition 2.4-2.8 are valid for Hamiltonians from Theorem 3.1. In particular, for $t \le t_0$, $|x - x_0| \le r$ with small enough r, t_0 there exists a unique solution to the boundary value problem for the Hamiltonian system with Hamiltonian H that provides the global minimum for the corresponding problem of the calculus of variations.*

Further we shall need the estimates for the function z from Theorem 5.2 and its derivatives.

Proposition 5.7. *Let H belong to the class of Hamiltonians described in Theorem 3.1 and z be the corresponding mapping from Theorem 5.2. If $v = (x - x_0)/t$ does not approach the origin and $|x - x_0| \le rc$, then*

$$p_0 - \frac{\partial L}{\partial v}(x_0, v) = O(|x - x_0|), \quad \frac{\partial L}{\partial v}(x_0, v) - \frac{\partial L}{\partial v}(x_0, z(t, v, x_0)) = O(|x - x_0|),$$

$$(5.31)$$

and

$$\frac{\partial \omega}{\partial v} = O(t), \quad \frac{\partial \omega}{\partial t} = O(|v|), \quad \frac{\partial^2 \omega}{\partial v^2} = O(t)|v|^{-1}|\log t|. \quad (5.32)$$

Proof. Plainly for any w

$$\frac{\partial L}{\partial v}(x_0, w) - \frac{\partial L}{\partial v}(x_0, v) = \int_0^1 \frac{\partial^2 L}{\partial v^2}(x_0, v + s(w - v)) \, ds(w - v),$$

$$\le \max_{s \in (0,1)} \left(\frac{\partial^2 H}{\partial p^2}\left(x_0, \frac{\partial L}{\partial v}(x_0, v + s(w - v)))\right) \right)^{-1} (w - v).$$

First let $w = \dot{x}_0 = \frac{\partial H}{\partial p}(x_0, p_0)$. From the first equation in (5.6) and the estimates for g in Theorem 5.3 it follows that

$$|\dot{x}_0 - v| = O(t)|g(x, p)| = O(t)|p|^{-1}H^2(x, p)$$

$$= O(t)|\log t|^{-1}H^2(x, p) = O(t^{-1})|x - x_0|^2|\log t|^{-1}.$$

Therefore, due to (5.22), one has

$$\left| \frac{\partial L}{\partial v}(x_0, \dot{x}_0) - \frac{\partial L}{\partial v}(x_0, v) \right| = O(|\log t|)\frac{t}{|x - x_0|}\frac{|x - x_0|^2}{t}|\log t|^{-1} = O(|x - x_0|),$$

i.e. the first inequality in (5.31). Now let $w = z(t, v, x_0)$. In that case it follows from (5.16) that $|z(t, v, x_0) - v| = O(t^{-1})|x - x_0|^2$ only, i.e. without an additional multiplier of the order $|\log t|$ as in the previous situation. Therefore, the previous arguments would lead here to an additional multiplier of the order $|\log t|$ on the r.h.s. of the second inequality (5.31). Hence, one needs here a more careful consideration. Namely, as it was noted in the proof of Theorem 5.3, the matrix $\frac{\partial^2 H}{\partial p^2}(x_0, p)$ has the maximal eigenvalue of the order $H(x_0, p)$ and the corresponding eigenvector is asymptotically (for large p) proportional to $\xi_0 = \xi_0(x_0, \bar{p})$. Other eigenvalues are already of the order $|p|^{-1}H(x, p)$. Therefore, in order to obtain the required estimate it is enough to show that the ratio of the projection of the vector $z(t, v, x_0) - v$ on the direction of $\xi_0(x_0, \bar{p})$ (for p around p_0) and the projection of $z(t, v, x_0) - v$ on the perpendicular hyperplane is not less than of the order $|\log t|$. But the vector $z(t, v, x_0) - v$ is proportional to $v = (x - x_0)/t$, which in its turn is close to $\frac{\partial H}{\partial p}(x_0, p_0)$. Hence, one must prove that the vector $\frac{\partial H}{\partial p}(x_0, p_0)$ lies essentially in the direction of $\xi_0(x_0, \bar{p}_0)$. But this is surely true, because due to (5.24) the principle term of the asymptotics of $\frac{\partial H}{\partial p}(x_0, p_0)$ is proportional to ξ_0 and next terms differs exactly by the multiplier of the order $|\log t|^{-1}$.

Let us turn now to the proof of (5.32). Differentiating (5.17) with respect to x yields

$$\frac{\partial S}{\partial x} = (1 + \omega)\frac{\partial L}{\partial v}(x_0, z(t, \frac{x - x_0}{t}, x_0))D_t + \left(\frac{\partial L}{\partial v}(x_0, z(t, \frac{x - x_0}{t}, x_0)), D_t v\right)\frac{\partial \omega}{\partial v}.$$

Hence

$$\frac{\partial \omega}{\partial v} = \frac{\frac{\partial S}{\partial x} - (1 + \omega)\frac{\partial L}{\partial v}(x_0, z(t, \frac{x - x_0}{t}, x_0))D_t}{\left(\frac{\partial L}{\partial v}(x_0, z(t, \frac{x - x_0}{t}, x_0)), D_t v\right)}. \tag{5.33}$$

To estimate the denominator in this formula notice that

$$\left(\frac{\partial L}{\partial v}(x_0, z(t, \frac{x - x_0}{t}, x_0)), D_t v\right) = \left(\frac{\partial L}{\partial v}(x_0, z(t, \frac{x - x_0}{t}, x_0)), z\right)(1 + O(|x - x_0|)$$

$$\geq H(x_0, \frac{\partial L}{\partial v}(x_0, z(t, \frac{x - x_0}{t}, x_0)))(1 + O(|x - x_0|)),$$

which is of the order $|x - x_0|/t$. Turning to the estimate of the nominator we present it as the sum of two terms, first being the difference between the final momentum $\frac{\partial S}{\partial x}$ and the initial momentum p_0 on a trajectory, which is of the order $O(|x - x_0|)$ due to Lemma 5.1, and the second being the difference between p_0 and $\frac{\partial L}{\partial v}(x_0, z)$, which is of the order $O(|x - x_0|)$ due to (5.31). Consequently, taking into account the estimates for the nominator and denominator gives the first estimate in (5.32).

Differentiating (5.17) with respect to t yields

$$\frac{\partial S}{\partial t} = L(x_0, z) + t\frac{\partial L}{\partial v}D_t\left(\frac{\partial \omega}{\partial t}v - \frac{1}{t}(1 + \omega)\frac{\partial z}{\partial v}v\right)$$

up to a nonessential smaller term. Since S satisfies the Hamilton-Jacobi equation it follows that

$$\frac{\partial \omega}{\partial t} = \frac{(1 + \omega)\left(\frac{\partial L}{\partial v}, \frac{\partial z}{\partial v}v\right) - H(x, \frac{\partial S}{\partial x}) - L(x_0, z)}{t\frac{\partial L}{\partial v}(x_0, z)v}.$$

The nominator is of the order $O(|x - x_0|^2)/t$, because the main term has the form

$$\frac{\partial L}{\partial v}(x_0, z)z - L(x_0, z) - H(x, \frac{\partial S}{\partial x})$$

$$= \left(H(x_0, \frac{\partial L}{\partial v}) - H(x, \frac{\partial L}{\partial v})\right) + \left(H(x, \frac{\partial L}{\partial v}) - H(x, \frac{\partial S}{\partial x})\right),$$

which is of the order $O(|x - x_0|^2)t^{-1}$, due to the estimates of the first order derivatives of H. The denominator is of the order $|x - x_0|$, which proves the second formula in (5.32).

Differentiating (5.17) two times in x yields

$$t\frac{\partial^2 S}{\partial x^2} - (1+\omega)\frac{\partial^2 L}{\partial v^2} - \frac{\partial \omega}{\partial v} \otimes \frac{\partial^2 L}{\partial v^2}v - \left(\frac{\partial L}{\partial v} \otimes \frac{\partial \omega}{\partial v} + \frac{\partial \omega}{\partial v} \otimes \frac{\partial L}{\partial v}\right) = \left(\frac{\partial L}{\partial v}, v\right)\frac{\partial^2 \omega}{\partial v^2}.$$
(5.34)

The coefficient at $\frac{\partial^2 \omega}{\partial v^2}$ in this formula was already found to be of the order $O(|x - x_0|/t)$. Thus one needs to show that the l.h.s. of (5.34) has the order $t|\log t|$. All tensor products on the l.h.s. of (5.34) certainly have this order due to the first estimate in (5.32). Next,

$$t\frac{\partial^2 S}{\partial x^2} = \left(\frac{\partial^2 H}{\partial p^2}\right)^{-1}(x_0, p_0)(1 + O(|x - x_0|)) = \left(\frac{\partial^2 H}{\partial p^2}\right)^{-1}(x_0, p_0) + O(t)|\log t|.$$

Therefore, it remains to show that

$$\left(\frac{\partial^2 H}{\partial p^2}\right)^{-1}(x_0, p_0) - \left(\frac{\partial^2 H}{\partial p^2}\right)^{-1}\left(x_0, \frac{\partial L}{\partial v}(x_0, z(v))\right) = O(t)|\log t|. \quad (5.35)$$

Using (5.31) and the mean value theorem for the difference on the l.h.s. of (5.35) leads directly to the estimate $O(t)|\log t|^2$ for this difference. But slightly more careful considerations, similar to those used in the proof of the second inequality (5.31) allow to decrease the power of $|\log t|$, which gives the required estimate.

Proposition 5.8. *Let H, z and v be the same as in Proposition 5.7. Then:*

$$\frac{\partial z}{\partial v}(t, v, x_0) = 1 + O(|x - x_0|), \qquad \frac{\partial^2 z}{\partial v^2}(t, v, x_0) = O(t)(1 + \log^+\frac{|x - x_0|}{t}), \quad (5.36)$$

where we used the usual notation $\log^+ M = \max(0, \log M)$.

Proof. For v outside a neighbourhood of the origin, it follows directly from Proposition 5.7. Notice only that for brevity we used always the estimate $|\log t|$ for $|p_0|$, but in formula (5.36) we have restored a more precise estimate for $|p_0|$ from (5.27). For the bounded v formulas (5.36) follow from Proposition 5.3 and explicit formulas for z from the proof of Lemma E2.

6. Complex Hamiltonians and calculus of variations for saddle-points

Here we discuss the solutions to the boundary-value problem for complex Hamiltonians depending quadratically on momenta. As we have seen in Sect.1, in the case of real Hamiltonians, a solution to the Hamiltonian system furnishes a minimum (at least locally) for a corresponding problem of calculus of variations. It turns out that for complex Hamiltonians the solutions of Hamiltonian equations have the property of a saddle-point. Let $x = y + iz \in C^m$, $p = \xi + i\eta \in C^m$ and

$$H = \frac{1}{2}(G(x)p, p) - (A(x), p) - V(x), \quad (6.1)$$

where

$$G(x) = G_R(x) + iG_I(x), \quad A(x) = A_R(x) + iA_I(x), \quad V(x) = V_R(x) + iV_I(x)$$

are analytic matrix-, vector-, and complex valued functions in a neighbourhood of the real plane. Moreover, G is non-degenerate, G_R, G_I are symmetric and G_R is non-negative for all x. Under these assumptions, one readily sees that all the results and proofs of Lemma 2.1, Lemma 2.2, Theorem 2.1 and Propositions 2.1-2.3 on the existence of the solutions of the Hamiltonian system and the asymptotic expansions for the two-point function are valid for this complex situation, where x_0 is considered to be any complex number in the domain where G, A, V are defined, and initial momenta are complex as well. In particular, one gets therefore the existence of the family $\Gamma(x_0)$ of complex characteristics joining x_0 with any point x from some complex domain $D(x_0)$ in time $t \leq t_0$, and the corresponding complex momentum field $p(t,x) = (\xi + i\eta)(t, y, z)$, which in its turn defines the complex invariant Hilbert integral (1.6). Let us clarify what optimisation problem is solved by the complex characteristics of the family $\Gamma(x_0)$.

Notice first that

$$Re\, H = \frac{1}{2}(G_R\xi, \xi) - \frac{1}{2}(G_R\eta, \eta) - (G_I\xi, \eta) - (A_R, \xi) + (A_I, \eta) - V_R, \quad (6.2)$$

$$Im\, H = \frac{1}{2}(G_I\xi, \xi) - \frac{1}{2}(G_I\eta, \eta) + (G_R\xi, \eta) - (A_I, \xi) - (A_R, \eta) - V_I, \quad (6.3)$$

and if $(x(s), p(s))$ is a complex solution to (1.1), then the pairs (y, ξ), (y, η), (z, ξ), (z, η) are real solutions to the Hamiltonian systems with Hamiltonians $ReH, ImH, ImH, -ReH$ respectively. Next, if G_R^{-1} exists, then

$$(G_R + iG_I)^{-1} = (G_R + G_I G_R^{-1} G_I)^{-1} - iG_R^{-1} G_I (G_R + G_I G_R^{-1} G_I)^{-1},$$

and if G_I^{-1} exists, then

$$(G_R + iG_I)^{-1} = G_I^{-1} G_R (G_I + G_R G_I^{-1} G_R)^{-1} - i(G_I + G_R G_I^{-1} G_R)^{-1}.$$

Therefore, since G_R, G_I are symmetric, $(G^{-1})_R, (G^{-1})_I$ are also symmetric. Moreover, $G_R > 0$ is equivalent to $(G^{-1})_R > 0$, and $G_I > 0$ is equivalent to $(G^{-1})_I < 0$.

By definition, the Lagrangian corresponding to the Hamiltonian H is

$$L(x, \dot{x}) = (p\dot{x} - H(x, p))|_{p=p(x)} \quad (6.4)$$

with $p(x)$ uniquely defined from the equation $\dot{x} = \frac{\partial H}{\partial p}(x, p)$. Therefore, the formula for L is the same as in the real case, namely

$$L(x, \dot{x}) = \frac{1}{2}(G^{-1}(\dot{x} + A(x)), \dot{x} + A(x)) + V(x). \quad (6.4')$$

Consequently

$$ReL(y, z, \dot{y}, \dot{z}) = \xi(y, z, \dot{y}, \dot{z})\dot{y} - \eta(y, z, \dot{y}, \dot{z})\dot{z} - ReH(y, z, \xi(y, z, \dot{y}, \dot{z}), \eta(y, z, \dot{y}, \dot{z})),$$
(6.5)
$$ImL(y, z, \dot{y}, \dot{z}) = \eta(y, z, \dot{y}, \dot{z})\dot{y} + \xi(y, z, \dot{y}, \dot{z})\dot{z} - ImH(y, z, \xi(y, z, \dot{y}, \dot{z}), \eta(y, z, \dot{y}, \dot{z})),$$
(6.6)

where $(\xi, \eta)(y, z, \dot{y}, \dot{z})$ are defined from the equations

$$\dot{y} = \frac{\partial ReH}{\partial \xi} = G_R \xi - G_I \eta - A_R, \quad \dot{z} = -\frac{\partial ReH}{\partial \eta} = G_R \eta + G_I \xi - A_I. \quad (6.7)$$

Proposition 6.1 *For all ξ, η*

$$\xi\dot{y} - \eta(y, z, \dot{y}, \dot{z})\dot{z} - ReH(y, z, \xi, \eta(y, z, \dot{y}, \dot{z})) \le ReL(y, z, \dot{y}, \dot{z})$$

$$\le \xi(y, z, \dot{y}, \dot{z})\dot{y} - \eta\dot{z} - ReH(y, z, \xi(y, z, \dot{y}, \dot{z}), \eta)), \quad (6.8)$$

or equivalently

$$ReL(y, z, \dot{y}, \dot{z}) = \max_{\xi} \min_{\eta} (\xi\dot{y} - \eta\dot{z} - ReH(x, p)) = \min_{\eta} \max_{\xi} (\xi\dot{y} - \eta\dot{z} - ReH(x, p)).$$
(6.8')

In other words, $ReL(y, z, \dot{y}, \dot{z})$ is a saddle-point for the function $(\xi\dot{y} - \eta\dot{z} - ReH(x, p))$. Moreover, ReL is convex with respect to $\dot{y}$ and concave with respect to $\dot{z}$ (strictly, if $G_R > 0$ strictly). Furthermore, if $G_I \ge 0$, then

$$ImL(y, z, \dot{y}, \dot{z}) = \max_{\xi} \min_{\eta} (\xi\dot{z} + \eta\dot{y} - ImH(x, p)) = \min_{\eta} \max_{\xi} (\xi\dot{z} + \eta\dot{y} - ImH(x, p)),$$
(6.9)

i.e. $Im(y, z, \dot{y}, \dot{z})$ is a saddle-point for the function $\xi\dot{z} + \eta\dot{y} - ImH(x, p)$. Moreover, if $G_I \ge 0$, then ImL is convex with respect to $\dot{z}$ and concave with respect to $\dot{y}$ (strictly, if $G_I > 0$ strictly).

Proof. Formula (6.7) is obvious, since the function $\xi\dot{y} - \eta\dot{z} - ReH(x, p)$ is concave with respect to ξ and convex with respect to η. Furthermore, $\frac{\partial^2 L}{\partial \dot{x}^2} = (G^{-1})(x)$ due to (6.4'). Therefore

$$\frac{\partial^2 ReL}{\partial \dot{y}^2} = Re\frac{\partial^2 L}{\partial \dot{x}^2} = (G^{-1})_R(x), \quad \frac{\partial^2 ReL}{\partial \dot{z}^2} = -Re\frac{\partial^2 L}{\partial \dot{x}^2} = -(G^{-1})_R(x),$$

which proves the required properties of ReL, because $(G^{-1})_R \ge 0$. The statements about ImL are proved similarly.

Consider now the complex-valued functional

$$I_t(x(.)) = \int_0^t L(x(\tau), \dot{x}(\tau)) \, d\tau,$$

defined on piecewise-smooth complex curves $x(\tau)$ joining x_0 and x in time t, i.e. such that $x(0) = x_0, x(t) = x$. As in Section 1, we define $S(t, x; x_0) = I_t(X(.))$, where $X(s)$ is the (unique) characteristic of the family $\Gamma(x_0)$ joining x_0 and x in time t.

Proposition 6.2. *The characteristic* $X(s) = Y(s) + iZ(s)$ *of the family* $\Gamma(x_0)$ *joining* x_0 *and* x *in time* t *is a saddle-point for the functional* ReI_t, *i.e. for all real piecewise smooth* $y(\tau), z(\tau)$ *such that* $y(0) = y_0$, $z(0) = z_0$, $y(t) = y$, $z(t) = z$ *and* $y(\tau) + iZ(\tau)$, $Y(\tau) + iz(\tau)$ *lie in the domain* $D(x_0)$

$$ReI_t(Y(.) + iz(.)) \le ReI_t(Y(.) + iZ(.)) = Re\,S(t, x, x_0) \le ReI_t(y(.) + iZ(.)). \tag{6.10}$$

In particular,

$$ReI_t(Y(.) + iZ(.)) = \min_{y(.)} \max_{z(.)} ReI_t(y(.) + iz(.)) = \max_{z(.)} \min_{y(.)} ReI_t(y(.) + iz(.)). \tag{6.11}$$

If $G_I(x) \ge 0$, *then similar fact holds for* ImI_t, *namely*

$$ImI_t(Y(.) + iZ(.)) = \min_{z(.)} \max_{y(.)} ImI_t(y(.) + iz(.)) = \max_{y(.)} \min_{z(.)} ImI_t(y(.) + iz(.)). \tag{6.12}$$

Proof. Let us prove, for example, the right inequality in (6.10). Notice

$$ReI_t(y(.), Z(.)) = \int_0^t (\xi(y, Z, \dot{y}, \dot{Z})\dot{y} - \eta(y, Z, \dot{y}, \dot{Z})\dot{Z}$$

$$- ReH(y, Z, \xi(y, Z, \dot{y}, \dot{Z}), \eta(y, Z, \dot{y}, \dot{z})))(\tau)\,d\tau$$

$$\ge \int_0^t (\xi(y, Z)\dot{y} - \eta(y, Z, \dot{y}, \dot{Z})\dot{Z} - ReH(y, Z, \xi(y, Z), \eta(y, Z, \dot{y}, \dot{Z}))(\tau)\,d\tau,$$

due to the left inequality in (6.8). The last expression can be written in equivalent form as

$$\int_0^t [\xi(y, Z)\dot{y} - \eta(y, Z)\dot{Z} - ReH(y, Z, \xi(y, Z), \eta(y, Z)](\tau)\,d\tau$$

$$+ \int_0^t [(\eta(y, Z) - \eta(y, Z, \dot{y}, \dot{Z}))\dot{Z}$$

$$+ ReH(y, Z, \xi(y, Z), \eta(y, Z)) - ReH(y, Z, \xi(y, Z), \eta(y, Z, \dot{y}, \dot{Z}))](\tau)\,d\tau.$$

Let us stress (to avoid ambiguity) that in our notation, say, $\eta(y, z)(\tau)$ means the imaginary part of the momentum field in the point $(\tau, (y + iz)(\tau)$ defined by the family $\Gamma(x_0)$, and $\eta(y, z, \dot{y}, \dot{z})$ means the solution of equations (6.7). Now notice that in the last expression the first integral is just the real part of the invariant Hilbert integral and consequently one can rewrite the last expression in the form

$$ReS(t, x; x_0) - \int_0^t [ReH(y, Z, \xi(y, Z), \eta(y, Z, \dot{y}, \dot{Z})) - ReH(y, Z, \xi(y, Z), \eta(y, Z))$$

$$- \left(\eta(y, Z, \dot{y}, \dot{Z}) - \eta(y, Z), \frac{\partial ReH}{\partial \eta}(y, Z, \xi(y, Z), \eta(y, Z)) \right) \right] d\tau.$$

The function under the second integral is negative (it is actually the real part of the Weierstrass function), since with respect to η the function ReH is concave. It follows that

$$ReI_t(y(.), Z(.)) \geq ReS(t, x : x_0) = ReI_t(Y(.), Z(.)).$$

Further on we shall deal mostly with a particular case of Hamiltonian (6.1), namely with the case of vanishing A and a constant G.

Proposition 6.3 *If the drift A vanishes and the diffusion matrix G is constant, then formula (2.19)-(2.21) hold. More exact formulas can be written as well:*

$$\frac{\partial^2 S}{\partial x^2} = \frac{1}{t} G^{-1} (1 + \frac{1}{3} t^2 \frac{\partial^2 V}{\partial x^2}(x_0) G + O(t^2 c)),$$

$$\frac{\partial^2 S}{\partial x_0^2} = \frac{1}{t} G^{-1} (1 + \frac{1}{3} t^2 \frac{\partial^2 V}{\partial x^2}(x_0) G + O(t^2 c)), \tag{6.13}$$

$$\frac{\partial^2 S}{\partial x \partial x_0} = -\frac{1}{t} G^{-1} (1 - \frac{1}{6} t^2 \frac{\partial^2 V}{\partial x^2}(x_0) G + O(t^2 c)). \tag{6.14}$$

and,

$$\frac{\partial X}{\partial p_0} = tG(1 + \frac{1}{6} t^2 \frac{\partial^2 V}{\partial x^2}(x_0) G + O(t^2 c)), \quad \frac{\partial P}{\partial p_0} = 1 + \frac{1}{2} t^2 \frac{\partial^2 V}{\partial x^2}(x_0) G + O(t^2 c),$$
$$\tag{6.15}$$

where c is from Theorem 2.1.

Proof. Under the assumptions of the Proposition

$$X(t, x_0, p_0) = x_0 + tGp_0 + \frac{1}{2} G \frac{\partial V}{\partial x}(x_0) t^2 + \frac{t^3}{6} G \frac{\partial^2 V}{\partial x^2} Gp_0 + O(t^4 p_0^2),$$

$$P(t, x_0, p_0) = p_0 + \frac{\partial V}{\partial x}(x_0) t + \frac{t^2}{2} \frac{\partial^2 V}{\partial x^2} Gp_0 + O(t^3 p_0^2).$$

This implies (6.15) and also the estimate

$$\frac{\partial X}{\partial x_0} = 1 + \frac{1}{2} t^2 G \frac{\partial^2 V}{\partial x^2}(x_0) + O(t^2 c).$$

These estimates imply (6.13), (6.14) due to (1.19)-(1.21).

In the theory of semiclassical approximation , it is important to know whether the real part of the action S is nonnegative.

Proposition 6.4. *(i) If G_R is strictly positive for all x, then $ReS(t, x; x_0)$ restricted to real values x, x_0 is nonnegative and convex for small enough t and $x - x_0$.*

(ii) Let G_R and the drift A vanish for all x, and let G_I be a constant positive matrix, which is proportional to the unit matrix. Then ReS restricted to real values x, x_0 is nonnegative and convex for small enough $t, x - x_0$ iff V_R is nonnegative and strictly convex with respect to $y = Re\,x$.

Proof. (i) Follows directly from representation (2.17).

(ii) It follows from (6.13), (6.14) that (under assumptions (ii)) $Re\,S(t, x, x_0)$ is convex in x and x_0 for real x and x_0 whenever V_R is convex for real V. Consequently, to prove the positivity of $Re\,S$ it is enough to prove the positivity of $S(t, \tilde{x}, x_0)$ for all x_0, because, this is a minimum of S, as a function of x_0. Using expansion (2.17) yields

$$S(t, \tilde{x}, x_0) = tV(x_0) + O(t^3)\|\frac{\partial V}{\partial x}\|^2.$$

Since $V(x)$ is nonnegative, it follows that $V(\hat{x}_0) \geq 0$ at the point $\hat{x}_0$ of its global minimum. The previous formula implies directly that $S(t, \tilde{x}, x_0)$ is positive (for small t at least) whenever $V(\hat{x}_0) > 0$. If $V(\hat{x}_0) = 0$, then S is clearly nonnegative outside a neighbourhood of $\hat{x}_0$. Moreover, in the neighbourhood of $\hat{x}_0$, it can be written in the form

$$S(t, \tilde{x}, x_0) = \frac{t}{2}\left(\frac{\partial^2 V}{\partial x^2}(\hat{x}_0)(x_0 - \hat{x}_0), x_0 - \hat{x}_0\right) + O(t|x_0 - \hat{x}_0|^3) + O(t^3|x_0 - \hat{x}_0|^2),$$

which is again non-negative for small t.

7. Stochastic Hamiltonians

The theory developed in the previous Sections can be extended to cover the stochastic generalisations of Hamiltonian systems, namely the system of the form

$$\begin{cases} dx = \frac{\partial H}{\partial p}\, dt + g(t, x) \circ dW \\ dp = -\frac{\partial H}{\partial x}\, dt - (c'(t, x) + pg'(t, x)) \circ dW, \end{cases} \tag{7.1}$$

where $x \in \mathcal{R}^n$, $t \geq 0$, $W = (W^1, ...W^m)$ is the standard m-dimensional Brownian motion ($\circ$, as usual, denotes the Stratonovich stochastic differential), $c(t, x)$ and $g(t, x) = g_{ij}(t, x)$ are given vector-valued and respectively $(m \times n)$-matrix-valued functions and the Hamiltonian $H(t, x, p)$ is convex with respect to p. Stochastic Hamiltonian system (7.1) correspond formally to the singular Hamiltonian function

$$H(t, x, p) + (c(t, x) + pg(t, x))\dot{W}(t),$$

where $\dot{W}$ is the white noise (formal derivative of the Wiener process). The corresponding stochastic Hamilton-Jacobi equation clearly has the form

$$dS + H(t, x, \frac{\partial S}{\partial x})\, dt + (c(t, x) + g(t, x)\frac{\partial S}{\partial x}) \circ dW = 0. \tag{7.2}$$

To simplify the exposition we restrict ourselves to the most important particular case, when $g = 0$ in (7.1) and the functions H and c do not depend explicitly on t. Namely, we shall consider the stochastic Hamiltonian system

$$\begin{cases} dx = \frac{\partial H}{\partial p}\, dt \\ dp = -\frac{\partial H}{\partial x}\, dt - c'(x)\, dW. \end{cases} \tag{7.3}$$

and the stochastic Hamilton-Jacobi equation

$$dS + H(x, \frac{\partial S}{\partial x})\, dt + c(x)\, dW = 0. \tag{7.4}$$

In that case the Ito and the Stratonovich differentials coincide. The generalisation of the theory to (7.1) and (7.2) is almost straightforward. As the next stage of simplification we suppose that the matrix of the second derivative of H with respect to all its arguments is uniformly bounded. An example of this situation is given by the standard quantum mechanical Hamiltonian $p^2 - V(x)$. In that important for the application case one can get rather nice results on the existence of the solution to the boundary-value problem uniform with respect to the position of the boundary values x_0, x. However, the restriction to this type of Hamiltonians is by no means necessary. More general Hamiltonians that was discussed in Sections 2-6 can be considered in this framework similarly and the result are similar to those obtained for the deterministic Hamiltonian systems of Sections 2-6.

Theorem 7.1 [K1], [K2]. *For fixed $x_0 \in R^n$ and $t > 0$ let us consider the map $P : p_0 \mapsto X(t, x_0, p_0)$, where $X(\tau, x_0, p_0)$, $P(\tau, x_0, p_0)$ is the solution to (7.3) with initial values (x_0, p_0). Let all the second derivatives of the functions H and c are uniformly bounded, the matrix $Hess_p H$ of the second derivatives of H with respect to p is uniformly positive (i.e. $Hess_p H \geq \lambda E$ for some constant λ), and for any fixed x_0 all matrices $Hess_p H(x_0, p)$ commute. Then the map P is a diffeomorphism for small $t \leq t_0$ and all x_0.*

Proof. Clear that the solution of the linear matrix equation

$$dG = B_1 G\, dt + B_2(t)\, dW, \quad G|_{t=0} = G_0, \tag{7.5}$$

where $B_j = B_j(t, [W])$ are given uniformly bounded and non-anticipating functionals on the Wiener space, can be represented by the convergent series

$$G = G_0 + G_1 + G_2 + \ldots \tag{7.6}$$

with

$$G_k = \int_0^t B_1(\tau) G_{k-1}(\tau)\, d\tau + \int_0^t B_2(\tau) G_{k-1}(\tau)\, dW(\tau). \tag{7.7}$$

Differentiating (7.3) with respect to the initial data (x_0, p_0) one gets that the matrix

$$G = \frac{\partial(X, P)}{\partial(x_0, p_0)} = \begin{pmatrix} \frac{\partial X}{\partial x_0} & \frac{\partial X}{\partial p_0} \\ \frac{\partial P}{\partial x_0} & \frac{\partial P}{\partial p_0} \end{pmatrix} (x(\tau, [W]), p(\tau, [W]))$$

satisfies a particular case of (7.5):

$$dG = \begin{pmatrix} \frac{\partial^2 H}{\partial p \partial x} & \frac{\partial^2 H}{\partial p^2} \\ -\frac{\partial^2 H}{\partial x^2} & -\frac{\partial^2 H}{\partial x \partial p} \end{pmatrix} (X,P)(t)\, G\, dt - \begin{pmatrix} 0 & 0 \\ c''(X(t)) & 0 \end{pmatrix} G\, dW \qquad (7.8)$$

with G_0 being the unit matrix. Let us denote by $\tilde{O}(t^\alpha)$ any function that is of order $O(t^{\alpha-\epsilon})$ for any $\epsilon > 0$, as $t \to 0$. Applying the log log law for stochastic integrals [Ar] first to the solutions of system (7.3) and then calculating G_1 by (7.7) we obtain

$$G_1 = \left(t \begin{pmatrix} \frac{\partial^2 H}{\partial p \partial x} & \frac{\partial^2 H}{\partial p^2} \\ -\frac{\partial^2 H}{\partial x^2} & -\frac{\partial^2 H}{\partial x \partial p} \end{pmatrix} (x_0,p_0) + \begin{pmatrix} 0 & 0 \\ c''(x_0)\tilde{O}(t^{1/2}) & 0 \end{pmatrix} \right) \begin{pmatrix} E & 0 \\ 0 & E \end{pmatrix}$$

up to a term of order $\tilde{O}(t^{3/2})$. Application of the log log law to the next terms of series (7.5) yields for the remainder $G - G_0 - G_1$ the estimate $\tilde{O}(t^{3/2})$. Thus, we have the convergence of series (7.5) for system (7.8) and the following approximate formula for its solutions:

$$\frac{\partial X}{\partial x_0} = E + t\frac{\partial^2 H}{\partial p \partial x}(x_0,p_0) + \tilde{O}(t^{3/2}), \quad \frac{\partial X}{\partial p_0} = t\frac{\partial^2 H}{\partial p^2}(x_0,p_0) + \tilde{O}(t^{3/2}), \quad (7.9)$$

$$\frac{\partial P}{\partial x_0} = \tilde{O}(t^{1/2}), \quad \frac{\partial P}{\partial p_0} = E + t\frac{\partial^2 H}{\partial x \partial p}(x_0,p_0) + \tilde{O}(t^{3/2}). \qquad (7.10)$$

These relations imply that the map $P : p_0 \mapsto X(t,x_0,p_0)$ is a local diffeomorphism and is globally injective. The last statement follows from the formula

$$X(t,x_0,p_0^1) - X(t,x_0,p_0^2) = t(1+O(t))(p_0^1 - p_0^2),$$

which one gets by the same arguments as in the proof of Theorem 2.1. Moreover, from this formula it follows as well that $x(t,p_0) \to \infty$, as $p_0 \to \infty$ and conversely. From this one deduces that the image of the map $P : p_0 \mapsto X(t,x_0,p_0)$ is simultaneously closed and open and therefore coincides with the whole space, which completes the proof of the Theorem.

Let us define now the two-points function

$$S_W(t,x,x_0) = \inf \int_0^t (L(y,\dot{y})\, d\tau - c(y)\, dW), \qquad (7.11)$$

where inf is taken over all continuous piecewise smooth curves $y(\tau)$ such that $y(0) = x_0$, $y(t) = x$, and the Lagrangian L is, as usual, the Legendre transform of the Hamiltonian H with respect to its last argument.

Theorem 7.2. *Under the assumptions of Theorem 7.1*

$$(i) \quad S_W(t,x,x_)) = \int_0^t (p\, dx - H(x,p)\, dt - c(x)\, dW), \qquad (7.12)$$

where the integral is taken along the solution $X(\tau), P(\tau)$ of system (7.3) that joins the points x_0 and x in time t (and which exists and is unique due to Theorem 7.1),

$$(ii) \qquad P(t) = \frac{\partial S_W(t, x, x_0)}{\partial x}, \qquad p_0 = -\frac{\partial S_W(t, x, x_0)}{\partial x_0},$$

(iii) S satisfies equation (7.4), as a function of x,
(iv) $S(t, x, x_0)$ is convex in x and x_0.

Proof. The proof can be carried out by rather long and tedious direct differentiations with the use of the Ito formula. But fortunately, we can avoid it by using the following well known fact [SV, Su, WZ]: if we approximate the Wiener trajectories W in some (ordinary) stochastic Stratonovich equation by a sequence of smooth functions

$$W_n(t) = \int_0^t q_n(s)\, ds \qquad (7.13)$$

(with some continuous functions q_n), then the solutions of the corresponding classical (deterministic) equations will tend to the solution of the given stochastic equation. For functions (7.13), equation (7.4) as well as system (7.3) become classical and results of the Theorem become well known (see, for instance, [MF1],[KM1]). In Section 1.1 we have presented these result for the case of Hamiltonians which do not depend explicitly on t, but this dependence actually would change nothing in these considerations. By the approximation theorem mentioned above the sequence of corresponding diffeomorphisms P_n of Theorem 7.1 converges to the diffeomorphism P, and moreover, due to the uniform estimates on their derivatives (see (7.9),(7.10)), the convergence of $P_n(t, x_0, p_0)$ to $P(t, x_0, p_0)$ is locally uniform as well as the convergence of the inverse diffeomorphisms $P_n^{-1}(t, x) \to P^{-1}(t, x)$. It implies the convergence of the corresponding solutions S_n to function (2.2) together with their derivatives in x. Again by the approximation arguments we conclude that the limit function satisfies equation (7.4). Let us note also that the convex property of S is due to equations (1.19),(1.20),(7.9),(7.10).

By similar arguments one gets the stochastic analogue of the classical formula to the Cauchy problem for Hamilton-Jacobi equation, namely the following result

Theorem 7.3 [TZ1],[K1]. *Let $S_0(x)$ is a smooth function and for all $t \le t_0$ and $x \in \mathcal{R}^n$ there exists a unique $\xi = \xi(t, x)$ such that $x(t, \xi) = x$ for the solution $x(\tau, \xi), p(\tau, \xi)$ of system (7.3) with initial data $x_0 = \xi, p_0 = (\partial S_0/\partial x)(\xi)$. Then*

$$S(t, x) = S_0(\xi) + \int_0^t (p\, dx - H(x, p)\, dt - c(x)\, dW) \qquad (7.14)$$

(where the integral is taken along the trajectory $x(\tau, \xi), p(\tau, \xi))$ is a unique classical solution of the Cauchy problem for equation (1.4) with initial function $S_0(x)$.

Theorems 7.1, 7.2 imply simple sufficient conditions for the assumptions of Theorem 2.3 to be true. The following result is a direct corollary of Theorem 7.2.

Theorem 7.4. *Under the assumptions of Theorem 7.1 let the function $S_0(x)$ is smooth and convex. Then for $t \leq t_0$ there exists a unique classical (i.e. everywhere smooth) solution to the Cauchy problem of equation (7.4) with initial function $S_0(x)$ and it is given by equation (7.14) or equivalently by the formula*

$$R_t S_0(x) = S(t, x) = \min_\xi (S_0(\xi) + S_W(t, x, \xi)). \qquad (7.15)$$

One can directly apply the method of constructing generalised solution to deterministic Bellman equation from [KM1],[KM2] to the stochastic case, which gives the following result (details in [K1],[K2]):

Theorem 7.5. *For any bounded from below initial function $S_0(x)$ there exists a unique generalised solution of the Cauchy problem for equation (7.3) that is given by formula (7.15) for all $t \geq 0$.*

Let us mention for the conclusion that the results of Propositions 2.4-2.8 can be now obtained by the similar arguments for the stochastic Hamiltonians. Furthermore, most of the results of the previous Section can be obtained also for complex stochastic Hamiltonian system of form (7.3). For example, let us formulate one of the results.

Theorem 7.6. *Consider a complex stochastic Hamilton-Jacobi equation of form (7.4), with H of form (6.1) supposing that G, A, V, c are analytic in the band $\{|Im\, x| \leq 2\epsilon\}$ with some $\epsilon > 0$ and such that G, A, V'', c' are uniformly bounded there together with all their derivatives. Then there exist $\delta > 0$, $t_0 > 0$ and a smooth family Γ of solutions of the corresponding Hamiltonian system (7.3) joining uniquely in time $t \leq t_0$ any two points x_0, x such that $|x - x_0| \leq \delta$, $|Im\, x| \leq \epsilon$, $|Im\, x_0| \leq \epsilon$. Moreover, all trajectories from Γ are saddle-points for the corresponding functional $\int_0^t (L(y, \dot{y})\, d\tau - c(y)\, dW)$ (in the sense of Section 2.6), and the corresponding random two-point function $S_W(t, x, x_0)$ satisfies (almost surely) equation (7.4).*

Chapter 3. SEMICLASSICAL APPROXIMATION FOR REGULAR DIFFUSION

1. Main ideas of the WKB-method with imaginary phase

In this chapter we construct exponential WKB-type asymptotics for solutions of equations of type

$$h\frac{\partial u}{\partial t} = \frac{h^2}{2}tr\left(G(x)\frac{\partial^2 u}{\partial x^2}\right) + h\left(A(x), \frac{\partial u}{\partial x}\right) - V(x), \qquad (1.1)$$

where $t \leq 0, x \in \mathcal{R}^m$, V, A and G are smooth real, vector-valued, and matrix-valued functions on $\mathcal{R}^m$ respectively, $G(x)$ is symmetric non-negative, and h is a positive parameter. Equivalently, one can write equation (1.1) in the "pseudo-differential form"

$$h\frac{\partial u}{\partial t} = H\left(x, -h\frac{\partial}{\partial x}\right)u = Lu \qquad (1.2)$$

with the Hamiltonian function

$$H(x,p) = \frac{1}{2}(G(x)p,p) - (A(x),p) - V(x). \qquad (1.3)$$

Our main aim will be the construction of the *Green function* of equation (1.1), i.e. of the solution $u_G(t,x,x_0)$ with Dirac initial data

$$u_G(t,x,x_0) = \delta(x - x_0). \qquad (1.4)$$

The solution of the Cauchy problem for equation (1.1) with general initial data $u_0(x)$ can be then given by the standard integral formula

$$u(t,x) = \int u_G(t,x,x_0)u_0(x_0)\,dx_0. \qquad (1.5)$$

In this introductory section we describe the main general steps of the construction of the formal asymptotic solution for the problem given by (1.1) and (1.4), presenting in a compact but systematic way rather well-known ideas (see, e.g. [MF1], [M1],[M2], [KM2]), which were previously used only for non-degenerate diffusions, i.e. when the matrix G in (1.1) was non-degenerate (and usually only for some special cases, see [MC1],[DKM1], [KM2]). Here we shall show that these ideas can also be applied for the case of regular (in particular degenerate) Hamiltonians of type (1.3) introduced and discussed in the previous chapter from the point of view of the calculus of variations. In fact, the results of the previous chapter form the basis that allows us to carry out successfully (effectively and rigorously) the general steps described in this section. Moreover, it seems that regular Hamiltonians form the most general class, for which it can be done in this way. As we shall see in Section 3.6, for non-regular Hamiltonians, the procedure must be modified essentially, even at the level of formal expansions, if one is interested in small time asymptotics, but for small h asymptotics (with fixed t), this procedure seems to lead to correct results even for non-regular degenerate diffusions. We shall construct two types of asymptotics for (1.1), (1.4), namely, small time asymptotics, when $t \to 0$ and h is fixed, say $h = 1$, and (global) small diffusion asymptotics, when t is any finite number and $h \to 0$.

Step 1. One looks for the asymptotic solution of (1.1), (1.4) for small h in the form

$$u_G^{as}(t, x, x_0, h) = C(h)\phi(t, x, x_0) \exp\{-S(t, x, x_0)/h\}, \tag{1.6}$$

where S is some non-negative function called the action or entropy, and $C(h)$ is a normalising coefficient.

In the standard WKB method traditionally used in quantum mechanics to solve the Schrödinger equation, one looks for the solutions in "oscillatory form"

$$C(h)\phi(t, x, x_0) \exp\{-\frac{i}{h}S(t, x, x_0)\} \tag{1.7}$$

with real functions ϕ and S called the amplitude and the phase respectively. For this reason one refers sometimes to the ansatz (1.6) as to the WKB method with imaginary phase, or as to exponential asymptotics, because it is exponentially small outside the zero-set of S. The difference between the asymptotics of the types (1.6) and (1.7) is quite essential. On the one hand, when justifying the standard WKB asymptotics of type (1.7) one should prove that the exact solution has the form

$$C(h)\phi(t, x, x_0) \exp\{-\frac{i}{h}S(t, x, x_0)\} + O(h) \tag{1.8}$$

(additive remainder), which can be proved under rather general conditions by L^2 methods of functional analysis [MF1]. For asymptotics of form (1.6) this type of justification would make no sense, because the expression (1.6) is exponentially small outside the zero-set of S. Thus, to justify (1.6) one should instead prove that the exact solution has the form

$$C(h)\phi(t, x, x_0) \exp\{-S(t, x, x_0)/h\}(1 + O(h)), \tag{1.9}$$

which must be carried out by some special pointwise estimates. Because of the multiplicative remainder in (1.9) one calls asymptotics of this type multiplicative. On the other hand, essential difference between (1.6) and (1.7) lies in the fact that if one adds different asymptotic expressions of form (1.9), then, unlike the case of the asymptotics (1.8), in the sum only the term with the minimal entropy survives at each point (because other terms are exponentially small in compared with this one), and therefore for the asymptotics (1.9) the superposition principle transforms into the idempotent superposition principle $(S_1, S_2) \mapsto \min(S_1, S_2)$ at the level of actions. For a detailed discussion of this idempotent superposition principle and its applications see [KM1],[KM2].

It seems that among parabolic differential equations only second order equations can have asymptotics of the Green function of form (1.9). Considering more general pseudo-differential equations one gets other classes, which enjoy this property, for example, the so called tunnel equations introduced in [M1], [M2] (see Chapter 6).

Inserting (1.6) in (1.1) yields

$$h\left(\frac{\partial\phi}{\partial t} - \frac{1}{h}\phi\frac{\partial S}{\partial t}\right) = \frac{h^2}{2}tr\, G(x)\left(\frac{\partial^2\phi}{\partial x^2} - \frac{\phi}{h}\frac{\partial^2 S}{\partial x^2}\right) + h\left(A(x), \frac{\partial\phi}{\partial x} - \frac{1}{h}\phi\frac{\partial S}{\partial x}\right)$$

$$+ \frac{1}{2}\left(G(x)\frac{\partial S}{\partial x}, \frac{\partial S}{\partial x}\right)\phi - h\left(G(x)\frac{\partial S}{\partial x}, \frac{\partial\phi}{\partial x}\right) - V(x). \tag{1.10}$$

Comparing coefficients of h^0 yields the Hamilton-Jacobi equation

$$\frac{\partial S}{\partial t} + H\left(x, \frac{\partial S}{\partial x}\right) = 0 \tag{1.11}$$

corresponding to the Hamiltonian function (1.3), or more explicitly

$$\frac{\partial S}{\partial t} + \frac{1}{2}\left(G(x)\frac{\partial S}{\partial x}, \frac{\partial S}{\partial x}\right) - \left(A(x), \frac{\partial S}{\partial x}\right) - V(x) = 0. \tag{1.12}$$

Comparing coefficients of h one gets the so called transport equation

$$\frac{\partial\phi}{\partial t} + \left(\frac{\partial\phi}{\partial x}, \frac{\partial H}{\partial p}(x, \frac{\partial S}{\partial x})\right) + \frac{1}{2}tr\left(\frac{\partial^2 S}{\partial x^2}\frac{\partial^2 H}{\partial p^2}(x, \frac{\partial S}{\partial x})\right)\phi(x) = 0, \tag{1.13}$$

or more explicitly

$$\frac{\partial\phi}{\partial t} - \left(A(x), \frac{\partial\phi}{\partial x}\right) + \left(G(x)\frac{\partial S}{\partial x}, \frac{\partial\phi}{\partial x}\right) + \frac{1}{2}tr\left(G(x)\frac{\partial^2 S}{\partial x^2}\right)\phi = 0. \tag{1.13'}$$

Therefore, if S and ϕ satisfy (1.12), (1.13), then the function u of form (1.6) satisfies equation (1.1) up to a term of order h^2, i.e.

$$h\frac{\partial u_G^{as}}{\partial t} - H\left(x, -h\frac{\partial}{\partial x}\right)u_G^{as} = -\frac{h^2}{2}C(h)\,tr\left(G(x)\frac{\partial^2\phi}{\partial x^2}\right)\exp\{-\frac{S(t,x)}{h}\}. \tag{1.14}$$

As is well known and as was explained in the previous chapter, the solutions of the Hamilton-Jacobi equation (1.12) can be expressed in terms of the solutions of the corresponding Hamiltonian system

$$\begin{cases} \dot{x} = -\frac{\partial H}{\partial p} = G(x)p - A(x) \\ \dot{p} = \frac{\partial H}{\partial x} = \frac{\partial}{\partial x}(A(x), p) - \frac{1}{2}\frac{\partial}{\partial x}(G(x)p, p) + \frac{\partial V}{\partial x} \end{cases} \cdot \tag{1.15}$$

Step 2. If we had to solve the Cauchy problem for equation (1.1) with a smooth initial function of form (1.6), then clearly in order to get an asymptotic solution in form (1.6) we would have to solve the Cauchy problem for the Hamilton-Jacobi equation (1.11). The question then arises, what solution of

(1.11) (with what initial data) one should take in order to get asymptotics for the Green function. The answer is the following. If the assumptions of Theorem 2.1.1 hold, i.e. there exists a family $\Gamma(x_0)$ of characteristics of (1.15) going out of x_0 and covering some neighbourhood of x_0 for all sufficiently small t, then one should take as the required solution of (1.15) the two-point function $S(t, x, x_0)$ defined in the previous chapter (see formulas (2.1.5), (2.1.6) for local definition and Proposition 2.2.8 for global definition). As was proved in the previous chapter (see Propositions 2.2.8 and 2.3.7), for regular Hamiltonians this function is an almost everywhere solution of the Hamilton-Jacobi equation. One of the reasons for this choice of the solution of (1.12) lies in the fact that for the Gaussian diffusion described in the first chapter this choice of S leads to the exact formula for Green function. Another reason can be obtained considering the Fourier transform of equation (1.1). Yet another explanation is connected with the observation that when considering systematically the idempotent superposition principle on actions as described above one finds that the resolving operator for the Cauchy problem (of generalised solutions) for the nonlinear equation (1.11) is "linear" with respect to this superposition principle and the two-point function $S(t, x, x_0)$ can be interpreted as well as "the Green function" for (1.11) (see details in [KM1],[KM2]). Therefore, by the "correspondence principle", the Green function for (1.1) should correspond to "the Green function" for (1.11). All this reasoning are clearly heuristic, and the rigorous justification of asymptotics constructed in this way needs to be given independently.

Step 3. This is to construct solutions of the transport equation (1.13). The construction is based on the well known (and easily proved) Liouville theorem, which states that if the matrix $\frac{\partial x}{\partial \alpha}$ of derivatives of the solution of an m-dimensional system of ordinary differential equations $\dot{x} = f(x, \alpha)$ with respect to any m-dimensional parameter α is non-degenerate on some time interval, then the determinant J of this matrix satisfies the equation $\dot{J} = J \operatorname{tr} \frac{\partial f}{\partial x}$. Let us apply the Liouville theorem to the system

$$\dot{x} = \frac{\partial H}{\partial p}\left(x, \frac{\partial S}{\partial x}(t, x, x_0)\right),$$

which is the first equation of (1.15), whose momentum p is expressed in terms of the derivatives of the two-point function according to Proposition 1.1 of the second chapter. Considering the initial momentum p_0 as the parameter α one gets in this way that on the characteristic $X(t, x_0, p_0)$ the determinant $J = \det \frac{\partial X}{\partial p_0}$ satisfies the equation

$$\dot{J} = J \operatorname{tr}\left(\frac{\partial^2 H}{\partial p \partial x} + \frac{\partial^2 H}{\partial p^2}\frac{\partial^2 S}{\partial x^2}\right),$$

or more explicitly (using (1.4))

$$\dot{J} = J \operatorname{tr}\left(G\frac{\partial^2 S}{\partial x^2} - \frac{\partial A}{\partial x} + \frac{\partial G}{\partial x}p\right),$$

which yields the equation for $J^{-1/2}$

$$(J^{-1/2})^{\cdot} = -\frac{1}{2}J^{-1/2}\,tr\left(\frac{\partial^2 H}{\partial p \partial x} + \frac{\partial^2 H}{\partial p^2}\frac{\partial^2 S}{\partial x^2}\right), \tag{1.16}$$

or more explicitly

$$(J^{-1/2})^{\cdot} = -\frac{1}{2}J^{-1/2}tr\left(G\frac{\partial^2 S}{\partial x^2} - \frac{\partial A}{\partial x} + \frac{\partial G}{\partial x}p\right). \tag{1.16'}$$

Now consider the behaviour of the function ϕ satisfying the transport equation (1.13) along the characteristic $X(t, x_0, p_0)$. The total time derivative is $\frac{d}{dt} = \frac{\partial}{\partial t} + \dot{x}\frac{\partial}{\partial x}$. Consequently denoting this total time derivative by a dot above a function and using (1.15) one can rewrite (1.13) as

$$\dot{\phi} + \frac{1}{2}\phi\,tr\left(\frac{\partial^2 H}{\partial p^2}(x, \frac{\partial S}{\partial x})\frac{\partial^2 S}{\partial x^2}\right) = 0, \tag{1.17}$$

or more explicitly

$$\dot{\phi} + \frac{1}{2}\phi\,tr\left(G(x)\frac{\partial^2 S}{\partial x^2}\right) = 0. \tag{1.17'}$$

This is a first order linear equation, whose solution is therefore unique up to a constant multiplier. Introducing the new unknown function α by $\phi = J^{-1/2}\alpha$ one gets the following equation for α using (1.16),(1.17):

$$\dot{\alpha} = \frac{1}{2}\alpha\,tr\,\frac{\partial^2 H}{\partial p \partial x}(x, \frac{\partial S}{\partial x}),$$

whose solution can be expressed in terms of the solutions of (1.15). Thus one finds a solution for (1.13) in the form

$$\phi(t, x, x_0) = J^{-1/2}(t, x, x_0)\exp\left\{\frac{1}{2}\int_0^t tr\,\frac{\partial^2 H}{\partial p \partial x}(x(t))p(t))\,d\tau\right\}, \tag{1.18}$$

or more explicitly

$$\phi(t, x, x_0) = J^{-1/2}(t, x, x_0)\exp\left\{\frac{1}{2}\int_0^t tr\left(\frac{\partial G}{\partial x}(x(t))p(t)) - \frac{\partial A}{\partial x}(x(t))\right)d\tau\right\}, \tag{1.18'}$$

where the integral is taken along the solution $(X, P)(t, x_0, p_0(t, x, x_0))$ of (1.15) joining x_0 and x in time t.

Notice that $J^{-1/2}(t, x, x_0)$ and therefore the whole function (1.18) are well defined only at regular points (see the definitions at the end of Section 2.2), because at these points the minimising characteristic joining x_0 and x in time t is unique and J does not vanish there. This is why in order to get a globally defined function of form (1.6) (even for small t) one should introduce a molyfier.

Namely, let for $t \in (0, t_0]$ and x in some domain $D = D(x_0)$, all points (t, x) are regular (such t_0 and domain $D(x_0)$ exist for regular Hamiltonians again due to the results of the previous chapter) and let χ_D be a smooth function such that χ_D vanishes outside D, is equal to one inside D except for the points in a neighbourhood of the boundary ∂D of D, and takes value in $[0, 1]$ everywhere. Then the function

$$u_G^{as} = C(h)\chi_D(x - x_0)\phi(t, x, x_0) \exp\{-S(t, x, x_0)/h\} \tag{1.19}$$

with $\phi(t, x; x_0)$ of form (1.18) is globally well defined for $t \le t_0$ and (by (1.14)) satisfies the following equation:

$$h\frac{\partial u_G^{as}}{\partial t} - H\left(x, -h\frac{\partial}{\partial x}\right)u_G^{as} = -h^2 F(t, x, x_0), \tag{1.20}$$

where F (which also depends on h and D) is equal to

$$F = C(h)\left[\frac{1}{2}tr\left(G(x)\frac{\partial^2 \phi}{\partial x^2}\right)\chi_D(x - x_0) - h^{-1}f(x)\right] \times \exp\left\{-\frac{S(t, x, x_0)}{h}\right\}, \tag{1.21}$$

where $f(x)$ has the form $O(\phi)(1 + O(\frac{\partial S}{\partial x})) + O(\frac{\partial \phi}{\partial x})$ and is non vanishing only in a neighbourhood of the boundary ∂D of the domain D.

Step 4. All the constructions that we have described are correct for regular Hamiltonians by the results of the previous chapter. The only thing that remains in the formal asymptotic construction is to show that u_G^{as} as defined by (1.19) satisfies the Dirac initial condition (1.4). But for regular Hamiltonians this is simple, because, as we have seen in the previous chapter and as we shall show again by another method in the next section, the main term (for small t and $x - x_0$) of the asymptotics of the two-point function is the same as for its quadratic or Gaussian diffusion approximation, and one can simply refer to the results of the first chapter. Alternatively, having the main term of the asymptotics of $S(t, x, x_0)$, one proves the initial condition property of (1.19) (with appropriate coefficient $C(h)$) by means of the asymptotic formula for Laplace integrals, see e.g. Appendix B. Consequently, the function (1.19) is a formal asymptotic solution of the problem given by (1.1), (1.4) in the sense that it satisfies the initial conditions (1.4), and satisfies equation (1.1) approximately up to order $O(h^2)$. Moreover, as we shall see further in Section 4, the exact Green function will have the form (1.9) with the multiplicative remainder $1 + O(h)$ having the form $1 + O(ht)$, which will imply that we have got automatically also the multiplicative asymptotics for the Green function for small times and fixed h, say $h = 1$. The same remark also applies to the next terms of the asymptotics which are described below.

Step 5. Till now we have constructed asymptotic solutions to (1.1) up to terms of the order $O(h^2)$. In order to construct more precise asymptotics (up to order $O(h^k)$ with arbitrary $k > 2$), one should take instead of the ansatz (1.6) the expansion

$$u_G^{as} = C(h)(\phi_0(t, x) + h\phi_1(t, x) + \dots + h^k \phi_k(t, x)) \exp\{-S(t, x)/h\}. \tag{1.22}$$

Inserting this in (1.1) and comparing the coefficients of h^j, $j = 0, 1, ..., k+1$, one sees that

$$h\frac{\partial u_G^{as}}{\partial t} - H\left(x, -h\frac{\partial}{\partial x}\right) u_G^{as} = \frac{h^{k+2}}{2} tr\left(G(x)\frac{\partial^2 \phi_k}{\partial x^2}\right) \exp\{-\frac{S(t, x)}{h}\}, \quad (1.23)$$

if (1.12), (1.13) hold for S and ϕ_0 and the following recurrent system of equations (higher order transport equations) hold for the functions ϕ_j, $j = 1, ..., k$:

$$\frac{\partial \phi_j}{\partial t} - \left(A(x), \frac{\partial \phi_j}{\partial x}\right) + \left(G(x)\frac{\partial S}{\partial x}, \frac{\partial \phi}{\partial x}\right) + \frac{1}{2}tr\left(G(x)\frac{\partial^2 S}{\partial x^2}\right)\phi_j$$

$$= \frac{1}{2}tr\left(G(x)\frac{\partial^2 \phi_{j-1}}{\partial x^2}\right),$$

which takes the form

$$\dot{\phi}_j + \frac{1}{2}tr\left(G(x)\frac{\partial^2 S}{\partial x^2}\right)\phi_j = \frac{1}{2}tr\left(G(x)\frac{\partial^2 \phi_{j-1}}{\partial x^2}\right) \quad (1.24)$$

in terms of the total derivative along the characteristics. The change of unknown $\phi_k = \phi_0 \psi_k$, $k = 1, 2, ...$, yields

$$\dot{\psi}_j = \frac{1}{2}\phi_0^{-1}tr\left(G(x)\frac{\partial^2 (\psi_{j-1}\phi_0)}{\partial x^2}\right)$$

and the solution to this equation with vanishing initial data can be found recursively by the integration

$$\psi_j = \frac{1}{2}\int_0^t \phi_0^{-1}tr\left(G(x)\frac{\partial^2 (\psi_{j-1}\phi_0)}{\partial x^2}\right)(x(\tau))\, d\tau. \quad (1.25)$$

By this procedure one gets a function of form (1.22), which is a formal asymptotic solution of (1.1) and (1.4) of order $O(h^{k+2})$, i.e. it satisfies the initial condition (1.4) exactly and satisfies equation (1.1) approximately up to order $O(h^{k+2})$, or more precisely, since each ψ_j is obtained by integration, for small times t the remainder is of the form $O(t^k h^{k+2})$.

Example. To conclude this section, let us show how the method works on the simple example of Gaussian diffusions presenting the analytic proof of formula (1.1.4). Consider the Hamiltonian

$$H = -(Ax, p) + \frac{1}{2}(Gp, p) \quad (1.26)$$

with constant matrices A, G, and the corresponding equation (1.2):

$$\frac{\partial u}{\partial t} = \frac{h}{2}tr\left(G\frac{\partial^2 u}{\partial x^2}\right) + \left(Ax, \frac{\partial u}{\partial x}\right). \quad (1.27)$$

Let us use formulas (2.1.5), (2.1.4) to calculate the two-point function $S(t, x, x_0)$. For this purpose we need to solve the boundary value problem for the corresponding Hamiltonian system

$$\begin{cases} \dot{x} = -Ax + Gp \\ \dot{p} = A'p. \end{cases} \tag{1.28}$$

The solution of (1.28) with initial data x_0, p_0 has the form

$$\begin{cases} P = e^{A't}p_0 \\ X = e^{-At}x_0 + \int_0^t e^{-A(t-\tau)}Ge^{A'\tau}\, d\tau\, p_0, \end{cases}$$

and therefore the function $p_0(t, x, x_0)$ defined by (2.1.2) is (globally) well defined if the matrix $E(t)$ of form (1.3) is non-degenerate, and is given by

$$p_0(t, x, x_0) = E^{-1}(t)(e^{At}x - x_0).$$

Therefore from (2.1.4), (2.1.5) one gets

$$S(t, x, x_0) = \frac{1}{2} \int_0^t (Gp(\tau), p(\tau))\, d\tau = \frac{1}{2} \int_0^t (Ge^{A'\tau}p_0, e^{A'\tau}p_0)\, d\tau$$

$$= \frac{1}{2}(E(t)p_0, p_0) = \frac{1}{2}(E^{-1}(t)(x_0 - e^{At}x), x_0 - e^{At}x)$$

and from (1.18)

$$\phi = (\det \frac{\partial X}{\partial p_0})^{-1/2} e^{t\,\mathrm{tr}\,A/2} = (\det E(t))^{-1/2}.$$

It follows from (1.14) that since ϕ does not depend on x, the r.h.s. of (1.14) vanishes, i.e. in the situation under consideration the asymptotic solution of form (1.6) constructed is in fact an exact solution and is defined globally for all t and x. Therefore one gets the formula for the Green function

$$(2\pi h)^{-m/2}(\det E(t))^{-1/2} \exp\{-\frac{1}{2h}(E^{-1}(x_0 - e^{At}x), x_0 - e^{At}x)\}, \tag{1.29}$$

the coefficient $C(h)$ being chosen in the form $(2\pi h)^{-m/2}$ in order to meet the initial condition (1.4), which one verifies directly. Expression (1.29) coincide with (1.1.4) for $h = 1$.

2. Calculation of the two-point function
for regular Hamiltonians

The most important ingredient in the asymptotics of second order parabolic equations is the two-point function. It was investigated in the previous chapter in the case of regular Hamiltonians; it was proved that this function is smooth and satisfies the Hamilton-Jacobi equation almost everywhere, and a method

of calculation of its asymptotic for small times and small distances was proposed: by means of the asymptotic solutions of the boundary value problem for
corresponding Hamiltonian system. In this section we describe an alternative,
more direct method of its calculation for small times and small distances. This
method seems to be simpler for calculations but without the rigorous results of
the previous chapter, the proof of the correctness of this method seems to be
rather difficult problem (especially when the coefficients of the Hamiltonian are
not real analytic).

In the case of a non-degenerate matrix G in (1.12), one can represent the
two-point function in the form (2.2.17) for small t and $x - x_0$. Substituting
(2.2.17) in (1.12) yields recursive formulas, by which the coefficients of this
expansion can be calculated to any required order. These calculations are widely
represented in the literature, and therefore we omit the details here. We shall
deal more carefully with degenerate regular Hamiltonians of form (2.3.1), where
the corresponding Hamilton-Jacobi equation has the form

$$\frac{\partial S}{\partial t} + \frac{1}{2}\left(g(x)\frac{\partial S}{\partial y}, \frac{\partial S}{\partial y}\right) - a(x,y)\frac{\partial S}{\partial x} - b(x,y)\frac{\partial S}{\partial y} - V(x,y) = 0. \qquad (2.1)$$

A naive attempt to try to solve this equation (following the procedure of the
non-degenerate case) by substituting into this equation an expression of the
form $Reg(t, x - x_0, y - y_0)/t$ with Reg being a regular expansion with respect
to its argument (i.e. as a non-negative power series), or even more generally
substituting $t^{-l}Reg$ with some $l > 0$, does not lead to recurrent equations but
to a difficult system, for which neither uniqueness nor existence of the solution
is clear even at the formal level. In order to get recurrent equations one should
chose the arguments of the expansion in a more sophisticated way. Corollary to
Proposition 2.1.3 suggests that it is convenient to make the (non-homogeneous)
shift of the variables, introducing a new unknown function

$$\sigma(t, x, y) = S(t, x + \tilde{x}(t), y + \tilde{y}(t), x_0, y_0), \qquad (2.2)$$

where $(\tilde{x}, \tilde{y}, \tilde{p}, \tilde{q})(t)$ denote the solution of the corresponding Hamiltonian system
with initial conditions $(x_0, y_0, 0, 0)$. In terms of the function σ the Hamilton-
Jacobi equation (2.1) takes the form

$$\frac{\partial \sigma}{\partial t} - (a(x+\tilde{x}(t), y+\tilde{y}(t)) - a(\tilde{x}(t), \tilde{y}(t)))\frac{\partial \sigma}{\partial x} - (b(x+\tilde{x}(t), y+\tilde{y}(t)) - b(\tilde{x}(t), \tilde{y}(t)))\frac{\partial \sigma}{\partial y}$$

$$-g(\tilde{x}(t), \tilde{y}(t))\tilde{q}(t)\frac{\partial \sigma}{\partial y}$$

$$+\frac{1}{2}\left(g(x + \tilde{x}(t), y + \tilde{y}(t))\frac{\partial \sigma}{\partial y}, \frac{\partial \sigma}{\partial y}\right) - V(x + \tilde{x}(t), y + \tilde{y}(t)) = 0. \qquad (2.3)$$

The key idea (suggested by Proposition 2.3.7) for the asymptotic solution of
this equation in the case of general regular Hamiltonian (2.4.1) is to make the

change of variables $(x^0, ..., x^M, y) \mapsto (\xi^0, ..., \xi^M, y)$ defined by the formula $x^I = t^{M-I+1}\xi^I$, $I = 0, ..., M$. Introducing the new unknown function by the formula

$$\Sigma(t, \xi^0, ..., \xi^M, y) = \sigma(t, t^{M+1}\xi^0, ..., t\xi^M, y) \tag{2.4}$$

and noting that

$$\frac{\partial \Sigma}{\partial \xi^I} = \frac{\partial \sigma}{\partial x^I} t^{M-I+1}, \quad I = 0, ..., M,$$

$$\frac{\partial \Sigma}{\partial t} = \frac{\partial \sigma}{\partial t} + (M+1)t^M \xi^0 \frac{\partial \sigma}{\partial x^0} + \cdots + \xi^M \frac{\partial \sigma}{\partial x^M},$$

one write down equation (2.3) for the case of Hamiltonian (2.4.1) in terms of the function Σ:

$$\frac{\partial \Sigma}{\partial t} + \frac{1}{2}\left(g(t^{M+1}\xi^0 + \tilde{x}^0(t))\frac{\partial \Sigma}{\partial y}, \frac{\partial \Sigma}{\partial y}\right)$$

$$-\left(g(\tilde{x}^0(t))\tilde{q}(t), \frac{\partial \Sigma}{\partial y}\right) - R_{2(M+2)}(x(t, \xi) + \tilde{x}(t), y + \tilde{y})$$

$$-\frac{t^M \xi^0 + R_1(x(t, \xi) + \tilde{x}(t), y + \tilde{y}(t)) - R_1(\tilde{x}(t), \tilde{y}(t))}{t^{M+1}}\frac{\partial \Sigma}{\partial \xi^0} - \cdots$$

$$-\frac{\xi^M + R_{M+1}(x(t, \xi) + \tilde{x}(t), y + \tilde{y}(t)) - R_{M+1}(\tilde{x}(t), \tilde{y}(t))}{t}\frac{\partial \Sigma}{\partial \xi^M}$$

$$-(R_{M+2}(x(t, \xi) + \tilde{x}(t), y + \tilde{y}(t)) - R_{M+2}(\tilde{x}(t), \tilde{y}(t)))\frac{\partial \Sigma}{\partial y}. \tag{2.5}$$

It turns out that by expanding the solution of this equation as a power series in its arguments one does get uniquely solvable recurrent equations. In this procedure lies indeed the source of the main definition of regular Hamiltonians, which may appear to be rather artificial at first sight. This definition insures that after expansion of obtained formal power series solution of (2.5) in terms of initial variables (t, x, y) one gets the expansion of S in the form $t^{-(2M+3)}Reg(t, x - \tilde{x}, y - \tilde{y})$ (where Reg is again a regular, i. e. positive power series, expansion with respect to its arguments), and not in the form of a Laurent type expansion with infinitely many negatives powers. More precisely, the following main result holds.

Theorem 2.1. *Under the assumptions of the main definition of regular Hamiltonians (see Sect. 4 in Chapter 2), there exists a unique solution of equation (2.5) of the form*

$$\Sigma = \frac{\Sigma_{-1}}{t} + \Sigma_0 + t\Sigma_1 + t^2\Sigma_2 + \cdots \tag{2.6}$$

such that Σ_{-1} and Σ_0 vanish in the origin, Σ_{-1} is strictly convex in a neighbourhood of the origin, and all Σ_j are regular power series in (ξ, y). Moreover, in this solution *(i)* Σ_{-1} *is the quadratic form with the matrix* $\frac{1}{2}(E^0)^{-1}(1)$, *where* $E^0(t)$ *and its inverse are as in Lemmas 1.2.2, 1.2.4,*

(ii) all Σ_j *are polynomials in* (ξ, y) *such that the degrees of* $\Sigma_{-1}, \Sigma_0, ...,$ Σ_{M-1} *do not exceed 2, the degrees of* $\Sigma_M, ..., \Sigma_{2M}$ *do not exceed 3 and so on, i.e. the degree of* $\Sigma_{k(M+1)-1+j}$ *does not exceed* $k+2$ *for any* $j = 0, ..., M$ *and* $k = 0, 1, ...$

Corollary. *For the function* σ *corresponding to* Σ *according to (2.4) (and to the two-point function* S *according to (2.2)) one obtains an expansion in the form*

$$\sigma(t, x, y) = t^{-(2M+3)} \sum_{j=0}^{\infty} t^j P_j(x, y), \qquad (2.7)$$

where each P_j *is a polynomial in* $(x^1, ..., x^{M+1})$ *of degree* $\deg_M P_j \leq j$ *with the coefficients being regular power series in* x^0. *Moreover,* P_0 *and its derivative vanish at the origin.*

Remark 1. Clearly, in order to have the complete series (2.6) or (2.7) one must assume all coefficients of the Hamiltonian to be infinitely differentiable functions of x^0. More generally, one should understand the result of Theorem 2.1 in the sense that if these coefficients have continuous derivatives up to the order j, then the terms Σ_l of expansion (2.6) can be uniquely defined for $l = -1, 0, 1, ..., j-1$ and the remainder can be estimated by means of the Theorem 3.2 of the previous chapter.

Remark 2. One can see that if the conditions on the degrees of the coefficients of the regular Hamiltonians (RH) 2.4.1 are not satisfied, then after the series expansion of the solution (2.6) with respect to t, x, y one necessarily gets infinitely many negative powers of t and therefore the definition of RH gives a necessary and sufficient condition for a second order parabolic equation to have the asymptotics of the Green function in the form (0.6) with ϕ and S being regular power series in $(x - x_0), t$ (up to a multiplier $t^{-\alpha}$).

This section is devoted to the constructive proof of Theorem 2.1, which is based essentially on Proposition 7.1 obtained in the special section at the end of this chapter. For brevity, we confine ourselves to the simplest nontrivial case, when $M = 0$, i.e. to the case, which was considered in detail in Section 2.3, and we shall use the notations of Section 2.3. The general case is similar, but one should use the canonical coordinate system in a neighbourhood of x_0, y_0, which was described in Chapter 2. In the case of Hamiltonian (2.3.4), equation (2.5) takes the form

$$\frac{\partial \Sigma}{\partial t} - \frac{\xi + (a(t\xi + \tilde{x}) - a(\tilde{x})) + \alpha(t\xi + \tilde{x})(y + \tilde{y}) - \alpha(\tilde{x})\tilde{y}}{t} \frac{\partial \Sigma}{\partial \xi}$$

$$- [b(t\xi + \tilde{x}) - b(\tilde{x})) + (\beta(t\xi + \tilde{x})(y + \tilde{y}) - \beta(\tilde{x})\tilde{y})$$

$$+ \frac{1}{2}(\gamma(t\xi + \tilde{x})(y + \tilde{y}), y + \tilde{y}) - \frac{1}{2}(\gamma(\tilde{x})\tilde{y}, \tilde{y})] \frac{\partial \Sigma}{\partial y}$$

$$- \left(g(\tilde{x})\tilde{q}, \frac{\partial \Sigma}{\partial y} \right) + \frac{1}{2} \left(g(t\xi + \tilde{x}) \frac{\partial \Sigma}{\partial y}, \frac{\partial \Sigma}{\partial y} \right) - V(t\xi + \tilde{x}, y + \tilde{y}) = 0, \qquad (2.8)$$

where

$$\Sigma(t, \xi, y) = \sigma(t, t\xi, y) = S(t, t\xi + \tilde{x}, y + \tilde{y}; x_0, y^0). \qquad (2.9)$$

We are looking for the solution of (2.8) of form (2.6) with additional conditions as described in Theorem 2.1. Inserting (2.6) in (2.8) and comparing coefficients of t^{-2} yields

$$-\Sigma_{-1} - (\alpha_0 y + \xi) \frac{\partial \Sigma_{-1}}{\partial \xi} + \frac{1}{2} \left(g_0 \frac{\partial \Sigma_{-1}}{\partial y}, \frac{\partial \Sigma_{-1}}{\partial y} \right) = 0. \qquad (2.10)$$

Actually we already know the solution. It is the quadratic form (2.3.49) with $t = 1$ and $\delta = 0$. However, let us prove independently in the present setting a stronger result, which includes the uniqueness property.

Proposition 2.1. *Under the additional assumption that Σ_{-1} vanishes and is strictly convex at the origin, equation (2.10) has a unique analytic solution (in fact a unique solution in the class of formal power series) and this solution is the quadratic form (2.3.49), where $t = 1$ and $\delta = 0$.*

Proof. We consider only the case $k = n$, i.e. a quadratic non-degenerate matrix $\alpha(x_0)$. The general case similar, but one needs the decomposition of Y as the direct sum of the kernel of $\alpha(x_0)\sqrt{g(x_0)}$ and its orthogonal complement (as in Section 2.3), which results in more complicated expressions. Thus, let α_0 be quadratic and non-degenerate. It follows then directly that $\Sigma_{-1}(0,0) = 0$ implies that $\frac{\partial \Sigma_{-1}}{\partial y}(0,0) = 0$. Furthermore, differentiating (2.10) with respect to y yields

$$- \frac{\partial \Sigma_{-1}}{\partial y} - \alpha_0 \frac{\partial \Sigma_{-1}}{\partial \xi} - (\alpha_0 y + \xi) \frac{\partial^2 \Sigma_{-1}}{\partial y \partial \xi} + \left(g_0 \frac{\partial \Sigma_{-1}}{\partial y}, \frac{\partial^2 \Sigma_{-1}}{\partial y^2} \right) = 0,$$

and using the non-degeneracy of α_0 one sees that $\frac{\partial \Sigma_{-1}}{\partial \xi}(0)$ also vanishes.

Let us now find the quadratic part

$$\frac{1}{2}(A\xi, \xi) + (B\xi, y) + \frac{1}{2}(Cy, y) \qquad (2.11)$$

of Σ_{-1} supposing that A is non-degenerate. From (2.10) one has

$$-\frac{1}{2}(A\xi, \xi) - (B\xi, y) - \frac{1}{2}(Cy, y)$$

$$-(\alpha_0 y + \xi)(A\xi + B'y) + \frac{1}{2}(g_0(B\xi + Cy), B\xi + Cy) = 0. \qquad (2.12)$$

This equation implies the 3 equations for the matrices A, B, C:

$$A = \frac{1}{3}B'g_0 B, \quad 2B + \alpha_0' A = Cg_0 B, \quad +\frac{1}{2}C + B\alpha_0 = \frac{1}{2}Cg_0 C. \qquad (2.13)$$

By the first equation of (2.13) and the non-degeneracy of A and g_0, the matrix B is non-degenerate as well. Therefore, inserting the first equation in the second one gets

$$2 + \frac{1}{3}\alpha_0' B' g_0 = C g_0 \quad \text{or} \quad g_0 B \alpha_0 = 3 g_0 C - 6.$$

Inserting this in the third equation (2.13) yields

$$(g_0 C)^2 - 7 g_0 C + 12 = 0,$$

and consequently $g_0 C$ is equal to 3 or 4 (here by 3, for example, we mean the unit matrix multiplied by 3).

Supposing $g_0 C = 3$ gives

$$C = 3 g_0^{-1}, \quad B = 3 g_0^{-1} \alpha_0^{-1}, \quad A = 3(\alpha_0^{-1})' g_0^{-1} \alpha_0^{-1},$$

and the quadratic form (2.11) is then equal to

$$\frac{3}{2} \left[(g_0^{-1}\alpha_0^{-1}\xi, \alpha_0^{-1}\xi) + 2(g_0^{-1}\alpha_0^{-1}\xi, y) + (g_0^{-1}y, y) \right] = \frac{3}{2} \left[g_0^{-1/2}(\alpha_0^{-1}\xi + y) \right]^2,$$

so that this form would not be strictly positive. Thus, we must take $g_0 C = 4$, and therefore

$$C = 4 g_0^{-1}, \quad B = 6 g_0^{-1} \alpha_0^{-1}, \quad A = 12(\alpha_0^{-1})' g_0^{-1} \alpha_0^{-1},$$

and the quadratic form (2.11) is

$$6(g_0^{-1}\alpha_0^{-1}\xi, \alpha_0^{-1}\xi) + 6(g_0^{-1}\alpha_0^{-1}\xi, y) + 2(g_0^{-1}y, y). \tag{2.14}$$

Therefore, the quadratic part of Σ_{-1} is as required. It remains for us to prove that all terms σ_j, $j > 2$, vanish in the expansion $\Sigma_{-1} = \sigma_2 + \sigma_3 + \dots$ of Σ_{-1} as a series of homogeneous polynomials σ_j. For σ_3 one gets the equation

$$-\sigma_3 - (\alpha_0 y + \xi)\frac{\partial \sigma_3}{\partial \xi} + \left(g_0 \frac{\partial \sigma_2}{\partial \xi}, \frac{\partial \sigma_3}{\partial y} \right) = 0$$

or, by (2.14),

$$-\sigma_3 - (\alpha_0 y + \xi)\frac{\partial \sigma_3}{\partial \xi} + (6\alpha_0^{-1}\xi + 4y)\frac{\partial \sigma_3}{\partial y} = 0.$$

This implies that $\sigma_3 = 0$ by Proposition 7.1, since for the case under consideration the matrix A in (7.1) is

$$\begin{pmatrix} -1 & -\alpha_0 \\ 6\alpha_0^{-1} & 4 \end{pmatrix} \tag{2.15}$$

and has eigenvalues 1 and 2 for any invertible α_0. (The simplest way to prove the last assertion is to use (2.3.41) when calculating the characteristic polynomial for the matrix (2.15).) One proves similarly that all σ_j, $j > 2$, vanish, which completes the proof of Proposition 2.1.

Therefore, Σ_{-1} is given by (2.14). For example, in a particular important case with unit matrices g_0 and α_0, one has

$$\Sigma_{-1} = 6\xi^2 + 6(\xi, y) + 2y^2. \tag{2.16}$$

Furthermore, comparing the coefficients of t^{-1} in (2.8) and using (2.3.24), (2.3.25) yields

$$-(\alpha_0 y + \xi)\frac{\partial \Sigma_0}{\partial \xi} + \left(g_0 \frac{\partial \Sigma_{-1}}{\partial y}, \frac{\partial \Sigma_0}{\partial y}\right) - \left(\beta_0 y + \frac{1}{2}(\gamma_0 y, y), \frac{\partial \Sigma_{-1}}{\partial y}\right)$$

$$- \left(\frac{\partial a}{\partial x}(x_0)\xi + \frac{\partial \alpha}{\partial x}(x_0)\xi(y + y^0) - \frac{\partial \alpha}{\partial x}(x_0)(a_0 + \alpha_0 y^0)y, \frac{\partial \Sigma_{-1}}{\partial \xi}\right) = 0. \tag{2.17}$$

This equation is of the type (7.1) with F being a sum of polynomials of degree 2 and 3, and the matrix A given by (2.15). Thus, by Proposition 7.1, the solution (with the additional condition $\Sigma_0(0,0) = 0$) is defined uniquely and is a polynomial of degree 2 and 3 in ξ, y. The statement of Theorem 2.1 about the other Σ_j can be obtained by induction using Proposition 7.1, because comparing the coefficients of $t^{-(k+1)}$ in (2.8) always yields equation of the type (7.1) on Σ_k with the same matrix A. Thus, we have completed the proof of Theorem 2.1.

3. Asymptotic solution of the transport equation

After the asymptotics of the two-point function has been carried out, the next stage in the construction of the exponential multiplicative asymptotics of the Green function for a second order parabolic equation is the asymptotic solution of the transport equation (1.13). On the one hand, it can be solved using (1.18) and the results of the previous chapter. We shall use this representation of the solution in the next section to get the estimates for the remainder (1.21). On the other hand, when one is interested only in small time and small distances from initial point, one can solve the transport equation by formal expansions similarly to the construction of the two-point function in the previous section. We shall now explain this formal method in some detail. In the non-degenerate case one simply looks for the solution of (1.13) in the form of a regular power series in t and $(x - x_0)$ with a multiplier $(2\pi h t)^{-m/2}$. It is a rather standard procedure and we omit it. Let us consider the case of a degenerate regular Hamiltonian (2.4.1). Proceeding as in the previous section one first make a shift in the variables introducing the function

$$\psi(t, x, y) = \phi(t, x + \tilde{x}(t), y + \tilde{y}(t), x_0, y_0), \tag{3.1}$$

and then one must make the change of the variables $x^I = t^{M-I+1}\xi^I$ as in (2.4). A new feature in comparison with (2.4) consists in the observation that in the case of the transport equation one also needs "the explicit introduction of the normalising constant", i.e. one defines the new unknown function by the formula

$$\Psi(t, \xi^0, ..., \xi^M, y) = t^\alpha \psi(t, t^{M+1}\xi^0, ..., t\xi^M, y),\tag{3.2}$$

where α is some positive constant (which is to be calculated). For the Hamiltonian (2.4.1) written in terms of the function Ψ equation (1.13) takes the form

$$\frac{\partial \Psi}{\partial t} - \frac{\alpha}{t}\Psi + \frac{1}{2}\Psi\, tr\left(g(t^{M+1}\xi^0 + \tilde{x}^0(t))\frac{\partial^2 \Sigma}{\partial y^2}\right)$$

$$+ \left(g(t^{M+1}\xi^0 + \tilde{x}^0(t))\frac{\partial \Psi}{\partial y}, \frac{\partial \Sigma}{\partial y}\right) - \left(g(\tilde{x}^0(t))\tilde{q}(t), \frac{\partial \Psi}{\partial y}\right)$$

$$-\frac{t^M\xi^0 + R_1(x(t,\xi) + \tilde{x}(t), y + \tilde{y}(t)) - R_1(\tilde{x}(t), \tilde{y}(t))}{t^{M+1}}\frac{\partial \Psi}{\partial \xi^0} - \cdots$$

$$-\frac{\xi^M + R_{M+1}(x(t,\xi) + \tilde{x}(t), y + \tilde{y}(t)) - R_{M+1}(\tilde{x}(t), \tilde{y}(t))}{t}\frac{\partial \Psi}{\partial \xi^M}$$

$$-(R_{M+2}(x(t,\xi) + \tilde{x}(t), y + \tilde{y}(t)) - R_{M+2}(\tilde{x}(t), \tilde{y}(t)))\frac{\partial \Psi}{\partial y}.\tag{3.3}$$

The main result of this section is the following.

Theorem 3.1. *There exists a unique $\alpha > 0$, in fact this α is given by (1.2.11), such that there exists a solution of (3.3) in the form*

$$\Psi = \Psi + t\Psi_1 + t^2\Psi_2 + ...\tag{3.4}$$

with each Ψ_j being a regular power series in (ξ, y) and Ψ_0 being some constant. Moreover, this solution is unique up to a constant multiplier and all Ψ_j turn out to be polynomials in ξ, y such that the degree of $\Psi_{k(M+1)-1-j}$, $j = 0, ..., M$, does not exceed $k - 1$ for any $k = 1, 2, ...$

Proof. Inserting (3.4) in (3.3) and using the condition that Ψ_0 is a constant one gets comparing the coefficients of t^{-1}:

$$\alpha\Psi_0 + \frac{1}{2}\Psi_0\, tr\left(g(x_0^0)\frac{\partial^2 \Sigma_{-1}}{\partial y^2}(x_0, y_0)\right) = 0,$$

and therefore

$$\alpha = \frac{1}{2}tr\left(g(x_0^0)\frac{\partial^2 \Sigma_{-1}}{\partial y^2}(x_0, y_0)\right).\tag{3.5}$$

Clearly α is positive. Using the canonical coordinates of Lemmas 1.2.2, 1.2.4, one proves that (3.5) coincides with (1.2.11). The remaining part of the proof of the theorem is the same as the proof of Theorem 2.1. Comparing the coefficients of t^q, $q = 0, 1, ...$, one get a recurrent system of equations for Ψ_q of the form

(2.8) with polynomials F_q of the required degree. Proposition 2.1 completes the proof.

Corollary. *The function $\psi(t,x,y)$ corresponding to the solution ϕ of (1.13) via (3.1) has the form of a regular power expansion in (t,x,y) with multiplier $Ct^{-\alpha}$, where C is a constant.*

This implies in particular that the solution ϕ of the transport equation also has the form of a regular power expansion in $t, x - x_0, y - y_0$ with the same multiplier. Comparing the asymptotic solution constructed with the exact solution for Gaussian approximation, one sees that in order to satisfy the initial condition (1.4) by the function u_G^{as} of form (1.19), where S and ϕ are constructed above, one must take the constant C such that $Ct^{-\alpha}$ is equal to the amplitude (pre-exponential term) in formula (1.2.10) multiplied by $h^{-m/2}$. With this choice of C the dominant term of the asymptotic formula (1.19) will coincide with the dominant term of the asymptotics (1.2.10) for its Gaussian approximation, which in its turn by Theorem 1.2.2 coincides with the exact Green function for the "canonical representative" of the class of Gaussian diffusions defined by the corresponding Young scheme.

4. Local asymptotics of the Green function for regular Hamiltonians

In Sect. 1.1 we have described the construction of the asymptotic solution (1.19) for problem (1.1), (1.4) and in Sect. 1.2, 1.3 we have presented an effective procedure for the calculation of all its elements. In this section we are going to justify this asymtptotical formula, i.e. to prove that the exact Green function can be presented in form (1.9). Roughly speaking, the proof consists in two steps. One should obtain an appropriate estimate for the remainder (1.21) and then use it in performing a rather standard procedure (based on Du Hammel principle) of reconstructing the exact Green function by its approximation. When one is interested only in asymptotics for small times and small distances, it is enough to use only the approximations for S and ϕ obtained in two previous sections (a good exposition of this way of justification for non-degenerate diffusion see e.g. in [Roe]). But in order to be able to justify as well the global "small diffusion" asymptotics, as we intend to do in the next section, one has to use the exact global formulas (2.5), (1.18) for S and ϕ. We shall proceed systematically with this second approach. The starting point for justification is the estimate of the r.h.s. in (1.14), when ϕ is given by (1.18).

Proposition 4.1. *If ϕ is given by (1.18), then the r.h.s. of (1.14) (in a neighbourhood of x_0, where (1.18) is well defined) has the form*

$$O(h^2)C(h)t^{2M+2}\exp\{-\frac{S(t,x,x_0)}{h}\}\phi(t,x;x_0) = O(h^2 t^{2M+2} u_G^{as}),$$

where as always $M+1$ is the rank of the regular Hamiltonian under consideration.

Proof. We omit the details concerning the simplest case of non-degenerate diffusion ($M+1=0$) and reduce ourselves to the degenerate regular case, when

$M \geq 0$ and therefore the Hamiltonian is defined by (2.4.1). Then clearly the first term under the integral in (1.18) vanishes and it is enough to prove that

$$\frac{\partial \nu}{\partial y^j} J^{-1/2}(t, x, x_0) = O(t^{\nu(M+1)}) J^{-1/2}(t, x, x_0), \quad \nu = 1, 2, \qquad (4.1)$$

$$\frac{\partial \nu}{\partial y^j} \exp\left\{-\frac{1}{2} \int_0^t tr \frac{\partial A}{\partial x}(X(\tau)) \, d\tau\right\}$$

$$= O(t^{\nu(M+1)}) \exp\left\{-\frac{1}{2} \int_0^t tr \frac{\partial A}{\partial x}(X(\tau)) \, d\tau\right\}, \quad \nu = 1, 2, \qquad (4.2)$$

where $X(\tau) = X(\tau, x_0, p_0(t, x; x_0))$. We have

$$\frac{\partial}{\partial x^I} J^{-1/2} = -\frac{1}{2} J^{-1/2} \left(J^{-1} \frac{\partial J}{\partial x^I}\right) = -\frac{1}{2} J^{-1/2} tr \left(\frac{\partial}{\partial x^I} \left(\frac{\partial X}{\partial p_0}\right) \left(\frac{\partial X}{\partial p_0}\right)^{-1}\right)$$

$$= -\frac{1}{2} J^{-1/2} \frac{\partial^2 X^K}{\partial p_0^L \partial p_0^N} \left(\frac{\partial X}{\partial p_0}\right)^{-1}_{LI} \left(\frac{\partial X}{\partial p_0}\right)^{-1}_{NK}, \qquad (4.3)$$

and by estimates (2.4.3)-(2.4.5) it can be presented in the form

$$J^{-1/2} O \left(t^{6+4M-K-L-N} t^{-(2M+3-L-I)} t^{-(2M+3-N-K)}\right) = J^{-1/2} O(t^I).$$

For the derivatives with respect to $y = x^{M+1}$, one has $I = M + 1$ and one gets (4.1) with $\nu = 1$. Differentiating the r.h.s. in (4.3) once more and again using (2.4.3)-(2.4.5) one gets (4.1) for $\nu = 2$. Let us turn now to (4.2). Let us prove only one of these formula, namely that with $\nu = 1$, the other being proved similarly. Note that due to the main definition of RH, the function under the integral in (4.2) depends only on x^0 and x^1, because it is a polynomial $Q_1(x)$ in $x^1, ..., x^{M+1}$ of M-degree ≤ 1. Therefore, one should prove that

$$\frac{\partial}{\partial y} \exp\left\{-\frac{1}{2} \int_0^t Q_1(X(\tau)) \, d\tau\right\} = O(t^{M+1}) \exp\left\{-\frac{1}{2} \int_0^t Q_1(X(\tau)) \, d\tau\right\}. \qquad (4.4)$$

One has

$$\frac{\partial}{\partial y} \exp\left\{-\frac{1}{2} \int_0^t Q_1(X(\tau)) \, d\tau\right\}$$

$$= \int_0^t \sum_{I=0}^1 \frac{\partial Q_1}{\partial x^I} \frac{\partial X^I}{\partial p_0^K}(\tau) \left(\frac{\partial X}{\partial p_0}\right)^{-1}_{K, M+1}(t) \, d\tau \exp\left\{-\frac{1}{2} \int_0^t Q_1(X(\tau)) \, d\tau\right\},$$

and using again (2.4.3)-(2.4.5) one sees that the coefficient before the exponential in the r.h.s. of this expression has the form

$$\int_0^t O \left(\tau^{2M+3-1-K}\right) O \left(t^{-(2M+3-K-M-1)}\right) d\tau = O(t^{M+1}),$$

which proves (4.4) and thus completes the proof of Proposition 4.1.

Consider now the globally defined function (1.19). For RH (2.4.1) it is convenient to take the polydisc $D_t^r = B_r(x_0^0) \times B_{r/t}(x_0^1) \times \ldots \times B_{r^{M+1}/t^{M+1}}(x_0^{M+1})$ as the domain D. The following is the direct consequence of the previous result.

Proposition 4.2. *For the remainder F in (1.21) one has the estimate*

$$F = O(t^{M+1})u_G^{as} + O\left((\exp\{-\frac{\Omega}{ht^{2M+3}}\})\right)$$

with some positive Ω.

Now, in order to prove the representation (1.9) for the exact solution of (1.1),(1.4) we shall use the following classical general method. Due to the Du Hammel principle (the presentation of the solutions of a non-homogeneous linear equation in terms of the general solution of the corresponding homogeneous one), the solution u_G^{as} of problem (1.21), (1.4) is equal to

$$u_G^{as}(t, x; x_0) = u_G(t, x; x_0) - h \int_0^t \int_{\mathcal{R}^m} u_G(t - \tau, x, \eta) F(\tau, \eta, x_0)\, d\eta d\tau, \quad (4.6)$$

where u_G is the exact Green function for equation (1.1). It is convenient to rewrite (4.6) in the abstract operator form

$$u_G^{as} = (1 - h\mathcal{F}_t)u_G, \quad (4.7)$$

with $\mathcal{F}_t$ being the integral operator

$$(\mathcal{F}_t\phi)(t, x, \xi) = (\phi \otimes F)(t, x, \xi) \equiv \int_0^t \int_{\mathcal{R}^m} \phi(t - \tau, x, \eta) F(\tau, \eta, \xi)\, d\eta d\tau, \quad (4.8)$$

where we have denoted by $\phi \star F$ the (convolution type) integral in the r.h.s. of (4.8). It follows from (4.7) that

$$u_G = (1 - h\mathcal{F}_t)^{-1}u_G^{as} = (1 + h\mathcal{F}_t + h^2\mathcal{F}_t^2 + \ldots)u_G^{as}$$

$$= u_G^{as} + hu_G^{as} \otimes F + h^2 u_G^{as} \otimes F \otimes F + \ldots. \quad (4.9)$$

Therefore, in order to prove the representation (1.9) for u_G one ought to show the convergence of series (4.9) and its presentation in form (1.9). This is done in the following main Theorem of this section.

Theorem 4.1. *For small t, the Green function of equation (1.2) with regular Hamiltonian (2.4.1), whose coefficients can increase polynomially as $x \to \infty$, has the form*

$$u_G = u_G^{as}(1 + O(ht)) + O(\exp\{-\frac{\Omega}{ht}\}), \quad (4.10)$$

where u_G^{as} is given by (1.19) with the domain D defined in Proposition 4.2 above, the functions S and ϕ defined by formulas (2.1.5) and (1.18), and calculated *asymptotically in Sect. 1.2, 1.3. Moreover, the last term in (4.10) is an integrable function of x, which is exponentially small as $x \to \infty$.*

Remark 1. The result of this theorem is essentially known for the case of non-degenerate diffusion. Let us note however that usually in the literature one obtains separately and by different methods the small time and small distance asymptotics, either without a small parameter h (see e.g. a completely analytical exposition in [Roe], [CFKS]), or with a small parameter (see e.g [Var1],[MC1],[Mol], which are essentially based on the probabilistic approach), and global estimates often given for bounded coefficients and without a small parameter. (see e.g. [PE], [Da1], where completely different technique is used). Therefore, the uniform analytic exposition of all these facts together as given here can be perhaps of interest even in non-degenerate situation.

Remark 2. In our proof of the Theorem we obtain first for the case of bounded coefficients the estimate for the additive remainder in (4.10) in the form $O(e^{-|x|})$, which allows afterwards to extend the result to the case of polynomially increasing coefficients. More elaborate estimate of the series (4.9) in the case of bounded coefficients gives for the additive remainder in (4.10) more exact estimate $O(\exp\{-\Omega|x|^2/ht\})$, which allows to generalise the result of the Theorem to the case of the unbounded coefficients increasing exponentially as $x \to \infty$.

Remark 3. In the previous arguments, namely in formula (4.6), we have supposed the existence of the Green function for (1.1), which follows surely from general results on parabolic second order equations, see e.g. [IK]. But this assumption proves to be a consequence of our construction. In fact, when the convergence of series (4.9) and its representation in form (1.9) is proved, one verifies by simple direct calculations that the sum of series (4.9) satisfies equation (1.1).

Remark 4. When the Theorem is proved, the justification of more exact asymptotics as constructed in Sect. 1, Step 5, can be now carried out automatically. In fact, if

$$h\frac{\partial u_G^{as}}{\partial t} - H(x, -h\frac{\partial}{\partial x})u_G^{as} = O(t^j h^k)u_G^{as},$$

then from (4.6)

$$u_G = u_G^{as} + \int_0^t \int_{\mathcal{R}^m} u_G(t - \tau, x, \eta)O(t^j h^k)u_G^{as}(\tau, \eta, x_0)\,d\eta d\tau,$$

and due to (4.10) and the semigroup property of u_G one concludes that

$$u_G = u_G^{as}(1 + O(t^{j+1}h^k)) + O(\exp\{-\frac{\Omega}{ht}\}).$$

Proof. Though in principle the convergence of (4.9) is rather clear from the estimate of the first nontrivial term by the Laplace method using Proposition

2.1.4, the rigorous estimate of the whole series involves the application of the Laplace method infinitely many times, where one should keep control over the growth of the remainder in this procedure, which requires a "good organisation" of the recursive estimates of the terms of (4.9). Let us present the complete proof in the simplest case of the non-degenerate diffusion, the general case being carried out similarly due to Proposition 4.2, but requires the consideration of polydisks instead of the disks, which makes all expressions much longer. Consider first the diffusion with bounded coefficients. In non-degenerate case one can take the ball $B_r(x_0)$ as the domain D for the molifier χ_D. Let

$$\Omega = \min\{tS(t, x, \xi) : |x - \xi| = r - \epsilon\}. \tag{4.11}$$

For given $\delta > 0, h_0 > 0$ one can take t_0 such that for $t \le t_0$, $h \le h_0$

$$(2\pi ht)^{-m/2} \le \exp\{\frac{\delta}{th}\}, \quad \frac{1}{ht} \le \exp\{\frac{\delta}{th}\}. \tag{4.12}$$

To write the formulas in a compact form let us introduce additional notations. Let

$$f(t, x, x_0) = \Theta_r(|x - x_0|)(2\pi ht)^{-m/2} \exp\{-S(t, x, x_0)/h\}, \tag{4.13}$$

$$g_k(t, x, x_0) = \Theta_{kr}(|x - x_0|) \exp\{-\frac{\Omega - \delta}{ht}\}. \tag{4.14}$$

From Proposition 4.2

$$F(t, x, x_0) = O(u_G^{as}(t, x, x_0)) + O((ht)^{-1} \exp\{-S(t, x, x_0)/h\}),$$

where the second function has a support in the ring $B_{x_0}(r) \setminus B_{x_0}(r - \epsilon)$. Using these formulas and the estimates for the solution ϕ of the transport equation one can choose a constant $C > 1$ such that

$$\frac{1}{C}\chi_D(x - x_0)f(t, x, x_0) \le u_G^{as}(t, x, x_0) \le Cf(t, x, x_0), \tag{4.15}$$

$$F(t, x, x_0) \le C(f(t, x, x_0) + g_1(t, x, x_0)). \tag{4.16}$$

To estimate the terms $u_G^{as} \otimes F \otimes F...$ in series (4.9) we shall systematically use the simple estimate of the Laplace integral with convex phase, namely the formula (B3) from Appendix B. Let us choose d such that

$$\frac{\partial^2 S}{\partial x^2}(t, x, \xi) \ge \frac{d}{t}, \quad \frac{\partial^2 S}{\partial \xi^2}(t, x, \xi) \ge \frac{d}{t} \tag{4.17}$$

for $|x - \xi| \le r$ and $t \le t_0$. Such d exists due to the asymptotic formula for S given in Sect. 2.2 or 3.3. We claim that the following inequalities hold (perhaps, for a smaller t_0):

$$f \otimes f \le td^{-m/2}(f + g_2), \quad f \otimes g_k \le td^{-m/2}g_{k+1}, \quad g_k \otimes g_1 \le td^{-m/2}g_{k+1}. \tag{4.18}$$

In fact, since

$$\min_{\eta}(S(t-\tau, x, \eta) + S(\tau, \eta, x_0)) = S(t, x, x_0) \tag{4.19}$$

and the minimum point η_0 lies on the minimal extremal joining x_0 and x in time t, it follows from (B3) that

$$(f \otimes f)(t, x, x_0) \leq \theta_{2r}(x - x_0) \int_0^t (2\pi h(t-\tau))^{-m/2} (2\pi h\tau)^{-m/2}$$

$$\times \exp\{-S(t, x, x_0)/h\} \left(\frac{d}{t} + \frac{d}{t-\tau}\right)^{-m/2} (2\pi h)^{m/2} d\tau$$

$$= \Theta_{2r}(|x - x_0|) d^{-m/2} t (2\pi ht)^{-m/2} \exp\{-S(t, x, x_0)/h\}.$$

To estimate this function outside $B_r(x_0)$ we use (4.12) and thus get the first inequality in (4.18). Furthermore,

$$(f \otimes g_k)(t, x, x_0)$$

$$\leq g_{k+1}(x - x_0) \int_0^t \int_{\mathcal{R}^d} (2\pi h(t-\tau))^{-m/2} \exp\{-S(t-\tau, x, \eta)/h\} d\eta d\tau.$$

Since $S \geq 0$ and due to (B3), this implies the second inequality in (4.18). At last, obviously, $g_k \star g_l \leq t b_m r^m g_{k+1}^2$, where b_m denotes the volume of the unit ball in $\mathcal{R}^m$. This implies the last inequality in (4.18) for small enough t_0.

It is easy now to estimate the terms of series (4.18):

$$h u_G^{as} \otimes F \leq hC^2 (f \otimes f + f \otimes g_1) = hC^2 t d^{-m/2} (f + 2g_2),$$

$$h^2 (u_G^{as} \otimes F) \otimes f \leq h^2 C^3 t d^{-m/2} (f + 2g_2)(f + g_1) \leq C(Chtd^{-m/2})^2 (f + 2g_2 + 4g_3),$$

and by induction

$$h^{k-1} \mathcal{F}^{k-1} u_G^{as} \leq C[Chtd^{-m/2}]^{k-1} (f + 2g_2 + 4g_3 + \ldots + 2^{k-1} g_k).$$

Since $2g_2 + 4g_3 + \ldots + 2^{k-1} g_k \leq 2^k g_k$, one has for $k > 1$

$$h^{k-1} \mathcal{F}^{k-1} u_G^{as} \leq 2C(2Cthd^{-m/2})^{k-1} (f + g_k).$$

Therefore, series (4.9) is convergent (uniformly on the compacts); outside the ball $B_{rk}, k \geq 1$, it can be estimated by

$$2C \sum_{l=k}^{\infty} (2Cthd^{-m/2})^l \exp\{-\frac{\Omega - \delta}{th}\} \leq \frac{2C(2Cthd^{-m/2})^k}{1 - 2Ct_0 h_0 d^{-m/2}} \exp\{-\frac{\Omega - \delta}{th}\}, \tag{4.20}$$

and inside the ball B_r, $|u_G - u_G^{as}|$ does not exceed

$$\frac{4C^2 thd^{-m/2}}{1 - 2Ct_0 h_0 d^{-m/2}} \left[(2\pi ht)^{-m/2} \exp\{-S(t, x, x_0)/h\} + \exp\{-\frac{\Omega - \delta}{th}\}\right]. \tag{4.21}$$

Notice that the number k in (4.20) is of the order $|x - x_0|/r$ and therefore the coefficient t^k can be estimated by a function of the form $e^{-\kappa|x-x_0|}$ with some $\kappa > 0$. The statement of the theorem follows now from (4.16),(4.20),(4.21).

Now let the functions A, G, V are not bounded but can increase polynomially as $x \to \infty$ (the uniform boundedness of V from below is supposed always). This case can be reduced to the case of bounded coefficients in following way. Let x_0 be given and let $\tilde{G}(x), \tilde{A}(x), \tilde{V}(x)$ be the functions which are uniformly bounded and coincide with $G(x), A(x), V(x)$ respectively in a neighbourhood of x_0. Taking the Green function $\tilde{u}_G$ of the diffusion equation with coefficients $\tilde{G}, \tilde{A}, \tilde{V}$ as the first approximation to the Green function $u_G(t, x, x_0)$ yields for u_G the series representation

$$u_G = \tilde{u}_G + \tilde{u}_G \otimes F + \tilde{u}_G \otimes F \otimes F + \dots$$

with

$$F = \frac{1}{2}tr(\tilde{G} - G)\frac{\partial^2 \tilde{u}_G}{\partial x^2} - (\tilde{A} - A)\frac{\partial \tilde{u}}{\partial x} - (\tilde{v} - V)\tilde{u}.$$

Due to the exponential decrease (of type $e^{-\kappa|x-x_0|}$ with some $\kappa > 0$) of $\tilde{u}$ (and, as one shows similarly, the same rate of decrease holds for the first and second derivatives of $\tilde{u}$) all terms of this series are well defined and it is convergent, which completes the proof.

Let us note that by passing we have proved a convergent series representation for the Green function, i.e. the following result, which we shall use in Chapter 9.

Proposition 4.3. *Under the assumptions of Theorem 4.1, the Green function of equation (1.2) can be presented in the form of absolutely convergent series (4.9), where*

$$u_G^{as} = C(h)\chi_D(x - x_0)\phi(t, x, x_0) \exp\{-S(t, x, x_0)/h\},$$

F is defined by (1.21) and the operation $\otimes$ is defined by (4.8).

Theorem 4.1 gives for the Green function of certain diffusion equations the multiplicative asymptotics for small times and small distances, but only a rough estimate for large distances. In the next section we shall show how to modify this asymptotics in order to have an asymptotic representation valid for all (finite) distances. There is however a special case when the global asymptotic formula is as simple as the local one (at least for small times), this is the case of diffusion equations with constant drift and diffusion coefficients. This special case is important for the study of the semi-classical spectral asymptotics of the Shrödinger equation and therefore we formulate here the corresponding result obtained in [DKM1].

Theorem 4.2. *Let $V(x)$ be a positive smooth function in $\mathcal{R}^m$ with uniformly bounded derivatives of the second, third and fourth order. Then there exists $t_0 > 0$ such that for $t \leq t_0$ the boundary value problem for the corresponding Hamiltonian system with the Hamiltonian $H = \frac{1}{2}p^2 - V(x)$ is uniquely* *solvable for all x, x_0 and t_0 and the Green function for the equation*

$$h\frac{\partial u}{\partial t} = \frac{h^2}{2}\Delta u - V(x)u \tag{4.22}$$

has the form

$$u_G(t, x, x_0, h) = (2\pi h)^{-m/2} J(t, x, x_0)^{-1/2} \exp\{-S(t, x, x_0)/h\}(1 + O(ht^3)) \tag{4.23}$$

with $O(ht^3)$ being uniform with respect to all x, x_0.

Remark. Notice the remainder is of the order t^3, which holds only in the described situation.

5. Global small diffusion asymptotics and large deviations

In the previous section we have written the exponential asymptotics of the Green function of equation (1.1) for small times. Moreover, for large distances it states only that the Green function is exponentially small. In order to obtain the asymptotics for any finite t and also give a more precise formula for large $x - x_0$, one can use the semigroup property of the Green function. Namely, for any $n \in \mathcal{N}$, $\tau > 0$, and $x \in \mathcal{R}^m$

$$u_G(n\tau, x, x_0)$$

$$= \int_{\mathcal{R}^m} \dots \int_{\mathcal{R}^m} u_G(\tau, x, \eta_1) u_G(\tau, \eta_1, \eta_2) \dots u_G(\tau, \eta_{n-1}, x_0)\, d\eta_1 \dots d\eta_{n-1}. \tag{5.1}$$

Notice that according to the result of sect. 2.2, 2.3, the set $Reg(x_0)$ of regular pairs (t, x), i.e such pairs for which there exists a unique minimising extremal joining x_0 and x in time t, is open and everywhere dense in $\mathcal{R}_+ \times \mathcal{R}^m$ for any regular Hamiltonian. Therefore, formula (1.19) without the multiplier χ_D is correctly defined globally (for all x and $t > 0$) almost everywhere, because $S(t, x, x_0)$ is defined globally for all t, x as the minimum of the functional (2.1.12) and for regular points S, ϕ are given by (2.24),(2.2.5), and (1.18) with integrals in both formulas taken along the unique minimising extremal. It turns out that though this function does not give a correct asymptotics to the Green function in a neighbourhood of a non-regular point, its convolution with itself already does. We restrict ourselves to the approximations of the first order. Next orders can be obtained by formulas (1.22)- (1.25).

Theorem 5.1. *For any t, x and $\tau < t$*

$$u_G(t, x; x_0) = (2\pi h)^{-m}(1 + O(h)) \int_{\mathcal{R}^m} \phi(t - \tau, x, \eta)\phi(\tau, \eta, x_0)$$

$$\times \exp\left\{ -\frac{S(t - \tau, x, \eta) + S(\tau, \eta, x_0)}{h} \right\} d\eta. \tag{5.2}$$

In particular, for any $(t, x) \in Reg(x_0)$

$$u_G(t, x, x_0) = (2\pi h)^{-m/2} \phi(t, x, x_0)(1 + O(h)) \exp\{-S(t, x, x_0)/h\}, \qquad (5.3)$$

Proof. First let us show that the integral in (5.2) is well defined. To this end, we use the Cauchy inequality and (1.18) to estimate (5.2) by

$$\int_{\mathcal{R}^m} J^{-1}(t - \tau, x, \eta) \exp\left\{-\frac{2S(t - \tau, x, \eta)}{h}\right\} d\eta$$

$$\times \int_{\mathcal{R}^m} J^{-1}(\tau, \eta, x_0) \exp\left\{-\frac{2S(\tau, \eta, x_0)}{h}\right\} d\eta$$

We shall estimate the second integral in this product, the first one being dealt in the completely similar way. Let us make in this second integral the change of the variable of integration $\eta \mapsto p_0(\tau, \eta, x_0)$, where p_0 is the initial momentum of the minimising extremal joining x_0 and η in time τ. The mapping $\eta \mapsto p_0$ is well defined for regular, and thus for almost all η. Thus the second integral can be estimated by the integral

$$\int_{\mathcal{R}^m} \exp\left\{-\frac{2S(\tau, \eta(\tau, x_0, p_0), x_0)}{h}\right\} dp_0,$$

which already does not contain any singularities. To see that this integral converges one only need to observe that due to the estimates of the two-point function S, the function under the integral here is decreasing exponentially as $p_0 \to \infty$.

We have shown that integral (5.2) is well defined. It follows from the Laplace method and Proposition 2.1.4 that that for regular t, x formula (5.2) can be written in form (5.3). To show that (5.2) presents the global asymptotics for the Green function, let us start first with the simplest case of equation (4.22), where the asymptotics of the Green function for small times is proved to be given by (4.23). For any t there exists $n \in \mathcal{N}$ such that $t/n < t_0$ and one can present the Green function for the time t in form (5.1) with function (4.23) instead of u_G. If the point (t, x) is regular, then there is only one and non-degenerate minimum point $(\eta_1, ..., \eta_{n-1})^{min}$ of the "compound action"

$$S(\tau, x, \eta_1) + S(\tau, \eta_1, \eta_2) + ... + S(\tau, \eta_{n-1}, x_0)$$

given by the formula

$$\eta_j^{min} = X(j\tau, x_0, p_0(t, x, x_0)),$$

i.e. all η_j^{min} lie on the unique minimising extremal. Using Proposition 2.1.4 and the Laplace method one gets (5.3). Alternatively, one can use the induction in n. Thus we have proved (5.3) for all regular points. Now, for any non-regular (t, x) let us use (5.1) with $n = 2$. It follows from the Laplace method that only those

η contribute to the first order asymptotics of this integral that lie on minimising extremals (which may not be unique now) joining x_0 and x in time t. But by the Jacobi theory these points η are regular with respect to x_0 and x and hence around this point one can use formula (5.3) for u_G. Therefore the asymptotics of this integral is the same as that of (5.2).

In the case of a general regular Hamiltonian, there is only one additional difficulty. Namely, if one calculates the asymptotic of the Green function for a regular point using (5.1) with the function (4.10) instead of u_G, then for sufficiently large n and $x - x_0$ the exponentially small remainders in (4.10) begin to spoil the correct phase of the corresponding Laplace integral. Hence, in order to get the correct asymptotics for (t, x) from any fixed compact one should improve (4.10) respectively. Namely, in proving theorem 4.1, one must take instead of the approximation (1.19), its convolution with itself of the form

$$\tilde{u}_G^{as}(n\tau, x, x_0)$$

$$= \int_{\mathcal{R}^m} \cdots \int_{\mathcal{R}^m} u_G^{as}(\tau, x, \eta_1) u_G^{as}(\tau, \eta_1, \eta_2) \ldots u_G^{as}(\tau, \eta_{n-1}, x_0)\, d\eta_1 \ldots d\eta_{n-1}.$$

Formulas (4.9) and (4.10) are then modified respectively. Increasing n to infinity, one increases to infinity the range of x for which (5.2) is valid. This argumentation completes the proof.

The global integral formula (valid for regular and non-regular points) for the asymptotics of the Green function for non-degenerate diffusion was first written by Maslov [M2] by means of his tunnel canonical operator. We have given here an equivalent but essentially more simple formula (5.2) thus avoiding the beautiful but rather sophisticated definition of the tunnel canonical operator (see [M2], [DKM1]). Moreover, we have presented the rigorous proof including a large class of degenerate diffusions.

For some non-regular points, the integral (5.2) can still be calculated explicitly. The two important cases are the following.

(i) There exist a finite number of non-degenerate extremals joining x_0 and x in time t. Then (5.2) is equal to the sum of expressions (5.3) corresponding to each extremal.

(ii) There is a (non-degenerate) closed manifold of extremals (as for instance is often the case for geodesics on symmetric spaces) joining x_0 and x in time t. Then one integrates (5.2) by means of the modified Laplace method (see, e.g. [Fed1], [K3]) standing for the case of the whole manifold of minimal points of the phase.

For general non-regular points one can write down explicitly only the logarithmic asymptotics of the solution. To this end, let us recall first a general result on logarithmic asymptotics of Laplace integrals.

Proposition 5.1 [MF2],[Fed1]. *Let the functions f and ϕ in the integral*

$$F(h) = \int_{\mathcal{R}^m} \phi(x) \exp\{-\frac{S(x)}{h}\}\, dx \qquad (5.4)$$

be continuous, ϕ having finite support $\operatorname{supp}\phi$. Let $M(S)$ denote the set of all $x \in \operatorname{supp}\phi$, where S is equal to its minimum M, and let $M_c(S) \subset \operatorname{supp}\phi$ be the set of x, where $S(x) \leq M + c$. Denote by $V(c)$ the volume of the set $M_c(S)$. Suppose $\phi \geq \delta$ in a neighbourhood of $M(S)$ for some positive constant δ.

(i) Then

$$\lim_{h \to 0} h \log F(h) = -M. \tag{5.5}$$

(ii) If

$$\lim_{c \to 0} \frac{\log V(c)}{\log c} = \alpha > 0, \tag{5.6}$$

then

$$\log F(h) = -\frac{M}{h} + \alpha \log h + o(\log h). \tag{5.7}$$

(iii) If $V(0) = 0$ and (5.7) holds, then (5.6) holds as well.

(iv) If $M(S)$ consists of a unique point and S is real analytic in a neighbourhood of this point, then the limit (5.6) exists and (5.7) holds.

The last statement is in fact a consequence of a theorem from [BG], see e.g. [At].

Theorem 5.1 and Proposition 5.1 imply the principle of large deviation for the Green function of regular diffusions.

Proposition 5.2. *For all t, x*

$$\lim_{h \to 0} h\, u_G(t, x, x_0) = -S(t, x, x_0). \tag{5.8}$$

This principle for regular points of non-degenerate diffusion was obtained by Varadhan, see [Var1]-[Var4]. In some cases, one can calculate the logarithmic asymptotic more precisely. For instance, Theorem 5.1 and Proposition 5.1 imply the following result.

Proposition 5.3. *If there exists a unique minimising extremal joining x_0 and x in time t (generally speaking, degenerate, i.e. the points x_0 and x can be conjugate along this extremal) and if $S(t, y, x_0)$ is real analytic in a neighbourhood of this extremal, then there exists $\alpha > 0$ such that for small h*

$$\log h\, u_G(t, x, x_0) = -\frac{S(t, x, x_0)}{h} + \alpha \log h + o(\log h). \tag{5.9}$$

Formula (5.9) for the case of non-degenerate diffusion was first written in [MC1] (and proved there under some additional assumptions), where α was called the invariant of the degeneracy of the extremal.

Let us present the solution of the general large deviation problem for regular diffusions. If in (1.1) the last term $V(x)$ vanishes, then the corresponding second order equation describes the evolution of the expectations (and its adjoint operator - the probability density) of the diffusion process defined by the stochastic equation

$$dX = A(X)\, dt + h\sqrt{G(x)}\, dW. \tag{5.10}$$

One is especially interested in the solution of (1.1) with the discontinuous initial function

$$u_0(x) = \begin{cases} 1, & \text{if } x \in D \\ 0, & \text{otherwise,} \end{cases} \tag{5.11}$$

where D is some closed bounded domain in $\mathcal{R}^m$. This solution corresponds to the diffusion starting in D. The problem of large deviation is to find the small h asymptotics of this solution on large distances from D. The solution of (1.1), (5.11) is given by formula (1.5),(5.3). To simplify it, one can use the Laplace method. As in the case of the Green function, for a general non-regular point only the logarithmic limit can be found explicitly. Namely, the following result is the direct consequence of formulas (1.5), (5.2), (5.5).

Proposition 5.4 (Large deviation principle for regular diffusions). *For the solution $u(t,x)$ of the problem (1.2), (5.11) with a regular Hamiltonian, one has*

$$\lim_{h \to 0} h \log u(t,x) = -S(t,x),$$

where $S(t,x)$ is the generalised solution of the Cauchy problem for Hamilton-Jacobi equation (1.11) with initial data

$$S(0,x) = \begin{cases} 0, & \text{if } x \in D \\ +\infty, & \text{otherwise,} \end{cases}$$

i.e.. $S(t,x)$ is given by the formula

$$S(t,x) = \min_\xi (S(t,x,\xi) + S(0,\xi)) = \min_{\xi \in D} S(t,x,\xi), \tag{5.12}$$

where $S(t,x;\xi)$ denotes as always the corresponding two-point function.

The explicit formula (without integration) for the asymptotics of $u(t,x)$ exists on the open everywhere dense set of regular (with respect to the domain D) points, where the critical point of the phase used in the integration of (1.5) by the Laplace method is unique and non-degenerate. On the complement to this set the asymptotics can be only written in the integral form similar to (5.3). Let us give the precise results, which follow more or less straightforwardly from (1.5), (5.2) and the Laplace method. For the case of the equation of form (4.22) this result was proved in [DKM1]. Let Hamiltonian (1.3) be regular with vanishing V, i.e for some Young scheme $\mathcal{M}$ it has the form (2.4.1) with vanishing $R_{2(M+1)}$. Let $Y(t,y_0)$ denote the solution of the system

$$\dot{y} = -A(y) \tag{5.13}$$

with initial value $y(0) = y_0$. Note that the solution of (5.13) is in fact the characteristic of the Hamiltonian system on which the momentum vanishes identically (the insertion of the vanishing momentum in the Hamiltonian system does not lead to a contradiction due to the assumption of vanishing V). Let D_t denote the smooth manifold with boundary, which is the image of $D = D_0$ with respect to the mapping $y_0 \mapsto Y(t,y_0)$.

Proposition 5.5. *On the set* $Int\,D_t$ *of the internal points of the domain* D_t *the solution* $u(t,x)$ *of problem (1.1), (5.11) can be presented in the form of regular series in* h. *More precisely, if* $x \in Int\,D_t$, *then*

$$u(t,x) = \left(\det \frac{\partial Y}{\partial y_0}(t,y_0)\right)^{-1/2}(1 + h\phi_1 + \ldots + h^k\phi_k + O(h^{k+1})),$$

where ϕ_j *can be found by the formulas similar to (1.24), (1.25).*

As we have mentioned, the most interesting is the problem of calculating the solution far away from D, in particular outside D_t. To formulate the result of the calculation of the Laplace integral (1.5) in that case, we need some other notations. Consider the m-dimensional manifold $\Lambda_0 = \partial D \times \mathcal{R}_+$ with coordinates $(\alpha_1, \ldots, \alpha_{m-1}, s)$, where $s \geq 0$ and $\alpha = (\alpha_1, \ldots, \alpha_{m-1})$ are some orthogonal coordinates on ∂D. Let $n(\alpha)$ denote the unit vector of the external normal to ∂D at the point α and let $\Gamma(D)$ be the family of characteristics $X(t, \alpha, sn(\alpha))$ with initial conditions $x_0 = \alpha, p_0 = sn(\alpha)$. For any t and $x \in \mathcal{R}^m \setminus D_t$, there exists a pair $(\alpha, s) = (\alpha, s)(t, x)$ (perhaps not unique) such that the characteristic $X(t, \alpha, sn(\alpha))$ comes to x at the time t and $S(t, x)$ (as defined in (5.12)) is equal to the action (2.1.4) along this characteristic. In fact, α is the coordinate of the point $\xi \in D$ that furnishes minimum in (5.8). Clear that p_0 is perpendicular to ∂D at ξ and thus has the form $p_0 = sn(\alpha)$ for some $s > 0$. Let Reg_D denote the set of pairs (t, x) such that $(\alpha, s)(t, x)$ is unique and moreover, the Jacobian $J(t, x) = \det \frac{\partial X}{\partial(\alpha, s)}$ does not vanish. Similarly to the proof of Proposition 2.2.7, 2.3.7 one shows that the set Reg_D is open and everywhere dense in the outside of the set $\{(t, x \in D_t)\}$.

Proposition 5.6. *For* $(t, x) \in Reg_D$ *the solution* $u(t, x)$ *of problem (1.2), (5.11) has the following asymptotics for small* h:

$$u(t,x) = \left(\det \frac{\partial X}{\partial(\alpha, s)}\Big|_{(\alpha,s)=(\alpha,s)(t,x)}\right)^{-1/2}$$

$$\times \exp\{-\frac{S(t,x)}{h}\}(1 + \ldots + h^k\phi_k + O(h^{k+1})).$$

The asymptotics of the global representation (1.5), (5.3) can be also calculated explicitly for some classes of non-regular points described similarly to the case of the Green function (see (i), (ii) after Theorem 5.2 and Theorem 5.4).

6. Non-regular degenerate diffusions: an example

In this Chapter we have constructed the theory of global semi-classical asymptotics and large deviations for a class of degenerate diffusions that were called regular. This class is characterised in particular by regular asymptotic representation (0.6) of the Green function. It seems however that the global small h asymptotics are valid actually for a larger class of degenerate diffusions. We present here an example of a non-regular diffusion for which small h and small time asymptotics can be calculated explicitly, so to say, by hands, and shall see that the small h asymptotics can be obtained as well by a formal application of the formulas of Section 6. We consider the equation

$$h\frac{\partial u}{\partial t} = \frac{h^2}{2}\frac{\partial^2 u}{\partial y^2} + \frac{h}{2}y^2\frac{\partial u}{\partial x}, \tag{6.1}$$

which corresponds to the simple non-regular Hamiltonian $H = (q^2 - y^2p)/2$ discussed at the end of Section 2.3, where we have noted that for this Hamiltonians the boundary value problem is not solvable if $x > x_0$ for any time. We are going to construct the Green function u_G for this equation corresponding to the initial point $(0,0)$, i.e. the solution with initial condition

$$u_G(t,x,y)|_{t=0} = \delta(x)\delta(y).$$

Proposition 6.1. *(i) The Green function u_G vanishes for $x \geq 0$,*
(ii) If $x < 0$ and $y = 0$, then

$$u_G(t,x,0) = \frac{1}{\sqrt{|x|}ht^{3/2}} \exp\{-\frac{\pi^2|x|}{ht^2}\}(1 + O(\frac{ht^2}{|x|})), \tag{6.2}$$

(iii) if $x < 0$ and $y < 0$, then there exists a unique real solution $\lambda(t,x,y) > -\pi^2$ of the equation

$$\frac{4x}{ty^2} = \frac{1}{\sinh^2\sqrt{\lambda}} - \frac{\cot\sqrt{\lambda}}{\sqrt{\lambda}}, \tag{6.3}$$

and

$$u_G = \frac{1}{2\pi ht^2\sqrt{|S''(\lambda(t,x,y))|}}\sqrt{\frac{\sqrt{\lambda(t,x,y)}}{\sinh\sqrt{\lambda(t,x,y)}}}$$

$$\times \exp\{-\frac{1}{ht}S(\lambda(t,x,y))\}(1 + O(ht)), \tag{6.4}$$

where the function S is defined by the formula

$$S(\lambda;t,x,y) = \frac{\lambda x}{t} + \frac{\sqrt{\lambda}}{2}y^2\coth\sqrt{\lambda}; \tag{6.5}$$

moreover $\lambda(t,x,y) \in (-\pi^2,0)$, $\lambda = 0$, $\lambda > 0$ respectively for $x < -ty^2/6$, $x = -ty^2/6$, $x > -ty^2/6$ and in the first case $\sqrt{\sqrt{\lambda}/\sinh\sqrt{\lambda}}$ should be understood as $\sqrt{\sqrt{|\lambda|}/\sin\sqrt{|\lambda|}}$; at last, for small

$$\epsilon = \frac{x}{ty^2} + \frac{1}{6}$$

the function $\lambda(t, x, y)$ can be presented as the convergent power series in ϵ: $\lambda = 45\epsilon + O(\epsilon^2)$, so that for small ϵ

$$u_G = \frac{3\sqrt{5}}{2\pi h t^2 |y|}(1 + O(\epsilon))(1 + O(th))\exp\{-\frac{y^2}{2ht}(1 + O(\epsilon))\}. \qquad (6.6)$$

Sketch of the proof. It is done essentially by direct calculations using the Fourier transform and the saddle-point method. Namely, carrying out the h-Fourier transform F_h of equation (6.1) with respect to the variable y one finds for $\tilde{u}(t, x, p) = (F_h u)(t, x, p)$ the equation

$$h\frac{\partial \tilde{u}}{\partial t} = \frac{h^2}{2}\frac{\partial^2 \tilde{u}}{\partial y^2} + \frac{i}{2}y^2\tilde{u},$$

which is actually the equation of the evolution of the quantum oscillator in imaginary time and with the complex frequency $\sqrt{-ip} = \sqrt{|p|}\exp\{-i\pi\, sgnp/4\}$. Since the Green function for such equation is well known, one obtains for u_G the following integral representation

$$u_G = -\frac{i}{(2\pi h)^{3/2}t^{5/2}}\int_{-i\infty}^{i\infty}\sqrt{\frac{\sqrt{\lambda}}{\sinh\sqrt{\lambda}}}\exp\{-\frac{1}{ht}S(\lambda(t, x, y))\}\,d\lambda. \qquad (6.7)$$

Notice that the function under the integral in this representation is regular everywhere except for singularities at points $-k^2\pi^2$, $k = 1, 2, ...$, and is a one-valued analytic function on the complex plane cut along the line $(-\infty, \pi^2)$. For $\lambda = Re^{i\phi}$ with $|\phi| < \pi$, one has

$$|\sinh\sqrt{\lambda}|^2 = \sinh^2(\sqrt{R}\cos\frac{\phi}{2}) + \sin^2(\sqrt{R}\sin\frac{\phi}{2}),$$

$$ReS(\lambda; t, x, y)) = \frac{x}{t}R\cos\phi$$

$$+\frac{y^2\sqrt{R}\left[\cos\frac{\phi}{2}\sinh(2\sqrt{R}\cos\frac{\phi}{2}) + \sin\frac{\phi}{2}\sin(2\sqrt{R}\sin\frac{\phi}{2})\right]}{2\left[\cosh(2\sqrt{R}\cos\frac{\phi}{2}) - \cos(2\sqrt{R}\sin\frac{\phi}{2})\right]}.$$

It implies that for $x \geq 0$ one can close the contour of integration by a semi-circle on the right half of the complex plane, which by Cauchy theorem gives the statement (i) of the Proposition. Let $x < 0$ and $y = 0$. Then one can transform the contour of integration in (6.7) to the contour C which goes from $-\infty$ to $-\pi^2$ along the lower edge of the half-line $(-\infty, -\pi^2)$ and then returns to the $-\infty$ along the upper edge of this half-line (notice that all singularities at $\lambda = -k^2\pi^2$ are of the type $z^{-1/2}$ and are therefore integrable). The simple analysis of the argument of $\sqrt{\sinh\sqrt{\lambda}}$ shows that the values of the integrand in (6.7) on the upper edge of the cut coincides (respectively differs by the sign) with its corresponding values on the lower edge on the intervals $((2k\pi)^2, (2k + 1)^2\pi^2)$

(resp. on the intervals $((2k-1)^2\pi^2, (2k\pi)^2))$, which yields (after the change $\lambda = v^2$) the formula

$$u_G = \frac{4}{(2\pi h)^{3/2} t^{5/2}} \sum_{k=1}^{\infty} \int_{\pi(2k-1)}^{2k\pi} (-1)^{k-1} v \exp\{-\frac{v^2|x|}{ht^2}\} \sqrt{\frac{v}{|\sin v|}}\, dv$$

$$= \frac{\sqrt{2}}{(\pi h)^{3/2} t^{5/2}} \int_0^\pi \sum_{k=1}^{\infty} (-1)^{k-1} \frac{(u+\pi(2k-1))^{3/2}}{\sqrt{\sin u}} \exp\{-\frac{|x|(u+\pi(2k-1))^2}{ht^2}\}\, dv$$

For large $|x|/(ht^2)$ all terms in this sum are exponentially small as compared with the first one. Calculating this first term for large $|x|/(ht^2)$ by the Laplace method yields (6.2).

Consider now the main case $x < 0, y < 0$. To calculate this integral asymptotically for small h one can use the saddle-point method. The equation $S'(\lambda) = 0$ for saddle points is just equation (6.3), and simple manipulations show the properties of the solution $\lambda(t,x,y)$ given in the formulation of the Proposition. Now the application of the saddle-point method to the integral in (6.7) amounts to the shift of the contour of integration on $\lambda(t,x,y)$ and the following calculation of thus obtained integral by means of the Laplace method, which yields (6.4), (6.6) and thus completes the proof.

Notice now that as we have mentioned above, the formal application of semi-classical formulas of Sections 1 or 5, i.e. of representation (1.19) with S being the two point function corresponding to Hamiltonian $H = (q^2 - y^2 p)/2$ and ϕ being given by (1.18), would give the same result as we have obtained above using the explicit expression for the h-Fourier transform of u_G. In fact, the Cauchy problem for the Hamiltonian system with the Hamiltonian H and initial conditions $(0,0,p_0,q_0)$ has the explicit solution

$$y = \frac{q_0}{\sqrt{p_0}} \sinh(\sqrt{p_0} t), \quad x = \frac{q_0^2}{4p_0}(t - \frac{\sinh(2\sqrt{p_0}t)}{2\sqrt{p_0}}),$$

and the problem of finding the solution to the boundary value problem with $x(t) = x, y(t) = y$ reduces to the solution of equation (6.3) for $\lambda = pt^2$. For $x > 0$ there is no solution to this boundary value problem, i.e. S is infinity and the u_G should vanish. Similarly for $x < 0$ one finds that semi-classical formulas (5.2), (5.1) yield (6.2) and (6.4). That is where the natural question arises, which we pose for the conclusion. For what class of non-regular Hamiltonians, to begin with those given by (2.4.7), one can justify asymptotic representations of type (1.19) or (5.2) for the Green function with the two-point function as the phase? Notice that unless $f = y^2$ as in the example before, exact representation of type (6.7) does not exist, and since these Hamiltonians are not regular (unless $f = y$) the machinery presented above in Sections 3-5 also does not work.

7. Analytic solutions to some linear PDE

In this short section we collect some general facts on analytic (or even formal power series) solutions to linear first order partial differential equations of the form

$$\lambda S + \left(Ax, \frac{\partial S}{\partial x} \right) = F(x), \tag{7.1}$$

where $x \in \mathcal{R}^d$, $F(x)$ is a polynomial, λ is a constant and A is a matrix with strictly positive eigenvalues $a_1 \leq ... \leq a_d$ which are simple, i.e. there exists an invertible matrix C such that $C^{-1}AC = D$ is the diagonal matrix $diag(a_1, ..., a_d)$. These facts are used in the asymptotic calculations of the two point function and of the solutions to the transport equation, which are carried out in Sections 3.2, 3.3 and Chapter 4.

Let p be the smallest non-negative integer such that $\lambda + pa_1 > 0$. In the most of examples $\lambda > 0$, and therefore $p = 0$.

Proposition 7.1.(i) *Let $F(x)$ be a homogeneous polynomial of degree $q \geq p$. Then there exists a solution S of (7.1) which is a polynomial of degree q with coefficients defined by:*

$$\frac{\partial^q S}{\partial x_{i_1}...\partial x_{i_q}} = \frac{1}{\lambda + a_{j_1} + ... + a_{j_q}}(C^{-1})_{j_1 i_1}...(C^{-1})_{j_q i_q}\frac{\partial^q F}{\partial x_{l_1}...\partial x_{l_q}}C_{l_1 j_1}...C_{l_q j_q}. \tag{7.2}$$

This solution is unique in the class of real analytic functions (in fact, even in the class of formal power series) under the additional assumption that all its derivatives at the origin up to order $p - 1$ vanish (this additional assumption is void in the main case $p = 0$).

(ii) *Let F be a sum $F = \sum_{q=p}^{m} F_q$ of homogeneous polynomials F_q of degree q. If $m = \infty$, let us suppose that this sum is absolutely convergent in a ball B_R (R may be finite or not). Then the analytic solution of (7.2) exists and is again unique under the condition above, and is given by the sum $\sum_{q=p}^{m} S_q$ of the solutions corresponding to each F_q. If $m = \infty$, this sum is convergent in the same ball B_R, as the sum presenting the function F.*

Proof. The change of variables $x = Cy$ transforms (7.1) to

$$\lambda \tilde{S} + \sum_{m=1}^{d} a_m y_m \frac{\partial \tilde{S}}{\partial y_m} = \tilde{F}(y), \tag{7.3}$$

where $\tilde{S}(y) = S(Cy)$ and $\tilde{F}(y) = F(Cy)$. Differentiating this equation $k \geq p$ times yields

$$(\lambda + a_{i_1} + ... + a_{i_k})\frac{\partial^k \tilde{S}}{\partial y_{i_1}...\partial y_{i_k}} + \sum a_m y_m \frac{\partial^{k+1} \tilde{S}}{\partial y_m \partial y_{i_1}...\partial y_{i_k}} = \frac{\partial^k \tilde{F}}{\partial y_{i_1}...\partial y_{i_k}}.$$

It follows that under the conditions of (i) only the derivatives of order q at the origin do not vanish, and for the derivatives of order q one gets

$$\frac{\partial^q \tilde{S}}{\partial y_{i_1}...\partial y_{i_q}}(0) = \frac{1}{\lambda + a_{i_1} + ... + a_{i_q}}\frac{\partial^q \tilde{F}}{\partial y_{i_1}...\partial y_{i_q}}(0). \tag{7.4}$$

Returning to the original variables x yields (7.2), because

$$\frac{\partial^q S}{\partial x_{i_1}...\partial x_{i_q}}(0) = \frac{\partial^q \tilde{S}}{\partial y_{j_1}...\partial y_{j_q}}(0)(C^{-1})_{j_1 i_1}...(C^{-1})_{j_q i_q},$$

$$\frac{\partial^q \tilde{F}}{\partial y_{j_1}...\partial y_{j_q}}(0) = \frac{\partial^q F}{\partial x_{l_1}...\partial x_{l_q}}(0)C_{l_1 j_1}...C_{l_q j_q}.$$

Similar arguments prove (ii) for finite m. If $m = \infty$, the convergence of the series representing the solution S (and thus the analyticity of S in the ball B_R) follows from (7.4), because this equations imply that

$$\left|\frac{\partial^q \tilde{S}}{\partial y_{i_1}...\partial y_{i_q}}(0)\right| = \frac{O(1)}{q}\left|\frac{\partial^k \tilde{F}}{\partial y_{i_1}...\partial y_{i_q}}(0)\right|.$$

We are going now to present an equivalent form of formula (7.2), which is more convenient for calculations. This formula will be used only in Section 4.3.

Cosider the graph Γ_q with vertices of two kinds such that there are exactly d^q vertices of each kind, and the vertices of the first kind (resp. second kind) are labeled by the sequences $(l_1,...,l_q)_x$ (resp. $(l_1,...,l_q)_y$) with each $l_j \in \{1,...,d\}$. The graph Γ_q is considered to be a complete oriented bipartite graph, which means that any pair of the vertices of different kind are connected by a (unique) arc, and the vertices of the same kind are not connected. Let us define the weights of the arcs by the formulas

$$W[(l_1,...,l_q)_x \to (j_1,...,j_q)_y] = C_{l_1 j_1}...C_{l_q j_q},$$

$$W[(j_1,...,j_q)_y \to (l_1,...,l_q)_x] = (C^{-1})_{j_1 l_1}...(C^{-1})_{j_q l_q}.$$

Furthermore, let us consider the weight of any vertex of the first kind to be one, and the weights of the vertices $(j_1,...,j_q)_y$ of the second kind being equal to $(\lambda + a_{j_1} + ... + a_{j_q})^{-1}$. The weight of any path in the graph Γ_q will be equal (by definition) to the product of the weights of its arcs and vertices. In particular, the weight of a two-step path is given by the formula

$$W[(l_1,...,l_q)_x \to (j_1,...,j_q)_y \to (i_1...i_q)_x]$$

$$= W[(l_1,...,l_q)_x \to (j_1,...,j_q)_y]\frac{1}{\lambda + a_{j_1} + ... + a_{j_q}}W[(j_1,...,j_q)_y \to (i_1,...,i_q)_x].$$

$$(7.5)$$

The fooliowing statement is a direct consequence of Proposition 7.1 and the definition of the bipartite weighted graph Γ_q.

Coroolary. *Formula (7.2) can be written in the following geometric form*

$$\frac{\partial^q S}{\partial x_{i_1}...\partial x_{i_q}}$$

$$= \sum_{(l_1,...,l_q)_x} \sum_{(j_1,...,j_q)_y} \frac{\partial^q F}{\partial x_{l_1}...\partial x_{l_q}} W[(l_1,...,l_q)_x \to (j_1,...,j_q)_y \to (i_1...i_q)_x]. \quad (7.6)$$

Let us discuss now the computational aspects of this formula for a special type of equation (7.1), which appears in the calculation (in Chapter 4) of the trace of the Green function of regular invariant diffusions corresponding to the stochastic geodesic flows. This equation has additional symmetries, which allow to reduce a large number of calculations encoded in formulas (7.2) or (7.6). The equation we are going to discuss, has the form

$$\lambda f - \left(\xi + y, \frac{\partial f}{\partial \xi} \right) + \left(6\xi + 4y, \frac{\partial f}{\partial y} \right) = F, \quad (7.7)$$

where ξ and y belong to $\mathcal{R}^k$, λ is a positive integer, and F is a polynomial in ξ, y. By Proposition 2.1 it is enough to be able to calculate the solutions corresponding to homogeneous polynomials F of each degree q. The solution is then given by the polynomial of degree q whose coefficients are calculated from (7.6). In the case of equation (7.7), A is the block-diagonal $2k \times 2k$-matrix with 2×2-blocks $\begin{pmatrix} -1 & -1 \\ 6 & 4 \end{pmatrix}$ on its diagonal. The corresponding matrix C is then also block-diagonal with 2×2-blocks $\begin{pmatrix} 1 & -1 \\ -2 & 3 \end{pmatrix}$ on its diagonal. It means that the change of the variables, which was used in the proof of Proposition 7.1 is now $(\xi, y)^i \mapsto (\eta, z)^i$, $i = 1, ..., k$, with

$$\begin{pmatrix} \xi^i \\ y^i \end{pmatrix} = \begin{pmatrix} 1 & -1 \\ -2 & 3 \end{pmatrix} \begin{pmatrix} \eta^i \\ z^i \end{pmatrix} = C \begin{pmatrix} \eta^i \\ z^i \end{pmatrix},$$

$$\begin{pmatrix} \eta^i \\ z^i \end{pmatrix} = \begin{pmatrix} 3 & 1 \\ 2 & 1 \end{pmatrix} \begin{pmatrix} \xi^i \\ y^i \end{pmatrix} = C^{-1} \begin{pmatrix} \xi^i \\ y^i \end{pmatrix}, \quad (7.8)$$

and thus equation (7.3) is

$$\tilde{f} + \eta^i \frac{\partial \tilde{f}}{\partial \eta^i} + 2z^i \frac{\partial \tilde{f}}{\partial z^i} = \tilde{F}(\eta, z),$$

which implies

$$\frac{\partial^{q+p} \tilde{f}}{\partial \eta^{i_1}...\partial \eta^{i_q} \partial z^{j_1}...\partial z^{j_p}} = \frac{1}{\lambda + q + 2p} \frac{\partial^{q+p} \tilde{F}}{\partial \eta^{i_1}...\partial \eta^{i_q} \partial z^{j_1}...\partial z^{j_p}}. \quad (7.9)$$

Due to the special block-diagonal form of C one sees that in the sum (7.2) consisting of $(2k)^{2q}$ terms only 2^{2q} terms do not necessarily vanish. Moreover, there is a large amount of symmetry, since A has only two different eigenvalues. Using simple combinatorial considerations we shall obtain now the following result.

Proposition 7.2. *If F is a polynomial of degree q and λ is positive number, then the unique analytic solution of (7.5) is the homogeneous polynomial of degree q in ξ, y with derivatives of the order q at the origin given by the formula*

$$\frac{\partial^q f}{\partial \xi^I \partial y^J} = \sum_{\bar{I} \subset I} \sum_{\bar{J} \subset J} A_{|\bar{I}|,|\bar{J}|}^{|I|,|J|} \frac{\partial^q F}{\partial \xi^{\bar{I}} \partial y^{I \backslash \bar{I}} \partial \xi^{J \backslash \bar{J}} \partial y^{\bar{J}}}, \tag{7.10}$$

where I and J are arbitrary sequences of indices from $\{1, ..., k\}$ such that $|I| + |J| = q$, and the coefficients A are given by the formula

$$A_{\mu\nu}^{\sigma\kappa} = \sum_{l=0}^{\mu} \sum_{m=0}^{\sigma-\mu} \sum_{n=0}^{\kappa-\nu} \sum_{p=0}^{\nu} C_{\mu}^l C_{\sigma-\mu}^m C_{\kappa-\nu}^n C_{\nu}^p$$

$$\times \frac{1}{\lambda + 2q - l - m - n - p} (-1)^{m-n+\kappa} 2^{p-l} (-3)^{l-p+\nu-\mu} 6^{\sigma}, \tag{7.11}$$

where C_i^j are the binomial coefficients.

Proof. In the case under consideration the vertices of the first kind (resp. of the second) of the graph Γ_q can be labeled by sequences ξ^I, y^J (resp. η^I, z^J), where both I and J are sequences of numbers from the set $1, ..., k$ of the lengths $|I|$ and $|J|$ respectively with $|I| + |J| = q$. Consequently, formula (7.6) takes the form

$$\frac{\partial^q f}{\partial \xi^I \partial y^J} = \sum_{\bar{I} \subset I} \sum_{\bar{J} \subset J} \frac{\partial^q F}{\partial \xi^{\bar{I}} \partial y^{I \backslash \bar{I}} \partial \xi^{J \backslash \bar{J}} \partial y^{\bar{J}}}$$

$$\times \sum_{\omega} W[(\xi^{\bar{I}}, y^{I \backslash \bar{I}}, \xi^{J \backslash \bar{J}}, y^{\bar{J}}) \to (\omega^{\bar{I}}, \omega^{I \backslash \bar{I}}, \omega^{J \backslash \bar{J}}, \omega^{\bar{J}}) \to (\xi^{\bar{I}}, \xi^{I \backslash \bar{I}}, y^{J \backslash \bar{J}}, y^{\bar{J}})], \tag{7.12}$$

where each ω can be either η or z with the corresponding index. Therefore we obtained (7.10), and it remains only to obtain formula for the weights in (7.12). To this end, we denote by l m, n and p the number of varibles η in $\omega^{\bar{I}}$, $\omega^{I \backslash \bar{I}}$, $\omega^{J \backslash \bar{J}}$ and $\omega^{\bar{J}}$ respectively, and we have

$$A_{\mu\nu}^{\sigma\kappa} = \sum_{l=0}^{\mu} \sum_{m=0}^{\sigma-\mu} \sum_{n=0}^{\kappa-\nu} \sum_{p=0}^{\nu} C_{\mu}^l C_{\sigma-\mu}^m C_{\kappa-\nu}^n C_{\nu}^p W_{l,m,n,p}^{\mu,\sigma-\mu,k-\nu,\nu},$$

where $W_{l,m,n,p}^{\mu,\sigma-\mu,k-\nu,\nu}$ is the weight of an arc having l transactions of the type $\xi \to \eta \to \xi$, $\mu - l$ transactions of the type $\xi \to z \to \xi$, m transactions of the type $y \to \eta \to \xi$, $\sigma - \mu - m$ transactions of the type $y \to z \to \xi$, n transactions of the type $\xi \to \eta \to y$, $k - \nu - n$ transactions of the type $\xi \to z \to y$, p transactions of the type $y \to \eta \to y$, and $\nu - p$ transactions of the type $y \to z \to y$. Due to (7.8), the weights of the transactions $\eta \to \xi$, $\eta \to y$, $z \to \xi$, $z \to y$ equal to $1, -2, -1, 3$ respectively, and the weights of the transactions $\xi \to \eta$, $\xi \to z$, $y \to \eta$, $y \to z$ equal to $3, 2, 1, 1$ respectively. Multiplying the corresponding weights yields

$$W_{l,m,n,p}^{\mu,\sigma-\mu,k-\nu,\nu} = \frac{1}{\lambda + 2q - l - m - n - p}$$

$$\times 3^l 2^{\mu-l}(-1)^{\mu-l}(-2)^m(-1)^{\sigma-\mu-m}3^n(-2)^n 2^{k-\nu-n}3^{k-\nu-n}(-2)^p 3^{\nu-p},$$

which implies (7.11), and the Proposition is proved.

Notice that it follows from (7.9) that

$$A_{00}^{\sigma\kappa} = (-6)^{\sigma-\kappa} A_{00}^{\kappa\sigma}. \tag{7.13}$$

In the next chapter, we shall need to solve equation (7.7) for the polynomials F of the fourth order. Actually, we shall need the full solution for the case of polynomials of order 2, and only a part of it for orders 3 and 4. We obtain now the neccesary formulas as an example of the application of Proposition 7.2.

Proposition 7.3. *Let f_j be the solution of equation (7.7) with $\lambda = 1$ and the r.h.s. F_j being a homogeneous polynomials in ξ, y of degree j. Then $f_0 = F_0$ and*

$$f_1 = \left(\frac{5}{6}\frac{\partial F_1}{\partial \xi^i} - \frac{\partial F_1}{\partial y_i}\right)\xi^i + \frac{1}{6}\frac{\partial F_1}{\partial \xi^i}y^i, \tag{7.14}$$

$$f_2 = \frac{1}{2}\left[\frac{6}{5}\frac{\partial^2 F_2}{\partial y^i \partial y^j} - \frac{9}{10}\left(\frac{\partial^2 F_2}{\partial y^i \partial \xi^j} + \frac{\partial^2 F_2}{\partial y^j \partial \xi^i}\right) + \frac{4}{5}\frac{\partial^2 F_2}{\partial \xi^i \partial \xi^j}\right]\xi^i \xi^j$$

$$+ \left[-\frac{1}{5}\frac{\partial^2 F_2}{\partial y^i \partial \xi^j} + \frac{3}{20}\frac{\partial^2 F_2}{\partial \xi^i \partial \xi^j} + \frac{1}{10}\frac{\partial^2 F_2}{\partial y^i \partial y^j} + \frac{1}{20}\frac{\partial^2 F_2}{\partial \xi^i \partial y^j}\right]\xi^i y^j$$

$$+ \frac{1}{2}\left[\frac{1}{30}\frac{\partial^2 F_2}{\partial \xi^i \partial \xi^j} - \frac{1}{60}\left(\frac{\partial^2 F_2}{\partial \xi^i \partial y^j} + \frac{\partial^2 F_2}{\partial \xi^j \partial y^i}\right) + \frac{2}{15}\frac{\partial^2 F_2}{\partial y^i \partial y^j}\right]y^i y^j. \tag{7.15}$$

Moreover,

$$\frac{\partial^3 f_3}{\partial \xi_i \partial \xi_j \partial \xi_k} = -\frac{54}{35}\frac{\partial^3 F_3}{\partial y_i \partial y_j \partial y_k} + \frac{39}{35}\left(\frac{\partial^3 F_3}{\partial \xi_i \partial y_j \partial y_k} + \frac{\partial^3 F_3}{\partial y_i \partial \xi_j \partial y_k} + \frac{\partial^3 F_3}{\partial y_i \partial y_j \partial \xi_k}\right)$$

$$- \frac{61}{70}\left(\frac{\partial^3 F_3}{\partial \xi_i \partial \xi_j \partial y_k} + \frac{\partial^3 F_3}{\partial \xi_i \partial y_j \partial \xi_k} + \frac{\partial^3 F_3}{\partial y_i \partial \xi_j \partial \xi_k}\right) + \frac{113}{140}\frac{\partial^3 F_3}{\partial \xi_i \partial \xi_j \partial \xi_k}, \tag{7.16}$$

and

$$\frac{\partial^4 f_4}{\partial \xi_i \partial \xi_j \partial \xi_k \partial \xi_l} = \frac{72}{35}\frac{\partial^4 F_4}{\partial y_i \partial y_j \partial y_k \partial y_l} + \frac{263}{315}\frac{\partial^4 F_4}{\partial \xi_i \partial \xi_j \partial \xi_k \partial \xi_l}$$

$$- \frac{51}{35}\left(\frac{\partial^4 F_4}{\partial \xi_i \partial y_j \partial y_k \partial y_l} + \frac{\partial^4 F_4}{\partial y_i \partial \xi_j \partial y_k \partial y_l} + \frac{\partial^4 F_4}{\partial y_i \partial y_j \partial \xi_k \partial y_l} + \frac{\partial^4 F_4}{\partial y_i \partial y_j \partial y_k \partial \xi_l}\right)$$

$$- \frac{92}{105}\left(\frac{\partial^4 F_4}{\partial y_i \partial \xi_j \partial \xi_k \partial \xi_l} + \frac{\partial^4 F_4}{\partial \xi_i \partial y_j \partial \xi_k \partial \xi_l} + \frac{\partial^4 F_4}{\partial \xi_i \partial \xi_j \partial y_k \partial \xi_l} + \frac{\partial^4 F_4}{\partial \xi_i \partial \xi_j \partial \xi_k \partial y_l}\right)$$

$$+ \frac{38}{35}\sum_{I \subset \{i,j,k,l\} : |I|=2}\frac{\partial^4 F_4}{\xi^I \partial y^{\{i,j,k,l\}\setminus I}}. \tag{7.17}$$

Proof. Using (7.10) yields

$$\frac{\partial f}{\partial \xi_i} = A_{00}^{10}\frac{\partial F}{\partial y_i} + A_{10}^{10}\frac{\partial F}{\partial \xi_i},$$

$$\frac{\partial f}{\partial y_i} = A_{01}^{00}\frac{\partial F}{\partial \xi_i} + A_{01}^{01}\frac{\partial F}{\partial y_i},$$

and from (7.11) one obtains

$$A_{00}^{10} = \sum_{m=0}^{1}\frac{6}{3-m}(-1)^m = 6\left(\frac{1}{3}-\frac{1}{2}\right) = -1,$$

$$A_{10}^{10} = \sum_{l=0}^{1}\frac{6}{3-l}2^{-l}(-3)^{l-1} = 6\left(-\frac{1}{9}+\frac{1}{4}\right) = \frac{5}{6},$$

$$A_{00}^{01} = \sum_{n=0}^{1}\frac{1}{3-n}(-1)^{1-n} = -\frac{1}{3}+\frac{1}{2} = \frac{1}{6},$$

$$A_{01}^{01} = \sum_{p=0}^{1}\frac{1}{3-p}(-1)2^p(-3)^{1-p} = 1-1 = 0,$$

which implies (7.14).

To get (7.15) one first uses (7.10) to obtain the formulas

$$\frac{\partial^2 f}{\partial \xi_i \partial \xi_j} = A_{00}^{20}\frac{\partial^2 F}{\partial y_i \partial y_j} + A_{10}^{20}\left(\frac{\partial^2 F}{\partial \xi_i \partial y_j} + \frac{\partial^2 f}{\partial y_i \partial \xi_j}\right) + A_{20}^{20}\frac{\partial^2 F}{\partial \xi_i \partial \xi_j},$$

$$\frac{\partial^2 f}{\partial \xi_i \partial y_j} = A_{00}^{11}\frac{\partial^2 F}{\partial y_i \partial \xi_j} + A_{10}^{11}\frac{\partial^2 F}{\partial \xi_i \partial \xi_j} + A_{01}^{11}\frac{\partial^2 F}{\partial y_i \partial y_j} + A_{11}^{11}\frac{\partial^2 F}{\partial \xi_i \partial y_j},$$

$$\frac{\partial^2 f}{\partial y_i \partial y_j} = A_{00}^{02}\frac{\partial^2 F}{\partial \xi_i \partial \xi_j} + A_{01}^{02}\left(\frac{\partial^2 F}{\partial \xi_i \partial y_j} + \frac{\partial^2 F}{\partial y_i \partial \xi_j}\right) + A_{02}^{02}\frac{\partial^2 F}{\partial y_i \partial y_j}. \qquad (7.18)$$

Using (7.11) one calculates

$$A_{00}^{20} = \sum_{m=0}^{2}C_2^m 6^2 \frac{1}{5-m}(-1)^m = 36(\frac{1}{5}-\frac{1}{2}+\frac{1}{3}) = \frac{6}{5},$$

$$A_{10}^{20} = \sum_{l=0}^{1}\sum_{m=0}^{1}6^2\frac{1}{5-l-m}(-1)^{m+l-1}2^{-l}3^{l-1} = 36(-\frac{1}{15}+\frac{1}{12}+\frac{1}{8}-\frac{1}{6}) = -\frac{9}{10},$$

$$A_{20}^{20} = \sum_{l=0}^{2}C_2^l 6^2 \frac{1}{5-l}2^{-l}(-3)^{l-2} = 36(\frac{1}{45}-\frac{1}{12}+\frac{1}{12}) = \frac{4}{5},$$

$$A_{01}^{02} = \sum_{n=0}^{1}\sum_{p=0}^{1}\frac{1}{5-n-p}(-1)^{-n-p+3}2^p 3^{-p+1} = -\frac{3}{5}+\frac{3}{4}+\frac{2}{4}-\frac{2}{3} = -\frac{1}{60},$$

$$A_{02}^{02} = \sum_{p=0}^{2}C_2^p \frac{1}{5-p}(-1)^p 2^p 3^{2-p} = \frac{9}{5}-3+\frac{4}{3} = \frac{2}{15},$$

$$A_{00}^{02} = \sum_{n=0}^{2} C_2^n \frac{1}{5-n}(-1)^{-n+2} = \frac{1}{5} - \frac{1}{2} + \frac{1}{3} = \frac{1}{30},$$

the last coefficient could be also obtain using the previous calculations and formula (7.13). Furthermore,

$$A_{00}^{11} = 6 \sum_{m=0}^{1} \sum_{n=0}^{1} \frac{1}{5-m-n}(-1)^{m-n+1} = 6\left(-\frac{1}{5} + \frac{1}{2} - \frac{1}{3}\right) = -\frac{1}{5},$$

$$A_{01}^{11} = 6 \sum_{m=0}^{1} \sum_{p=0}^{1} \frac{1}{5-m-p}(-1)^{m-p}2^p3^{-p+1} = 6\left(\frac{3}{5} - \frac{3}{4} - \frac{2}{4} + \frac{2}{3}\right) = \frac{1}{10},$$

$$A_{10}^{11} = 6 \sum_{l=0}^{1} \sum_{n=0}^{1} \frac{1}{5-l-n}(-1)^{l-n}2^{-l}3^{l-1} = 6\left(\frac{1}{15} - \frac{1}{12} - \frac{1}{8} + \frac{1}{6}\right) = \frac{3}{20},$$

$$A_{11}^{11} = 6 \sum_{l=0}^{1} \sum_{p=0}^{1} \frac{1}{5-l-p}(-1)^{l-p+1}2^{p-l}3^{l-p} = 6\left(-\frac{1}{5} + \frac{1}{6} + \frac{3}{8} - \frac{1}{3}\right) = \frac{1}{20}.$$

Substituting these formulas in (7.18) yields (7.15).

To obtain (7.16) one uses (7.10) to write

$$\frac{\partial^3 f_3}{\partial \xi_i \partial \xi_j \partial \xi_k} = A_{00}^{30} \frac{\partial^3 F_3}{\partial y_i \partial y_j \partial y_k} + A_{10}^{30}\left(\frac{\partial^3 F_3}{\partial \xi_i \partial y_j \partial y_k} + \frac{\partial^3 F_3}{\partial y_i \partial \xi_j \partial y_k} + \frac{\partial^3 F_3}{\partial y_i \partial y_j \partial \xi_k}\right)$$

$$+A_{20}^{30}\left(\frac{\partial^3 F_3}{\partial \xi_i \partial \xi_j \partial y_k} + \frac{\partial^3 F_3}{\partial \xi_i \partial y_j \partial \xi_k} + \frac{\partial^3 F_3}{\partial y_i \partial \xi_j \partial \xi_k}\right) + A_{30}^{30}\frac{\partial^3 F_3}{\partial \xi_i \partial \xi_j \partial \xi_k}. \qquad (7.19)$$

Then one uses formula (7.11) to calculate

$$A_{00}^{30} = 6^3 \sum_{m=0}^{3} C_3^m \frac{1}{7-m}(-1)^m = 6^3\left(\frac{1}{7} - \frac{1}{2} + \frac{3}{5} - \frac{1}{4}\right) = -\frac{54}{35},$$

$$A_{10}^{30} = 6^3 \sum_{l=0}^{1} \sum_{m=0}^{2} C_2^m \frac{1}{7-l-m}(-1)^{m+l-1}2^{-l}3^{l-1}$$

$$= 6^3\left(-\frac{1}{21} + \frac{1}{9} - \frac{1}{15} + \frac{1}{12} - \frac{1}{5} + \frac{1}{8}\right) = \frac{39}{35},$$

$$A_{20}^{30} = 6^3 \sum_{l=0}^{2} \sum_{m=0}^{1} C_2^l \frac{1}{7-l-m}(-1)^{m+l}2^{-l}3^{l-2}$$

$$= 6^3\left(\frac{1}{9}(\frac{1}{7} - \frac{1}{6}) + \frac{1}{3}(\frac{1}{5} - \frac{1}{6}) + \frac{1}{4}(\frac{1}{5} - \frac{1}{4})\right) = -\frac{61}{70},$$

$$A_{30}^{30} = 6^3 \sum_{l=0}^{3} C_3^l \frac{1}{7-l} 2^{-l}(-3)^{l-3} = 6^3 \left(-\frac{1}{189} + \frac{1}{36} - \frac{1}{20} + \frac{1}{32} \right) = \frac{113}{140}.$$

Substituting these coefficients in (7.19) yields (7.16). Similarly, to get (7.17) one needs the following coefficients (which are again obtained from the general formula (7.11):

$$A_{40}^{40} = 6^4 \sum_{l=0}^{4} C_4^l \frac{1}{9-l} 2^{-l}(-3)^{l-4} = 6^4 \left(\frac{1}{9} \frac{1}{3^4} - \frac{1}{4} \frac{1}{27} + \frac{1}{42} - \frac{1}{36} + \frac{1}{80} \right) = \frac{263}{315},$$

$$A_{30}^{40} = 6^4 \sum_{l=0}^{3} \sum_{m=0}^{1} C_3^l \frac{1}{9-l-m} (-1)^{m+l-3} 2^{-l} 3^{l-3} = -\frac{92}{105},$$

$$A_{20}^{40} = 6^4 \sum_{l=0}^{2} \sum_{m=0}^{2} C_2^l C_2^m \frac{1}{9-l-m} (-1)^{m+l-2} 2^{-l} 3^{l-2} = \frac{38}{35},$$

$$A_{10}^{40} = 6^4 \sum_{l=0}^{1} \sum_{m=0}^{3} C_3^m \frac{1}{9-l-m} (-1)^{m+l-1} 2^{-l} 3^{l-1} = -\frac{51}{35},$$

$$A_{00}^{40} = 6^4 \sum_{m=0}^{4} C_4^m \frac{1}{9-m} (-1)^m = 6^4 \left(\frac{1}{9} - \frac{1}{2} + \frac{6}{7} - \frac{2}{3} + \frac{1}{5} \right) = \frac{72}{35}.$$

Proposition is proved.

Chapter 4. INVARIANT DEGENERATE DIFFUSION ON COTANGENT BUNDLES

1. Curvilinear Ornstein-Uhlenbeck process and stochastic geodesic flow

In this chapter we apply the theory developed in the previous chapter to the investigation of invariant degenerate diffusions on manifolds. We confine ourselves to the case of a regular degenerate diffusion of rank one. Since in the conditions of the regularity of a Hamiltonian the linearity of some coefficient in the second variable y is included, one has to suppose when constructing an invariant object that this second variable lives in a linear space. Therefore, an invariant operator ought to be defined on a vector bundle over some manifold: coordinates y in fibres and coordinate x on a base. We reduce ourselves to the most commonly used vector bundle, namely to the cotangent bundle T^*M of a compact n-dimensional manifold M. In local coordinates, a regular Hamiltonian H of a degenerate diffusion of rank one has form (2.3.4), where the matrix g is positive definite and α is non-degenerate. The corresponding diffusion equation (3.1.2) has the form

$$h\frac{\partial u}{\partial t} = Lu = H\left(x, y, -h\frac{\partial}{\partial x}, -h\frac{\partial}{\partial y}\right) = \frac{h^2}{2}g_{ij}\frac{\partial^2 u}{\partial y_i \partial y_j}$$

$$+h(a^i(x)+\alpha^{ij}(x)y_j)\frac{\partial u}{\partial x^i} + h(b_i(x)+\beta_i^j(x)y_j+\frac{1}{2}\gamma_i^{jl}(x)y_jy_l)\frac{\partial u}{\partial y_i} - V(x,y)u. \quad (1.1)$$

In this section, we give the complete description of the invariant operators of that kind on T^*M. Let us recall that a tensor γ of type (q,p) on a manifold M is by definition a set of n^{p+q} smooth functions $\gamma_{j_1\ldots j_q}^{i_1\ldots i_p}(x)$ on x that under the change of coordinates $x \mapsto \tilde{x}$ changes by the law

$$\tilde{\gamma}_{j_1\ldots j_q}^{i_1\ldots i_p}(\tilde{x}) = \gamma_{l_1\ldots l_q}^{k_1\ldots k_p}(x)\frac{\partial \tilde{x}^{i_1}}{\partial x^{k_1}}\cdots\frac{\partial \tilde{x}^{i_p}}{\partial x^{k_p}}\frac{\partial x^{l_1}}{\partial \tilde{x}^{j_1}}\cdots\frac{\partial x^{l_q}}{\partial \tilde{x}^{j_q}}.$$

To each tensor of the type $(0,p)$ corresponds the polylinear function on the cotangent bundle T^*M defined by the formula $\gamma(x,y) = \gamma^{i_1\ldots i_p}(x)y_{i_1}\cdots y_{i_p}$.

Theorem 1.1 *Suppose the following objects are given on M:*

(i) Riemanian metric, which in local coordinates x on M is given by a positive definite matrix $g(x)$, $x \in M$;

(ii) non-degenerate tensor $\alpha = \{\alpha^{ij}(x)\}$ of the type $(0,2)$ (non-degeneracy means that the matrix α is non-degenerate everywhere) and a tensor $a = \{a^i(x)\}$ of the type $(0,1)$ (i.e. a vector field); these tensors obviously define a quadratic function $f(x,y) = \alpha^{ij}(x)y_iy_j + a^i(x)y_i$ on T^*M;

(iii) tensors b, β, γ of the types $(1,0), (1,1), (1,2)$ respectively;

(iv) the sum V of tensors of the types $(0,0), (0,1), (0,2), (0,3), (0,4)$, which defines a bounded from below function $V(x,y)$ on T^*M.

Then the second order differential operator

$$L = \frac{1}{2}g_{ij}(x)\frac{\partial^2}{\partial y_i \partial y_j} + \frac{\partial f}{\partial y_i}(x,y)\frac{\partial}{\partial x_i} - \frac{\partial f}{\partial x_i}(x,y)\frac{\partial}{\partial y_i}$$

$$+ \left(b_i(x) + \beta_i^j(x)y_j + \frac{1}{2}\gamma_i^{kl}(x)y_k y_l\right)\frac{\partial}{\partial y_i} - V(x,y) \tag{1.2}$$

is an invariant operator on T^*M , which is a regular diffusion of the rank one. Conversely, each such operator has this form.

Proof. Under the change of the variables $x \mapsto \tilde{x}(x)$ the moments change by the rule $\tilde{y} = y\frac{\partial x}{\partial \tilde{x}}$. Therefore,

$$\frac{\partial u}{\partial y_i} = \frac{\partial u}{\partial \tilde{y}_j}\frac{\partial \tilde{y}_j}{\partial y_i}, \quad \frac{\partial^2 u}{\partial y_i \partial y_j} = \frac{\partial^2 u}{\partial \tilde{y}_k \partial \tilde{y}_m}\frac{\partial \tilde{y}_m}{\partial y_j}\frac{\partial \tilde{y}_k}{\partial y_i}, \tag{1.3}$$

$$\frac{\partial u}{\partial x^i} = \frac{\partial u}{\partial \tilde{x}^j}\frac{\partial \tilde{x}^j}{\partial x^i} + \frac{\partial u}{\partial \tilde{y}_j}\frac{\partial \tilde{y}_j}{\partial x^i}, \quad \frac{\partial \tilde{y}_k}{\partial y_j} = \frac{\partial x^j}{\partial \tilde{x}^k}. \tag{1.4}$$

It follows, in particular, that under the change $(x,y) \mapsto (\tilde{x},\tilde{y})$, the second order part of (1.1), the first order part of (1.1), and the zero order part of (1.1) transforms to second order, first order, and zero order operators respectively, and consequently, if the operator (1.1) is invariant, then its second order part, its first order part, and its zero order part must be invariant. In order that the zero order term $V(x,y)u$ was invariant it is necessary and sufficient that $V(x,y)$ is invariant and therefore $V(x,y)$ is a function. From the invariance of the second order part one has

$$g_{ij}(x)\frac{\partial^2 u}{\partial y_i \partial y_j} = g_{ij}(x)\frac{\partial^2 u}{\partial \tilde{y}_k \partial \tilde{y}_m}\frac{\partial \tilde{y}_m}{\partial y_j}\frac{\partial \tilde{y}_k}{\partial y_i}$$

$$= g_{ij}(x)\frac{\partial^2 u}{\partial \tilde{y}_k \partial \tilde{y}_m}\frac{\partial x^j}{\partial \tilde{x}^m}\frac{\partial x^i}{\partial \tilde{x}^k} = \tilde{g}_{km}(\tilde{x})\frac{\partial^2 u}{\partial \tilde{y}_k \partial \tilde{y}_m},$$

and consequently, the invariance of the second order part is equivalent to the requirement that g is a tensor, and therefore defines a riemannian metric. Let us write now the condition of the invariance of the first order part of operator (1.1). Changing the variable $(x,y) \mapsto (\tilde{x},\tilde{y})$ in the first order part of (1,1) one has

$$\left(a^i(x) + \alpha^{ij}(x)y_j\right)\frac{\partial u}{\partial x^i} + \left(b_i(x) + \beta_i^j(x)y_j + \frac{1}{2}\gamma_i^{jl}(x)y_j y_l\right)\frac{\partial u}{\partial y_i}$$

$$= \left(a^i(x) + \alpha^{ij}(x)\tilde{y}_m\frac{\partial \tilde{x}^m}{\partial x^j}\right)\left(\frac{\partial u}{\partial \tilde{x}^l}\frac{\partial \tilde{x}^l}{\partial x^i} + \frac{\partial u}{\partial \tilde{y}_l}\frac{\partial \tilde{y}_l}{\partial x^i}\right)$$

$$+ \left(b_i(x) + \beta_i^j(x)\tilde{y}_m\frac{\partial \tilde{x}^m}{\partial x^j} + \frac{1}{2}\gamma_i^{jl}(x)\tilde{y}_m\tilde{y}_p\frac{\partial \tilde{x}^m}{\partial x^j}\frac{\partial \tilde{x}^p}{\partial x^l}\right)\frac{\partial u}{\partial \tilde{y}_q}\frac{\partial \tilde{y}_q}{\partial y_i}.$$

Therefore, the invariance of this first order part is equivalent to the following two equations:

$$\tilde{a}^i(\tilde{x}) + \tilde{\alpha}^{ij}(\tilde{x})\tilde{y}_j = \left(a^l(x) + \alpha^{lj}(x)\tilde{y}_m\frac{\partial\tilde{x}^m}{\partial x^j}\right)\frac{\partial\tilde{x}^i}{\partial x^l}, \tag{1.5}$$

and

$$\left(\tilde{b}_i(\tilde{x}) + \tilde{\beta}_i^j(\tilde{x})\tilde{y}_j + \frac{1}{2}\tilde{\gamma}_i^{jl}(\tilde{x})\tilde{y}_j\tilde{y}_l\right) = \left(a^l(x) + \alpha^{lj}(x)\tilde{y}_m\frac{\partial\tilde{x}^m}{\partial x^j}\right)\frac{\partial\tilde{y}_i}{\partial x^l}$$

$$+ \left(b_q(x) + \beta_q^j(x)\tilde{y}_m\frac{\partial\tilde{x}^m}{\partial x^j} + \frac{1}{2}\gamma_q^{jl}(x)\tilde{y}_m\tilde{y}_p\frac{\partial\tilde{x}^m}{\partial x^j}\frac{\partial\tilde{x}^p}{\partial x^l}\right)\frac{\partial\tilde{y}_i}{\partial y_q}. \tag{1.6}$$

From (1.5) one obtains that a and α are tensors, as is required. Next,

$$\frac{\partial\tilde{y}_i}{\partial x^l} = -\tilde{y}_p\frac{\partial\tilde{x}^p}{\partial x^l\partial x^m}\frac{\partial x^m}{\partial\tilde{x}^i},$$

Therefore, equating in (1.6) the terms which do not depend on $\tilde{y}$, the terms depending on $\tilde{y}$ linearly, and the terms depending on $\tilde{y}$ quadratically, one gets that b is a tensor of the type $(1,0)$, and that the law of the transformation of β and γ has the form

$$\tilde{\gamma}_i^{mp}(\tilde{x}) = \gamma_q^{jl}(x)\frac{\partial\tilde{x}^m}{\partial x^j}\frac{\partial\tilde{x}^p}{\partial x^l}\frac{\partial x^q}{\partial\tilde{x}^i} - 2\alpha^{lj}(x)\frac{\partial\tilde{x}^m}{\partial x^j}\frac{\partial^2\tilde{x}^p}{\partial x^l\partial x^q}\frac{\partial x^q}{\partial\tilde{x}^i},$$

$$\tilde{\beta}_i^p(\tilde{x}) = \beta_q^j(x)\frac{\partial\tilde{x}^p}{\partial x^j}\frac{\partial x^q}{\partial\tilde{x}^i} - a^l(x)\frac{\partial^2\tilde{x}^p}{\partial x^l\partial x^m}\frac{\partial x^m}{\partial\tilde{x}^i}.$$

Since

$$\frac{\partial\tilde{a}^p}{\partial\tilde{x}^i}(\tilde{x}) = \frac{\partial a^j}{\partial x^q}\frac{\partial x^q}{\partial\tilde{x}^i}\frac{\partial\tilde{x}^p}{\partial x^j} + a^l(x)\frac{\partial^2\tilde{x}^p}{\partial x^l\partial x^m}\frac{\partial x^m}{\partial\tilde{x}^i},$$

$$\frac{\partial\tilde{\alpha}^{mp}}{\partial\tilde{x}^i}(\tilde{x}) = \frac{\partial\alpha^{lj}}{\partial x^q}\frac{\partial x^q}{\partial\tilde{x}^i}\frac{\partial\tilde{x}^m}{\partial x^l}\frac{\partial\tilde{x}^p}{\partial x^j} + 2\alpha^{lj}(x)\frac{\partial^2\tilde{x}^p}{\partial x^q\partial x^l}\frac{\partial x^q}{\partial\tilde{x}^i}\frac{\partial\tilde{x}^m}{\partial x^j},$$

it follows that $\{\gamma_i^{mp} + \frac{\partial\alpha^{mp}}{\partial x^i}\}$ and $\{\beta_i^p + \frac{\partial a^p}{\partial x^i}\}$ are tensors of the types $(1,2)$ and $(1,1)$ respectively. Denoting these tensors again by γ and β respectively, yields representation (1.2). The proof is complete.

Let us write the stochastic differential equation for the diffusion process corresponding to the operator (1.2) with vanishing V. Let $r : M \mapsto \mathcal{R}^N$ be an embedding of the Riemanian manifold M in the Euclidean space (as is well known, such embedding always exists). The operator (1.2) stands for the diffusion on T^*M defined by the stochastic system

$$\begin{cases} dx = \frac{\partial f}{\partial y}\,dt \\ dy_i = -\frac{\partial f}{\partial x^i}\,dt + (b_i(x) + \beta_i^j(x)y_j + \frac{1}{2}\gamma_i^{kl}(x)y_ky_l)\,dt + \frac{\partial r^j}{\partial x^i}\,dw_j, \end{cases} \tag{1.7}$$

where w is the standard N-dimensional Wiener process. This statement follows from the well known formula for the Riemanian metric

$$g_{ij}(x) = \sum_{k=1}^{N} \frac{\partial r^k}{\partial x^i} \frac{\partial r^k}{\partial x^j}$$

and the Ito formula. It is interesting to note that though system (1.7) depends explicitly on the embedding r, the corresponding operator L defining the transition probabilities for diffusion process (1.7) depends only on the Riemanian structure.

One sees that system (1.7) describes a curvilinear version of the classical Ornstein-Uhlenbeck process (see e.g. [Joe] for an invarian definition) defined originally (see, e.g.[Nel1]) by the system ($x, y \in \mathcal{R}^n$)

$$\begin{cases} \dot{x} = y \\ dy = -\frac{\partial V}{\partial x} dt - \beta y \, dt + dw(t) \end{cases} \tag{1.8}$$

as a model of Brownian motion, where $\beta \geq 0$ is some constant and $V(x)$ is some (usually bounded from below) function (potential). System (1.8) defines a Newton particle (Hamiltonian system with the Hamiltonian $V(x) + y^2/2$ disturbed by the friction force βy and by the white noise random force dw. System (1.7) describes a Hamiltonian system (defined by the Hamiltonian function f which is quadratic in momentum but with varying coefficients) with additional deterministic force (defined by the 1-form b), the friction $\beta_i^j(x)y_j + \frac{1}{2}\gamma_i^{kl}(x)y_k y_l$ (which can depend on the first and second degree of the velocity) and the white noise force depending on the position of the particle.

In the case of vanishing b, β, γ system (1.7) is a stochastic Hamiltonian system with non-homogeneous singular random Hamiltonian $f(x, y) + r(x)\dot{w}$, which describes the deterministic Hamiltonian flow disturbed by the white noise force:

$$\begin{cases} dx = \frac{\partial f}{\partial y} dt \\ dy = -\frac{\partial f}{\partial x} dt + \frac{\partial}{\partial x}(r, dw). \end{cases} \tag{1.9}$$

The "plane" stochastic Hamiltonian systems, i.e. (1.9) for $M = \mathcal{R}^n$, were investigated recently in connection with their application to the theory of stochastic partial differential equation, see [K1], [TZ1], [TZ2].

The mostly used example of the Hamiltonian system on the cotangent bundle T^*M of a Riemanian manifold is of course the geodesic flow, which stands for the Hamiltonian function $f = (G(x)y, y)/2$, where $G(x) = g^{-1}(x)$. For this f, system (1.9) takes the form

$$\begin{cases} \dot{x} = G(x)y \\ dy = -\frac{1}{2}\frac{\partial}{\partial x}(G(x)y, y) \, dt + \frac{\partial}{\partial x}(r, dw), \end{cases} \tag{1.10}$$

This system was called in [K1] the stochastic geodesic flow. The investigation of its small time asymptotics was begun in [AHK2]. Corresponding Hamiltonian (2.3.4) of the stochastic geodesic flow is

$$H = \frac{1}{2}(g(x)q, q) - (G(x)y, p) + \frac{1}{2}\left(\frac{\partial}{\partial x}(G(x)y, y), q\right) \tag{1.11}$$

and the invariant diffusion equation is

$$\frac{\partial u}{\partial t} = Lu = \frac{h}{2} \, tr \left(g(x) \frac{\partial^2 u}{\partial y_i \partial y_j} \right) + \left(G(x)y, \frac{\partial u}{\partial x} \right) - \frac{1}{2} \left(\frac{\partial}{\partial x}(G(x)y, y), \frac{\partial u}{\partial y} \right).$$
(1.12)

It depends only on the Riemanian structure and therefore its property should reflect the geometry of M, which explain more explicitly in the next sections.

2. Small time asymptotics for stochastic geodesic flow

The stochastic geodesic flow is a good example for performing the general results of the previous chapter. Using these results we present now the calculation of the main terms of the small time asymptotics for the Green function of equation (1.12), i. e. its solution with the initial data

$$u_G(0, x, y; x_0, y^0) = \delta(x - x_0)\delta(y - y^0)$$
(2.1)

in a neighbourhood of the point $(x_0, y^0) \in T^*M$.

All calculations will be carried out in normal coordinates around x_0 (see, e.g. [CFKS]), in which $x_0 = 0$,

$$g_{ij}(x) = \delta_i^j + \frac{1}{2} g_{ij}^{kl} x^k x^l + O(|x|^3),$$
(2.2)

and $\det g(x) = 1$ identically. These conditions imply that

$$\sum_{i=1}^{n} g_{ii}^{kl} = 0 \quad \forall k, l$$
(2.3)

and that the Gaussian (or scalar) curvature in x_0 is equal to

$$R = \sum_{i,k} g_{ik}^{ik}.$$
(2.4)

Remark. Some authors do not include the requirement $\det g = 1$ in the definition of normal coordinates. Notice however that if a system of coordinates x on a n-dimensional riemanian manifold M satisfies all other conditions of normality but for the condition $\det g = 1$, then the coordinates $\tilde{x}$ defined by the formula

$$\tilde{x}^1 = \int_0^{x^1} \sqrt{g}(s, x^2, ..., x^n) \, ds, \quad \tilde{x}^i = x^i, \, i \geq 2,$$

satisfies all the conditions of normality given above, as one checks easily (see [CFKS]).

Moreover, from (2.2) one gets obviously the expansions

$$G_{ij} = \delta_i^j - \frac{1}{2} g_{ij}^{kl} x^k x^l + O(|x|^3),$$
(2.5)

for the inverse matrix $G(x) = g^{-1}(x)$, and also

$$\frac{\partial G_{ij}}{\partial x^k}(x) = -g_{ij}^{kl} + O(|x|^2). \tag{2.6}$$

To find the asymptotics of the two-point function one should solve the main equation (3.2.12), which for the case of Hamiltonian (1.11) takes the form

$$\frac{\partial \Sigma}{\partial t} - \frac{\xi + G(t\xi + \tilde{x})(y + \tilde{y}) - G(\tilde{x})\tilde{y}}{t}\frac{\partial \Sigma}{\partial \xi} - \left(g(\tilde{x})\tilde{q}, \frac{\partial \Sigma}{\partial y}\right)$$

$$+ \frac{1}{2}\left[\left(\frac{\partial G}{\partial x}(t\xi + \tilde{x})(y + \tilde{y}), y + \tilde{y}\right) - \left(\frac{\partial G}{\partial x}(\tilde{x})\tilde{y}, \tilde{y}\right)\right]\frac{\partial \Sigma}{\partial y}$$

$$+ \frac{1}{2}\left(g(t\xi + \tilde{x})\frac{\partial \Sigma}{\partial y}, \frac{\partial \Sigma}{\partial y}\right) = 0. \tag{2.7}$$

Using (2.2), (2.3) one concludes that

$$\tilde{x} = x_0 - ty^0 + O(t^3), \quad \tilde{y} = y^0 + O(t^2), \quad \tilde{q} = O(t^2), \tag{2.8}$$

and then one rewrites (2.7) in the coordinate form (using now low indices for both ξ and y:

$$\frac{\partial \Sigma}{\partial t} - \frac{(\xi + y)_i - \frac{t^2}{2}g_{ij}^{kl}[(\xi_k - y_k^0)(\xi_l - y_l^0)(y_j + y_j^0) - y_k^0 y_l^0 y_j^0] + O(t^3)}{t}\frac{\partial \Sigma}{\partial \xi_i}$$

$$- \left[\frac{t}{2}g_{ij}^{kl}[(\xi_l - y_l^0)(y_i + y_i^0)(y_j + y_j^0) + y_i^0 y_j^0 y_l^0] + O(t^2)\right]\frac{\partial \Sigma}{\partial y_k}$$

$$+ \frac{1}{2}(1 + \frac{t^2}{2}g_{ij}^{kl}(\xi_k - y_k^0)(\xi_l - y_l^0) + O(t^3))\frac{\partial \Sigma}{\partial y_i}\frac{\partial \Sigma}{\partial y_j} = 0. \tag{2.9}$$

Following the arguments of Sect.2 of the previous chapter one looks for the solution of this equation in form (3.2.6), where Σ_{-1} is a positive quadratic form and $\Sigma_0(0,0) = 0$. For Σ_{-1} one gets equation (3.2.10) with α_0 and g_0 being unit matrices. Its solution is given by (3.2.16). For Σ_0 one finds then the equation

$$-(y + \xi)\frac{\partial \Sigma_0}{\partial \xi} + (6\xi + 4y)\frac{\partial \Sigma_0}{\partial y} = 0,$$

whose solution vanishes, due to Proposition 3.7.1. Furthermore, for Σ_1 one obtains the equation

$$\Sigma_1 - (y + \xi)_i\frac{\partial \Sigma_1}{\partial \xi_i} + (6\xi + 4y)_i\frac{\partial \Sigma_1}{\partial y_i}$$

$$+ g_{ij}^{kl}[(\xi_k - y_k^0)(\xi_l - y_l^0)(y_j + y_j^0) - y_k^0 y_l^0 y_j^0](6\xi + 3y)_i$$

$$-g_{ij}^{kl}[(\xi_l - y_l^0)(y_i + y_i^0)(y_j + y_j^0) + y_i^0 y_j^0 y_l^0](3\xi + 2y)_k$$

$$+g_{ij}^{kl}(\xi_k - y_k^0)(\xi_l - y_l^0)(3\xi + 2y)_i(3\xi + 2y)_j = 0.$$

Opening the brackets one presents this equation in the form

$$\Sigma_1 - (y + \xi)_i \frac{\partial \Sigma_1}{\partial \xi_i} + (6\xi + 4y)_i \frac{\partial \Sigma_1}{\partial y_i} = F(\xi, y), \tag{2.10}$$

where F is the sum $F_2 + F_3 + F_4$ of the homogeneous polynomials of degree 2,3,4 given by the formulas

$$F_2 = g_{ij}^{kl}[(12\xi_i\xi_k - 4y_iy_k)y_j^0 y_l^0 - (18\xi_iy_j + 7y_iy_j + 9\xi_i\xi_j)y_k^0 y_l^0 + (3\xi_k\xi_l + 2\xi_ky_l)y_i^0 y_j^0], \tag{2.11}$$

$$F_3 = g_{ij}^{kl}[(-6\xi_i\xi_k\xi_l + 3\xi_k\xi_ly_i + 4\xi_ky_iy_l)y_j^0$$

$$+(36\xi_i\xi_ky_j + 11\xi_ky_iy_j - 2y_iy_jy_k + 18\xi_i\xi_j\xi_k)y_l^0], \tag{2.12}$$

$$F_4 = g_{ij}^{kl}[2\xi_ky_iy_jy_l - 4\xi_k\xi_ly_iy_j - 18\xi_i\xi_k\xi_ly_j - 9\xi_i\xi_j\xi_k\xi_l]. \tag{2.13}$$

The solution of this equation is the sum of the solutions Σ_1^2, Σ_1^3, Σ_1^4 corresponding to F_2, F_3, and F_4 in the r.h.s. of (2.10). These solutions can be calculated by formula (3.7.10). For instance, Σ_1^2 is given by (3.7.15) with F_2 being equal to (2.11). These calculations are rather long, but the form of the solution is clear:

$$\Sigma_1 = g_{ij}^{kl} R_{ijkl}(\xi, y, y^0), \tag{2.14}$$

where R_{ijkl} are homogeneous polynomials of degree 4 in the variables ξ, y, y^0. Similarly one sees that the other terms Σ_j are homogeneous polynomials in ξ, y, y^0 of degree $j + 3$, which is important to know when making the estimates uniform in y^0.

Let us find now the first nontrivial term of the asymptotic solution of the transport equation. In the case of Hamiltonian (1.11), the general equation (3.3.3) takes the form

$$\frac{\partial \Psi}{\partial t} - \frac{\alpha}{t}\Psi - \frac{\xi + G(\xi t + \tilde{x})(y + \tilde{y}) - G(\tilde{x})\tilde{y}}{t} \frac{\partial \Psi}{\partial \xi}$$

$$+\frac{1}{2}\left[\left(\frac{\partial G}{\partial x}(t\xi + \tilde{x})(y + \tilde{y}), y + \tilde{y}\right) - \left(\frac{\partial G}{\partial x}(\tilde{x})\tilde{y}, \tilde{y}\right)\right]\frac{\partial \Psi}{\partial y}$$

$$-\left(g(\tilde{x})\tilde{q}, \frac{\partial \Psi}{\partial y}\right) + \left(g(t\xi + \tilde{x})\frac{\partial \Sigma}{\partial y}, \frac{\partial \Psi}{\partial y}\right) + \frac{1}{2}\Psi \, tr\left(g(t\xi + \tilde{x})\frac{\partial^2 \Sigma}{\partial y^2}\right) = 0, \tag{2.15}$$

where

$$\Psi(t, \xi, y) = t^\alpha \phi(t, t\xi + \tilde{x}, y + \tilde{y}; 0, y^0). \tag{2.16}$$

From (3.3.5) one finds $\alpha = 2n$. Looking for the solution of (2.15) in the form

$$\Psi = 1 + t\Psi_1 + t^2\Psi_2 + \dots$$

one gets comparing the terms at t^0 the following equation (since $\Sigma_0 = 0$):

$$\Psi_1 - \left(\xi + y, \frac{\partial \Psi_1}{\partial \xi}\right) + \left(6\xi + 4y, \frac{\partial \Psi_1}{\partial y}\right) = 0.$$

Due to Proposition 3.7.1, Ψ_1 vanishes. Comparing the coefficients at t yields

$$\Psi_2 - \left(\xi + y, \frac{\partial \Psi_2}{\partial \xi}\right) + \left(6\xi + 4y, \frac{\partial \Psi_2}{\partial y}\right)$$

$$+ tr\left(\frac{1}{2}\frac{\partial^2 \Sigma_1}{\partial y^2} + g^{kl}(\xi_k - y_k^0)(\xi_l - y_l^0)\right) = 0. \tag{2.17}$$

It is again the equation of type (2.10) with the polynomials of degree 0,1,2 in the r.h.s. The solution of this equation is therefore given by Proposition 3.7.3. Again the calculations are rather long but the form of the solution is clear:

$$\Psi_2 = \sum_i g_{kl}^{ii} P_{kl} + g_{il}^{ik} Q_{kl} + G_{ii}^{kl} R_{kl}, \tag{2.18}$$

where P_{kl}, Q_{kl}, R_{kl} are some homogeneous polynomials in ξ, y, y^0 of degree 2.

3. The trace of the Green function and geometric invariants

It turns out that similarly to the case of non-degenerate diffusion on a compact manifold (see, e.g. [Gr],[Roe]), the resolving operator for the Cauchy problem for equation (1.12) belongs to the trace class, i.e. the trace

$$tr e^{-tL} = \int_{T^*M} u_G(t, x, y; x, y) \, dx dy \tag{3.1}$$

exists. Moreover, this integral can be developed in asymptotic power series in t with coefficients being the invariants of the Riemanian manifold. For brevity, let us put $h = 1$. The following result was announced in [AHK2] and its complete proof will be published elsewhere. We shall sketch here only the main line of necessary calculations using the technique developed in Section 3.7.

Theorem 3.1. *Integral (3.1) exists and has the asymptotical expansion for small time in the form*

$$(2\pi t^3)^{-n/2}(Vol\,M + a_3 t^3 + a_4 t^4 + ...),$$

the first nontrivial coefficient a_3 being proportional to the Gaussian curvature $G(M) = \int_M R\,dx$ of M and $Vol\,M = \int_M dx$ being the Riemanian volume.

Sketch of the Proof. The existence of the expansion follows from the asymptotic formula for the Green function obtained above. Let us show how to prove the last statement, indicating as well the main steps of the exact calculation of a_3. From (3.2.2),(3.2.4) it follows that

$$S(t, x_0, y^0; x_0, y^0) = \Sigma(t, \frac{x_0 - \tilde{x}}{t}, y^0 - \tilde{y}).$$

Therefore in normal coordinate around the point $x_0 = 0$ one has

$$S = \frac{1}{t}\left(6\frac{\tilde{x}^2}{t^2} - 6\frac{\tilde{x}}{t}(y^0 - \tilde{y}) + 2(y^0 - \tilde{y})^2\right) + t\Sigma_1\left(-\frac{\tilde{x}}{t}, y^0 - \tilde{y}\right) + O(t^2).$$

Using (2.3.5), (2.3.14) and expansion (2.2), (2.5),(2.6) let us make formulas (2.8) more precise:

$$\begin{cases} \tilde{x}^i = -ty_i^0 + \frac{1}{6}t^3(g_{ij}^{kl} - \frac{1}{2}g_{kl}^{ij})y_j^0 y_k^0 y_l^0 + O(t^4) \\ \tilde{y}_i = y_i^0 + \frac{1}{4}t^2 g_{kl}^{ij} y_j^0 y_k^0 y_l^0 + O(t^3). \end{cases} \tag{3.3}$$

Therefore

$$S = \frac{6}{t}\sum_i (y_i^0 - \frac{1}{6}t^2(g_{ij}^{kl} - \frac{1}{2}g_{kl}^{ij})y_j^0 y_k^0 y_l^0)^2 - \frac{3t}{2}y_i^0 g_{kl}^{ij} y_k^0 y_l^0 y_j^0 + t\Sigma_1(y^0, 0) + O(t^2).$$

Consequently,

$$S = \frac{6}{t}(y^0, y^0) - \frac{5}{2}t g_{ij}^{kl} y_i^0 y_j^0 y_k^0 y_l^0 + t\Sigma_1(y^0, 0) + O(t^2). \tag{3.4}$$

Therefore, to get the first nontrivial term of the expansion of S one needs the solution of (2.10) at $y = 0, \xi = y^0$.

Similarly, we have

$$\phi(t, 0, y^0; 0, y^0) = t^{-2n}\Psi(t, -\frac{\tilde{x}}{t}, y^0 - \tilde{y})$$

$$= t^{-2n}(1 + t^2\Psi_2(-\frac{\tilde{x}}{t}, y^0 - \tilde{y}) + O(t^3)) = t^{-2n}(1 + t^2\Psi_2(y^0, 0) + O(t^3)), \tag{3.5}$$

and therefore we need the solution of (2.17) also only at $y = 0, \xi = y^0$. From (2.14) and (2.18) it follows that

$$\Sigma_1(y^0, 0) = \sigma g_{ij}^{kl} y_i^0 y_j^0 y_k^0 y_l^0, \tag{3.6}$$

$$\Psi_2(y^0, 0) = \sum_k (\beta g_{kk}^{ij} + \gamma g_{ij}^{kk} + \delta g_{jk}^{ik}) y_i^0 y_j^0 \tag{3.7}$$

with some constants $\sigma, \beta, \gamma, \delta$.

The key point in the proof of the theorem is the following fact.

Lemma 3.1. *In formula (3.6), one has $\sigma = \frac{5}{2}$.*

Proof. To simplify calculations let us first note that formula (7.31) will not change if we take instead of the tensor g_{ij}^{kl} its symmetrisation, and therefore, when calculating $\Sigma_1(y^0, 0)$ from equation (7.18) we can consider the coefficients g_{ij}^{kl} in the expression for F to be completely symmetric (with respect to any change of the order of its indices i, j, k, l. In particular, it means that instead of F_2 and F_3 from (2.11), (2.12) we can take

$$\tilde{F}_2 = (6\xi_i\xi_j - 16\xi_i y_j - 11 y_i y_j)g_{ij}^{kl} y_k^0 y_l^0, \tag{3.8}$$

$$\tilde{F}_3 = (12\xi_i\xi_j\xi_k + 39\xi_i\xi_k y_j + 15\xi_i y_j y_k - 2 y_i y_j y_k)g_{ij}^{kl} y_l^0. \tag{3.9}$$

Next, clearly

$$\Sigma_1(y^0,0) = \frac{1}{2}\frac{\partial^2 \Sigma_1^2}{\partial \xi_i \partial \xi_j}y_0^i y_0^j + \frac{1}{3!}\frac{\partial^3 \Sigma_1^3}{\partial \xi_i \partial \xi_j \partial \xi_k}y_0^i y_0^j y_0^k + \frac{1}{4!}\frac{\partial^4 \Sigma_1^4}{\partial \xi_i \partial \xi_j \partial \xi_k \partial \xi_l}y_0^i y_0^j y_0^k y_0^l, \tag{3.10}$$

where Σ_1^p, $p = 2, 3, 4$, denote the corresponding homogeneous part of Σ_1. Now taking into consideration the assumed symmetricity of the coefficients of g_{ij}^{kl} one gets from (3.7.15) and (3.8) that

$$\frac{1}{2}\frac{\partial^2 \Sigma_1^2}{\partial \xi_i \partial \xi_j}(y^0,0) = \left(-\frac{6}{5}\times 11 + \frac{9}{10}\times 16 + \frac{4}{5}\times 6\right)g_{ij}^{kl} y_k^0 y_l^0, = 6g_{ij}^{kl} y_k^0 y_l^0,$$

from (3.7.16) and (3.9) that

$$\frac{1}{3!}\frac{\partial^3 \Sigma_1^3}{\partial \xi_i \partial \xi_j \partial \xi_k} = \left(\frac{54}{35}\times 2 + \frac{39}{35}\times 15 - \frac{61}{70}\times 39 + \frac{113}{140}\times 12\right)g_{ij}^{kl} y_l^0 = -\frac{9}{2}g_{ij}^{kl} y_l^0,$$

and from (3.7.17) and (2.13) that

$$\frac{1}{4!}\frac{\partial^4 \Sigma_1^4}{\partial \xi_i \partial \xi_j \partial \xi_k \partial \xi_l} = \left(-\frac{263}{315}\times 9 - \frac{51}{35}\times 2 + \frac{92}{105}\times 18 - \frac{38}{35}\times 4\right)g_{ij}^{kl} = g_{ij}^{kl}.$$

Subsituting these formulas to (3.10) yields

$$\Sigma_1(y^0,0) = (6 - \frac{9}{2} + 1)g_{ij}^{kl} y_i^0 y_j^0 y_k^0 y_l^0 = \frac{5}{2}g_{ij}^{kl} y_i^0 y_j^0 y_k^0 y_l^0,$$

and the Lemma is proved.

End of the proof of the Theorem. Due to the Lemma, the sum of the second and third terms in the expression (3.4) for S vanishes. Therefore, due to (3.4), (3.6), (3.7), and to the fact that the odd degrees of y^0 do not contribute to the integral, one concludes that the integral $\int u(t, 0, y; 0, y)\, dy$ is equal to

$$\left(\frac{\sqrt{3}}{\pi t^2}\right)^n \int e^{-6y^2/t}[1 + t^2 \sum_k(\beta g_{kk}^{ij} + \gamma g_{ij}^{kk} + \delta g_{jk}^{ik})y_i y_j + O(t^3|y|^6) + O(t^4|y|^4)]\, dy.$$

Due to (2.3), (2.4), this is equal to

$$= \left(\frac{\sqrt{3}}{\pi t^2}\right)^n \left(\frac{t\pi}{6}\right)^{n/2}[1 + \frac{1}{12}t^3 \delta R + O(t^4)].$$

Integrating this expression over M obviously gives (3.2) with $a_3 = \delta G(M)/12$.

Chapter 5. TRANSITION PROBABILITY DENSITIES
FOR STABLE JUMP-DIFFUSIONS

1. Asymptotic properties of one-dimensional stable laws

This chapter is devoted to a study of the transition probability densities for stable jump-diffusions and its natural modifications such as truncated stable jump-diffusions and stable-like diffusions. In the last section, some applications to the study of the sample path properties of these processes are presented.

In this introductory section we recall the well known asymptotical expansions of one-dimensional stable densities More circumstantial exposition of the theory of one-dimensional stable laws and their applications can be found e.g. in [Lu] or [Zo]. Let us comment only that the first term of the large distance asymptotics for stable laws seemed first to appear in [Pol], and the whole expansions was obtained in [Fel]. The characteristic function of the general (up to a shift) one-dimensional stable law with the index of stability $\alpha \in (0,2), \alpha \neq 1$, is

$$\exp\{-\sigma|y|^\alpha e^{i\frac{\pi}{2}\gamma \, sgn \, y}\} \tag{1.1}$$

(see e.g. Appendix C), where the parameter γ (which measures the skewness of the distribution) satisfies the conditions $|\gamma| \leq \alpha$, if $0 < \alpha < 1$, and $|\gamma| \leq 2 - \alpha$, if $1 < \alpha < 2$. Parameter $\sigma > 0$ is called the scale. For $\alpha = 1$ only in symmetric case, i.e. for $\gamma = 0$, the characteristic function can be written in form (1.1). In order to have unified formulas we exclude the non-symmetric stable laws with the index of stability $\alpha = 1$ from our exposition and will always consider $\gamma = 0$ whenever $\alpha = 1$. The probability density corresponding to characteristic function (1.1) is

$$S(x; \alpha, \gamma, \sigma) = \frac{1}{2\pi} \int_{-\infty}^{+\infty} \exp\{-ixy - \sigma|y|^\alpha e^{i\frac{\pi}{2}\gamma \, sgn \, y}\} \, dy. \tag{1.2}$$

Due to the evident relations

$$S(-x; \alpha, \gamma, \sigma) = S(x; \alpha, -\gamma, \sigma), \tag{1.3}$$

$$S(x; \alpha, \gamma, \sigma) = \sigma^{-1/\alpha} S(x\sigma^{-1/\alpha}; \alpha, \gamma, 1), \tag{1.4}$$

it is enough to investigate the properties of the normalised density $S(x; \alpha, \gamma, 1)$ for positive values of x. Clearly for these x

$$S(x; \alpha, \gamma, 1) = \frac{1}{\pi} Re \int_0^\infty \exp\{-ixy - y^\alpha e^{i\frac{\pi}{2}\gamma}\} \, dy. \tag{1.5}$$

It follows that all S are infinitely differentiable and bounded

$$|S(x; \alpha, \gamma, \sigma)| \leq \frac{1}{\alpha\pi}(\sigma \cos \frac{\pi}{2}\gamma)^{-1/\alpha}\Gamma(1/\alpha).$$

Using a linear change of the variable in (1.5) yields for $x > 0$

$$S(x; \alpha, \gamma, 1) = \frac{1}{\pi x} Re \int_0^\infty \exp\{-\frac{y^\alpha}{x^\alpha} e^{i\frac{\pi}{2}\gamma}\} e^{-iy} \, dy. \qquad (1.6)$$

Proposition 1.1. *For small $x > 0$ and any $\alpha \in (0, 2)$, the function $S(x; \alpha, \gamma, 1)$ has the following asymptotic expansion*

$$S(x; \alpha, \gamma, 1) \sim \frac{1}{\pi x} \sum_{k=1}^\infty \frac{\Gamma(1 + k/\alpha)}{k!} \sin \frac{k\pi(\gamma - \alpha)}{2\alpha} (-x)^k. \qquad (1.7)$$

Moreover, for $\alpha \in (1, 2)$ (resp. for $\alpha = 1$), the series on the r.h.s. of (1.7) is absolutely convergent for all x (resp. for x from a neighbourhood of the origin) and its sum is equal to $S(x; \alpha, \gamma, 1)$. The asymptotic expansion can be differentiated infinitely many times.

Proof. Expanding the function e^{-ixy} in (1.5) in the power series yields for $S(x; \alpha, \gamma)$ the expression

$$\frac{1}{\pi} Re \int_0^\infty \exp\{-y^\alpha e^{i\pi\gamma/2} \left(1 - ixy + \ldots + \frac{(-ixy)^k}{k!} + \theta \frac{(xy)^{k+1}}{(k+1)!}\right) dy$$

with $|\theta| \leq 1$. Since

$$\int_0^\infty y^{\beta-1} \exp\{-\lambda y^\alpha\} \, dy = \alpha^{-1} \lambda^{-\beta/\alpha} \Gamma(\beta/\alpha), \quad Re\,\lambda > 0$$

(and these integrals are absolutely convergent for $Re\,\lambda > 0$), one obtains

$$S(x; \alpha, \gamma, 1) = \frac{1}{\pi\alpha} Re \sum_{m=0}^k \exp\{-i\frac{\pi\gamma(m+1)}{2\alpha}\} \frac{(-ix)^m}{m!} \Gamma(\frac{m+1}{\alpha}) + R_{k+1}$$

with

$$|R_{k+1}| \leq \Gamma(\frac{k+2}{\alpha}) \frac{|x|^{k+1}}{(k+1)!}.$$

Therefore, we have got an asymptotic expansion for S. It is convenient to rewrite this expansion in the form

$$S(x; \alpha, \gamma, 1) \sim \frac{1}{\pi x \alpha} Re - \sum_{k=1}^\infty (-x)^k \frac{\Gamma(k/\alpha)}{(k-1)!} \exp\{-i\frac{\pi}{2}(\frac{\gamma}{\alpha} k - k + 1)\}.$$

Using the formula $\Gamma(k/\alpha) = \Gamma(1 + k/\alpha)\alpha/k$ and taking the real part yields (1.7). The statement about convergence follows from the asymptotic formula for Γ-function (Stirling formula), which implies that the radius of convergence of series (1.7) is equal to infinity, is finite, or is zero, respectively if $\alpha \in (1, 2)$, $\alpha = 1$, or $\alpha \in (0, 1)$.

We are going to discuss now the behaviour of stable densities for large x.

Proposition 1.2 (Zolotarev's identity). *If $x > 0$ and $\alpha \in (\frac{1}{2}, 1)$ or $\alpha \in (1, 2)$, then*

$$S(x; \alpha, \gamma, 1) = x^{-(1+\alpha)} S\left(x^{-\alpha}; \frac{1}{\alpha}, \frac{1}{\alpha}(\gamma + 1) - 1, 1\right). \qquad (1.8)$$

Proposition 1.3. *For any $\alpha \in (0, 2)$ and $x \to \infty$, the function $S(x; \alpha, \gamma)$ has the following asymptotic expansion:*

$$S(x; \alpha, \gamma, 1) \sim \frac{1}{\pi x} \sum_{k=1}^{\infty} \frac{\Gamma(1 + k\alpha)}{k!} \sin \frac{k\pi(\gamma - \alpha)}{2} (-x^{-\alpha})^k. \qquad (1.9)$$

Moreover, for $\alpha \in (0, 1)$ (resp. $\alpha = 1$, $\gamma = 0$), the series on the r.h.s. of (1.9) is absolutely convergent for all finite $x^{-\alpha}$ (resp. for $x^{-\alpha}$ in a neighbourhood of the origin) and its sum is equal to $S(x; \alpha, \gamma)$. Asymptotic expansion (1.9) can be differentiated infinitely many times.

Proof of Propositions 1.2, 1.3. First let $\alpha \in (0, 1]$. Due to the Cauchy theorem, one can change the path of integration in (1.6) to the negative imaginary axes, i.e.

$$S(x; \alpha, \gamma) = \frac{1}{\pi x} Re \int_0^{-i\infty} \exp\{-\frac{y^\alpha}{x^\alpha} e^{i\frac{\pi}{2}\gamma}\} e^{-iy} \, dy, \qquad (1.10)$$

because the magnitude of the integral along the arch $l = \{y = re^{-i\phi}, \phi \in [0, \frac{\pi}{2}]\}$ does not exceed

$$\int_0^{\pi/2} r \exp\{-r \sin \phi - \frac{r^\alpha}{x^\alpha} \cos(\alpha\phi - \frac{\pi}{2}\gamma)\} \, d\phi,$$

and tends to zero as $r \to \infty$, due to the assumptions on α and γ. Changing now the variable $y = ze^{-i\pi/2}$ in (1.10) yields

$$S(x; \alpha, \gamma) = Re - \frac{i}{\pi x} \int_0^\infty \exp\{-z - \frac{z^\alpha}{x^\alpha} e^{-i\frac{\pi}{2}(\gamma - \alpha)}\} \, dz.$$

Expanding $\exp\{-\frac{z^\alpha}{x^\alpha} e^{-i\frac{\pi}{2}(\gamma - \alpha)}\}$ in power series and evaluating the standard integrals one gets

$$S(x; \alpha, \gamma) = Re - \frac{i}{\pi x} \sum_{k=1}^{\infty} \frac{\Gamma(1 + k\alpha)}{k!} (-x^{-\alpha})^k \exp\{i\frac{k\pi}{2}(\alpha - \gamma)\},$$

which implies (1.9). As in the proof of Proposition 1.1, one sees from the asymptotic formula for Γ-function that the radius of convergence of series (1.9) is equal to infinity for $\alpha \in (0, 1)$ and is finite non-vanishing for $\alpha = 1$. Therefore, we have proved (1.9) for $\alpha \in (0, 1]$. Comparing formulas (1.9) for $\alpha \in (1/2, 1)$ and (1.7) for $\alpha \in (1, 2)$ one gets Zolotarev's identity (1.8). Using this identity and asymptotic expansion (1.7) for $\alpha \in (\frac{1}{2}, 1)$ one obtains asymptotic formula (1.9) for $\alpha \in (1, 2)$. Surely one can easily justify asymptotic expansion (1.9) for $\alpha \in (1, 2)$ independently from Zolotarev's identity, see proof of Proposition 2.2 below.

2. Asymptotic properties of finite dimensional stable laws

Here we generalise the results of the previous section to the case of finite dimensional symmetric stable densities, and then deduce some estimates for its derivatives, which will be used in the following sections.

Let us start with some bibliographical comments on the subject of this section. The results of Proposition 2.1 are rather trivial but I am not aware whether they appeared somewhere. The results of Proposition 2.2, 2.3 are partially known. Namely, the first term of the large distance asymptotic expansion of stable laws with the uniform spectral measure was obtained in general form in [BG], (see also a different proof in [Ben1]), though some particular cases were known in physics essentially earlier, see e.g. [Cha]. Some generalisations of these results to the infinite dimensional situation can be found in [Ben2]. On the other hand, the existence of an asymptotic expansion in powers of $|x|^{-1}$ was proved for more general Fourier integrals in [Fed2]. In our Propositons 2.2, 2.3, we present explicit formulas for asymptotic expansions of general finite dimensional stable laws, also taking care of the estimates of the remainder, which is of vital importance for our purposes. Further on we give the asymptotic expansions and global estimates for the derivatives of stable densities and for some relevant functions. Some estimates for these functions follow from more general estimates obtained in [Koch], but in [Koch] these functions are estimated in terms of some rational expressions, and our estimates are given in terms of the stable densities themselves, which bacomes possible when using the unimodality property of stable laws (see Proposition 2.4), which could not be used in a more general situation considered in [Koch].

The general symmetric stable density (up to a shift) has the form

$$S(x; \alpha, \sigma\mu) = \frac{1}{(2\pi)^d} \int_{\mathcal{R}^d} \exp\{-\sigma|p|^\alpha \int_{S^{d-1}} |(\bar{p}, s)|^\alpha \mu(ds)\} e^{-ipx} \, dp, \qquad (2.1)$$

where the measure μ on S^{d-1} is called the spectral measure, and where we have written explicitly a parameter σ, the scale (which is normally included in μ), having in mind the future applications to stable motions, where σ plays the role of the time.

We shall denote by $\bar{p}$ the unit vector in the direction of p, i.e. $\bar{p} = p/|p|$. Using for $\bar{p}$ spherical coordinates (θ, ϕ), $\theta \in [0, \pi]$, $\phi \in S^{d-2}$ with the main axis directed along x and then changing the variable θ to $t = \cos\theta$ yields

$$S(x; \alpha, \sigma\mu) = \frac{1}{(2\pi)^d} \int_0^\infty d|p| \int_{-1}^1 dt \int_{S^{d-1}} d\phi$$

$$\exp\{-\sigma|p|^\alpha \int_{S^{d-1}} |(\bar{p}, s)|^\alpha \mu(ds)\} \cos(|p||x|t)|p|^{d-1}(1 - t^2)^{(d-3)/2}. \qquad (2.2)$$

Changing the variable of integration $|p|$ to $y = |p||x|$ one can write it in the equivalent form

$$S(x; \alpha, \sigma\mu) = \frac{1}{(2\pi|x|)^d} \int_0^\infty dy \int_{-1}^1 dt \int_{S^{d-1}} d\phi$$

$$\exp\{-\sigma \frac{y^\alpha}{|x|^\alpha} \int_{S^{d-1}} |(\bar{p}, s)|^\alpha \mu(ds)\} \cos(yt)y^{d-1}(1 - t^2)^{(d-3)/2}. \qquad (2.3)$$

Proposition 2.1. *If*

$$C_1 \le \int_{S^{d-1}} |(\bar{p}, u)|^\alpha \mu(du) \le C_2 \qquad (2.4)$$

for all $\bar{p}$ and some positive constants $C_1 \le C_2$, then for small $|x|/\sigma^{1/\alpha}$ the density S has the asymptotic expansion

$$S(x; \alpha, \sigma\mu) \sim \frac{1}{(2\pi\sigma^{1/\alpha})^d} \sum_{k=0}^\infty \frac{(-1)^k}{(2k)!} a_k(\bar{x}) \left(\frac{|x|}{\sigma^{1/\alpha}}\right)^{2k} \qquad (2.5)$$

with

$$a_k(\bar{x}) = \int_0^\infty d|p| \int_{-1}^1 dt \int_{S^{d-1}} d\phi$$

$$\times \exp\{-|p|^\alpha \int_{S^{d-1}} |(\bar{p}, s)|^\alpha \mu(ds)\}|p|^{2k+d-1}(1 - t^2)^{(d-3)/2}. \qquad (2.6)$$

These coefficients satisfy the estimates

$$C_2^{-(d+2k)/\alpha} \le \frac{a_k(\bar{x})\alpha}{A_{d-2}\Gamma\left(\frac{2k+d}{\alpha}\right) B\left(k + \frac{1}{2}, \frac{d-1}{2}\right)} \le C_1^{-(d+2k)/\alpha}, \qquad (2.7)$$

where $A_0 = 2$, A_{d-2} for $d > 2$ denotes the area of the sphere S^{d-2}, and $\Gamma(p), B(p, q)$ denote the Euler Gamma and Beta functions respectively; in particular, if $C_1 = C_2 = 1$ (the case of the uniform spectral measure), the coefficients a_k do not depend on $\bar{x}$ and

$$a_k = \alpha^{-1} A_{d-2}\Gamma\left(\frac{2k + d}{\alpha}\right) B\left(k + \frac{1}{2}, \frac{d-1}{2}\right). \qquad (2.8)$$

The modulus of each term in expansion (2.5) serves also as an estimate of the remainder in this asymptotic representation, i.e. for each m, $S(x; \alpha, \sigma\mu)$ equals

$$\frac{1}{(2\pi\sigma^{1/\alpha})^d} \left(\sum_{k=0}^m \frac{(-1)^k}{(2k)!} a_k(\bar{x}) \left(\frac{|x|}{\sigma^{1/\alpha}}\right)^{2k} + \theta \frac{a_{m+1}(\bar{x})}{(2m + 1)!} \left(\frac{|x|}{\sigma^{1/\alpha}}\right)^{2m+1}\right) \qquad (2.9)$$

with $|\theta| \leq 1$. *Finally, if $\alpha > 1$ (resp. $\alpha = 1$), the series on the r.h.s. of (2.5) is absolutely convergent for all $|x|$ (resp. in a neighbourhood of the origin) and equals $S(x; \alpha, \sigma\mu)$.*

Proof. This is rather trivial and uses no new ideas as compared with the one-dimensional case. Let first $C_1 = C_2 = 1$. Expanding the function $\cos(|p||x|)$ in (1.2) in the power series and integrating in ϕ, yields

$$S(x; \alpha, \sigma) = \frac{A_{d-2}}{(2\pi)^d} \int_0^\infty d|p| \int_{-1}^1 dt \, \exp\{-\sigma|p|^\alpha\}|p|^{d-1}(1-t^2)^{(d-3)/2}$$

$$\left(\sum_{m=0}^k (-1)^m \frac{(|p||x|t)^{2m}}{(2m)!} + \theta \frac{(|p||x|t)^{2m+2}}{(2m+2)!} \right)$$

with $|\theta| \leq 1$. Since

$$\int_{-1}^1 t^{2m}(1-t^2)^{(d-3)/2} \, dt = B(m + \frac{1}{2}, \frac{d-1}{2}),$$

and

$$\int_0^\infty y^{\beta-1} \exp\{-\sigma y^\alpha\} \, dy = \alpha^{-1}\sigma^{-\beta/\alpha}\Gamma(\beta/\alpha), \quad Re\beta > 0,$$

one can integrate in $|p|$ and t to obtain for $S(x; \alpha, \sigma)$ the expression

$$\frac{A_{d-2}}{(2\pi)^d \alpha \sigma^{d/\alpha}} \left[\sum_{m=0}^k \frac{(-1)^m}{(2m)!} \left(\frac{|x|}{\sigma^{1/\alpha}} \right)^{2m} \Gamma\left(\frac{2m+d}{\alpha} \right) B\left(m + \frac{1}{2}, \frac{d-1}{2} \right) \right.$$

$$\left. + \frac{\theta}{(2m+1)!} \left(\frac{|x|}{\sigma^{1/\alpha}} \right)^{2m+1} \Gamma\left(\frac{2m+2+d}{\alpha} \right) B\left(m + \frac{3}{2}, \frac{d-1}{2} \right) \right]$$

with $|\theta| \leq 1$. Consequently one obtains the required expansion with a_k given in (2.8). The statement about the convergence of the series for $\alpha \geq 1$ follows from the Stirling formula for the Γ function and the well known expression of the function B in terms of Γ. The case of general μ is more or less the same: one expands $\cos(|p||x|t)$ in (2.2) in the power series and then changes the variable of integration $\sigma^{1/\alpha}|p|$ to $|p|$ in each term. Assumption (2.4) ensures firstly the existence of the integrals in (2.6) (in fact, only the left part of (2.4) is necessarily for that) and secondly it allows us to estimate $a_k(\bar{x})$ by means of corresponding coefficients (2.8).

We shall consider now two approaches to the construction of the asymptotic expansion of S in a more involved case, namely for large distances. The first of them, will be applied only to the case of the uniform spectral measure, but it gives explicit formulae for the coefficients in terms of special functions. To explain this method, let us recall first some facts on the Bessel and Whittaker functions (see [WW]). For any complex z that is not a negative real and any real

$n > 1/2$ the Bessel function $J_n(z)$ and the Whittaker function $W_{0,n}(z)$ can be defined by the integral formulae

$$J_n(z) = \frac{(z/2)^n}{\Gamma(n+1/2)\sqrt{\pi}} \int_{-1}^{1} (1-t^2)^{n-1/2} \cos(zt)\, dt,$$

$$W_{0,n}(z) = \frac{e^{-z/2}}{\Gamma(n+1/2)} \int_0^{\infty} [t(1+t/z)]^{n-1/2} e^{-t}\, dt,$$

where $arg\, z$ is understood to take its principle value, i.e. $|arg\, z| < \pi$. Furthermore, for these n and z these functions are connected by the formula

$$J_n(z) = \frac{1}{\sqrt{2\pi z}} [\exp\{\frac{1}{2}\left(n+\frac{1}{2}\right)\pi i\} W_{0,n}(2iz)$$

$$+ \exp\{-\frac{1}{2}\left(n+\frac{1}{2}\right)\pi i\} W_{0,n}(-2iz)],$$

which for real positive z implies

$$J_n(z) = 2Re[\frac{1}{\sqrt{2\pi z}} \exp\{\frac{1}{2}(n+\frac{1}{2})\pi i\} W_{0,n}(2iz)]. \tag{2.10}$$

If $n = m + 1/2$ with nonnegative integer m, then $W_{0,n}$ can be expressed in elementary functions

$$W_{0,n}(z) = e^{-z/2}\big(1 + \frac{n^2 - (1/2)^2}{z} + \frac{(n^2 - (1/2)^2)(n^2 - (3/2)^2)}{2z^2} + \dots$$

$$+ \frac{(n^2 - (1/2)^2)\dots(n^2 - (m-1/2)^2)}{m! z^m}\big)]. \tag{2.11}$$

In particular, $W_{0,1/2}(z) = e^{-z/2}$. More generally, for any $n > 1/2$ one has the following asymptotic expansion as $z \to \infty$, $|arg\, z| \le \pi - \epsilon$ with some $\epsilon > 0$:

$$W_{0,n}(z) \sim e^{-z/2} \left[1 + \frac{n^2 - (1/2)^2}{z} + \frac{(n^2 - (1/2)^2)(n^2 - (3/2)^2)}{2z^2} + \dots\right]. \tag{2.12}$$

Proposition 2.2. *Let the spectral measure μ of a stable law be uniform. If, in particular, (2.4) holds with $C_1 = C_2 = 1$, we shall denote $S(x; \alpha, \sigma\mu)$ by $S(x; \alpha, \sigma)$. In that case, for $|x|/\sigma^{1/\alpha} \to \infty$, one has the asymptotic expansion*

$$S(x; \alpha, \sigma) \sim \frac{1}{(2\pi|x|)^d} \sum_{k=1}^{\infty} \frac{a_k}{k!} (\sigma|x|^{-\alpha})^k \tag{2.13}$$

with

$$a_k = (-1)^{k+1} A_{d-2} 2^{-\alpha k-1} \sin(\frac{\pi}{2}k\alpha) \int_0^{\infty} \xi^{\alpha k+(d-1)/2} W_{0,\frac{d}{2}-1}(\xi)\, d\xi. \tag{2.14}$$

In particular, a_1 is positive for all d, and for odd dimensions $d = 2m+3$, $m \geq 0$,

$$a_k = (-1)^{k+1} A_{2m+1} \sin(\frac{\pi}{2} k\alpha) \Gamma(m + 2 + \alpha k)$$

$$\times \left(2^{m+1} + \frac{(m + \frac{1}{2})^2 - (\frac{1}{2})^2}{2(m + 1 + \alpha k)} 2^m + \frac{((m + \frac{1}{2})^2 - (\frac{1}{2})^2)((m + \frac{1}{2})^2 - (\frac{3}{2})^2)}{3!(m + \alpha k + 1)(m + \alpha k)} 2^{m-1}\right.$$

$$\left. + ... + \frac{((m + \frac{1}{2})^2 - (\frac{1}{2})^2)...((m + \frac{1}{2})^2 - (m - \frac{1}{2})^2)}{m!(m + \alpha k + 1)(m + \alpha k)...(2 + \alpha k)}\right). \tag{2.15}$$

Moreover, for $\alpha \in (0,1)$ (resp. $\alpha = 1$) this series is convergent for all $|x|^{-1}$ (resp. in a neighbourhood of the origin) and its sum is equal to $S(x; \alpha, \sigma)$. Furthermore, as in the case of the expansion of Proposition 2.1, each term in (2.13) serves also as an estimate for the remainder, in the sense that the difference between S and the sum of the $(k-1)$ terms of the expansion does not exceed in magnitude the magnitude of the k-th term.

Proof. Due to (2.3), (2.4) with $C_1 = C_2 = 1$ and the definition of the Bessel functions,

$$S(x; \alpha, \sigma) = \frac{A_{d-2}}{(2\pi|x|)^d} \int_0^\infty 2^{d/2-1} \Gamma(\frac{d-1}{2}) \sqrt{\pi} J_{\frac{d}{2}-1}(y) y^{d/2} \exp\{-\sigma \frac{y^\alpha}{|x|^\alpha}\} dy. \tag{2.16}$$

The key point in the proof is to use (2.10) and rewrite the last expression in the form

$$S(x; \alpha, \sigma) = \frac{A_{d-2}}{(2\pi|x|)^d} Re \int_0^\infty \Gamma(\frac{d-1}{2})$$

$$\times W_{0, \frac{d}{2}-1}(2iy)(2y)^{(d-1)/2} \exp\{-\sigma \frac{y^\alpha}{|x|^\alpha}\} e^{(d-1)\pi i/4} dy. \tag{2.17}$$

Suppose now that $\alpha \in (0, 1]$. From the asymptotic formula (2.9) it follows that one can justify the change of the variable of the path of integration in (2.17) to the negative imaginary half line. Taking this new path of integration and then changing the variable of integration $y = -i\xi$ yields

$$S(x; \alpha, \sigma) = \frac{A_{d-2}}{(2\pi|x|)^d} Re - i \int_0^\infty \Gamma(\frac{d-1}{2})$$

$$\times W_{0, \frac{d}{2}-1}(2\xi)(2\xi)^{(d-1)/2} \exp\{-\sigma \frac{\xi^\alpha}{|x|^\alpha} e^{-i\alpha\pi/2}\} d\xi. \tag{2.18}$$

Expanding the exponent under this integral in the power series and taking the real part yields (2.13),(2.14). Estimating coefficients (2.14) using the asymptotic formula (2.12) and the fact that $z^{m-1/2} W_{0,m}(z)$ is continuous for $z \geq 0$ (which follows from the definition of $W_{0,n}$ given above) one easily gets the convergence of the series (2.13) and the estimate $a_1 > 0$. In the case of odd dimensions one calculates coefficients (2.14) explicitly using (2.11).

Let $\alpha \in (1,2)$. In this case one cannot rotate the contour of integration in (2.17) through the whole angle $\pi/2$, but one can rotate it through the angle $\pi/(2\alpha)$. This amounts to the possibility of making the change of the variable in (2.17) $y = ze^{-i\pi/2\alpha}$ and then considering z to be again real, which gives

$$S(x;\alpha,\sigma) = \frac{A_{d-2}}{(2\pi|x|)^d} Re\Big[\int_0^\infty \Gamma(\frac{d-1}{2})W_{0,\frac{d}{2}-1}\Big(2z\exp\{\frac{i\pi(\alpha-1)}{2\alpha}\}\Big)$$

$$\times (2z)^{(d-1)/2}\exp\{i\sigma\frac{z^\alpha}{|x|^\alpha} + \frac{i}{4}(d-1)\pi - i\frac{\pi(d+1)}{4\alpha}\}\,dz\Big].$$

Using the Taylor formula for $\exp\{i\sigma\frac{z^\alpha}{|x|^\alpha}\}$ yields

$$S(x;\alpha,\sigma) = \frac{A_{d-2}}{(2\pi|x|)^d} Re\exp\{\frac{\pi i}{4\alpha}(\alpha(d-1) - (d+1))\}\Gamma\left(\frac{d-1}{2}\right)$$

$$\times \int_0^\infty (2z)^{\frac{d-1)}{2}}\Big[1 + \sum_{k=1}^m \frac{(i\sigma z^\alpha)^k}{x^{\alpha k}k!}$$

$$+\frac{\theta}{(m+1)!}\frac{(\sigma z^\alpha)^{m+1}}{x^{\alpha(m+1)}}\Big]W_{0,\frac{d}{2}-1}\Big[2z\exp\{\frac{i\pi(\alpha-1)}{2\alpha}\}\Big]\,dz$$

with $|\theta| \le 1$. It implies the asymptotic expansion (2.13) with

$$a_k = A_{d-2}Re\exp\{\frac{\pi i}{4\alpha}(\alpha(d-1) - (d+1))\}\Gamma\left(\frac{d-1}{2}\right)2^{(d-1)/2}i^k$$

$$\times \int_0^\infty z^{\alpha k+(d-1)/2}W_{0,\frac{d}{2}-1}\Big(2z\exp\{\frac{i\pi(\alpha-1)}{2\alpha}\}\Big)\,dz.$$

To simplify this expression, one makes here a new rotation of the path of integration, which amounts to the change of the variable $\xi = 2z\exp\{\frac{i\pi(\alpha-1)}{2\alpha}\}$ and again considering ξ to be real. After simple manipulations one obtains the same formula (2.14) as for the case $\alpha \in (0,1)$.

Consider now the general case.

Proposition 2.3. *Let the spectral measure μ of a stable law satisfy the r.h.s inequality in (2.4) and moreover, let μ has a smooth density with respect to Lebesgue measure. Then for large $|x|/\sigma^{1/\alpha}$ the density $S(x;\alpha,\sigma\mu)$ has an asymptotic expansion of type (2.13) with some $a_k = a_k(\bar{x})$ depending continuously on α, μ and $\bar{x}$ and with a_1 being positive.*

Proof. Let $\epsilon \in (0,1/2)$ and let $\chi(t)$ be a smooth even function $\mathcal{R} \to [0,1]$ that equals one (resp. zero) for $|t| \le 1 - 2\epsilon$ (resp. for $|t| \ge 1 - \epsilon$). Denote

$$g_\mu(t,\phi) = g_\mu(t,\phi;\alpha,\bar{x}) = \int_{S^{d-1}} |(\bar{p},s)|^\alpha \mu(ds).$$

Notice that g_μ depends on $\bar{x}$ because the choice of polar coordinates (t, ϕ) for $\bar{p}$ depends on $\bar{x}$. The existence of a smooth density for μ implies that g_μ is differentiable with respect to t. Let

$$f_1(t) = (1 - t^2)^{(d-3)/2}\chi(t), \quad f_2(t) = (1 - t^2)^{(d-3)/2}(1 - \chi(t)),$$

and let us present density (2.3) as the sum $S_1 + S_2$ with

$$S_j = \frac{1}{(2\pi|x|)^d} \int_0^\infty dy \int_{-1}^1 dt \int_{S^{d-1}} d\phi \exp\{-\sigma\frac{y^\alpha}{|x|^\alpha}g_\mu(t, \phi)\} \cos(yt)y^{d-1} f_j(t).$$

Expanding the exponent in the expression for S_1 in the power series leads straightforwardly to the asymptotic expansion

$$S_1 \sim \frac{1}{(2\pi|x|)^d} \sum_{k=0}^\infty \frac{1}{k!} b_k(\bar{x}) \left(\frac{\sigma}{|x|^\alpha}\right)^k, \tag{2.19}$$

where

$$b_k(\bar{x}) = (-1)^k \int_0^\infty F_k(y)y^{k\alpha+d-1}\, dy \tag{2.20}$$

and

$$F_k(y) = Re \int_{-\infty}^\infty \int_{S^{d-1}} e^{-iyt}f_1(t)g_\mu^k(t, \phi)\, dt d\phi.$$

Since $f_1(t) \int g_\mu^k(t, \phi)\, d\phi$ is a smooth function of t with a compact support, its Fourier transform F_k belongs to the Schwartz space on $\mathcal{R}$. Hence all coefficients (2.20) are well defined, and (2.19) presents an asymptotic expansion. More precisely, in order to be able to represent S_1 as the sum of k terms of this expansion with the estimate of the remainder of the form $O((\sigma/|x|^\alpha)^{m+1})$, it is sufficient to assume the existence of $l > k\alpha + d$ bounded derivatives of the density of the measure μ.

Next, in the expression for S_2 the variable t does not approach zero, and consequently, to expand S_2 one can use for each t, ϕ the method used for expanding one-dimensional densities. Consider, for instance , the case $\alpha \leq 1$. Clearly

$$S_2 = \frac{2}{(2\pi|x|)^d}Re \int_0^\infty dy \int_{1-2\epsilon}^1 dt \int_{S^{d-1}} d\phi \exp\{-\frac{\sigma y^\alpha}{|x|^\alpha}g_\mu(t, \phi)\}e^{-iyt}y^{d-1} f_2(t).$$

For any t, ϕ one can rotate the contour of integration in y to the negative imaginary axe. Changing then y to $y = -iz$ yields

$$S_2 = \frac{2}{(2\pi|x|)^d}Re \int_0^\infty dy \int_{1-2\epsilon}^1 dt \int_{S^{d-1}} d\phi$$

$$\times \exp\{-\frac{\sigma z^\alpha}{|x|^\alpha}g_\mu(t, \phi)e^{-i\alpha\pi/2}\}e^{-zt}(-iz)^{d-1} f_2(t).$$

Expanding the first exponent in the power series and taking standard integrals over z yields the asymptotic expansion

$$S_2 \sim \frac{1}{(2\pi|x|)^d} \sum_{k=0}^{\infty} \frac{1}{k!} c_k(\bar{x}) \left(\frac{\sigma}{|x|^\alpha}\right)^k \qquad (2.21)$$

with

$$c_k(\bar{x}) = 2Re \int_{1-2\epsilon}^1 dt \int_{S^{d-1}} d\phi (-g_\mu(t,\phi))^k (-i)^d f_2(t) e^{-i\pi\alpha k/2} t^{-(\alpha k+d)} \Gamma(\alpha k + d),$$
$$(2.22)$$

where again the modulus of each term serves also as an estimate to the remainder. The sum of expansions (2.19) and (2.21) gives the expansions for S. To prove the assertion it remains to show that the first coefficient $b_0 + c_0$ in this expansion vanishes. The simplest way to see it is to refer to Proposition 2.2. Namely, due to the construction, the first coefficient $b_0 + c_0$ does not depend on the spectral measure μ, and due to Proposition 2.2, it vanishes when the spectral measure is uniform. Hence, it vanishes for any μ. We can also prove this directly. Let us prove it, for example, for the case of an odd dimension $d = 2m + 1$. In that case, c_0 vanishes (because the integral in (2.22) is purely imaginary in this case) and we must show that

$$b_0 = \int_0^\infty F_0(y) y^{2m} \, dy$$

vanishes. But F_0 is the Fourier transform of the function $f_1(t) = (1-t^2)^{m-1}\chi(t)$. Hence, $b_0 = f_1^{(2m)}(0)$, which obviously vanishes. The case of $\alpha \geq 1$ is considered similarly, only one should rotate the contour of integration in the expression for S_2 through the angle $\pi/(2\alpha)$, as in the proof of Proposition 2.2.

Proposition 2.4. *For any $K > 1$ there exists $C > 1$ such that $C^{-1}|x|^{-d} \leq S(x; \alpha, \sigma\mu) \leq C|x|^{-d}$ whenever $K^{-1} \leq |x|/\sigma^{1/\alpha} \leq K$ uniformly for all spectral measures satisfying (2.4) and all α from any compact subinterval of the interval $(0, 2)$.*

Proof. Due to the small distance and large distance asymptotics and the property of unimodality of symmetric stable laws (see Appendix F), it follows that the stable densities are always (strictly) positive. On the other hand, it follows from (2.3) that $|x|^d S(x, \alpha, \sigma\mu)$ is a continuous function of $|x|/\sigma^{1/\alpha}$, g_μ, and $\bar{x}$. Since on any compact set it achieves its minimal and maximal values, which are both positive, the statement of the Proposition readily follows.

In the following sections, we shall need also estimates for the derivatives of stable densities with respect to x or σ, and also for more general relevant functions of the form

$$\phi_{b,\nu}(x; \alpha, \sigma\mu)$$

$$= \frac{1}{(2\pi)^d} \int_{\mathcal{R}^d} |p|^b g(\bar{p}, b, \nu) \exp\{-\sigma|p|^\alpha \int_{S^{d-1}} |(\bar{p}, s)|^\alpha \mu(ds)\} e^{-ipx} \, dp, \qquad (2.23)$$

where the real parameter b is supposed to be such that $b > -1$, $b \neq 0$, the measure ν on S^{d-1} may be not necessarily positive, but having finite $|\nu|$, and

$$g_\nu(\bar{p}; b) = \int_{S^{d-1}} |(\bar{p}, s)|^b \nu(ds).$$

The study of these functions can be carried out in the same way as that of $S(x; \alpha, \sigma\mu)$. For instance, the function $\phi_{b,\nu}$ can be presented in form (2.2) with the additional multiplier $|p|^b g_\nu(\bar{p}; b)$ under the integral, which yields for small $x\sigma^{-1/\alpha}$ (whenever inequalities (2.4) hold) the asymptotic representation

$$\phi_{b,\nu}(x; \alpha, \sigma\mu) \sim \frac{1}{(2\pi\sigma^{1/\alpha})^d \sigma^{b/\alpha}} \sum_{k=0}^{\infty} \frac{(-1)^k}{(2k)!} a_k(\bar{x}) \left(\frac{|x|}{\sigma^{1/\alpha}} \right)^{2k} \tag{2.24}$$

with a_k satisfying the estimates

$$a_k(\bar{x}) \leq \frac{A_{d-2}}{\alpha} \Gamma\left(\frac{2k+d+b}{\alpha} \right) B\left(k + \frac{1}{2}, \frac{d-1}{2} \right) C_1^{-(d+2k+b)/\alpha} \sup_{\bar{p}} |g(\bar{p}, b, \nu)|. \tag{2.25}$$

In particular, if $C_1 = C_2 = 1$ and $\nu = \mu$,

$$a_k = \alpha^{-1} A_{d-2} \Gamma\left(\frac{2k+d+b}{\alpha} \right) B\left(k + \frac{1}{2}, \frac{d-1}{2} \right).$$

As in the case of expansion of S, the modulus of each term in expansion (2.24) serves also as an estimate for the remainder.

Turning to the estimate of ϕ for large x, consider first the case when $g_\nu(\bar{p}; b) = g_\mu(\bar{p}; \alpha) = 1$. For such $\phi_{b,\nu}$, which we shall denote for brevity $\phi_b = \phi_b(x; \alpha, \sigma)$, one obtains representations (2.16),(2.17) with the additional multiplier $(y/|x|)^b$ under the integral, and representation (2.18) with the multiplier $(-i\xi/|x|)^b$ under the integral. Consequently, in that case, for large $|x|/\sigma^{1/\alpha}$, one obtains for ϕ_b the asymptotic expansion similar to (2.13), namely

$$\phi_b(x; \alpha, \sigma) \sim \frac{1}{(2\pi|x|)^d |x|^b} \sum_{k=0}^{\infty} \frac{a_k}{k!} (\sigma|x|^{-\alpha})^k \tag{2.26}$$

with

$$a_k = (-1)^{k+1} A_{d-2} 2^{-\alpha k - 1 - b} \sin(\frac{\pi}{2}(k\alpha + b)) \int_0^\infty \xi^{\alpha k + b + (d-1)/2} W_{0, \frac{d}{2}-1}(\xi) \, d\xi. \tag{2.27}$$

Notice that $a_0 \neq 0$ in the expansion for ϕ_b unlike the case of expansion (2.13). In particular, for all x

$$\phi_b(x; \alpha, \sigma) = -\frac{A_{d-2} \sin(\pi b/2)}{(2\pi)^d 2^{1+b} |x|^{d+b}} \int_0^\infty \xi^{b+(d-1)/2} W_{0, \frac{d}{2}-1}(\xi) \, d\xi (1 + \frac{\omega\sigma}{|x|^\alpha}) \tag{2.28}$$

with $\omega < |a_1|/|a_0|$. For general μ, ν one uses the approach from Proposition 2.3 to obtain the corresponding expansion for ϕ. All facts about ϕ that we shall need further are summarised in the following statement.

Proposition 2.5. *For any positive K there exists a positive C such that for all α from any compact subinterval of the interval $(0, 2)$, all b from a compact subset of the set $(-1, 0) \cup (0, \infty)$, all μ satisfying (2.4) and all uniformly bounded $|\nu|$ one has*

$$\phi_{b,\nu}(x; \alpha, \sigma\mu) \leq C\sigma^{-b/\alpha} S(x; \alpha, \sigma\mu) \tag{2.29}$$

or

$$\phi_{b,\nu}(x; \alpha, \sigma\mu) \leq C|x|^{\alpha-b}\sigma^{-1} S(x; \alpha, \sigma\mu) \tag{2.30}$$

respectively for $|x|\sigma^{-1/\alpha} \leq K$ or $|x|\sigma^{-1/\alpha} \geq K$. In particular,

$$\phi_{b,\nu}(x; \alpha, \sigma\mu) \leq C\sigma^{-1} S(x; \alpha, \sigma\mu),$$

if $\alpha \geq b$ and both $|x|$ and σ are bounded. Moreover, if additionally ν is non-negative and also satisfies (2.4), then

$$\phi_{b,\nu}(x, \alpha, \sigma\mu) = -\frac{1}{\sigma}S(x; b, \sigma\nu)(1 + O(\sigma|x|^{-\alpha}) + O(\sigma|x|^{-b})) \tag{2.31}$$

Furthermore,

$$\left|\frac{\partial\phi_{b,\nu}}{\partial b}(x; \alpha, \sigma\mu)\right| \leq C(1 + |\log\sigma|)\sigma^{-b/\alpha} S(x; \alpha, \sigma\mu) \tag{2.32}$$

or

$$\left|\frac{\partial\phi_{b,\nu}}{\partial b}(x; \alpha, \sigma\mu)\right| \leq C(1 + |\log|x|| + |\log\sigma|)|x|^{\alpha-b}\sigma^{-1} S(x; \alpha, \sigma\mu) \tag{2.33}$$

respectively for $|x|\sigma^{-1/\alpha} \leq K$ or $|x|\sigma^{-1/\alpha} \geq K$.

Proof. Comparing the asymptotic expansions of ϕ for small and large x with the corresponding expansions for S one obtains (2.29), (2.30) for small and large $|x|\sigma^{-1/\alpha}$. For finite $|x|\sigma^{-1/\alpha}$ these estimates are equivalent and they follow from Proposition 2.4. To estimate the derivative

$$\frac{\partial\phi_{b,\nu}(x; \alpha, \sigma\mu)}{\partial b} = \frac{1}{(2\pi)^d} \int_{\mathcal{R}^d} |p|^b \int_{S^{d-1}} |(\bar{p}, v)|^b (\log|p| + \log|(\bar{p}, v)|)\nu(dv)$$

$$\times \exp\{-\sigma|p|^\alpha \int_{S^{d-1}} |(\bar{p}, s)|^\alpha \mu(ds)\}e^{-ipx}\, dp, \tag{2.34}$$

or equivalently

$$\frac{\partial\phi_{b,\nu}(x; \alpha, \sigma\mu)}{\partial b} = \frac{1}{(2\pi)^d} \int_{\mathcal{R}^d} \frac{y^b}{|x|^b} \int_{S^{d-1}} |(\bar{p}, v)|^b (\log y - \log|x| + \log|(\bar{p}, v)|)\nu(dv)$$

$$\times \exp\{-\sigma \frac{y^\alpha}{|x|^\alpha} \int_{S^{d-1}} |(\bar{p}, s)|^\alpha \mu(ds)\} e^{-ipz}\, dp, \qquad (2.35)$$

one does in the same way using also the well known integral

$$\int_0^\infty x^{\beta-1} \log x \exp\{-\sigma x^\alpha\}\, dx = \alpha^{-2}\sigma^{-\beta/\alpha} \left[\Gamma'(\beta/\alpha) - \Gamma(\beta/\alpha)\log\sigma\right]$$

(which follows from the trivial formula $\int_0^\infty x^{\beta-1} \log x e^{-x}\, dx = \Gamma'(\beta)$). For example, for small $|x|\sigma^{-1/\alpha}$ one gets (from (2.34)) for $\frac{\partial \phi_{b,\nu}}{\partial b}$ a representation of type (2.2) with the additional multiplier $|p|^b g_\nu(\bar{p}; b)(\log|p| + \log|(\bar{p},\nu)|)$. The term $\log|(\bar{p},\nu)|$ is bounded from above and from below, and therefore the corresponding term is estimated in the same way as in the case of $\phi_{b,\nu}$. In its turn, the term $\log|p|$ will transform to $(\log|p| - \alpha^{-1}\log\sigma)$ after the change of the variable $\sigma^{1/\alpha}|p|$ to $|p|$, which gives the additional term with $\log\sigma$ in (2.32). For large $|x|\sigma^{-1/\alpha}$ one deals with representation (2.35) in the same way as in Proposition 3.2 to obtain (2.33).

To prove the important asymptotic equation (2.31) one must show that the major (non-vanishing) term of the expansion of the function $\phi_{b,\nu}(x;\alpha,\sigma\mu)$ as $|x|\sigma^{-1/\alpha} \to \infty$ coincides (up to the multiplier $-\sigma$) with the major term of the expansion of $S(x;b,\sigma\nu)$ as $|x|\sigma^{-1/b} \to \infty$. For the case of the uniform measures μ, ν, it follows from (2.28),(2.13), (2.14). In order to see this in the general case, one follows the line of the arguments of the proof of Proposition 2.3 and presents $\phi_{b,\nu}$ in the form $\psi_1 + \psi_2$ with

$$\psi_j = \frac{1}{(2\pi|x|)^d |x|^b} \int_0^\infty dy \int_{-1}^1 dt \int_{S^{d-1}} d\phi$$

$$\times \exp\{-\sigma \frac{y^\alpha}{|x|^\alpha} g_\mu(t,\phi)\} \cos(yt) y^{b+d-1} f_j(t) g_\nu(t,\phi),$$

where f_j are the same as in Proposition 2.3. The function ψ_2 is the rewritten in the form

$$\psi_2 = \frac{2}{(2\pi|x|)^d |x|^b} Re \int_0^\infty dy \int_{1-2\epsilon}^1 dt \int_{S^{d-1}} d\phi \exp\{-\frac{\sigma z^\alpha}{|x|^\alpha} g_\mu(t,\phi) e^{-i\alpha\pi/2}\}$$

$$\times e^{-zt}(-iz)^{b+d-1} f_2(t) g_\nu(t,\phi).$$

One sees now directly that the first terms (corresponding to $k = 1$) of the expansions of S_1 and S_2 from Proposition 2.3 coincide (up to the multiplier $-\sigma$) with the zero terms (corresponding to $k = 0$) of the corresponding expansions of ψ_1 and ϕ_2 respectively, and thus (2.31) follows.

Similarly one can estimate the partial derivatives $\frac{\partial S}{\partial x}$. Moreover, since $\phi_{\alpha,\mu}$ coincides with $-\frac{\partial S}{\partial \sigma}$ and

$$\frac{\partial S}{\partial \alpha}(x;\alpha,\sigma\mu) = -\sigma \left.\frac{\partial \phi_{b,\mu}(x;\alpha,\sigma\mu)}{\partial b}\right|_{b=\alpha}$$

one obtains the estimates for the derivative of S with respect to σ and α from Proposition 2.5. Thus one obtains the following result.

Proposition 2.6. *Suppose as usual that (2.4) holds. There exists a constant C such that for the derivative of S with respect to σ one has the asymptotic equation*

$$\frac{\partial S}{\partial \sigma}(x; \alpha, \sigma\mu) = \frac{1}{\sigma}S(x; \alpha, \sigma\mu)(1 + O(\sigma|x|^{-\alpha})). \tag{2.36}$$

and the global estimate

$$\left|\frac{\partial S}{\partial \sigma}(x; \alpha, \sigma\mu)\right| \leq \frac{C}{\sigma}S(x; \alpha, \sigma\mu). \tag{2.37}$$

Moreover, for the derivatives of S with respect to x and α one has the global estimates

$$\left|\frac{\partial S}{\partial x}(x; \alpha, \sigma\mu)\right| \leq C\min(\sigma^{-1/\alpha}, |x|^{-1})S(x; \alpha, \sigma\mu), \tag{2.38}$$

and

$$\left|\frac{\partial S}{\partial \alpha}(x; \alpha, \sigma\mu)\right| \leq C(1 + |\log \sigma| + \log\max(1, |x|))S(x; \alpha, \sigma\mu). \tag{2.39}$$

These estimates are uniform for α from any compact subinterval of the interval $(0, 2)$.

The results of Propositions 2.1, 2.3 can be easily generalised to the case of non-symmetric stable laws. On the other hand, the property of unimodality, which was crucial for Proposition 2.4 and the subsequent results, is not known for general stable laws. However, it is known for general one-dimensional stable laws, see Appendix F. Therefore, one obtains the following result.

Proposition 2.7. *The statements of Proposition 2.4 -2.6 hold also for general one-dimensional stable densities $S(x; \alpha, \gamma, \sigma)$ uniformly for $|\gamma| \leq \alpha - \epsilon$ (resp. $|\gamma| \leq 2 - \alpha - \epsilon$) with any positive ϵ whenever $\alpha < 1$ (resp. $\alpha > 1$).*

As a corollary of the formulas of this Section, let us present now some identities (which seem to be of independent interest) expressing the stable densities in odd dinensions in terms of the one-dimensional densities.

Proposition 2.8. *Let*

$$S_d(|x|; \alpha) = \frac{1}{(2\pi)^d}\int_{\mathcal{R}^d}\exp\{-i(p, x) - |p|^{-\alpha}\}\,dp$$

be the density of the d-dimensional stable law of the index α with the uniform spectral measure, which obviously depends on $|x|$ only. Let $S_d^{(k)}(|x|; \alpha)$ denote its k-th derivative with respect to $|x|$. Then

$$S_3(|x|; \alpha) = \frac{-1}{2\pi|x|}S_1'(|x|; \alpha),$$

$$S_5(|x|;\alpha) = \frac{A_3}{8\pi^4|x|^2}\left(S_1''(|x|;\alpha) - \frac{1}{4|x|}S_1'(|x|;\alpha)\right),$$

and in general for each positive integer m

$$S_{2m+3} = \frac{(-1)^{m+1}A_{2m+1}}{(2\pi)^{2m+2}}\Big[\frac{2^m}{|x|^{m+1}}S_1^{(m+1)} - \frac{2^{m-1}}{2!|x|^{m+2}}\left((m+\tfrac{1}{2})^2 - (\tfrac{1}{2})^2\right)S_1^{(m)}$$

$$+... + \frac{(-1)^m}{(m+1)!|x|^{2m+1}}\left((m+\tfrac{1}{2})^2 - (\tfrac{1}{2})^2\right)...\left((m+\tfrac{1}{2})^2 - (m-\tfrac{1}{2})^2\right)S_1'\Big].$$

Proof. To get this result, one can either compare the coefficients of the power expansions (for large $|x|$, if $\alpha \leq 1$, and for small $|x|$, if $\alpha \geq 1$) of the l.h.s. and the r.h.s. of these identities, or one can use formulas (2.17), (2.11) to express the l.h.s. as the one-dimensional Fourier transform of some elementary function and to compare this function with the corresponding function of the r.h.s.

3. Transition probability density for stable jump-diffusions

This section is devoted to the construction and asymptotic properties of the transition probability densities (Green functions) for stable processes with varying coefficients. In the case of $\alpha > 1$, the existence of the transition probability density for such a process was first proved in a more general framework in [Koch]. For general α it is proved in [Neg],[KN], but under the additional assumption that the coefficients (functions $A(x)$ and $\mu(x)$) are infinitely smooth. The arguments of paper [Koch] are based on the theory of hypersingular integrals developed essentially in [Sam1],[Sam2] (and based also on the regularisation procedure from [Las]), and the arguments of papers [Neg], [KN] are based on the classical theory of ΨDO with symbols $S_{\rho,\delta}^\sigma$ introduced by Hörmander, and its extension to the case of symbols with varying order. We prove the existence of this density by a different method. The main new result is to provide local multiplicative asymptotics and global in x both-sided estimates of this density for finite times, taking into account also the parameter h. For $h = 1$ these results were obtained in [K11]. We need nontrivial dependence on h in the next chapter, in order to be able to investigate the semiclassical asymptotics as $h \to 0$ of the Green functions constructed here. We use here standard notations of the theory of pseudo-differential equations that are collected in Appendix D.

Let us recall that the solution $u_G(t, x; x_0; h)$ to the Cauchy problem

$$h\frac{\partial u}{\partial t} = \Phi(x, -ih\nabla)u, \quad x \in \mathcal{R}^d, \quad t \geq 0, \tag{3.1}$$

$$u(x)|_{t=0} = \delta(x - x_0) \tag{3.2}$$

is called the Green function (or the fundamental solution) of equation (3.1). If the function Φ does not depend on x, the Green function (if it exists) can be written explicitly

$$u_G(t, x; x_0; h) = u_G(t, x - x_0; h) = (2\pi h)^{-d} \int_{\mathcal{R}^d} \exp\{\frac{\Phi(p)t + ip(x - x_0)}{h}\} \, dp,$$
(3.3)

because its h-Fourier transform obviously satisfies the equation $h(\partial \hat{u}/\partial t)(t, p) = \Phi(p)\hat{u}(t, p)$ with initial condition $\hat{u}(0, p) = (2\pi h)^{-d/2} \exp\{-ipx_0/h\}$.

To simplify the formulas, we restrict our consideration to the case of processes with the uniform spectral measure (but with varying scale), noting that due to the general formulas of the previous section, all results can be automatically generalised to the case of general one-dimensional stable diffusions, namely to the case when the skewness parameter γ depends nontrivially on x with the only restriction that if $\alpha < 1$ (resp. $\alpha > 1$), then $|\gamma(x)| \leq \alpha - \epsilon$ (resp. $|\gamma| \leq 2 - \alpha - \epsilon$) for all x and some positive ϵ, and to the case of general symmetric stable diffusions, having a varying spectral measure $\mu(x, ds)$ with the only restriction that it satisfies (2.4) uniformly for all x (see details concerning general spectral measures in [K11]).

As it follows from the theory of stochastic processes (see. e.g. [Ja] or Appendix C,D), the transition probability density for the stable processes with varying scale $G(x) > 0$ and shift $A(x)$ is the Green function u_G^{st} for equation (3.1) with the symbol

$$\Phi(x, p) = ipA(x) - G(x)\|p\|^\alpha.$$
(3.4)

Due to (3.3), if G and A are constants, the solution to (3.1), (3.2),(3.4) is equal to

$$u_\alpha(t, x - x_0; G, A, h) = (2\pi h)^{-d} \int_{\mathcal{R}^d} \exp\{\frac{-G\|p\|^\alpha t + ip(x + At - x_0)}{h}\} \, dp,$$
(3.5)

and therefore

$$u_\alpha(t, x - x_0, G, A, h) = u_\alpha(t, x + At - x_0, G, 0, h) = S(x_0 - At - x; \alpha, Gth^{\alpha-1}),$$
(3.6)

where the function $S(x; \alpha, \sigma)$ was introduced in Proposition 2.2. From (3.6) and the results of the previous sections we get directly the following estimates for u and its derivatives in t and x:

Proposition 3.1. *For any $B > 0$, $K > 0$, there exists a constant $C > 1$ such that for*

$$G \leq B, \quad G^{-1} \leq B, \quad \|A\| \leq B,$$
(3.7)

the following estimates hold uniformly for α from any compact subset of the interval $(0, 2)$:

(i) if $|x_0 - At - x| \leq K(th^{\alpha-1})^{1/\alpha}$, then

$$\frac{1}{C}(th^{\alpha-1})^{-d/\alpha} \leq u_\alpha(t, x - x_0; G, A, h) \leq C(th^{\alpha-1})^{-d/\alpha},$$
(3.8)

(ii) if $|x_0 - At - x| \geq K(th^{\alpha-1})^{1/\alpha}$, then

$$\frac{th^{\alpha-1}}{C|x_0 - At - x|^{d+\alpha}} \leq u_\alpha(t, x - x_0; G, A, h) \leq \frac{Cth^{\alpha-1}}{|x_0 - At - x|^{d+\alpha}}, \qquad (3.9)$$

(iii) for all t, x

$$\left|\frac{\partial u_\alpha}{\partial t}(t, x - x_0; G, A, h)\right| \leq \frac{C}{t} u_\alpha(t, x - x_0; G, A, h), \qquad (3.10)$$

$$\left|\frac{\partial u_\alpha}{\partial x}(t, x - x_0; G; A, h)\right| \leq \frac{C}{(th^{\alpha-1})^{1/\alpha}} u_\alpha(t, x - x_0; G, A, h), \qquad (3.11)$$

$$\left|\frac{\partial u_\alpha}{\partial x}(t, x - x_0, G; A, h)\right| \leq \frac{C}{|x_0 - At - x|} u_\alpha(t, x - x_0; G, A, h). \qquad (3.12)$$

Corollary. *If $\alpha \geq 1$ (resp. $\alpha \leq 1$), for any pairs $(G_1, A_1), (G_2, A_2)$ satisfying (3.7)*

$$u_\alpha(t, x - x_0; G_1, A_1, h) \leq C u_\alpha(t, x - x_0; G_2, A_2, h) \qquad (3.13)$$

with some constant C and all sufficiently small t/h (resp. h/t). If $A_1 = A_2$, the same holds for all h, t uniformly for α from any compact subset of the open interval $(0, 2)$.

Naturally, one expects that for small times the Green function of equation (3.1) with varying coefficients can be approximated by the Green function (3.6) of the corresponding problem with constant coefficients, i.e. by the function $u_\alpha(t, x - x_0; G(x_0), A(x_0), h)$. This is in fact true. To prove this we first prepare some estimates of the convolutions of u_α with itself and relevant functions.

Remark. In future we shall often omit for brevity some of the last arguments in the notation for u_α, when it will not lead to ambiguity.

Lemma 3.1. *Let (3.7) holds. (i) If $\alpha > 1$, then uniformly for $\delta \in [0, 1]$, $t \leq h \leq h_0$ with any h_0 and x_0*

$$\int \min(\delta, |\eta|) u_\alpha(t, \eta; G(x_0), A(x_0), h) \, d\eta = O((th^{\alpha-1})^{1/\alpha}), \qquad (3.14)$$

if $A \equiv 0$, this holds uniformly for $t \leq t_0, h \leq h_0, \delta \in [0, 1]$;
(ii) if $\alpha \leq 1$, $A \equiv 0$, then

$$\int \min(\delta, |\eta|) u_\alpha(t, \eta; G(x_0), A(x_0), h) \, d\eta = O(th^{\alpha-1}) \qquad (3.14')$$

uniformly for $\delta \in (0, 1)$, $t \leq h \leq h_0$ with any h_0.

Proof. We shall consider $A \equiv 0$ for brevity, because one sees from the proof, that if $\alpha > 1$ and $t \leq h$, the presence of A does not change anything. We decompose our integral in the sum $I_1 + I_2$ corresponding to the decomposition of the domain of integration in the union $D_1 \cup D_2$ with

$$D_1 = \{\eta : |\eta| \le (th^{\alpha-1})^{1/\alpha}\}, \quad D_2 = \{\eta : |\eta| \ge (th^{\alpha-1})^{1/\alpha}\}.$$

Then, due to (3.8), $I_1 = O(1)(th^{\alpha-1})^{1/\alpha}$, which is plainly of the form $O(th^{\alpha-1})$ for $\alpha \le 1$ and $t \le h$.

Next, if $\alpha > 1$,

$$I_2 = O(t)h^{\alpha-1} \int_{D_2} \frac{d\eta}{|\eta|^{\alpha+d-1}} = O(t)h^{\alpha-1} \int_{(th^{\alpha-1})^{1/\alpha}}^{\infty} |\eta|^{-\alpha}\, d\eta = O(1)(th^{\alpha-1})^{1/\alpha}.$$

If $\alpha \le 1$, we decompose $I_2 = I_2' + I_2''$ decomposing D_2 in the union $D_2' \cup D_2''$ with

$$D_2' = \{\eta : (th^{\alpha-1})^{1/\alpha} \le |\eta| \le 1\}, \quad D_2'' = \{\eta : |\eta| \ge 1\}.$$

We have

$$I_2' = O(t)h^{\alpha-1} \int_{D_2'} \frac{d\eta}{|\eta|^{\alpha+d-1}} = O(t)h^{\alpha-1} \int_0^1 |\eta|^{-\alpha}\, d|\eta| = O(t)h^{\alpha-1},$$

$$I_2'' = O(t)h^{\alpha-1} \int_{D_2''} \frac{d\eta}{|\eta|^{\alpha+d}} = O(t)h^{\alpha-1} \int_1^{\infty} |\eta|^{-1-\alpha} d|\eta| = O(t)h^{\alpha-1},$$

which completes the proof.

Proposition 3.2. *Under the same assumptions as in Lemma 3.1, one has*

$$\int_{\mathcal{R}^d} u_\alpha(t-\tau, x-\eta; G(\eta), A(\eta)) \min(\delta, |\eta - x_0|)\tau^{-1} u_\alpha(\tau, \eta - x_0; G(x_0), A(x_0))\, d\eta$$

$$\le b(t^{-1}\min(\delta, |x - x_0|) + \Delta(\tau, h))u_\alpha(t, x - x_0; G(x_0), A(x_0)) \qquad (3.15)$$

with some b, where

$$\Delta(t, h) = \begin{cases} (t/h)^{\alpha^{-1}-1}, & \alpha > 1, \\ h^{\alpha-1}, & \alpha \le 1. \end{cases} \qquad (3.16)$$

Proof. Again consider $A = 0$ for brevity. Consider separately two domains of the values of x:

$$M_1 = \{x : |x - x_0| \ge (th^{\alpha-1})^{1/\alpha}\}, \quad M_2 = \{x : |x - x_0| \le (th^{\alpha-1})^{1/\alpha}\}.$$

First let $x \in M_1$. Then

$$|x - x_0| \ge \frac{1}{2}((t-\tau)h^{\alpha-1})^{1/\alpha} + \frac{1}{2}(\tau h^{\alpha-1})^{1/\alpha}.$$

We decompose the integral on the l.h.s. of (3.15) into the sum $I_1 + I_2$ corresponding to the partition of the domain of integration in the union $D_1 \cup D_2$ with

$$D_1 = \{|x - \eta| \geq \frac{1}{2}|x - x_0|\}, \quad D_2 = \{|x - \eta| \leq \frac{1}{2}|x - x_0|\}.$$

In D_1

$$|x - \eta| \geq \frac{1}{2}|x - x_0| \geq \frac{1}{4}((t - \tau)h^{\alpha-1})^{1/\alpha},$$

and consequently

$$u_\alpha(t - \tau, x - \eta) = \frac{O(t - \tau)h^{\alpha-1}}{|x - \eta|^{d+\alpha}} = \frac{O(t - \tau)h^{\alpha-1}}{|x - x_0|^{d+\alpha}} = \frac{O(t - \tau)}{t} u_\alpha(t, x - x_0; G(x_0)).$$

Hence

$$I_1 = \frac{O(t - \tau)}{t} u_\alpha(t, x - x_0; G(x_0)) \int \frac{\min(\delta, |\eta - x_0|)}{\tau} u_\alpha(\tau, \eta - x_0; G(x_0)) \, d\eta, \tag{3.17}$$

and due to Lemma 3.1, one has

$$I_1 = \frac{O(t - \tau)}{t} u_\alpha(t, x - x_0, G(x_0)) \Delta(\tau, h).$$

Next, in D_2,

$$\frac{3}{2}|x - x_0| \geq |\eta - x_0| \geq \frac{1}{2}|x - x_0| \geq \frac{1}{4}(\tau h^{\alpha-1})^{1/\alpha},$$

and consequently there

$$\min(\delta, |\eta - x_0|)u_\alpha(\tau, \eta - x_0; G(x_0)) = O(1)\min(\delta, |x - x_0|)\frac{\tau}{t} u_\alpha(t, x - x_0; G(x_0)).$$

Hence

$$I_2 = O(t^{-1})\min(1, |x - x_0|)u_\alpha(t, x - x_0; G(x_0)).$$

Thus the required estimates are proved for $x \in M_1$. Next, let $x \in M_2$. Then $u_\alpha(t, x - x_0)$ is of the order $(th^{\alpha-1})^{-d/\alpha}$. If $t - \tau > \tau$, one can estimate $u_\alpha(t - \tau, x - \eta)$ by $((t - \tau)h^{\alpha-1})^{-d/\alpha}$, which is of the order $(th^{\alpha-1})^{-d/\alpha}$. Consequently, the integral on the l.h.s. of (3.15) can be estimated by

$$O(1)u_\alpha(t, x - x_0; G(x_0)) \int \frac{\min(\delta, |\eta - x_0|)}{\tau} u_\alpha(\tau, \eta - x_0; G(x_0)) \, d\eta,$$

which is again of the order $\Delta(\tau, h)u_\alpha(t, x - x_0; G(x_0))$. If $\tau > t - \tau$, one decomposes the integral in the sum $I_1 + I_2$ by making the partition of the domain of integration in the union $D_1 \cup D_2$ with

$$D_1 = \{|\eta - x_0| \leq 2(th^{\alpha-1})^{1/\alpha}\}, \quad D_2 = \{|\eta - x_0| \geq 2(th^{\alpha-1})^{1/\alpha}\}.$$

In D_1 one estimates $u_\alpha(\tau, \eta - x_0)$ by $(th^{\alpha-1})^{1/\alpha}$, and obtains for I_1 the estimate $\Delta(t,h)u_\alpha$. In D_2 one has plainly that $|\eta - x| \geq (th^{\alpha-1})^{1/\alpha}$, and therefore

$$u_\alpha(t - \tau, x - \eta) = \frac{O(t-\tau)}{t} th^{\alpha-1}(th^{\alpha-1})^{-(d+\alpha)/\alpha} = \frac{O(t-\tau)}{t} u_\alpha(t, x - x_0),$$

and one gets for I_2 the estimate (3.17). The proof is complete.

Proposition 3.3. *Under the assumptions of Lemma 3.1*

$$\int \min(\delta, |x - \eta|)u_\alpha(t - \tau, x - \eta; G(\eta), A(\eta))$$

$$\times (\tau)^{-1} \min(\delta, |\eta - x_0|)u_\alpha(\tau, \eta - x_0; G(x_0), A(x_0))\, d\eta$$

$$= O(1)\Delta(\tau, h)u_\alpha(t, x - x_0; G(x_0), A(x_0)) \tag{3.18}$$

uniformly for $\delta \in [0, 1]$, where $\Delta(\tau, h)$ is from (3.16).

Proof. It is quite similar to the proof of the previous Proposition. Consider again $A = 0$ and set M_1, M_2 as in the previous Proposition. In M_1 one makes the same decomposition in domains D_1 and D_2. In D_1

$$\min(\delta, |x - \eta|)u_\alpha(t - \tau, x - \eta; G(\eta)) = O(\delta)u_\alpha(t, x - x_0; G(x_0))(t - \tau)/t,$$

and consequently the integral on the l.h.s. of (3.18) has the form

$$O(1)u_\alpha(t, x - x_0; G(x_0))\frac{t - \tau}{t} \int \frac{\min(\delta, |\eta - x_0|)}{\tau} u_\alpha(t, \eta - x_0; G(x_0))$$

$$= O(1)\Delta(\tau, h)u_\alpha(t, x - x_0; G(x_0)).$$

In D_2 one estimates the integral by

$$O(1)\min(\delta, |x - x_0|)u_\alpha(t, x - x_0; G(x_0))\frac{1}{t} \int \min(\delta, |x - \eta|)u_\alpha(t, x - \eta; G(\eta)),$$

which gives nothing new. Similarly one analyses the case with $x \in M_2$, where we omit the details.

We shall need also to carry out the convolution of u_α with itself. If $A \equiv 0$ and $G(x)$ satisfies (3.7), or if A, G satisfy (3.7) and $t \leq h$, then plainly

$$\int_{\mathcal{R}^d} u_\alpha(t - \tau, x - \eta; G(\eta), A(\eta))u_\alpha(\tau, \eta - x_0; G(x_0), A(x_0))\, d\eta$$

$$\leq bu_\alpha(t, x - x_0; G(x_0), A(x_0)) \tag{3.19}$$

with some constant b, because of (3.13) and the semigroup identity

$$\int_{\mathcal{R}^d} u_\alpha(t - \tau, x - \eta; G(x_0), A(x_0))u_\alpha(\tau, \eta - x_0; G(x_0), A(x_0))\, d\eta$$

$$= u_\alpha(t, x - x_0; G(x_0), A(x_0))$$

for the Green function (3.5) of equation (3.1),(3.4) with constant coefficients. It can be shown also directly, in the same way as above (see e.g. Lemma 5.1 below). We shall show now that this estimate can be expanded beyond the restriction $t \le h$ even for non-vanishing A. This result can be used for the corresponding generalisations of the main Theorem 3.1 below.

Proposition 3.4. *Let $A(x)$ and $G(x)$ satisfy the assumptions (3.7) for all x and let $A(x)$ have a uniformly bounded derivative. If $\alpha > 1$, then (3.19) holds for all x, x_0 and small $t \le h^{(\alpha-1)/(2\alpha-1)}$.*

Proof. Due to the assumptions, one has

$$(t - \tau)^2 \le ((t - \tau)h^{\alpha-1})^{1/\alpha}$$

for all $\tau \in (0, t]$. To prove (3.19) we decompose the integral in the l.h.s. of (3.19) in the sum $I_1 + I_2$ of the integrals corresponding to the decomposition of the domain of integration into two sets

$$M_1 = \{\eta : |\eta - x| \le 2B(t - \tau)\}, \quad M_2 = \{\eta : |\eta - x| \ge 2B(t - \tau)\},$$

where the constant B is from estimate (3.7). In M_1

$$\eta - A(\eta)(t - \tau) - x = \eta - A(x)(t - \tau) - x + O(t - \tau)^2.$$

Therefore, making the shift of the variable of integration $\eta \mapsto \xi = \eta + A(x_0)\tau$ in the integral I_1 and using the semigroup identity yields the estimate

$$I_1 \le u_\alpha(t, x - x_0 + A(x_0)t + (A(x) - A(x_0))(t - \tau) + O(t - \tau)^2).$$

If $|x-x_0| \ge K(t-\tau)$ with large enough K, then $(A(x)-A(x_0))(t-\tau)+O(t-\tau)^2 < |x - x_0|/2$ and the last expression can be estimated by $O(u_\alpha(t, x - x_0)$ due to Proposition 3.1. On the other hand, if $|x-x_0| = O(t-\tau)$, then the last expression is of the form

$$u_\alpha(t, x + O(t - \tau)^2, x_0; G(x_0), A(x_0))$$
$$= u_\alpha(t, x + O((t - \tau)h^{\alpha-1})^{1/\alpha}, x_0; G(x_0), A(x_0)),$$

which again can be estimated by $u_\alpha(t, x - x_0)$ due to (3.8),(3.9). In order to estimate I_2, one notes that in M_2 the magnitudes $|\eta - x|$, $|\eta - A(\eta)(t - \tau) - x|$, and $|\eta - A(x_0)(t - \tau) - x|$ are of the same order, and therefore in M_2

$$u_\alpha(t - \tau, x - \eta; G(\eta), A(\eta), h) \le C^2 u_\alpha(t - \tau, x - \eta; G(x_0), A(x_0), h),$$

which completes the proof of the Proposition.

Now we can state the main result of this section.

Theorem 3.1. *Let $G(x), A(x)$ be functions on $\mathcal{R}^d$ such that $G^{-1}(x)$ is uniformly bounded and $G, A(x)$ have uniformly bounded derivatives up to and including the order $q, q \ge 2$. Suppose also that $A(x) \equiv 0$ if $\alpha \le 1$. Then the*

Green function $u_G^{st}(t, x, x_0, h)$ for equation (3.1),(3.4) exists, is continuous and differentiable in t for $t > 0$. For arbitrary $h_0 > 0$ and $t_0 > 0$ the following representation holds uniformly for $t \leq t_0$, $t \leq h \leq h_0$:

$$u_G^{st}(t, x, x_0, h) = u_\alpha(t, x - x_0; G(x_0), A(x_0), h)$$

$$\times (1 + O(1) \min(1, |x - x_0|) + O(t\Delta(t, h))), \tag{3.20}$$

and

$$\frac{\partial u_G^{st}}{\partial t}(t, x, x_0, h) = \frac{\partial u_\alpha}{\partial t}(t, x - x_0)$$

$$+ O(t^{-1}) u_\alpha(t, x - x_0)(\min(1, |x - x_0|) + O(t\Delta(t, h)), \tag{3.21}$$

where $\Delta(t, h)$ is given in (3.16) and u_α is defined in (3.5),(3.6). If $\alpha > 1$ (respectively $\alpha \leq 1$, u_G^{st} has continuous derivatives of all orders $l \leq q$ (respectively $l \leq q - 1$), and for these derivatives the following representations hold:

$$\frac{\partial^l u_G^{st}}{\partial x^l}(t, x, x_0, h) = \frac{\partial^l u_\alpha}{\partial x^l}(t, x - x_0)$$

$$+ O((th^{\alpha-1})^{-l/\alpha}) u_\alpha(t, x - x_0)(\min(1, |x - x_0|) + t\Delta(t, h)). \tag{3.22}$$

At last, if $\alpha > 1$ and $A \equiv 0$, then all this estimates holds for small enough t without the restriction $t \leq h$.

Remark 1. Notice that in case $\alpha \geq 1$ (resp. $\alpha < 1$ and $A \equiv 0$), a sufficient condition for the right hand side of (1.8) to be well defined is the existence of the second (resp. only the first) derivative of the function u. This is the reason for assuming $q \geq 2$ in Theorem 3.1. Under weaker assumptions, for example if A, μ are only Hölder continuous, one can still prove the convergence of series (3.20) below defining the Green function. But in that case one faces a quite non-trivial problem to define rigorously, in what (generalised) sense equation (1.8) is actually satisfied by the corresponding function u_G^{st}. This can be done (see e.g. [Koch] for the case $\alpha \geq 1$), but we are not going to discuss this problem here. Notice also that since $\partial u_\alpha/\partial t = O(t^{-1}) u_\alpha$, it follows that the second term in (3.21) is actually smaller than the principle term. The same remark concerns formula (3.22).

Remark 2. If t/h is not small, then more rough estimates may be available, which one can obtain using Proposition 3.4.

Proof. The method is based on the Du Hamel formula as in Sect 3.4. The function u_α given by (3.14), (3.5) satisfies the equation (as a function of (t, x))

$$\frac{\partial u}{\partial t}(t, x, x_0, h) = \frac{1}{h}\Phi(x, -ih\nabla)u - F(t, x, x_0, h) \tag{3.23}$$

with

$$F(t, x, x_0, h) = \frac{1}{h}(\Phi(x, -ih\nabla) - \Phi(x_0, -ih\nabla))u_\alpha(t, x, x_0; G(x_0), A(x_0), h)$$

$$= \frac{A(x) - A(x_0)}{(2\pi h)^d} \int_{\mathcal{R}^d} \frac{ip}{h} \exp\{\frac{-G(x_0)\|p\|^\alpha t + ip(x + A(x_0)t - x_0)}{h}\} \, dp$$

$$- \frac{G(x) - G(x_0)}{(2\pi h)^d} \int_{\mathcal{R}^d} \frac{\|p\|^\alpha}{h} \exp\{\frac{-G(x_0)\|p\|^\alpha t + ip(x + A(x_0)t - x_0)}{h}\} \, dp.$$

Equivalently,

$$F(t, x, x_0, h) = (A(x) - A(x_0))\frac{\partial u_\alpha}{\partial x}(t, x - x_0) + \frac{G(x) - G(x_0)}{G(x_0)}\frac{\partial u_\alpha}{\partial t}(t, x - x_0),$$
(3.24)

where by the assumptions of the Theorem, the first term is not vanishing only for $\alpha > 1$. Therefore, due to the Du Hamel principle, if u_G^{st} is the Green function for equation (3.1), (3.4), then

$$u_\alpha(t, x - x_0) = (u_G^{st} - \mathcal{F}u_G^{st})(t, x, x_0, h),$$

where the integral operator $\mathcal{F}$ is defined, as in Sect. 3.4, by the formula

$$(\mathcal{F}\phi)(t, x, \xi) = \int_0^t \int_{\mathcal{R}^d} \phi(t - \tau, x, \eta) F(\tau, \eta, \xi) \, d\eta d\tau \equiv \phi \otimes F. \qquad (3.25)$$

We use, as in the proof of Theorem 3.4.1, the special symbol $\phi \otimes F$ for the (convolution-type) integral in (3.25). Therefore,

$$u_G^{st} = (1 - \mathcal{F})^{-1}u_\alpha = (1 + \mathcal{F} + \mathcal{F}^2 + ...)u_\alpha, \quad \mathcal{F}^k u_\alpha = u_\alpha \otimes F \otimes ... \otimes F. \quad (3.26)$$

Consequently, one needs to prove the convergence of this series and the required estimate for its sum and its derivative in t. In fact, though in previous arguments we presupposed the existence of the Green function u_G, one verifies directly that the sum (3.26) satisfies equation (3.1) (whenever it converges together with its derivative).

From (3.10), (3.12) and the assumptions on G and A, it follows from (3.24) that there exists $a > 0$ such that

$$|F(t, x, x_0, h)| \leq a(u_\alpha(t, x - x_0) + \tilde{u}(t, x - x_0)), \qquad (3.27)$$

where

$$\tilde{u}(t, x - x_0) = t^{-1} \min(1, |x - x_0|)]u_\alpha(t, x - x_0).$$

To make the following formulas shorter it is convenient to introduce an additional notation. Namely, for any functions $u(t, x, x_0)$, $v(t, x, x_0)$ on $\mathcal{R}_+ \times \mathcal{R}^d \times \mathcal{R}^d$ let

$$(u \circ v)(t, \tau, x, x_0) = \int_{\mathcal{R}^d} u(t - \tau, x, \eta)v(\tau, \eta, x_0) \, d\eta$$

(if this integral exists, of course). Obviously, the operation $\otimes$ introduced earlier is connected with the operation $\circ$ by the formula

$$(u \otimes v)(t, x, x_0) = \int_0^t (u \circ v)(t, \tau, x, x_0) \, d\tau,$$

and moreover, formulas (3.15), (3.18), (3.19) can be rewritten now in the following concise form:

$$(u_\alpha \circ \tilde{u})(t, \tau, x, x_0) \leq c(\Delta(\tau, h)u_\alpha(t, x, x_0) + \tilde{u}(t, x, x_0)), \tag{3.15'}$$

$$((t\tilde{u}) \circ \tilde{u})(t, \tau, x, x_0) \leq c\Delta(\tau, h)u_\alpha(t, x, x_0), \tag{3.18'}$$

$$(u_\alpha \circ u_\alpha)(t, \tau, x, x_0) \leq cu_\alpha(t, x, x_0) \tag{3.19'}$$

with some constant $c > 0$ and all $0 < \tau < t \leq t_0$ with an arbitrary fixed t_0. For the function $v = u_\alpha + \tilde{u}$ it follows directly that

$$(u_\alpha \circ v)(t, \tau, x, x_0) \leq c(\Delta(\tau, h)u_\alpha(t, x, x_0) + v(t, x, x_0))$$

and therefore (taking into account that $t\tilde{u} \leq u_\alpha$) also that

$$((tv) \circ v)(t, \tau, x, x_0) \leq c(\Delta(\tau, h)u_\alpha(t, x, x_0) + tv(t, x, x_0))$$

again with some constant $c > 0$ (perhaps different from the previous one). Using these inequalities we are going now to prove the convergence of series (3.20). The case $\alpha \leq 1$ is simpler. Let us discuss it first. In that case $\Delta(t, h) = h^{\alpha-1}$ and from above formulas we obtain:

$$(u_\alpha \circ v) \leq c(h^{\alpha-1}u_\alpha + v), \quad [t(h^{\alpha-1} + v)] \circ v \leq ch^{\alpha-1}u_\alpha.$$

with some $c > 0$. Therefore, since $|F| \leq av$, one obtains

$$|u_\alpha \otimes F| \leq act(h^{\alpha-1}u_\alpha + v), \quad |u \otimes F \otimes F| \leq a^2c^2th^{\alpha-1}u_\alpha,$$

and generally (by trivial induction)

$$|u \otimes F^{\otimes(2k)}| \leq \frac{(a^2c^2th^{\alpha-1})^k}{(k!)^2}u_\alpha, \quad |u \otimes F^{\otimes(2k+1)}| \leq \frac{ac(a^2c^2th^{\alpha-1})^k}{k!(k+1)!}t(h^{\alpha-1}u_\alpha + v)$$

for all natural k. Therefore

$$u \otimes F^{\otimes 2} + u \otimes F^{\otimes 4} + \ldots \leq e^{a^2c^2t}u_\alpha,$$

$$u \otimes F^{\otimes 3} + u \otimes F^{\otimes 5} + \ldots \leq ace^{a^2c^2t}t(h^{\alpha-1}u_\alpha + v).$$

Consequently, series (3.26) is convergent in the required domain uniformly outside any neighbourhood of the set $\{t = 0, x = x_0\}$ (Notice that if one divides each element of this series on u_α, then the corresponding new series will converge uniformly for all x, x_0 and $t \leq \min(t_0, h)$.) Moreover, for its sum one gets the representation

$$u_G^{st} = u_\alpha(1 + O(t\Delta(t, h))) + (u_\alpha \otimes F)(1 + O(t\Delta(t, h))), \tag{3.28}$$

which clearly implies the required estimate.

The case $\alpha > 1$ requires only a bit more elaborate calculations. In that case one proves by induction that

$$|u \otimes F^{\otimes(2k)}| \leq \frac{1}{(k!)^2}(B^2 t\Delta(t,h))^k u_\alpha,$$

$$|u \otimes F^{\otimes(2k+1)}| \leq \frac{1}{(k!)^2}tB(B^2 t\Delta(t,h))^k(\Delta(t,h)u_\alpha + v)$$

for all natural k and some constant B. Thus, we obtain the convergence of series (3.26) for $\alpha > 1$ and again the representation of its sum in form (3.28).

Remark. Using a slight modification of Proposition 3.2, one can obtain the estimate $|F \otimes F| = O(t^{\omega-1})u_\alpha$, which can be used to simplify the above given proof of the convergence of (3.20).

It remains to prove (3.21), (3.22).

It remains to prove (3.16), (3.17). The difficulty that arises here is due to the observation that if one differentiates directly the terms of series (3.26) and uses the estimates (3.10), (3.11), one obtaines the expressions which are not defined (because τ^{-1} is not an integrable function for small τ). To avoid this difficulty, one needs to rearrange appropriately the variables of integration in (3.25), before using the estimates for the derivatives. To begin with, notice that due to (3.10), (3.11) and the assumption that $A \equiv 0$ for $\alpha \leq 1$ one obtains that

$$\frac{\partial F}{\partial t}(t,x,x_0) = O(t^{-1})(u_\alpha(t,x-x_0) + \tilde{u}(t,x-x_0)),$$

$$\frac{\partial F}{\partial x}(t,x,x_0) = O(th^{\alpha-1})^{-1/\alpha})(u_\alpha(t,x-x_0) + \tilde{u}(t,x-x_0)), \qquad (3.29)$$

if G and A have bounded derivatives.

Noticing that the convolution (3.25) after the change of the variable $\tau = st$ can be presented in the equivalent form

$$(\phi \otimes F)(t,x,\xi) = t \int_0^1 \phi(t(1-s),x,\eta)F(ts,\eta,\xi)\,d\eta ds$$

one can now estimate the derivative of the second term in (3.26) in the following way:

$$\frac{\partial}{\partial t}(u_\alpha \otimes F)(t,x,x_0) = \int_0^1 u_\alpha(t(1-s),x-\eta)F(ts,\eta,x_0)\,d\eta ds$$

$$+t\int_0^1 [(1-s)\frac{\partial u_\alpha}{\partial t}(t(1-s),x-\eta)F(ts,\eta,x_0)$$

$$+su_\alpha(t(1-s),x-\eta)\frac{\partial F}{\partial t}(ts,\eta,x_0)]\,d\eta ds,$$

and all three terms of this expression are of the order $O(t^{-1})(u_\alpha \star F)(t, x, x_0)$. Similarly one estimates the derivatives of the other terms in series (3.26), which gives (3.21).

Turning to (3.22) let us bound ourselves to the estimate of the first derivative only, higher derivatives being estimated similarly. The consideration of the case $\alpha > 1$ is trivial, because in that case $\tau^{-1/\alpha}$ is an integrable function for small τ, and consequently, differentiating expansion (3.26) term by term and using estimate (3.11) yields the required result straightforwardly. Suppose $\alpha \le 1$ (and $A \equiv 0$). To estimate the derivative of the second term in (3.26) let us rewrite it in the following form:

$$\mathcal{F}u_\alpha(t, x, x_0) = (u \otimes F)(t, x, x_0) = \int_0^{t/2} u_\alpha(t - \tau, x - \eta) F(\tau, \eta, x_0) \, d\eta \, d\tau$$

$$+ \int_{t/2}^t u_\alpha(t - \tau, \eta) F(\tau, x - \eta, x_0) \, d\eta \, d\tau.$$

Now, differentiating with respect to x and using (3.11) and (3.29) to estimate the first and the second term respectively, yields for the magnitude of the derivative of $\mathcal{F}u_\alpha$ the same estimate as for $\mathcal{F}u_\alpha$ itself but with an additional multuplier of the order $O(th^{\alpha-1})^{-1/\alpha})$. We are going to estimate similarly the derivative of the term $\mathcal{F}^k u_\alpha(t, x, x_0)$ in (3.26), which equals

$$\int_{\sigma_t} d\tau_1 ... d\tau_k \int_{\mathcal{R}^{kd}} d\eta_1 ... d\eta_k u_\alpha(t - \tau_1 - ... - \tau_k, x - \eta_1) F(\tau_1, \eta_1, \eta_2) ... F(\tau_k, \eta_k, x_0),$$

where we denoted by σ_t the simplex

$$\sigma_t = \{\tau_1 \ge 0, ..., \tau_k \ge 0 : \tau_1 + ... \tau_k \le t\}.$$

To this end, we make the partition of this simplex in the union of the $k + 1$ domains D_l, $l = 0, ..., k$, (which clearly have disjoint interiors) with $D_0 = \sigma_{t/2}$ and

$$D_l = \{(\tau_1, ..., \tau_k) \in \sigma_t \setminus \sigma_{t/2} : \tau_l = \max\{\tau_j, j = 1, ..., k\}\}, \quad l = 1, ..., k.$$

and then make a shift in the variables η to obtain $\mathcal{F}^k u_\alpha = \mathcal{F}_0^k + ... + \mathcal{F}_k^k$ with $\mathcal{F}_0^k(t, x, x_0)$ being equal to

$$\int_{\sigma_{t/2}} d\tau_1 ... d\tau_k \int_{\mathcal{R}^{kd}} d\eta_1 ... d\eta_k u(t - \tau_1 - ... - \tau_k, x - \eta_1) F(\tau_1, \eta_1, \eta_2) ... F(\tau_k, \eta_k, x_0)$$

and with $\mathcal{F}_l^k(t, x, x_0)$ being equal to

$$\int_{D_l} d\tau_1 ... d\tau_k \int_{\mathcal{R}^{kd}} d\eta_1 ... d\eta_k u(t - \tau_1 - ... - \tau_k, y_1) F(\tau_1, x - y_1, x - y_2) ...$$

$$\times F(\tau_{l-1}, x - y_{l-1}, x - y_l) F(\tau_l, x - y_l, \eta_{l+1})...F(\tau_k, \eta_k, x_0)$$

for $l = 1, ..., k$. Now, differentiating $\mathcal{F}_0^k$ with respect to x we use estimate (3.11), and differentiating $\mathcal{F}_l^k$, $l = 1, ..., k$, we use (3.29) for the derivative of $F(\tau_l, x - y_l, \eta_{l+1})$ and the estimate

$$\frac{\partial^l}{\partial x^l} F(t, x - \eta, x - \xi) = O(1)(u_\alpha(t, \eta - \xi) + \tilde{u}(t, \eta - \xi)), \quad l \leq q - 1, \quad (3.29')$$

(actually we need only $l = 1$ now) for the derivatives of other multipliers. The estimate (3.29') follows easily from (3.24). In this way, one obtains (noticing also that $\tau_l \geq t/(2k)$ in D_l) for the derivative of the term $\mathcal{F}^k u$ in (3.26) the same estimate as for $\mathcal{F}^k u$ itself, but with additional multiplier of the form

$$O(1) \left[(th^{\alpha-1})^{-1/\alpha} + k \left(\frac{th^{\alpha-1}}{k} \right)^{-1/\alpha} + \frac{k(k+1)}{2} \right] = O(1)(th^{\alpha-1})^{-1/\alpha} k^{1+1/\alpha}.$$

As it was proved above, the terms $\mathcal{F}^k$ are estimated by the expressions of the form $O(1)(Cth^{\alpha-1})^k/(k!)$ with some constant C. Multiplying these terms by k^q with any fixed positive q does not spoil the convergence of the series, which implies the required estimate for the derivative of u_G^{st}. The proof of Theorem 3.1 is therefore complete.

Remark. It follows from the theorem that if $A(x) \equiv 0$ and $\alpha > 1$, the asymptotics of the u_G^{st} for small h and small t are the same. For the case of constant coefficients it follows directly from (3.6). As also can be seen from (3.6), if $\alpha < 1$, small h and small t asymptotics are different already for constant coefficients.

Let us indicate shortly some consequences of this theorem related to the solutions of the Cauchy problem of equation (3.1). As usual, we shall denote by $C(\mathcal{R}^d)$ the Banach space of continuous bounded functions on $\mathcal{R}^d$ with the sup-norm, by $C_0(\mathcal{R}^d)$ we shall denote its closed subspace consisting of functions vanishing at infinity, and by $C^k(\mathcal{R}^d)$, k being a positive integer, we denote the Banach space of continuous functions having bounded derivatives up to and including the order k with the norm $\|f\| = \max_{l \leq k} \sup_x |f^{(l)}(x)|$.

For an arbitrary $f \in C(\mathcal{R}^d)$ and $t > 0$, let

$$(R_t f)(x) = \int_{\mathcal{R}^d} u_G^{st}(t, x, \xi) f(\xi) \, d\xi.$$

Proposition 3.5. *Suppose the assumptions of Theorem 3.1 are satisfied, and let T be an arbitrary positive number. Then*

(i) $(R_t f)(x)$ tends to $f(x)$ as $t \to 0$ for each x and any $f \in C(\mathcal{R}^d)$; moreover, if $f \in C_0(\mathcal{R}^d)$, then $R_t f$ tends to f uniformly, as $t \to 0$;

(ii) if $\alpha > 1$ (resp. $\alpha \leq 1$), then R_t is a continuous operator $C(\mathcal{R}^d) \mapsto C^l(\mathcal{R}^d)$ with the norm of the order $O(t^{-l/\alpha})$ for all $l \leq q$ (resp. $l \leq q - 1$);

(iii) R_t, $t \in (0, T]$, is a uniformly bounded family of operators $C^l(\mathcal{R}^d) \mapsto C^l(\mathcal{R}^d)$ for all $l \leq q - 1$;

(iv) if $f \in C(\mathcal{R}^d)$ and $t > 0$, the function $R_t f(x)$ satisfies equation (3.1);

(v) the Cauchy problem for equation (3.1) can have at most one solution in the class of continuous functions belonging to $C_0(\mathcal{R}^d)$ for each t; this solution is necessarily non-negative whenever the initial function is non-negative.

Proof. (i) follows from representation (3.20) and the fact that the same statement clearly holds, if one replaces u_G^{st} with u_α in the definition of R_t. (ii) follows again from (3.20) and estimate (3.11) together with its trivial generaliations for higher derivatives. To prove (iii), we rewrite the expression for R_t in the following equivalent form:

$$R_t f(x) = \int_{\mathcal{R}^d} u_G^{st}(t, x, x - y) f(x - y) \, dy,$$

and then use the estimates

$$\frac{\partial^l}{\partial x^l} u_G^{st}(t, x, x - y) = O(1) u_\alpha(t, y), \quad l = 1, ..., q - 1,$$

which follow from (3.20) and (3.29'). (iv) holds, because u_G^{st} satisfies equation (1.8). (v) follows from a general fact on the positivity of the solutions to pseudo-differential equations with a generator satisfying the positive maximum principle (see e.g. [K11]).

Corollary. *The Green function u_G^{st} is everywhere non-negative, satisfies the semigroup identity (the Chapman-Kolmogorov equation) and $\int u_G^{st}(t, x, \eta) \, d\eta = 1$ holds for all t, $x \in \mathcal{R}^d$. In particular, the semigroup defined by equation (3.1) is a Feller semigroup, which therefore corresponds to a certain Feller process.*

The first two statements on u_G^{st} follow directly from Proposition 3.5 (actually from statements (iv),(v)). The last statement can be deduced from it by means of a rather standard trick, see [Koch], where it is proved in the case $\alpha \geq 1$ (in general case it is proved exactly in the same way, whenever the existence result of Proposition 3.5 is established).

Before formulating the next result, let us prove a simple Lemma.

Lemma 3.2. *Let $G(x)$ and $A(x)$ satisfy the assumptions of Theorem 3.1. Then*

$$\int_{\{|x-\eta|\leq(th^{\alpha-1})^{1/(1+\alpha)}\}} u_\alpha(t, x - \eta; \mu(\eta), A(\eta)) \, d\eta = 1 - O((th^{\alpha-1})^{1/(1+\alpha)}),$$

$$\int_{\{|x-\eta|\geq(th^{\alpha-1})^{1/(1+\alpha)}\}} u_\alpha(t, x - \eta; \mu(\eta), A(\eta)) \, d\eta = O((th^{\alpha-1})^{1/(1+\alpha)}).$$

Proof. The second estimate follows directly from (3.9),(3.13). To prove the first inequality, notice that due to the mean-value theorem and Proposition 2.6, one has

$$|u_\alpha(t, x - \eta; G(\eta), A(\eta)) - u_\alpha(t, x - \eta; G(x), A(x))|$$

$$= (O(t) + O(|x - \eta|)u_\alpha(t, x - \eta; G(x), A(x))$$
$$= O((th^{\alpha-1})^{1/(1+\alpha)})u_\alpha(t, x - \eta; G(x), A(x))$$

for $|x - \eta| \le (th^{\alpha-1})^{1/(1+\alpha)}$. Since $\int u_\alpha(t, x - \eta; \mu(x), A(x))\, dx = 1$ (because u_α is a probability density), it implies the first inequality stated in the Lemma.

Theorem 3.2. *Under the assumptions of Theorem 3.1, for any h_0 there exists a constant K such that for all $t \le h \le h_0$ and all x, x_0, y*

$$K^{-1}u_\alpha(t, x - x_0, G(y), A(y), h) \le u_G^{st}(t, x - x_0, h) \le Ku_\alpha(t, x - x_0, G(y), A(y), h).$$

Proof. Obviously it is enough to prove the statement for small enough t, because then one can automatically expand this result to all (finite) t using the semigroup identities for u_G^{as} and u_G^{st}. Moreover, due to (3.13), it is enough to get the result for $y = x_0$. Next, the upper bound for u_G^{as} follows directly from (3.20). Due to (3.28), in order to get the lower bound, it is enough to prove that

$$u_\alpha + u_\alpha \otimes F \ge \epsilon u_\alpha \tag{3.30}$$

with some positive ϵ. Notice first of all that due to the estimates of $u_\alpha \otimes F$ given above, (3.24) holds trivially for small $|x - x_0|$ (and small t). Thus one needs to consider only the case of $|x - x_0| > \delta$ with any positive fixed δ. Next, the contribution in $u_\alpha \otimes F$ of the first term in (3.24) is of the form $O(t^\omega)u_\alpha$, and consequently in proving (3.30)) only the second term of (3.24) must be taken into account. Clearly, for any $\delta > 0$, the second term of (3.24) can be presented in the form

$$O(1) \min(\delta, |x - x_0|)t^{-1}u_\alpha(t, x - x_0)$$
$$+ \frac{G(x) - G(x_0)}{G(x_0)}(1 - \Theta_\delta(|x - x_0|))\frac{\partial u_\alpha}{\partial t}(t, x - x_0),$$

where $O(1)$ is uniform with respect to $\delta \in (0, 1)$. Due to Proposition 3.2, the contribution in $u_\alpha \otimes F$ of the first term in this expression has the order $(O(\delta) + O(t^\omega))u_\alpha$ and can be made smaller that ϵu_α by choosing small enough δ and t. Thus only the contribution of the second term of the above expression is of interest (and only for $|x - x_0| > \delta$). It follows from Propositions 2.2, 2.5 that for $|x - x_0| > \delta$ with a given δ and $t \to 0$

$$\frac{\partial u_\alpha}{\partial t}(t, x - x_0) = \frac{1}{t}u_\alpha(t, x - x_0)(1 + O(th^{\alpha-1})) = \frac{\gamma h^{\alpha-1}G(x_0)}{|x - x_0|^{d+\alpha}}(1 + O(th^{\alpha-1}))$$

with a positive γ (which equals actually to the coefficient of the first term in expansion (2.10). Consequently, up to nonessential terms,

$$u_\alpha \otimes F = \int_0^t d\tau \int d\eta$$

$$\times u_\alpha(t - \tau, x - \eta)(1 - \Theta_\delta(|\eta - x_0|))\frac{\gamma h^{\alpha-1}(G(\eta) - G(x_0))}{|\eta - x_0|^{d+\alpha}}(1 + O(\tau h^{\alpha-1})).$$

Now we decompose this integral into the sum $I_1 + I_2$ by decomposing the domain of integration with respect to η into the union $D_1 \cup D_2$ with

$$D_1 = \{\eta : |\eta - x| \geq (th^{\alpha-1})^{1/(1+\alpha)}\}, \quad D_2 = \{\eta : |\eta - x| \leq (th^{\alpha-1})^{1/(1+\alpha)}\}.$$

Clearly

$$I_1 = \int_0^t d\tau \int_{D_1} d\eta \, \frac{O(th^{\alpha-1})}{|x - \eta|^{d+\alpha}} \frac{1}{|\eta - x_0|^{d+\alpha}}.$$

Representing $D_1 = D_1' \cup D_1''$ with $D_1' = \{\eta \in D_1 : |x - \eta| \geq |x_0 - \eta|\}$ and $D_1'' = D_1 \setminus D_1'$ one has $|x - \eta| \geq |x - x_0|/2$ in D_1', and $|\eta - x_0| \geq |x - x_0|/2$ in D_1''. Therefore,

$$I_1 = O(th^{\alpha-1})u_\alpha(t, x - x_0) \int_{|y| \geq (th^{\alpha-1})^{1/(1+\alpha)}} \frac{dy}{|y|^{d+\alpha}}$$

$$= O((th^{\alpha-1})^{1/(1+\alpha)})u_\alpha(t, x - x_0).$$

Consequently, the integral I_1 is small as compared with the first term on the l.h.s. of (3.30). Turning to the estimate of I_2 notice that due to the first estimate of Lemma 3.2 one can write

$$I_2 = tf(x)(1 + O((th^{\alpha-1})^{1/(1+\alpha)}))$$

$$+ \int_0^t d\tau \int u_\alpha(t - \tau, x - \eta; G(\eta), A(\eta))(f(\eta) - f(x)) \, d\eta$$

with

$$f(\eta) = \gamma h^{\alpha-1}(1 - \Theta_\delta(|x - x_0|)) \frac{G(x) - G(x_0)}{|x - x_0|^{d+\alpha}}.$$

The second term in I_2 we estimate using the mean-value theorem for function f, which gives for this term an estimate of the same order as I_1. It remains only the first term in the expression for I_2. It follows that up to nonessential terms

$$u_\alpha \otimes F = (1 - \Theta_\delta(|x - x_0|)) \frac{t\gamma h^{\alpha-1}(G(x) - G(x_0))}{|x - x_0|}(1 + O(t))$$

$$= (1 - \Theta_\delta(|x - x_0|))(1 + O(t))$$

$$\times [u_\alpha(t, x - x_0; G(x), A(x), h) - u_\alpha(t, x - x_0, G(x_0), A(x_0), h)].$$

Clearly, this expression when added to $u_\alpha(t, x - x_0, G(x_0), A(x_0), h)$ is not less than $\epsilon u_\alpha(t, x - x_0)$ with some $\epsilon \in (0, 1)$ if t is small enough. The proof is therefore completed.

It is often useful to know that the solution to a Cauchy problem preserves a certain rate of decay at infinity. We present now a result of this kind for equation

(3.1), (3.4), which we shall use also in the next Section. Let for $\beta > 0, \epsilon > 0$ the functions f_β^d be defined on $\mathcal{R}^d$ by the formulas

$$f_\beta^d(x) = (1 + |x|^{\beta+d})^{-1}, \quad f_{\beta,\epsilon}^d(x) = f_\beta^d(x/\epsilon) \tag{3.31}$$

Proposition 3.6. *Under the assumptions of Theorem 3.1 there exists a constant B such that*

$$\int_{\mathcal{R}^d} u_G^{st}(t, x, \eta, h) f_{\alpha,\epsilon}^d(\eta - x_0) \, d\eta \leq B f_{\alpha,\epsilon}^d(x - x_0). \tag{3.32}$$

whenever $t \leq h \leq \epsilon$.

Proof. Obviously it is enough to prove the statement for $x_0 = 0$, and due to (3.20), with u_α instead of u_G^{st}. Furthermore, for $|\eta| > \frac{1}{2}|x|$ one can estimate

$$f_{\alpha,\epsilon}^d(\eta) \leq (1 + (|x|/2)^{\alpha+d})^{-1} \leq 2^{\alpha+d} f_{\alpha,\epsilon}^d(x),$$

and therefore for $x_0 = 0$, the integral in (3.31) over the set $\{|\eta| > \frac{1}{2}|x|\}$ does not exceed

$$\int_{\mathcal{R}^d} u_\alpha(t, x - \eta; G, A) 2^{\alpha+d} f_{\alpha,\epsilon}^d(x) \, d\eta \leq 2^{\alpha+d} f_{\alpha,\epsilon}^d(x).$$

Hence, it remains to show that

$$\int_{\{|\eta| \leq \frac{1}{2}|x|\}} u_\alpha(t, x - \eta; G, A) f_{\alpha,\epsilon}^d(\eta) \, d\eta \leq B f_{\alpha,\epsilon}^d(x). \tag{3.33}$$

Notice now that

$$\frac{1}{2} \leq f_{\alpha,\epsilon}^d(x) \leq 1, \tag{3.34}$$

$$\frac{1}{2}(\epsilon/|x|)^{d+\alpha} \leq f_{\alpha,\epsilon}^d(x) \leq (\epsilon/|x|)^{d+\alpha}, \tag{3.35}$$

if $|x| \leq \epsilon$ or $|x| \geq \epsilon$ respectively. Consider now separately two cases.

(i) If $|x| \leq (th^{\alpha-1})^{1/\alpha}$, then in particular $|x| \leq \epsilon$ and moreover, one can estimate u_α under the integral on the r.h.s. of (3.33) by $C(th^{\alpha-1})^{-d/\alpha}$, due to (3.8), and thus one gets to this integral the estimate

$$\frac{O(1)}{(th^{\alpha-1})^{-d/\alpha}} \int_{\{|\eta| \leq |x|\}} |\eta|^{d-1} f_{\alpha,\epsilon}^d(\eta) \, d|\eta| = \frac{O(1)\epsilon^d}{(th^{\alpha-1})^{d/\alpha}} \int_0^{(|x|/\epsilon)^d} \frac{dr}{1 + r^{1+\alpha/d}}, \tag{3.36}$$

where we have changed the variable of integration by the formula $r = (|\eta|/\epsilon)^d$. Since the integral in (3.36) has the form $O(1)(|x|/\epsilon)^d$, the whole expression (3.36) has the form

$$\frac{O(1)|x|^d}{(th^{\alpha-1})^{1/\alpha}} = O(1) = O(1) f_{\alpha,\epsilon}^d(x),$$

where we have used (3.34).

(ii) If $|x| \geq (th^{\alpha-1})^{1/\alpha}$, one estimates u_α using formula (3.9) and thus one estimates the integral in (3.33) by

$$\frac{O(1)\epsilon^d th^{\alpha-1}}{|x|^{d+\alpha}} \int_0^{(|x|/\epsilon)^d} \frac{dr}{1+r^{1+\alpha/d}}. \tag{3.37}$$

If $|x| \leq \epsilon$, we estimate the integral in (3.37) by $O(1)(|x|/\epsilon)^d$ as in case (i) and thus get to (3.37) the representation $O(1) = O(1)f^d_{\alpha,\epsilon}(x)$. If $|x| \geq \epsilon$, we estimate the integral in (3.37) by $O(1)$ and then using (3.35) present (3.37) in the form

$$O(1)th^{\alpha-1}\epsilon^{-d}f^d_{\alpha,\epsilon}(x),$$

which is again $O(1)f^d_{\alpha,\epsilon}(x)$ due to the assumption $t \leq h \leq \epsilon$.

The same arguments prove the following generalisation, which we shall use considering the stable-like processes in Sect. 5.

Proposition 3.7. *Under the assumptions of Proposition 3.5, for any $\beta > 0$ there exists a constant B such that for $t \leq h \leq \epsilon$*

$$\int_{\mathcal{R}^d} u_G^{st}(t,x,\eta,h)f^d_{\beta,\epsilon}(\eta - x_0)\, d\eta \leq Bf^d_{\min(\alpha,\beta),\epsilon}(x - x_0).$$

Let us note for conclusion that certain estimates for the transition probability densities of stable processes on compact Riemannian manifolds can be found in [Mol].

4. Stable jump-diffusions combined with compound Poisson processes

Because of the well-known "long-tail" property of the stable densities one can't expect the large deviation principle to hold for stable processes, say, with small scale (or large distances). However, as will be shown in the next chapter, if one combines a stable process with an appropriate compound Poisson process that "kills the tails", in other words, if one considers the so called truncated stable processes, the large deviation principle can be obtained, as well as more precise asymptotic expansions with respect to the varying small scale. Having this in mind, we present here the necessary generalisation of the result of the previous section. For the process considered in this Section, not only the estimates but even the existence of the transition probability density seems to be a new result.

Theorem 4.1 *Let in the pseudo-differential equation*

$$h\frac{\partial u}{\partial t} = \Phi(x,-ih\nabla)u \equiv \Phi_s(x,-ih\nabla)u + \Phi_p(x,-ih\nabla)u \tag{4.1}$$

the function Φ_s be given by (3.4) or

$$\Phi_s(x,p) = ipA(x) - \frac{1}{2}(G(x)p,p) \tag{4.2}$$

with positive matrix $G(x)$ and uniformly bounded G, G^{-1}, A, and

$$\Phi_p(x, p) = \int_{\mathcal{R}^d} (e^{i\xi p} - 1) f(x, \xi) \, d\xi \qquad (4.3)$$

with an integrable function f such that

$$|f(x, \xi)| \le a f_\beta^d(\xi), \qquad (4.4)$$

where $\beta \in (0.2)$, f_α^d is defined in (3.31), and a is a constant. Then the Green function u_G for for the equation (4.1) exists, is continuous and for small t, t/h, h has the representation

$$u_G(t, x; x_0, h) = u_G^{st}(t, x; x_0, h)(1 + O(t/h)) + O(t)h^{-d-1} f_{\min(\alpha,\beta)}^d((x - x_0)/h), \qquad (4.5)$$

where u_G^{st} is the Green function (3.20) of equation (3.1) with $\Phi = \Phi_s$.

Remark 1. Function (4.3) can be presented as the difference of the two functions of the same form but with positive "densities" f, each of these functions presenting therefore the characteristic function of a compound Poisson process, see e.g. Appendix C. Thus in notations Φ_s, Φ_p the indices s and p stand respectively for the stable and Poissonian parts of the ΨDO Φ.

Remark 2. Notice that equation (4.1), (4.2) with a suitable function f describes the truncated stable diffusions, i.e. such processes, whose Lévy measure coincides with a stable measure in a neighborhood of the origin and vanishes outside some ball. In the case of permanent coefficients, the corresponding probability density was investigated in [Is]. Our Theorems 4.1, 4.2 generalise some results from [Is] to the case of varying coefficients.

Proof. As the first approximation to the solution of (4.1), (3.2) we take the Green function of the corresponding problem with $\Phi_p = 0$. Proceeding as in the proof of Theorem 3.1 we present the exact Green function u_G by series (3.30) with u_G^{st} instead of u_α and with

$$F(t, x, x_0, h) = -\frac{1}{h} \Phi_p(x, -ih\nabla) u_G(t, x, x_0, h)$$

$$= \frac{1}{h} \int [u_G^{st}(t, x + h\xi, x_0, h) - u_G^{st}(t, x, x_0, h)] f(x, \xi) \, d\xi. \qquad (4.6)$$

Due to Proposition 3.7, $F = F_1 - F_2$ with

$$|F_1(t, x, x_0, h)| \le \frac{aB}{h^{d+1}} f_{\min(\alpha,\beta),h}^d(x - x_0), \quad |F_2(t, x, x_0)| \le \frac{aC}{h} u_G^{st}(t, x, x_0)$$

with some constant C. Consequently, using Proposition 3.7 one gets

$$|\mathcal{F} u_G^{st}| = |u_G^{st} \otimes (F_1 - F_2)| \le atB^2 h^{-d-1} f_{\min(\alpha,\beta),h}^d + atCh^{-1} u_G^{st}(t, x, x_0).$$

for $\alpha < 2$. In the case of Φ_s of form (4.2) (i.e. in the case of $\alpha = 2$) to get the same estimate for $\mathcal{F}u_G^{st}$ one needs the estimate

$$u_G^{dif} \otimes f_\beta^d \le B f_\beta^d$$

with some constant B, where u_{dif}^G is the Green function of the corresponding diffusion equation (equation (4.1) with $\Phi_p = 0$). This estimate is trivial, if u_G^{dif} is replaced by the Green function u_{free} of the corresponding equation with permanent coefficients, and in general is then obtained using the well known fact that u_G^{diff} can be estimate by u_{free} (see e.g. [Da1], [Da2]).

Using induction and the trivial estimate

$$\int f_{\alpha,\epsilon}^d(x - \eta) f_{\alpha,\epsilon}^d(\eta) \, d\eta \le \epsilon^d f_{\alpha,\epsilon}^d(x), \tag{4.7}$$

one estimates the other terms in series (3.26), which gives its convergence and the required representation.

In order to get rid of $f_{\alpha,h}^d$ in (4.5) we present the following estimate.

Lemma 4.1. *One has*

$$th^{-(1+d)} f_{\alpha,h}^d(x - x_0) = O(1) \min(1, (|x - x_0|/h)^{d+\alpha}) u_\alpha(t, x - x_0). \tag{4.8}$$

Proof. Due to (3.8) and (3.34), if $|x + A(x_0)t - x_0| < (th^{\alpha-1})^{1/\alpha}$, the l.h.s. in (4.8) has the form

$$O(t) h^{-1-d} (th^{\alpha-1})^{d/\alpha} u_\alpha = O(t/h)^{1+d/\alpha} u_\alpha.$$

Let $(th^{\alpha-1})^{1/\alpha} \le |x + A(x_0)t - x_0|$. Consider two cases. If $|x - A(x_0)t - x_0| = O(h)$, the l.h.s. in (4.8) has the form

$$O\left(\frac{t}{h^{1+d}}\right) = O\left(\frac{t}{h^{1+d}} \frac{|x - x_0|^{d+\alpha}}{th^{\alpha-1}}\right) u_\alpha = O\left(\frac{|x - x_0|}{h}\right)^{d+\alpha} u_\alpha. \tag{4.9}$$

due to (3.9),(3.34). If $|x - x_0| \ge h$, the l.h.s. in (4.8) has the form

$$O\left(\frac{t}{h^{1+d}}\right) \frac{h^{d+\alpha}}{|x - x_0|^{d+\alpha}} = O\left(\frac{t}{h^{1+d}} \frac{h^{d+\alpha}}{th^{\alpha-1}}\right) u_\alpha = O(u_\alpha),$$

due to (3.9), (3.35). The Lemma is proved.

Proposition 4.1. *If Φ_s is of form (3.4), the function $u_G(t, x, x_0, h)$ from (4.5) can be rewritten in the form*

$$[1 + O(t/h) + \Omega(t, h) + O(1) \min(1, (|x - x_0|/h)^{d+\alpha}, |x - x_0|)] u_\alpha(t, x - x_0). \tag{4.10}$$

Proof. Follows directly from Lemma 4.1 and Theorem 4.1.

The following result gives an alternative representation for the Green function of equation (4.1). Let $0 < c_1 < c_2$ and let $\chi(t, y)$ be a smooth function on $\mathcal{R}_+ \times \mathcal{R}_+ \mapsto [0, 1]$ that vanishes (resp. equals to one) for $y \geq c_2$ (resp. for $y \leq c_1$).

Theorem 4.2. *Under the assumptions of Proposition 4.1 the Green function for equation (4.1) can be alternatively written in forms (4.5), (4.10) but with*

$$u_0(t, x, x_0, h) = \frac{1}{(2\pi h)^d} \int_{\mathcal{R}^d} \exp\{\frac{\Phi(x_0, p)t + ip(x - x_0)}{h}\} \, dp. \qquad (4.11)$$

or with $\tilde{u}_0(t, x, x_0, h) = \chi(t, |x - x_0|) u_0(t, x, x_0, h)$ *instead of* u_G^{st}. *Moreover, the localised function* $\tilde{u}_0$ *satisfies the Dirac initial condition (3.2).*

Proof. Let us show that

$$u_0(t, x, x_0, h) = (1 + O(t/h)) u_\alpha(t, x - x_0) + O(th^{-1-d}) f_{\alpha, h}^d(x - x_0) \qquad (4.12)$$

with u_α given by (3.5). To this end, we use the presentation of Φ in the sum $\Phi_s + \Phi_p$ and expand the function $\exp\{\Phi_p(x_0, p)t\}$ in the power series. The first term in thus obtained series in (4.10) will be just u_α. Let us estimate the second term

$$(2\pi)^{-d} \int_{\mathcal{R}^d} \frac{t}{h} \Phi_p(x_0, ph) \exp\{\Phi_s(x_0, p)th^{\alpha-1} + ip(x - x_0)\} \, dp. \qquad (4.13)$$

Due to (4.3), $\Phi_p(x_0, hp)$ is the sum of the Fourier transform $h^{-d}\hat{f}_h(x_0, p)$ of the function $h^{-d}f(x_0, -\xi/h)$ and a constant. Substituting a constant instead of $\Phi_p(x_0, hp)$ in (4.13) gives $O(t/h)u_\alpha$. Substituting $h^{-d}\hat{f}_h(x_0, p)$ gives in (4.13) the inverse Fourier transform of the product of the two functions, which is equal therefore to the convolution of their inverse Fourier transforms: u_α and $h^{-d}f(x_0, -\xi/h)$ (notice that f and $\hat{f}$ belong to the space L_2 due to (4.4)). But this convolution, due to (4.4), was estimated in Proposition 3.5, which implies that this term of the sum in (4.13) can be estimated by the second term in (4.12). Other terms of the expansion in (4.13) are estimated in the same way, using also (4.7), which proves the convergence of the series and estimate (4.12). Due to Theorem 4.1 and formula (4.10), formula (4.12) implies the statement of the Theorem for the function u_0. Furthermore, since for $|x - x_0| \geq c$, u_α is of the same order as $th^{-1-d}f_{\alpha, h}^d$, one can use $\tilde{u}_0$ instead of u_0 without changing the estimates of the remainder in (4.5). At last, function $\tilde{u}_0$ satisfies (3.2), because on the one hand, u_α satisfies (3.2), and on the other hand

$$\lim_{t \to 0} \int_{|x-x_0| \geq t^{-\delta+1/\alpha}} u_\alpha(t, x - x_0) = 0$$

for any $\delta > 0$, due to (3.9).

We are going now to get rid of the assumption of smallness of time.

Proposition 4.2. *Under the assumptions of Theorem 4.1, for any h_0 there exists K such that for $t \le h \le h_0$*

$$u_G \le Ku_\alpha, \quad u_\alpha \le K(u_G + f^d_{\alpha,h}), \quad u_0 \le Ku_\alpha.$$

Proof. Since we have these inequalities for small t/h we readily obtain them for any finite t/h using the semigroup identity for u_G and the solutions of the equations with constant coefficients, and Proposition 3.4.

One sees in particular that though the function (4.11) gives an approximation for the Green function of equation (4.1) for any fixed $h > 0$, it seems to be not a good candidate for the small h asymptotics. As will be shown in the next chapter, in order to get the uniform small t and small h asymptotics one should modify essentially the phase in integral (4.11).

5. Stable-like processes

Here we are going to generalise the results of Section 3 to the case of the stable processes with the varying index of stability, i.e. to the so called stable-like processes. For brevity we shall consider A from (3.4) to be zeros. Thus we shall study the Green function for equation (3.1) with the symbol $\Phi(x,p) = -G(x)|p|^{\alpha(x)}$, i.e the equation

$$h\frac{\partial u}{\partial t} = -G(x)\| -ih\nabla\|^{\alpha(x)}u, \quad x \in \mathcal{R}^d, \quad t \ge 0. \tag{5.1}$$

Throughout this section we suppose that α take values in some fixed compact subinterval $[\alpha_d, \alpha_u]$ of the open interval $(0,2)$ and the scales G satisfy (3.7) with some fixed constant B.

Remark. Literally the same argument and results hold also in the case when a smooth drift $A(x)$ is present but $\alpha_d > 1$.

For brevity, we shall consider $h = 1$ in this section, noting that the presence of h presents no new difficulties and can be dealt with as in Section 3. It was proven in [KN] (see also some particular cases of this result in [Neg]) that if α and μ are infinitely smooth functions of x, the transition probability measures of stable-like processes are absolutely continuous with respect to Lebesgue measure, which implies the existence of a measurable densities for these transition probabilities. The arguments of [Neg] and [KN] are based on the far developed theory of ΨDO with symbols having a varying order. Here we obtain this result generalising the arguments of our Section 3 under weaker assumptions of only first order differentiability of μ and α. In fact, one can easily generalise the arguments to cover the case of only Hölder continuous μ and α. Moreover, this approach which allows us to obtain local multiplicative asymptotics and global estimates for these densities. Before formulating the main result, let us prepare some estimates of the convolution of the functions u_α with varying indices α generalising the corresponding estimates from Section 3.

Lemma 5.1. *(i) Uniformly for all x and small enough t, one has*

$$\int u_{\alpha(\eta)}(t, x - \eta, G(\eta))\, d\eta = O(1) t^{-d(\alpha_u - \alpha_d)/(\alpha_u \alpha_d)}. \tag{5.2}$$

(ii) if additionally the function $\alpha(\eta)$ is Hölder continuous, i.e. $|\alpha(x) - \alpha(y)| = O(|x - y|^\beta)$ uniformly with some positive β, then uniformly for all x and all $t \leq t_0$ with any fixed t_0 one has

$$\int u_{\alpha(\eta)}(t, x - \eta, G(\eta))\, d\eta = O(1). \tag{5.3}$$

Proof. Let $|x - \eta| \geq c$ with some fixed $c \in (0, 1]$. Then $|x - \eta| \geq t^{1/\alpha(\eta)}$ for all η, if $t \leq t_0$ with a small enough t_0. It follows from (3.6), (3.10) that for such η

$$u_\alpha(\eta)(t, x - \eta, G(\eta)) = \frac{O(t)}{|x - \eta|^{d + \alpha(\eta)}} = \frac{O(t)}{|x - \eta|^{d + \alpha_d}}.$$

Since this function is integrable outside any neighbourhood of the origin, it follows that the integral from the l.h.s. of (5.2) over the set $|x - \eta| \geq c$ is bounded. Next, let $t^{1/\alpha_u} \leq |x - \eta| \leq c \leq 1$. Again $|x - \eta| \geq t^{1/\alpha(\eta)}$ in this set, and therefore

$$u_\alpha(\eta)(t, x - \eta, G(\eta)) = \frac{O(t)}{|x - \eta|^{d + \alpha(\eta)}} = \frac{O(t)}{|x - \eta|^{d + \alpha_u}}.$$

Consequently, the integral over this set can be estimate in magnitude by the expression

$$O(t) \int_{t^{1/\alpha_u}}^\infty \frac{dy}{y^{1 + \alpha_u}} = O(t) t^{-1} = O(1).$$

At last one has

$$\int_{\{|x - \eta| \leq t^{1/\alpha_u}\}} u_\alpha(\eta)(t, x - \eta, G(\eta))\, d\eta = O(1) t^{-d/\alpha_d} \int_{\{|x - \eta| \leq t^{1/\alpha_u}\}} d\eta,$$

which clearly has the form of the r.h.s. of (5.2), and therefore (5.2) is proved. To prove (5.3) under assumption (ii) we need now to consider only the integral over the set $\{|x - \eta| \leq t^{1/\alpha_u}\}$. Denote

$$\alpha_u(x, b) = \max\{\alpha(y) : |y - x| \leq b\}, \quad \alpha_d(x, b) = \min\{\alpha(y) : |y - x| \leq b\} \tag{5.4}$$

Now, let us take $b = b(t) = t^{1/\alpha_u}$ and let

$$M(x) = \{\eta : t^{1/\alpha(x, b)} \leq |\eta - x| \leq t^{1/\alpha_u}\}. \tag{5.5}$$

Then the integral on the l.h.s. of (5.3) over the set $M(x)$ can be estimated by

$$O(t) \int_{M(x)} \frac{d\eta}{|x - \eta|^{d + \alpha_u(x, b)}} = O(t) \int_{t^{1/\alpha_u(x, b)}}^\infty \frac{dy}{y^{1 + \alpha_u(x, b)}} = O(t) t^{-1} = O(1).$$

At last, the integral on the l.h.s. of (5.3) over the set $\{|\eta - x| \le t^{1/\alpha_u(x,b)}$ can be estimated by

$$O(t^{-d/\alpha_d(x,b)})t^{d/\alpha_u(x,b)} = O(1)\exp\{-\frac{d(\alpha_u(x,b) - \alpha_d(x,b))}{\alpha_u(x,b)\alpha_d(x,b)}\log t\}$$

$$= O(1)\exp\{O(t^{\beta/\alpha_u}\log t\} = O(1),$$

which completes the proof of Lemma 5.1.

Lemma 5.2. *Under the assumptions of Lemma 5.1 (ii) one has*

$$\int u_{\alpha(\eta)}(t - \tau, x - \eta, G(\eta))u_{\alpha(x_0)}(\tau, \eta - x_0, G(x_0))\, d\eta$$

$$= O(1)u_{\alpha(x_0)}(t, x - x_0, G(x_0)) + O(t)f^d_{\alpha_d}(x - x_0). \tag{5.6}$$

Proof. 1) If $|x - x_0| \ge c$ with some fixed $c \in (0, 1]$, either $|x - \eta| \ge |x - x_0|/2 \ge c/2$ and then the first multiplier under the integral in (5.6) is estimated by $O(t)|x - x_0|^{-d - \alpha_d} = O(t)f^d_{\alpha_d}(x - x_0)$, or $|\eta - x_0| \ge |x - x_0|/2 \ge c/2$ and then the second multiplier under the integral in (5.6) is estimated by $O(t)f^d_{\alpha_d}(x - x_0)$. Therefore, in that case, due to Lemma 5.1, the integral on the l.h.s. of (5.6) has the form $O(t)f^d_{\alpha_d}(x - x_0)$.

2) Suppose now that $t^{1/\alpha_u} \le |x - x_0| \le c$. Let us decompose the integral from the l.h.s. of (5.6) in the sum $I_1 + I_2$ decomposing the domain of integration into the union $D_1 \cup D_2$ with $D_1 = \{\eta : |\eta - x_0| \ge |x - x_0|/2\}$ and D_2 being its complement. In D_1

$$u_{\alpha(x_0)}(\tau, \eta - x_0, G(x_0)) = \frac{O(\tau)}{|\eta - x_0|^{d + \alpha(x_0)}}$$

$$= \frac{O(t)}{|x - x_0|^{d + \alpha(x_0)}} = O(1)u_{\alpha(x_0)}(t, x - x_0, G(x_0)),$$

and therefore, due to Lemma 5.1, $I_1 = O(1)u_{\alpha(x_0)}(t, x - x_0, G(x_0))$. Next, in D_2

$$u_{\alpha(\eta)}(t - \tau, x - \eta, G(\eta)) = \frac{O(t - \tau)}{|x - \eta|^{d + \alpha(\eta)}} = \frac{O(t)}{|x - x_0|^{d + \alpha_u(x_0, |x - x_0|)}}$$

$$= \frac{O(t)}{|x - x_0|^{d + \alpha(x_0)}}|x - x_0|^{\alpha(x_0) - \alpha_u(x_0, |x - x_0|)}$$

$$= O(1)u_{\alpha(x_0)}(t, x - x_0, G(x_0))\exp\{O(|x - x_0|^\beta)\log|x - x_0|\} =$$

$$O(1)u_{\alpha(x_0)}(t, x - x_0, G(x_0)).$$

Therefore, $I_1 = O(1)u_{\alpha(x_0)}(t, x - x_0, G(x_0))$ as well.

3) Analogously one considers now the case $x \in M(x_0)$ with $M(x)$ defined in (5.5) and $b = b(t)$ being the same as in the proof of Lemma 5.1. One obtains for the integral in this case the same estimate $O((1)u_{\alpha(x_0)}(t, x - x_0, G(x_0))$.

4) Let $|x - x_0| \leq t^{1/\alpha_u(x_0, b)}$. Decompose our integral into the sum $I_1 + I_2$ decomposing the domain of integration into the union $D_1 \cup D_2$ with $D_1 = \{\eta : |\eta - x_0| \geq 2t^{1/\alpha_u(x_0, b)}\}$ and D_2 being its complement. In D_1 one has

$$u_{\alpha(x_0)}(\tau, \eta - x_0, G(x_0)) = \frac{O(\tau)}{|\eta - x_0|^{d + \alpha(x_0)}} = O(t)t^{-(d + \alpha(x_0))/\alpha_u(x_0, b)}$$

$$= O(t^{-d/\alpha(x_0)}) \exp\left\{\left[\frac{d}{\alpha(x_0)} - \frac{d}{\alpha_u(x_0, b)} + 1 - \frac{\alpha(x_0)}{\alpha_u(x_0, b)}\right] \log t\right\},$$

which is clearly of the order $O(1)u_{\alpha(x_0)}(t, x - x_0, G(x_0))$. Consequently, one obtains for I_1 the same estimate as in the case 2), again using Lemma 5.1. Turning to I_2 we distinguish two cases. If $\tau \geq t/2$ we estimate the second multiplier under the integral by

$$O(1)\tau^{-d/\alpha(x_0)} = O(1)t^{-d/\alpha(x_0)} = O(1)u_{\alpha(x_0)}(t, x - x_0, G(x_0)),$$

and if $t - \tau \geq t/2$ we estimate the first multiplier under the integral by

$$O(1)(t - \tau)^{-d/\alpha_d(x_0, b)} = O(1)t^{-d/\alpha(x_0)} = O(1)u_{\alpha(x_0)}(t, x - x_0, G(x_0)).$$

In both cases one obtains therefore the same estimate for the integral as in cases 2) and 3). The proof of the Lemma is completed.

Similarly one generalises the other estimates of Section 3 and obtains the following results

Lemma 5.3. *If the family $\alpha(x)$ satisfies the assumptions of Lemma 5.1 (ii), the results of Lemma 3.1 and of Proposition 3.2, 3.3 are still valid, if in all formulas one puts $\alpha(\eta)$ instead of the constant α.*

We can give now the main result of this section.

Theorem 5.1. *Let the functions $\alpha(x)$ and $G(x)$ have uniformly bounded continuous derivatives. Then the Green function u_G^{stl} for equation (5.1) exists, is continuous and differentiable in t for $t > 0$. For $t \leq t_0$ with small enough t_0*

$$u_G^{stl}(t, x, x_0) = u_{\alpha(x_0)}(t, x - x_0, G(x_0))$$

$$\times [1 + O(1) \min(1, (1 + |\log t|)|x - x_0|) + O(t^\omega)] + O(t)f_{\alpha_d}^d(x - x_0) \quad (5.7)$$

with some $\omega \in (0, 1)$.

Proof. We shall follow the arguments of the proof of Theorem 3.1. One finds readily that

$$\frac{\partial u_{\alpha(x_0)}}{\partial t} = \Phi(x, -i\nabla)u_{\alpha(x_0)}(t, x - x_0, G(x_0)) - F(t, x, x_0) - \tilde{F}(t, x, x_0), \quad (5.8)$$

where F is the same as in Theorem 3.1, i.e. it is given by formula (3.18) with $\alpha = \alpha(x_0)$, and

$$\tilde{F} = \phi_{\alpha(x_0), G(x)}(x - x_0; \alpha(x_0), tG(x_0)) - \phi_{\alpha(x), G(x)}(x - x_0; \alpha(x_0), tG(x_0)), \quad (5.9)$$

where the functions ϕ were defined in (2.23). Consider first the case when $|x - x_0| \leq c$ with some $c > 0$. In that case, due to the mean-value theorem, one has

$$\tilde{F} = O(|x - x_0|) \max_b \left| \frac{\partial \phi_{b,\mu(x_0)}}{\partial b}(x - x_0, \alpha(x_0), G(x_0)t) \right|,$$

where b takes values between $\alpha(x_0)$ and $\alpha(x)$. If $|x - x_0| \leq t^{1/\alpha(x_0)}$, one finds using (2.32) that

$$\tilde{F} = O(|x - x_0|)t^{-1}(1 + |\log t|)t^{-|\alpha(x)-\alpha(x_0)|/\alpha(x_0)}u_{\alpha(x_0)}$$

$$= O(|x - x_0|)t^{-1}(1 + |\log t|)u_{\alpha(x_0)}.$$

If $t^{\alpha(x_0)} \leq |x - x_0| \leq c$ with any $c > 0$, one finds using (2.33) that

$$\tilde{F} = O(|x - x_0|)t^{-1}(1 + |\log t|)|x - x_0|^{\alpha(x_0)-\alpha(x)}u_{\alpha(x_0)}$$

$$= O(|x - x_0|)t^{-1}(1 + |\log t|)u_{\alpha(x_0)}. \tag{5.10}$$

If $|x - x_0| > c$, let us estimate each term in (5.6) separately using (2.30) to obtain the estimate

$$|\tilde{F}(t, x, x_0)| = O(t^{-1})u_{\alpha(x)}(t, x - x_0, G(x)) + O(t^{-1})u_{\alpha(x_0)}(t, x - x_0, G(x)), \tag{5.11}$$

which is clearly of the order $O(1)f^d_{\alpha_d}(x - x_0)$. Therefore one has

$$|\tilde{F}(t, x, x_0)| \leq t^{-1} \min(a_1, a_2|x - x_0||\log t|)u_{\alpha(x_0)}(t, x - x_0; \mu(x_0)) + a_3 f^d_{\alpha_d}(x - x_0)$$

with some constants a_1, a_2, a_3. By Lemma 5.3 $|(u_\alpha \otimes \tilde{F})(t, x, x_0)|$ does not exceed

$$\leq a(\min(1, |x - x_0||\log t|) + t^\omega)u_{\alpha(x_0)}(t, x - x_0, G(x_0)) + atf^d_{\alpha_d}(x - x_0).$$

Afterwards one finishes the proof in the same way as that of Theorem 3.1 proving that formula (3.22) is still valid for u^{stl}_G.

Let us present a more rough version of formula (5.7). Namely, estimating the first term on the r.h.s. of (5.7) separately for $|x - x_0| \leq t^\beta$ and for $|x - x_0| \geq t^\beta$, one gets the following result.

Corollary 5.1. *Under the assumptions of Theorem 5.1*

$$u^{stl}_G(t, x, x_0) = u_\alpha(t, x - x_0; \mu(x_0))(1 + O(t^\beta)) + O(t^\omega)f^d_\alpha(x - x_0), \tag{5.12}$$

with any $\beta < 1/(d + \alpha)$ and $\omega < 1 - \beta(d + \alpha)$.

At last, we generalise the upper bound estimate from Theorem 3.2.

Theorem 5.2. *Under the assumptions of Theorem 5.1 and for any fixed t_0, the Green function $u^{stl}_G(t, x, x_0)$ does not exceed the expression*

$$K[\Theta_{t^\gamma}(|x - x_0|)u_{\alpha(x_0)}(t, x - x_0, \mu(x_0)) + (1 - \Theta_{t^\gamma})(|x - x_0|)u_{\alpha_d}(t, x - x_0, \mu(x_0))] \tag{5.13}$$

uniformly for all $t \leq t_0$, and x, x_0, where γ and K are some positive constants.

Proof. It follows from the proof of Theorem 5.1. In fact, due to (3.22), one only needs to obtain the required estimate for $u \otimes \tilde{F}$. To this end one estimates $\tilde{F}$ by (5.10) for $|x - x_0| \leq c = t^\gamma$ and by (5.11) for $|x - x_0| \geq c = t^\gamma$.

Notice for conclusion that the generalisations of Theorem 3.1 presented in Section 4 can be extended also to the case of the processes with varying index α, e.g. to the case of a stable-like jump-diffusion disturbed by a compound Poisson process. Finally, the statement of the Corollary to Proposition 3.5 holds also for the Green function u_G^{stl}, which one proves in the same way as in section 3.

6. Applications to the sample path properties of stable jump-diffusions

Let $X(t)$ (resp. $X(t, x_0)$) be the Feller process starting at the origin (resp. at x_0) and corresponding to the Feller semigroup defined by equation (3.1), (3.4) with $h = 1$. The existence of the Feller process corresponding to equations (3.1), (3.4), or more generally to equations (4.1)-(4.3) is well known, the probabilistic proof being first obtained in [Kom2]. Some particular cases and generalisaton can be found e.g. in [Kom1], [Ho], [Por], [PP]. Of course, the existence of such Feller process follows as well from the well-posedness of the Cauchy problem for equation (3.1), (3.4), which follows in its turn from the existence of the Green function constructed in previous sections (see Corollary to Proposition 3.5). In this section we are going to show how the analytic results of the previous sections can be used in studying the sample path properties of the Feller process X, i.e. of a stable diffusion. We obtain first some estimates for the distribution of maximal magnitude of stable jump-diffusion X generalising partially the corresponding well known estimates for stable processes (see e.g. [Bi]). Then we obtain the principle of approximate scaling and the principle of approximate independence of increments, and finally apply these results to the study of the lim sup behaviour of $|X(t)|$ as $t \to 0$. For simplicity we have reduced here the discussion to the study of the stable diffusions only (studied analytically in Section 3), and will not go into the details of modifications needed to cover more general process of Sections 4, 5. The investigations of the sample path properties of the Feller-Courrége processes with pseudodifferential generators of type (0.3) are becoming popular now, see e.g. [Schi] and references therein for some recent results in this direction.

For any event B, by $P(B)$ we shall denote the probability of B with respect to the probability space on which X is defined. From the general theory of Feller processes it follows (see e.g. [RY]) that (i) one can choose a modification of $X(t, x_0)$ with the cadlag property, i.e. such that almost surely the trajectories $X(t, x_0)$ are right continuous and have left-hand limits, (ii) the natural filtration of σ-algebras corresponding to $X(t)$ is right-continuous. We shall suppose from now on that these conditions are satisfied. Let $\alpha \in (0, 2)$ (the case $\alpha = 2$ is, on the one hand, well studied, and on the other hand, it displays essentially different properties to $\alpha \neq 2$). We suppose throughout this section that the assumptions of Theorem 3.1 are satisfied. Consequently, the Green function

$u_G^{st}(t, x, x_0)$ constructed in Theorem 3.1 defines the transition probability density (from x to x_0 in time t) for the process X. The connection between the analytic language and the probabilistic language that we shall use in this section can be expressed essentially by the formula

$$\int_A P(X(t, y) - y \in d\xi) f(\xi) = \int_A u_G^{st}(t, y, \xi) f(\xi) \, d\xi$$

for $A \subset \mathcal{R}^d$. Let

$$X^\star(t, x_0) = \sup\{|X(s, x_0) - x_0| : s \in [0, t]\}, \quad X^\star(t) = X^\star(t, 0).$$

Theorem 6.1. *For any t_0 there exist positive constants C, K such that for all $t \leq t_0$, x_0, and $\lambda > K t^{1/\alpha}$*

$$C^{-1} t \lambda^{-\alpha} \leq P(X^\star(t, x_0) > \lambda) \leq C t \lambda^{-\alpha}. \tag{6.1}$$

Proof. Since we shall use the uniform estimates from Theorem 3.2, it will be enough to consider only $x_0 = 0$ in (6.1). Plainly $P(X^\star(t) > \lambda) \geq P(|X(t)| > \lambda)$, and thus the left hand side inequality in (6.1) follows directly from Theorem 3.2. Turning to the proof of the right hand side inequality, denote by T_a the first time when the process $X(t)$ leaves the ball $B(a)$, i.e. $T_a = \inf\{t \geq 0 : |X(t)| > a\}$. Notice now that due to Theorem 3.2,

$$P(|X(s) - x_0| \geq \lambda/2) = O(s)\lambda^{-\alpha} = O(K^{-\alpha})$$

uniformly for all x_0 and $s \leq t$. Therefore, due to the homogeneity and the strong Markov property, one has that

$$P(|X(t)| > \lambda/2) \geq P((X^\star(t) > \lambda) \cap (|X(t)| > \lambda/2))$$

$$\geq \int_0^t P(T_\lambda \in ds) P(|X(t - T_\lambda, X(T_\lambda)) - X(T_\lambda)| \leq \lambda/2)$$

$$\geq (1 - O(K^{-\alpha})) \int_0^t P(T_\lambda \in ds) \geq (1 - O(K^{-\alpha})) P((X^\star(t) > \lambda).$$

It follows that

$$P((X^\star(t) > \lambda) \leq (1 - O(K^{-\alpha})) P(|X(t)| > \lambda/2),$$

which implies the r.h.s. inequality in (6.1), again due to Theorem 3.2.

Let us formulate now explicitly the main tools in the investigation of the sample path properties of stable diffusions that can be used as substitutes to the scaling property and the independence of increments, which constitute the main tools in studying the stable Lévy motions (i.e. stable diffusions with constant coefficients).

Proposition 6.1. Local principle of approximate scaling. *There exists C such that for $t \leq 1$ and all x, x_0*

$$C^{-1} u_G^{st}(1, xt^{1/\alpha}, x_0 t^{1/\alpha}) \leq u_G^{st}(t, x, x_0) \leq C u_G^{st}(1, xt^{1/\alpha}, x_0 t^{1/\alpha}). \qquad (6.2)$$

Proof. It follows directly from Proposition 3.1, its corollary and Theorem 3.2.

Remark. Practically it is more convenient to use the following equivalent form of (6.2):

$$C^{-1} u_\alpha(1, (x - x_0) t^{1/\alpha}, G(\eta), A(\eta)) \leq u_G^{st}(t, x, x_0)$$

$$\leq C u_\alpha(1, (x - x_0) t^{1/\alpha}, G(\eta), A(\eta)). \qquad (6.3)$$

which holds for all x, x_0, η and $t \leq t_0$.

Proposition 6.2. Local principle of approximate independence of increments. *For any t_0 there exists a constant C such that if $0 \leq s_1 < t_1 \leq s_2 < t_2 \leq t_0$, M_1, M_2 are any measurable sets in $\mathcal{R}^d$ and x_0 is any point in $\mathcal{R}^d$, then*

$$C^{-1} P(X(t_1, x_0) - X(s_1, x_0) \in M_1) P(X(t_2, x_0) - X(s_2, x_0) \in M_2)$$

$$\leq P((X(t_1, x_0) - X(s_1, x_0) \in M_1) \cap (X(t_2, x_0) - X(s_2, x_0) \in M_2))$$

$$\leq C P(X(t_1, x_0) - X(s_1, x_0) \in M_1) P(X(t_2, x_0) - X(s_2, x_0) \in M_2). \qquad (6.4)$$

Proof. Consider for brevity only the case $s_2 = t_1$, the case $s_2 > t_1$ being similar. Also by homogeneity one can set $s_1 = 0$ without the loss of generality. Then, due to Theorem 3.2 and the Markov property, one has

$$\int_{M_1} P(X(t_1, x_0) \in dy) P(X(t_2, x_0) - X(t_1, x_0) \in M_2)$$

$$\leq C P(X(t_2 - t_1) \in M_2) \times \int_{M_1} P(X(t_1, x_0) \in dy),$$

which implies the r.h.s. inequality in (6.4), again due to Theorem 3.2. Similarly one obtains the l.h.s. inequality in (6.4).

As an example of the application of these general results, let us prove now that the well known integral test (discovered first in [Kh]) on the limsup behaviour of the stable processes is valid also for stable diffusions.

Theorem 6.2. *Let $f : (0, \infty) \to (0, \infty)$ be an increasing function. Then $\limsup_{t \to 0}$ of the function $(|X(t)|/f(t))$ is equal to 0 or ∞ almost surely according as the integral $\int_0^1 f(t)^{-\alpha} dt$ converges or diverges.*

Proof. The proof generalises the arguments given in [Ber] for the proof of the corresponding result for one-dimensional stable Lévy motions. Suppose first that the integral converges. This implies in particular that $t^{1/\alpha} = o(f(t))$ as $t \to 0$, which means that $f(t/2) t^{-1/\alpha} > K$ for $t \leq t_0$ with small enough t_0, where K is the constant from Theorem 6.1. Consequently, for every positive integer n,

$$P(X^\star(2^{-n}) > f(2^{-n-1})) \leq C2^{-n}(f(2^{-n-1}))^{-\alpha}.$$

Since the series $\sum 2^{-n}(f(2^{-n-1}))^{-\alpha}$ converges, the Borel-Cantelli lemma yields $X^\star(2^{-n}) \leq f(2^{-n-1})$ for all n large enough, almost surely. It follows that $X^\star(t) \leq f(t)$ for all $t > 0$ small enough, almost surely (because, if $n = n(t)$ denotes the maximal natural number such that $t \leq 2^{-n}$, then $X^\star(t) \leq X^\star(2^{-n}) \leq f(2^{-n-1}) \leq f(t)$). Using the function ϵf instead of f in the above arguments yields $X^\star(t)/f(t) \leq \epsilon$ for any ϵ and all small enough t almost surely. Consequently, $\lim_{t\to 0}(X^\star(t)/f(t)) = 0$, almost surely.

Now let the integral diverge. Let $a > 0$. For any integer $n > 0$ consider the event $A_n = A'_n \cap B_n$ with

$$B_n = \{|X(2^{-n})| \leq a2^{-n/\alpha}\},$$

$$A'_n = \{|X(2^{-n-1}) - X(2^{-n})| \geq f(2^{-n+1}) + a2^{-n/\alpha}\}.$$

Due to the local principle of approximate scaling,

$$C^{-1}P(|X(1)| \leq a) \leq P(B_n) \leq CP(|X(1)| \leq a)$$

for all n uniformly. Consequently, by the Markov property, homogeneity and again approximate scaling, one has

$$P(A_n) \geq C^{-1}P(|X(1)| \leq a) \min_{|\xi| \leq a2^{-n}} P(|X(2^{-n}, \xi) - \xi| \geq f(2^{-n+1}) + a2^{-n/\alpha})$$

$$\geq C^{-2}P(|X(1)| \leq a)P(|X(1)| \geq 2^{n/\alpha}f(2^{-n+1}) + a)$$
$$\geq \tilde{C}P(|X(1)| \leq a)(2^{n/\alpha}f(2^{-n+1}) + a)^{-\alpha}.$$

with some positive $\tilde{C}$. The sum $\sum 2^{-n}(f(2^{-n+1}) + a2^{-n/\alpha})^{-\alpha}$ diverges, because it represents a Riemann sum for the integral of the function

$$(f(t) + a(t/2)^{1/\alpha})^{-\alpha} \geq 2\max(a2^{-1/\alpha}t^{-1}, f(t)^{-\alpha})$$

and consequently the sum $\sum P(A_n)$ is divergent. Notice now that though the events A_n are not independent, the events A'_n and A'_m are "approximately independent" in the sense of Proposition 6.2, if $n \neq m$. The same remark concerns the events B_n and A_n. Therefore, if $n \neq m$,

$$P(A_n \cap A_m) \leq P(A'_n \cap A'_m) = O(1)P(A'_n)P(A'_m)$$

$$= O(1)P(A_n)P(A_m)P(B_n)^{-1}P(B_m)^{-1} = O(1)P(A_n)P(A_m)P(|X(1)| \leq a)^{-2},$$

and one can use the well known generalisation of the second Borel-Cantelli lemma for dependent event (see e.g. [Sp]) to conclude that $\limsup_{t\to 0}(X(t)/f(t)) \geq 1$ with positive probability, and therefore almost surely, due to the Blumenthal zero-one law (see e.g. [RY]). Repeating the same arguments for the function $\epsilon^{-1}f$ instead of f one gets that $\limsup_{t\to 0}(X(t)/f(t)) = \infty$ almost surely, which completes the proof of the theorem.

Chapter 6. SEMICLASSICAL ASYMPTOTICS
FOR THE LOCALISED FELLER-COURRÈGE PROCESSES

1. Maslov's tunnel equations and the Feller-Courrège processes

In Chapter 3 we have constructed the asymptotics sa $h \to 0$ of the Green function for diffusion equation in form (3.1.9) for almost all t, x. Real function (3.1.9) is exponentially small as $h \to 0$ for almost all t, x (where $S(t, x) \neq 0$) and it satisfies "the large deviation principle":

$$\lim_{h \to 0} h \log u(t, x, x_0) = S(t, x, x_0)$$

is almost everywhere a positive finite function (one can consider also S to be infinite in some points to include for instance degenerate non-regular diffusions, see e.g. Sect. 3.6). Natural question arises, what is the class of differential or pseudo-differential equations of type (3.1.2) (see Appendix D for main notations of the theory of ΨDE) with real H, for which the Green function has the small h asymptotics of form (3.1.9). Looking for the answer to this question, V.P. Maslov gave in [M3] the following definition.

Definition. *Continuous function $H(x, p)$ on $\mathcal{R}^d \times \mathcal{C}^d$ is called a Hamiltonian of tunnel type, if it has the following properties*

(i) H is smooth in x and holomorphic in p for $p \in \mathcal{C}^d \setminus \{Re\, p = 0\}$,

(ii) for real p the Hamiltonian $H(x, p)$ is real, the Lagrangian $p\frac{\partial H}{\partial p}(x, p) - H(x, p)$ is non-negative, and $\det \frac{\partial^2 H}{\partial p^2}(x, p) \neq 0$,

(iii) for real p the function $H(x, ip)$ and all its derivatives in x and p increase at most polynomially as $x, p \to \infty$,

(iv) main tunnel condition is satisfied:

$$\max_{\eta \in \mathcal{R}^d} Re H(x, p + i\eta) = H(x, p), \quad p \in \mathcal{R}^d, p \neq 0.$$

This definition was generalised to the case of systems of ΨDO in [M3], where also many examples of important equations from physics and probability theory satisfying these conditions were given. Maslov conjectured that the asymptotic Green function of tunnel equations can be given by (3.1.9) and gave some heuristic arguments in support of this hypothesis. In fact, one naturally comes to the definition of tunnel equations if one considers first the equations with constant coefficients and tries to ensure the possibility to come to formula (3.1.6) by carrying out the Fourier transform method of obtaining the Green function for diffusion equation. Rigorously, the problem of describing the class of equation (3.1.2) with asymptotic Green function of form (3.1.9) is open. In order to give at least partial answer to this problem we are going first to restrict the class of tunnel equations approaching the question from another point of view.

Notice that due to the construction of asymptotic formula (3.1.9) for diffusion, the amplitude $\phi(t, x, x_0)$ there is everywhere positive (whenever it is finite), because the Jacobian $J(t, x, x_0)$ considered along an extremal can change his sign only at a focal point and there are no such points on a minimising extremal (see Chapter 2). Therefore, the generalisation of the construction of Chapter 3 leads necessarily to positive asymptotics of the Green functions. Therefore, natural candidates to the equations with an asymptotic Green function of form (3.1.9) are the equations preserving positivity. But the equations with this property are well known. Essentially they describe the evolutions of the Feller-Courrège processes (or general jump-diffusions), see e.g. Appendix D. We conclude that it is reasonable to look for the generators of Maslov's tunnel equations in the class of Lévy-Khintchine ΨDO with symbols of type (D3). However, since for carrying out the WKB construction for equation (D6),(D7), one needs the values of $H(x, p)$ for real p (or, equivalently, the values of the symbol $\Psi(x, p) = H(x, -ip)$ for imaginary p), it is necessarily to restrict the class of symbols (D3) to the case of the Lévy-Khintchine measures decreasing fast at infinity. We shall consider here the simplest case of the Lévy-Khintchine measures with a bounded support in $\mathcal{R}^d$. The class of symbols (D3) (and corresponding PDO, evolutionary PDE (D1) and Hamiltonians (D6)) with the Lévy measures having a bounded support will be called the localised Lévy-Khintchine symbols (ΨDO, ΨDE, Hamiltonians). The corresponding semigroups and stochastic processes will be called the localised Feller-Courrège semigroups and processes.

It turns out that the two approaches in the search of equations with the asymptotic Green functions of form (3.1.9) (tunnel condition that comes from the Fourier transform method and the arguments concerning the conservation of positivity) lead to close results, because as one readily verifies, the localised Lévy-Kchinchine Hamiltonians of form (D6) are entire analytic functions with respect to p and belong to the class of Maslov's tunnel Hamiltonians as defined above. On the other hand, under some additional assumptions on the regularity of the boundary of the support of $\nu(x, d\xi)$ (see Propositions 2.5.1, 2.5.2) the Lévy-Khintchine Hamiltonians $H(x, p)$ belong to the class of the Hamiltonians of the exponential (or even uniform exponential) growth as defined in Section 2.5 and one can use the theory from this Section to construct the solutions of the boundary-value problem for such Hamiltonians and thus to be able to carry out the construction of the semi-classical approximation for the Green function of the localised Lévy-Khintchine ΨDE following the steps described in Section 3.1.

It turns out however that the procedure of Chapter 3 can not be successfully applied directly to such ΨDE. Let us indicate, where the problem lies. Substituting a function of form (3.1.6) in equation (D6),(D7) and using the formulas of the commutation of a ΨDO with an exponential function (see e.g. (D11), (D12)) one finds similarly to the case of quadratic Hamiltonians that

$$h\frac{\partial u}{\partial t} - H(x, -h\nabla)u = h^2 F(t, x; x_0, h)$$

with some F, whenever S and ϕ satisfy the Hamilton-Jacobi and the transport equations (3.1.11), (3.1.13) respectively. The results of Section 2.5 ensure that the two-point function $S(t, x, x_0)$ and the amplitude (3.1.18) are smooth and well-defined in a sufficiently large neighbourhood of x_0. However, using the estimates for the Jacobian from Section 2.5 one sees that the remainder F is fast increasing as $t \to 0$ in such a way that already the first term in the series (3.4.9) giving the representation of the exact Green function may not exist. On the other hand, it follows directly from Theorem 5.4.2 that the WKB asymptotics of form (3.1.9) is not uniform for small h and small t, simply because the behaviour of u_G is different from (3.1.9) for small times t. By the way, the same thing happens for non-regular degenerate diffusion from Section 3.6.

The aim of this chapter is to present a method of the justification of the asymptotics of form (3.1.9) for the Green function of the localised Lévy-Khintchine ΨDE for finite, "not very small" times, more precisely, for $t \in [\delta, t_0]$ with some t_0 and any fixed positive $\delta < t_0$. It will imply the global large deviation principle for these Green function and the validity of representation (3.1.9) almost everywhere. The main ingredient in this method is the construction of the uniform small time and small h asymptotics that differs from (3.1.9). This asymptotic formula is obtained by sewing the formulas of type (5.4.11) and (3.1.9) and it turns to (3.1.9) for any fixed t. We shall not try here to describe the most general class of the localised Lévy measures in (D6) that allow to carry out this construction but will give the full proof for the class of $\nu(x, d\xi)$ corresponding to the localised stable processes (see e.g. Appendix C) with the uniform spectral measure disturbed by a compound Poisson process, i.e. for the equations of form (5.4.1). More precisely, we shall consider the equation

$$\frac{\partial u}{\partial t} = (A(x), \frac{\partial u}{\partial x}) + \frac{1}{h} \int_{\mathcal{R}^d} \left(u(x + h\xi) - u(x) - \frac{h}{1 + \xi^2}(\xi, \frac{\partial u}{\partial x}) \right) \nu(x, d\xi) \quad (1.1)$$

with

$$\nu(x, d\xi) = (G(x)\Theta_{a(x)}(|\xi|)|\xi|^{-(d+\alpha)} + g(x, \xi)) \, d\xi, \quad (1.2)$$

where the following assumptions are made:

(A1) $A(x), G(x), a(x)$ are smooth functions such that G and a are positive and $|A(x)|$, $G(x)$, $a(x)$, $G^{-1}(x)$, $a^{-1}(x)$ are uniformly bounded;

(A2) the nonnegative function $g(x, \xi)$ depends smoothly on x, is uniformly bounded and vanishes for $|\xi| \geq a(x)$;

(A3) $A(x) \equiv 0$ whenever $\alpha \leq 1$.

Similarly to equation (1.1) one can treat the case of the stable process of the index $\alpha = 2$ (i.e. diffusion) disturbed by a compound Poisson process, namely the equation of the form

$$\frac{\partial u}{\partial t} = (A(x), \frac{\partial u}{\partial x}) + \frac{h}{2} tr \, (G(x)\frac{\partial^2 u}{\partial x^2}) + \frac{1}{h} \int_{\mathcal{R}^d} (u(x + h\xi) - u(x))g(x, \xi) \, d\xi \quad (1.3)$$

with a bounded nonnegative integrable function g having a finite support with respect to the second argument.

The main result of this Chapter is the following.

Theorem 1.1. *If h_0 and t_0 are small enough, then in the domain $h \leq h_0$, $t \in [\delta, t_0]$ with any $\delta > 0$, the Green function $u_G = u_G(t, x, x_0, h)$ of equation (1.1), (1.2) has the form*

$$(2\pi h)^{-d/2} \Theta_{0,c}(|x - x_0|) \phi(t, x, x_0) \exp\{-\frac{S(t, x, x_0)}{h}\}(1 + O(h^\omega)) + O(\exp\{-\frac{\Omega}{h}\})$$

with some $c > 0$, $\Omega > 0$, $\omega \in (0,1)$, where $S(t, x, x_0)$ is the two-point function corresponding to the Hamiltonian of equation (1.1) and ϕ is the solution (1.18) of the corresponding transport equation, and the last term in (1.4) is an integrable function of x.

This fact follows from more general result proven in Section 6.3. As a consequence, one obtains all global formulas of Section 3.5 for the stable jump-diffusion defined by equation (1.1), (1.2). Since the whole construction is rather complicated, we devote a special Section 6.2 to present first a rough small t and small h asymptotics of the Green function, which does not give the correct amplitude for small h but captures only the correct logarithmic limit.

The idea to represent the uniform small time and small h asymptotics for the Green functions of general tunnel equations in the integral form of type (3.1) from Section 6.3 belongs to V.P.Maslov, see his heuristic arguments in [M2]. Notice for conclusion that one can try also to justify the asymptoics for a fundamental solution not in the pointwise sense, as above, but in the sense of distribution. An approach to the construction and justification of such "weak" semiclassical asymptotics for tunnel equation was proposed in [Dan], where the case of diffusion process perturbed by a compound Poisson process was considered (under additional, rather restrictive, assumptions).

2. Rough local asymptotics and local large deviations

This section is devoted to the construction of a rough local asymptotics (but uniform in small t and h) to the Green function of equation (1.1) under assumptions (A1)-(A3) that are supposed to be satisfied throughout the Section. Equation (1.1) is of form (D6) with the Hamiltonian

$$H(x, p) = -(A(x), p) + \int_{\mathcal{R}^d} \left(e^{-(p,\xi)} - 1 + \frac{(p, \xi)}{1 + \xi^2} \right) \nu(x, d\xi). \qquad (2.1)$$

Moreover, it is of form (5.4.1) and therefore it has a continuous Green function due to theorem 5.4.1.

On the other hand, Hamiltonian (2.1) belongs to the class of the Hamiltonians of the uniform exponential growth described in Theorem 2.5.3 with the function $a(x, \bar{p})$ from (2.5.19) not depending on $\bar{p}$ and being equal to $a(x)$ from (1.2). Therefore, Theorems 2.5.1, 2.5.2 and Propositions 2.5.5-2.5.7 hold for H. We shall use further the notations of Section 2.5. In particular, let c be chosen in such a way that the boundary value problem for the Hamiltonian system with the Hamiltonian H is uniquely (in the sense of Theorem 2.1 or Proposition 5.6) solvable for $|x - x_0| \leq 2c$ and $t \leq t_0$. Let $z(t, v, x_0)$ be the function defined in Theorem 2.5.2 and let

$$y(t, x, x_0) = tz(t, \frac{x - x_0}{t}, x_0). \qquad (2.2)$$

This function is well defined and smooth for $|x - x_0| < 2c$. Let $\chi(t, y)$ be the smooth molyfier (as the function of the second variable) of the form $\chi_{c_1}^{c_2}$ from Lemma E1 with $0 < c_1, c_2 < c$. Then the function

$$u_{rough}(t, x, x_0, h) = \frac{\chi(t, |x - x_0|)}{(2\pi h)^d} \int_{\mathcal{R}^d} \exp\{-\frac{ipy(t, x, x_0) - H(x_0, ip)t}{h}\} \, dp \qquad (2.3)$$

or equivalently

$$u_{rough}(t, x, x_0, h) = \frac{\chi(t, |x - x_0|)}{(2\pi h)^d} \int_{\mathcal{R}^d} \exp\{\frac{\Phi(x_0, p)t + ipy(t, x, x_0)}{h}\} \, dp \qquad (2.4)$$

with $\Phi(x, p) = H(x, -ip)$ is a well defined smooth function for all x and small enough t. The aim of this Section is to prove that this function presents a uniform small t and small h asymptotics to the Green function u_G of equation (1.1), (1.2). First of all, in the next two propositions, we shall show that for small t this function turns to the function u_G^{as} from Section 5.3 that presents a small time asymptotics for u_G, and for small h it has the form (3.1.9) with S being the two-point function of the variational problem corresponding to the Hamiltonian (2.1).

We start with a simple lemma. Recall that the function $u_\alpha(t, x - x_0; G, A, h)$ was defined in (5.3.5), (5.3.6). As in Chapter 3 we shall often omit the last arguments in this function writing $u_\alpha(t, x - x_0)$ for $u_\alpha(t, x - x_0; G(x_0), A(x_0), h)$.

Lemma 2.1. *Under the assumptions of Proposition 5.3.1 there exists a constant C such that*
(i) for any y and $|x + A(x_0)t - x_0| \leq (th^{\alpha-1})^{1/\alpha}$

$$u_\alpha(t, x + y - x_0) \leq Cu_\alpha(t, x - x_0)$$

(ii) for any y and $|x + A(x_0)t - x_0| \geq (th^{\alpha-1})^{1/\alpha}$

$$\frac{u_\alpha(t, x + y - x_0)}{u_\alpha(t, x - x_0)} \leq C \min\left(\frac{|x + A(x_0)t - x_0|^{d+\alpha}}{|x + A(x_0)t - x_0 + y|^{d+\alpha}}, \frac{|x + A(x_0)t - x_0|^{d+\alpha}}{(th^{\alpha-1})^{1+d/\alpha}}\right),$$

(iii) if $y/|x - x_0|$ is bounded, then for any x

$$u_\alpha(t, x + y - x_0) = u_\alpha(t, x - x_0)(1 + O\left(\frac{y}{|x - x_0|}\right)). \qquad (2.5)$$

Proof. Statements (i) and (ii) follow directly from Proposition 5.3.1. To get (2.5) one writes

$$u_\alpha(t, x + y - x_0) = u_\alpha(t, x - x_0) + \frac{\partial u_\alpha}{\partial x}(t, x + \theta y - x_0)y$$

with some $\theta \in (0,1)$, then one uses (5.3.11),(5.3.12) to estimate $\frac{\partial u_\alpha}{\partial x}(t, x+\theta y - x_0)$ by means of $u_\alpha(t, x + \theta y - x_0)$, and then uses the cases (i),(ii) to estimate $u_\alpha(t, x + \theta y - x_0)$ by means of $u_\alpha(t, x - x_0)$, which yields (2.5).

Proposition 2.1. *For any $h_0 > 0$ there exists $t_0 > 0$ such that uniformly in the domain $t \le t_0, h \le h_0, t/h \le t_0/h_0$*

$$u_{rough}(t, x, x_0, h) = \chi(t, |x - x_0|)$$

$$\times (1 + O(t/h) + O(1) \min(1, |x - x_0|, (|x - x_0|/h)^{d+\alpha})) u_G^{as}(t, x - x_0, h). \quad (2.6)$$

In particular, $u_{rough} = O(u_\alpha)$, and the function u_{rough} gives a local multiplicative asymptotics to the Green function u_G for small t and t/h in the sense that u_{rough} satisfies Dirac's initial conditions (5.3.2) and u_G can be presented in form (5.4.10) with u_{rough} instead of u_α.

Proof. If $y(t, x, x_0) = (x - x_0)$ in (2.5), then all statements were proved in Theorem 5.4.2. But due to (2.5.16),

$$y(t, x, x_0) = (x - x_0)(1 + O(|x - x_0|)).$$

Consequently, using $y(t, x, x_0)$ instead of $x - x_0$ in (2.3) means the shift in the argument of the function u_G^{as}, which, due to (2.5), amounts to the additional multiplier of form $1 + O(|x - x_0|) = 1 + O(c)$, which completes the proof of the Proposition.

For any real g, we shall denote $\max(0, g)$ by g^+.

Proposition 2.2. *Let either h/t be bounded or $h(1 + \log^+ \frac{|x - x_0|}{t})^3/|x - x_0|$ and t/h be bounded. Let $\tilde{h} = h/t$ or $\tilde{h} = h(1 + \log^+ \frac{|x - x_0|}{t})/|x - x_0|$ in the first or in the second case respectively. Then for small enough t in the first case and for small enough t and $\tilde{h}$ in the second one*

$$u_{rough}(t, x, x_0, h) = \chi(t, |x - x_0|) u_{rough}^{exp}(t, x, x_0, h)(1 + O(\tilde{h})) \quad (2.7)$$

with

$$u_{rough}^{exp}(t, x, x_0, h) = (2\pi h t)^{-d/2} \left(\det \frac{\partial^2 H}{\partial p^2}(x_0, \hat{p}) \right)^{-1/2} \exp\{-\frac{S(t, x, x_0)}{h}\},$$

$$(2.8)$$

where $\hat{p} = \hat{p}(t, x, x_0)$ is the (obviously unique) solution of the equation

$$\frac{\partial H}{\partial p}(x_0, \hat{p}(t, x, x_0)) = \frac{1}{t} y(t, x, x_0).$$

Moreover,

$$\hat{p}(t, x, x_0) = p_0(t, x, x_0) + O(|x - x_0|), \quad (2.9)$$

where $p_0(t, x, x_0)$ denotes, as in Section 2.1, the initial momentum on the trajectory of the Hamiltonian flow joining x_0 and x in time t.

Proof. To calculate the integral in (2.3) we shall use the method of the saddle-point. Namely, one readily sees that the phase

$$\Sigma(p, t, x, x_0) = ipy(t, x, x_0) - H(x_0, ip)t \qquad (2.10)$$

in (2.3) is an entire analytic function in p that has a unique imaginary saddle-point $-i\hat{p}(t, x, x_0)$. Formula (2.9) for $\hat{p}$ follows from (2.5.31). Let us shift the contour of integration in (2.3) by $\hat{p}$, which is possible due to the Cauchy theorem and Proposition C1 from Appendix C. This amounts to writing $\hat{p} + ip$ instead of ip in (2.3). Hence

$$u_{rough}(t, x, x_0, h) = \frac{\chi(t, |x - x_0|)}{(2\pi h)^d} \exp\{-\frac{S(t, x, x_0)}{h}\}$$

$$\times \int_{\mathcal{R}^d} \exp\left\{-\frac{(H(x_0, \hat{p}) - H(x_0, \hat{p} + ip))t + ipy(t, x, x_0)}{h}\right\} dp, \qquad (2.11)$$

where we have used also the formula

$$S(t, x, x_0) = \hat{p}y(t, x, x_0) - H(x_0, \hat{p})t, \qquad (2.12)$$

which is equivalent to (2.5.17).

To calculate the integral in (2.11) we apply Proposition B2 considering $\tilde{h}$ as a small parameter. We omit the details concerning a simpler case $h \leq t$ and consider only the situation with $t \leq h \leq |x - x_0|/(1 + \log^+ \frac{|x-x_0|}{t})$. The assumptions (1)-(5) of Proposition B2 are satisfied for the phase

$$\hat{\Sigma}(p) = \frac{(H(x_0, \hat{p}) - H(x_0, \hat{p} + ip))t + ipy(t, x, x_0)}{|x - x_0|/(1 + \log^+ \frac{|x-x_0|}{t})^3} \qquad (2.13)$$

of the Laplace integral (2.11), because on the one hand

$$Re\,\hat{\Sigma}(p) = \int_{\mathcal{R}^d} e^{-(\hat{p}, \xi)}(1 - \cos(p\xi))\nu(x_0, d\xi)\frac{t(1 + \log^+(|x - x_0|/t))^3}{|x - x_0|},$$

and thus $Re\,\hat{\Sigma}$ is positive everywhere except the origin where it vanishes, and on the other hand $Re\hat{\Sigma}(p) \to \infty$ as $|p| \to \infty$, due to Proposition C1. Moreover, from (1.2) it follows that for the phase $\hat{\Sigma}$ one can takes the constant r from assumptions (6)-(7) of Proposition B2 to be independent from $\hat{p}$ and x_0 (for instance, one can take $r = \pi/3 \max a(x)$) and such that

$$Re\,\frac{\partial^2 \hat{\Sigma}}{\partial p^2} \geq C_1(1 + \log\frac{|x - x_0|}{t})^2 \qquad (2.14)$$

with some positive constant C_1, because for $|p| < r$ one has

$$\frac{\partial^2}{\partial p^2}\int e^{-(\hat{p}, \xi)}(1 - \cos(p, \xi))\nu(x, d\xi) \geq \frac{1}{2}\frac{\partial^2}{\partial p^2}\int(e^{-(\hat{p}, \xi)} - 1 + \frac{(\hat{p}, \xi)}{1 + \xi^2})\nu(x, d\xi),$$

and the r.h.s. of this inequality was estimated in Theorem 2.5.3. Since the amplitude of our Laplace integral is a constant, only the first two terms in (B12) are relevant. Since the derivatives of $\hat{\Sigma}$ with respect to p are given in terms of the derivatives of H that satisfy the assumption (i) of Definition 2.5.1 and due to (2.14) one readily sees that the constant A from Proposition B1 is bounded in our situation.. Therefore, $\delta_2(\tilde{h})$ is also bounded. Similarly one gets that $\delta_1(\tilde{h})$ is also bounded for small t. It remains to compare $\delta_3(\tilde{h})$ with the principle term in (B13). Due to (2.14) and Proposition C1, $\delta_3(\tilde{h})$ does not exceed

$$\exp\left\{-\frac{r^2 C_1(1 + \log^+(|x - x_0|/t))^2}{\tilde{h}}\right\}$$

$$\times \int \exp\left\{-\frac{\sigma(x_0)|p|^\alpha t(1 + \log^+(|x - x_0|/t))^3}{\tilde{h}_0|x - x_0|}\right\} dp.$$

Suppose $\log(|x - x_0|/t) \geq 1$ (the case of bounded $|x - x_0|/t$ is simpler and is omitted). Carrying out the integration yields

$$\delta_3(\tilde{h}) = O(1)\left(\frac{t}{|x - x_0|}\right)^{r^2 C_1/\tilde{h}}\left(\frac{|x - x_0|\tilde{h}_0}{t(1 + \log^+(|x - x_0|/t))^3}\right)^{d/\alpha}.$$

To complete the proof of the Proposition, we need to show that

$$(2\pi h)^{-d}\delta_3(\tilde{h}) = O(\tilde{h})(2\pi h t)^{-d/2}\left(\det\frac{\partial^2 H}{\partial p^2}(x_0, \hat{p})\right)^{-1/2}. \qquad (2.15)$$

Due to (2.9) and Theorem 2.5.3,

$$\frac{\partial^2 H}{\partial p^2}(x_0, \hat{p}) = O(|x - x_0|)t^{-1}. \qquad (2.16)$$

Hence, due the expression for δ_3, in order to prove (2.15) it is enough to show that

$$\left(\frac{t}{|x - x_0|}\right)^{r^2 C_1/2\tilde{h}}\left(\frac{|x - x_0|}{t(1 + \log^+(|x - x_0|/t))^3}\right)^{d/\alpha}$$

$$\leq \left(\frac{2\pi h}{t}\right)^{d/2}\left(\frac{|x - x_0|}{t}\right)^{-d/2} = \left(\frac{2\pi h}{|x - x_0|}\right)^{d/2},$$

which holds obviously for small enough $\tilde{h}_0$, because t/h is bounded. Hence, formula (2.7) follows from Proposition B2 and the above obtained estimates for δ_1, δ_2, δ_3.

Since u_α presents a multiplicative asymptotics for u_{rough} in the domains $\{t < h, |x - x_0| < h\}$, one sees that, roughly speaking, the domain $\{\max(t, |x - x_0|/|\log t|) < h < |x - x_0|$ is the boundary layer between the asymptotic representations u_α and u_{rough}^{exp}. In this domain we know that u_{rough} is bounded by

u_α, but this estimate is too rough and will be improved later on. We turn now to the estimate of the result of the substitution of function (2.4) in equation (1.1), i.e. of the function

$$F(t, x, x_0, h) = \left[\frac{\partial}{\partial t} - \frac{1}{h}H(x, -h\nabla)\right] u_{rough}. \tag{2.17}$$

Again we consider separately the cases of small h and small t.

Recall that $\Theta_{b,c}$ denotes the characteristic function of the closed interval $[b, c]$, i.e. $\Theta_{b,c}(y)$ is equal to one (resp. to zero) for $y \in [b, c]$ (resp. otherwise)

Proposition 2.3. *Under the assumptions of Proposition 2.2 for small t and $\tilde{h}$*

$$F = O(u_{rough}^{exp})\left[1 + \left(\frac{|x - x_0|}{t}\right)^{1+O(|x-x_0|)}\right]. \tag{2.18}$$

Proof. Using the formula of the commutation of a Lévy-Khintchine ΨDO with an exponential function, see Proposition D2, yields

$$F = -\frac{1}{(2\pi h)^d h}\int_{\mathcal{R}^d}(f_1 + hf_2 + hf_3 + hf_4)\exp\{-\frac{ipy(t, x, x_0) - H(x_0, ip)t}{h}\}\,dp \tag{2.19}$$

with

$$f_1 = f_1(t, x, p) = \chi(t, |x - x_0|)\left(\frac{\partial\Sigma}{\partial t} + H(x, \frac{\partial\Sigma}{\partial x})\right)$$

$$= \chi(t, |x - x_0|)\left(ip\frac{\partial y}{\partial t} - H(x_0, ip) + H(x, ip\frac{\partial y}{\partial x}(t, x, x_0))\right),$$

$$f_2 = \frac{1}{h}\chi(t, |x - x_0|)\int\nu(x, d\xi)$$

$$\times\left(\exp\{-hi\int_0^1(1 - s)(p_j\frac{\partial^2 y_j}{\partial x^2}(x + sh\xi)\xi, \xi)\,ds\} - 1\right)\exp\{-i(p, \frac{\partial y}{\partial x}\xi)\},$$

$$f_3 = -\frac{\partial\chi}{\partial t} - \left(\frac{\partial\chi}{\partial x}, \frac{\partial H}{\partial p}(x, ip\frac{\partial y}{\partial x}(t, x, x_0))\right) + \int\nu(x, d\xi)$$

$$\times\left(\frac{\partial\chi}{\partial x}, \xi\right)\left(\exp\{\frac{h}{i}\int_0^1(1 - s)(p_j\frac{\partial^2 y_j}{\partial x^2}(x + sh\xi)\xi, \xi)\,ds\} - 1\right)\exp\{-i(p, \frac{\partial y}{\partial x}\xi)\},$$

$$f_4 = h\int_{\mathcal{R}^d}\left(\int_0^1(1 - s)ds\frac{\partial^2\chi}{\partial x^2}(x + sh\xi)\xi, \xi\right)$$

$$\times\exp\{-hi\int_0^1(1 - s)(p_j\frac{\partial^2 y_j}{\partial x^2}(x + sh\xi)\xi, \xi)\,ds\}\exp\{-i(p, \frac{\partial y}{\partial x}\xi)\}\nu(x, d\xi).$$

We make now the same shift of the contour of integration as in the proof of Proposition 2.2 to find that (2.19) can be rewritten as

$$-\frac{1}{(2\pi h)^d h}\exp\{-\frac{S(t, x, x_0)}{h}\}\int_{\mathcal{R}^d}(f_1 + hf_2 + hf_3 + hf_4)\exp\{-\frac{\hat{\Sigma}(p)}{h}\}\,dp \tag{2.20}$$

with f_1, f_2, f_3, f_4 being the same as above but with the argument $-i\hat{p}+p$ instead of p and with $\hat{\Sigma}(p)$ given by (2.13).

To estimate this integral we shall use Propositions B1, B2. As in the proof of Proposition 2.2 let us consider only the case $t \leq h \leq |x - x_0|/|\log t|$, where $\tilde{h} = h|\log t|/|x - x_0|$. Notice first that since all f_j, $j = 1, ..., 4$, increase as $p \to \infty$ at most polynomially, the estimate of the exponentially small remainder $\delta_2(\tilde{h})$ of form (B13) is carried out exactly in the same way as in the proof of Proposition 2.2. Hence we shall deal only with the remainder $\delta_1(\tilde{h})$ of form (B12) and shall consider r to be chosen as in the proof of Proposition 2.2.

Let us begin with the contribution of f_1 in (2.20). The first observation is that

$$f_1(t, x, -i\hat{p}(t, x, x_0)) = 0. \tag{2.21}$$

In fact, from (2.2) and (2.5.32) it follows that

$$tL(x_0, \frac{1}{t}y(t, x, x_0)) = S(t, x, x_0).$$

Differentiating in t yields

$$\frac{\partial S}{\partial t} = L(x_0, \frac{1}{t}y(t, x, x_0)) + \hat{p}\left(\frac{\partial y}{\partial t}(t, x, x_0) - \frac{1}{t}y(t, x, x_0)\right).$$

Since the two-point function S satisfies the Hamilton-Jacobi equation, and moreover,

$$\frac{\partial S}{\partial x} = \frac{\partial \hat{\Sigma}}{\partial x} = \hat{p}\frac{\partial y}{\partial x},$$

it follows that

$$L(x_0, \frac{1}{t}y(t, x, x_0)) + \hat{p}\frac{\partial y}{\partial t} - \frac{1}{t}\hat{p}y(t, x, x_0) + H(x, \hat{p}\frac{\partial y}{\partial x}) = 0,$$

which is equivalent to (2.21). Therefore, the main term in the asymptotic formula (B11) for the Laplace integral (2.20) with the amplitude f_1 vanishes. To estimate $\delta_1(\tilde{h})$ we need the estimates of $f_1(-i\hat{p}+p)$ and its first two derivatives for small p. One has

$$\frac{\partial f_1}{\partial p}(-i\hat{p} + p) = i\frac{\partial y}{\partial t} - i\frac{\partial H}{\partial p}(x_0, \hat{p} + ip) + i\frac{\partial y}{\partial x}\frac{\partial H}{\partial p}(x, (\hat{p} + ip)\frac{\partial y}{\partial x}),$$

$$\frac{\partial^2 f_1}{\partial p^2} = \frac{\partial^2 H}{\partial p^2}(x_0, \hat{p} + ip) - (\frac{\partial y}{\partial x})\frac{\partial^2 H}{\partial p^2}(x, (\hat{p} + ip)\frac{\partial y}{\partial x})\left(\frac{\partial y}{\partial x}\right)'.$$

Due to (2.5.36), (2.5.37),

$$\frac{\partial y}{\partial t} = O(|x - x_0|^2)t^{-1}, \quad \frac{\partial y}{\partial x} = 1 + O(|x - x_0|), \quad \frac{\partial^2 y}{\partial x^2} = O(1 + \log^+ \frac{|x - x_0|}{t}). \tag{2.22}$$

It follows that the last term in the above expressions for the derivatives of f_1 give the principle contribution, which is of the order $O(|x-x_0|/t)^{1+O(|x-x_0|)}$ (up to an additive constant). Consequently, all three coefficients F_0, F_1, F_2 from formula (B12) have the same order, and therefore, $\delta_1(\tilde{h})$ as well. Hence, using also Proposition 2.2, one concludes that the term in (2.20) containing f_1 contributes to the first term in (2.18).

To estimate the contributions of other terms it is convenient to take into consideration only the real part (2.16) of the phase, to estimate the magnitude of f_2, f_3, f_4 at $-i\hat{p}+p$ for small p, and then apply Proposition B1. Due to (2.22), one has

$$|\exp\{-((\hat{p}+ip), \frac{\partial y}{\partial x}\xi)\}| = O(\exp\{a(x)|\hat{p}|(1+O(|x-x_0|))\})$$

$$= O(1)\left(1+\left|\frac{x-x_0}{t}\right|^{1+O(|x-x_0|)}\right). \tag{2.23}$$

Next,

$$h\hat{p}_j\left|\frac{\partial^2 y_j}{\partial x^2}(x+sh\xi)\right| = O(h)\left(1+\log^+\left|\frac{x-x_0}{t}\right|\right)^2 = O(|x-x_0|)\log$$

is bounded. Writing down

$$\exp\{hi\left(p_j\int_0^1(1-s)\frac{\partial^2 y_j}{\partial x^2}(x+sh\xi)\,ds\,\xi,\xi\right)\}-1$$

$$= hi\left(p_j\int_0^1(1-s)\frac{\partial^2 y_j}{\partial x^2}(x+sh\xi)\,ds\,\xi,\xi\right)$$

$$\exp\{\theta hi\left(p_j\int_0^1(1-s)\frac{\partial^2 y_j}{\partial x^2}(x+sh\xi)\,ds\,\xi,\xi\right)\}. \tag{2.24}$$

one has

$$|f_2(t,x,-i\hat{p}+p)| = O(1)\left(1+\log^+\left|\frac{x-x_0}{t}\right|\right)^2\left(1+\frac{|x-x_0|}{t}\right)^{1+O(|x-x_0|)},$$

which again contributes to the first term in (2.18). The terms with f_3, f_4 do not give anything new.

Proposition 2.4. *Let t/h be bounded. If $\alpha > 1$, then*

$$F = O(u_\alpha)\left[1+\frac{|x-x_0|}{t}+\left(\frac{t}{h}\right)^{\alpha^{-1}-1}|\log t|\right.$$

$$+\Theta_{(th^{\alpha-1})^{1/\alpha},c}(|x-x_0|)\frac{|x-x_0|^2}{ht}\left(\frac{t}{h}\right)^{-1/\alpha}|\log t|\right]. \tag{2.25}$$

If $\alpha \leq 1$, let $\delta_1 = (1 - \alpha)^2 \alpha^{-1}(2 - \alpha)^{-1}$ and δ_2 be any number such that

$$\frac{1 - \alpha}{\alpha(2 - \alpha)} < \delta_2 < \frac{1}{\alpha(2 - \alpha)}.$$

Then

$$F = O(u_\alpha)\left[1 + |\log t|\frac{|x - x_0|}{t}\right.$$

$$+\Theta_{(th^{\alpha-1})^{1/\alpha}, t^{\delta_2}h^{-\delta_1}}(|x - x_0|)\frac{|x - x_0|^2}{ht}\left(\frac{t}{h}\right)^{-1/\alpha}|\log t| + \Theta_{t^{\delta_2}h^{-\delta_1}}(|x - x_0|)t^{-1}\right].$$

$$(2.25')$$

Proof. We use again formula (2.19). Let $H = H_s + H_p$ be the decomposition of H in the stable and Poissonian parts as in Theorem 4.1, i.e. $H_s(x, -ip) = \Phi_s(x, p)$ is given by (5.3.4) and $H_p(x, -ip) = \Phi_p(x, p)$ is given by (5.4.3). It is enough to consider the phase in integral (2.19) to be simply $ip(x - x_0) - H_s(x_0, ip)t$, because the use of the exact phase (2.12) will give the result that differs by a bounded multiplier (that one shows as in the proof of Proposition 2.1), which is not essential. Thus, rewriting also everything in terms of the symbol $\Phi(x, p) = H(x, -ip)$ we present the r.h.s. of (2.19) in the form

$$\frac{O(1)}{(2\pi h)^d}\int_{R^d}(\frac{f_1}{h} + f_2 + f_3 + f_4)(t, x, -p)\exp\{\frac{\Phi_s(x_0, p)t + ip(x - x_0)}{h}\}\,dp. \quad (2.26)$$

Consider first the contribution of the term with the amplitude f_1. Let us rewrite this function in the form

$$f_1(t, x, -p) = -ip\frac{\partial y}{\partial t} + (\Phi_s(x, p) - \Phi_s(x_0, p)) + (\Phi_s(x, p\frac{\partial y}{\partial x}) - \Phi_s(x, p))$$

$$+(\Phi_p(x, p) - \Phi_p(x_0, p)) + (\Phi_p(x, p\frac{\partial y}{\partial x}) - \Phi_p(x, p)). \quad (2.27)$$

We claim that the contribution of this term can be estimated by the expression on the r.h.s. of (5.3.27). Due to (2.22) one sees directly that it is the case for the contribution of the first term in (2.27). The contribution of the second term in (2.27) has the form

$$O(|x - x_0|)\left[\frac{\partial u_\alpha}{\partial x}(t, x - x_0) + \frac{\partial u_\alpha}{\partial t}(t, x - x_0, h)\right], \quad (2.28)$$

which is estimated exactly as (5.3.24) and is of form (5.3.27). Next, due to (2.23) and (5.3.4), the third term in (2.27) can be presented as $O(|(x - x_0)|)\Phi_s(x, p)$, which again has form (2.28). To estimate the contribution of the fourth term in (2.27) we write

$$\Phi_p(x, p) - \Phi_p(x_0, p) = \frac{\partial \Phi_p}{\partial x}(x_0 + \theta(x - x_0), p)(x - x_0).$$

Since $\frac{\partial \Phi_p}{\partial x}$ has the same form as Φ_p, i.e.

$$\frac{\partial \Phi_p}{\partial x}(x,p) = \int (e^{ip\xi} - 1)\frac{\partial f}{\partial x}(x,\xi)\,d\xi$$

with $|\frac{\partial f}{\partial x}| = O(f_\alpha^d(\xi))$ (actually $\frac{\partial f}{\partial x}(x,\xi)$ vanishes for $|\xi| \geq a(x)$), we deal with the contribution of this term as in the proof of Theorem 5.4.2 to get the estimate for it of the form

$$O(|x - x_0|)(h^{-1} + t^{-1})u_\alpha(t,x,x_0) = O(|x - x_0|/t)u_\alpha(t,x,x_0),$$

which again gives nothing new sa compared with (2.28).

Consider the last term in (2.27). It is convenient to decompose the integral (5.4.3) defining Φ_p into two integrals over $|\xi| \leq a(x)$ and $|\xi| \geq a(x)$. The first integral is differentiable in p and the corresponding difference is estimated as the fourth term in (2.27). In the second integral $f(x,\xi) = |\xi|^{-(d+\alpha)}$ and the corresponding term is estimated readily.

Consider now the contribution of the term with f_2 in (2.26). To begin with, consider the case when $h \leq \epsilon|\log t|^{-1}$ with small positive ϵ. Introducing the function

$$w(x,\xi)_j = \left[\xi\frac{\partial y}{\partial x}\right]_j + h\int_0^1 (1-s)\left(\frac{\partial^2 y_j}{\partial x^2}(x + sh\xi)\xi,\xi\right)ds$$

and using (2.24) we can present the contribution of the term with f_2 in the form

$$O(h)|\log t|\int \left|\frac{\partial u_\alpha}{\partial x}(t,x - x_0 + hw(x,\xi))\right| |\xi|^2\,\nu(x,d\xi).$$

Due to our assumption on h, $w(x,\xi) = \xi(1 + O(|x - x_0|) + O(h|\xi|^2)$, which means that for small enough ϵ, one can make the change of the variable of integration in the previous integral $\xi \to \eta = w(x,\xi)$ to write the contribution of the term with f_2 in the form

$$O(h)|\log t|\int \left|\frac{\partial u_\alpha}{\partial x}(t,x - x_0 + h\xi)\right| |\xi|^2\nu(x,d\xi). \tag{2.29}$$

First let $|x - x_0| \leq 2(th^{\alpha-1})^{1/\alpha}$. Present expression (2.29) as the sum of two terms $I_1 + I_2$ corresponding to the decomposition of the domain of integration over ξ in the union of the sets

$$M_1 = \{\xi : |x - x_0 + h\xi| \leq 4(th^{\alpha-1})^{1/\alpha}\}, \quad M_2 = \{\xi : |x - x_0 + h\xi| \geq 4(th^{\alpha-1})^{1/\alpha}\}.$$

In the first integral we estimate

$$\left|\frac{\partial u_\alpha}{\partial x}(t,x - x_0 + h\xi)\right| = \frac{O(1)}{(th^{\alpha-1})^{1/\alpha}}u_\alpha(t,x - x_0 + h\xi) = \frac{O(1)}{(th^{\alpha-1})^{1/\alpha}}u_\alpha(t,x - x_0).$$

Therefore

$$I_1 = \frac{O(h)|\log t|}{(th^{\alpha-1})^{1/\alpha}} u_\alpha(t, x - x_0) \int_{\{h|\xi| \le 6(th^{\alpha-1})^{1/\alpha}\}} |\xi|^{1-\alpha} \, d|\xi|$$

and consequently

$$I_1 = \left(\frac{t}{h}\right)^{(\alpha^{-1}-1)} |\log t| O(u_G^{as}), \quad I_1 = |\log t| O(u_\alpha) \tag{2.30}$$

respectively for $\alpha \ge 1$ or $\alpha \le 1$. In the second integral I_2 we estimate $|x - x_0 + h\xi| > h|\xi|/2$ and thus

$$\left|\frac{\partial u_\alpha}{\partial x}(x - x_0 + h\xi)\right| = \frac{O(1)}{|x + h\xi|} u_\alpha(x - x_0 + h\xi) = \frac{O(1)}{h|\xi|} u_\alpha(x).$$

Consequently

$$I_2 = O(u_G^{as})|\log t| \int_{(t/h)^{1/\alpha}}^{O(h^{-1})} |\xi|^{-\alpha} \, d|\xi|.$$

Hence, I_2 is the same as (2.30).

Let $2(th^{\alpha-1})^{1/\alpha} \le |x - x_0| = O(h)$. Let us decompose expression (2.29) into the sum $I_1 + I_2$ of two terms corresponding to the decomposition of the domain of integration over ξ in the union of the sets $M_1 = \{h|\xi| \le |x - x_0|/2\}$ and $M_2 = \{h|\xi| \ge |x - x_0|/2\}$. The first term is

$$\frac{O(h)|\log t|}{|x - x_0|} u_\alpha(t, x - x_0) \int_0^{|x-x_0|/2h} |\xi|^{1-\alpha} \, d|\xi|$$

$$= \left(\frac{|x - x_0|}{h}\right)^{1-\alpha} |\log t| O(u_\alpha(t, x - x_0)),$$

which is easily seen to be again of form (2.30), due to the assumptions on x. To estimate the second term one notes that for $h|\xi| \ge |x - x_0|/2$

$$|\xi|^2 \nu(x, d\xi) \le \left(\frac{h}{|x - x_0|}\right)^{d+\alpha-2} d\xi = \left|\frac{x - x_0}{h}\right|^2 \frac{h^\alpha d(h\xi)}{|x - x_0|^{d+\alpha}}, \tag{2.31}$$

and therefore one can estimate this second term by

$$I_2 = O(|\log t|) \left(\frac{t}{h}\right)^{-1/\alpha} \left|\frac{x - x_0}{h}\right|^2 \frac{h^\alpha}{|x - x_0|^{d+\alpha}} \int u_\alpha(x - x_0 + \eta) \, d\eta$$

$$= O(|\log t|) \left(\frac{t}{h}\right)^{-1/\alpha} \left|\frac{x - x_0}{h}\right|^2 \frac{h^\alpha}{|x - x_0|^{d+\alpha}}$$

$$= \left(\frac{t}{h}\right)^{-1/\alpha} \frac{|x-x_0|^2}{th} |\log t| O(u_\alpha(t, x-x_0)). \tag{2.32}$$

Now let $|x - x_0| \geq 2a(x)h$. Then for all $|\xi| \leq a(x)$

$$|\frac{\partial u_\alpha}{\partial x}(t, x - x_0 + h\xi)| \leq \frac{u_\alpha(t, x - x_0 + h\xi)}{|x - x_0 + h\xi|}$$

$$= O(1)\frac{u_\alpha(t, x - x_0)}{|x - x_0|} = O(h^{-1})u_\alpha(t, x - x_0).$$

Therefore, one can estimate expression (2.29) by $|\log t| O(u_G^{as})$, which does not give anything new.

Let us suppose now that $h \geq \epsilon |\log t|^{-1}$, i.e. t is exponentially small with respect to h. Integral over the range of ξ (in the expression for the contribution of the term with f_2) such that $h|\xi| \leq \epsilon |\log t|^{-1}$ with small enough ϵ can be obviously estimated in the same way as in the previous case. So, one needs to consider the integral only over the domain

$$D = \{\frac{\epsilon}{h|\log t|} \leq |\xi| \leq a(x)\}$$

with a sufficiently small ϵ. To estimate this integral consider both terms in formula for f_2 from (2.19) separately. The contribution of the first term can be written sa

$$h^{-1} \int_D u_{rough}(t, x + h\xi, x_0, h)\, \nu(x, d\xi),$$

and is the same to estimate

$$h^{-1} \int_D u_G^{as}(t, x + h\xi - x_0, h)\, \nu(x, d\xi), \tag{2.33}$$

The estimate of the contribution of the second term is reduced plainly to the estimate of the same integral. In D

$$\nu(x, d\xi) = \frac{O(1)\, d\xi}{|\xi|^{d+\alpha}} = O(h|\log t|)^{d+\alpha}d\xi = O(h^\alpha |\log t|^{d+\alpha}d(h\xi).$$

Hence, integral (2.33) can be estimated by

$$O(h^{\alpha-1})|\log t|^{d+\alpha} \int u_G^{as}(t, x + \eta - x_0, h)\, d\eta = O(h^{\alpha-1})|\log t|^{d+\alpha}.$$

If $|x - x_0| \leq (th^{\alpha-1})^{1/\alpha}$, this is estimated by u_G^{as} (because t is exponentially small with respect to h). If $|x - x_0| \geq (th^{\alpha-1})^{1/\alpha}$, the last expression can be written as

$$|x - x_0|^{d+\alpha}t^{-1}|\log t|^{d+\alpha}O(u_G^{sa}).$$

If $\alpha \geq 1$, this is obviously bounded by (2.31), which completes the consideration of the case $\alpha > 1$.

If $\alpha \leq 1$, we shall use estimate (2.31) only for

$$(th^{\alpha-1})^{1/\alpha} \leq |x - x_0| \leq t^{\delta_2} h^{-\delta_1},$$

which gives the second term in (2.25'). (Notice that due to our assumptions, $(th^{\alpha-1})^{1/\alpha} \leq t^{\delta_2} h^{-\delta_1}$.) If $|x - x_0| \geq t^{\delta_2} h^{-\delta_1}$, we shall use estimate (2.29) only when integrating over $|\xi h| \leq (th^{\alpha-1})^{1/\alpha}$, and integrating over the remaining set of ξ we shall use formula (2.33). Thus, for $|\xi h| \leq (th^{\alpha-1})^{1/\alpha}$, estimating

$$|\xi|^2 \nu(x, d\xi) \leq \left(\frac{t}{h}\right)^{1/\alpha} \left|\frac{x - x_0}{h}\right| \frac{h^{\alpha} d|h\xi|}{|x - x_0|^{d+\alpha}}$$

instead (2.31) yields for I_2 the estimate

$$I_2 = O(|\log t|) \frac{|x - x_0|}{t} u_G^{as},$$

instead of (2.32), which contributes to the first term in (2.25'). As we have mentioned, for $|\xi h| \geq (th^{\alpha-1})^{1/\alpha}$, we shall use formula (2.33) for the contribution of the term with f_2. To estimate this integral we estimate

$$\frac{1}{h} \nu(x, d\xi) = \frac{1}{h} \frac{d\xi}{|\xi|^{d+\alpha}} = \frac{1}{t} \frac{th^{\alpha-1} d(h\xi)}{|h\xi|^{d+\alpha}},$$

and therefore the integral (2.33) is estimated by

$$\int u_\alpha(t, x - x_0 + \eta) t^{-1} u_\alpha(t, \eta) = O(t^{-1}) u_\alpha(2t, x - x_0) = O(t^{-1}) u_\alpha(t, x - x_0),$$

which contributes to the last term in (2.25').

Turning to the contributions of f_3 and f_4 in (2.17) one sees readily that they give nothing new as compared to the terms considered above, which completes the proof of the Proposition. Notice only that considering the contribution of the second term in the expression for f_3 one presents $\frac{\partial H}{\partial p} = \frac{\partial H_s}{\partial p} + \frac{\partial H_p}{\partial p}$. Due to Proposition 5.2.4, the contribution of the first term is $O(|x - x_0|/t) u_\alpha \Theta_{c_1, c_2}(|x - x_0|)$, and the second is considered as in the proof of theorem 5.4.2 to obtain for its contribution the estimate $O(1/t) u_\alpha \Theta_{c_1, c_2}(|x - x_0|)$, which is actually the same as the first one.

We shall give now an estimate for u_{rough} in the boundary layer $\{h \leq |x - x_0| \leq h|\log t|^k\}$.

Proposition 2.5. *For any $K > 0$ and $k > 0$ there exists $\kappa > 0$ such that*

$$u_{rough}(t, x, x_0, h) = O(1) \left(1 + \log^+ \frac{|x - x_0|}{t}\right)^\kappa \chi(t, |x - x_0|) u_{rough}^{exp}(t, x, x_0, h)$$

uniformly in the domain

$$\{\frac{1}{K} \le \frac{|x - x_0|}{h} \le K\left(1 + \log^+ \frac{|x - x_0|}{t}\right)^k, \quad t \le Kh\}.$$

Proof. As in the proof of Proposition 2.2 we shall use representation (2.11), but shall estimate it in a different way. Using formulas (C20)-(C22) and making the change of the variable $p \mapsto ph$ we rewrite (2.11) in the form

$$u_{rough}(t, x, x_0, h) = O(1)\chi(t, |x - x_0|) \exp\{-\frac{S(t, x, x_0)}{h}\} \exp\{-\frac{t}{h}H(x_0, \hat{p})\}I,$$
$$(2.34)$$

where I equals

$$\frac{1}{(2\pi)^d} \int \exp\{-\sigma|p|^\alpha th^{\alpha-1} + \frac{t}{h}\int e^{-ip\xi}(f_1 + f_2)(x, \xi/h)\,d(\xi/h) + ip(x - x_0)\}\,dp$$
$$(2.35)$$

or

$$I = \frac{1}{(2\pi)^d} \int \exp\{-\sigma|p|^\alpha th^{\alpha-1}\}$$

$$\times \exp\{-i(\tilde{p}, p)\sigma\alpha|p|^{\alpha-2}th^{\alpha-2} + \frac{t}{h}\int e^{-ip\xi}(f_1 + f_2)(x, \xi/h)\,d(\xi/h) + ip(x - x_0)\}\,dp$$
$$(2.36)$$

respectively for $\alpha < 1$ or $\alpha \ge 1$. Here

$$|f_1(\eta)| \le \exp\{(\hat{p}, \eta)\}f_\alpha^d(\eta)$$
$$(2.37)$$

(the function f_α^d being defined in (5.3.31)), and

$$|f_2(\eta)| \le \Theta_b(|\eta|)|\hat{p}| \exp\{\hat{p}\eta\}|\eta|^{-(d+\alpha-1)},$$

$$|f_2(\eta)| \le \Theta_b(|\eta|)|\hat{p}|^2 \exp\{\hat{p}\eta\}|\eta|^{-(d+\alpha-2)}$$
$$(2.38)$$

resp. for $\alpha < 1$ or $\alpha \ge 1$ with b being any constant less than $a(x)$.

Consider first the case $\alpha < 1$. We shall estimate (2.35) as in the proof of Theorem 5.4.2 by expanding the exponent $\exp\{\frac{t}{h}\int e^{-ip\xi}(f_1 + f_2)(x, \xi/h)\,d(\xi/h)\}$ in (2.35) in the power series. The first term in thus obtained series in (2.35) is just u_α, and it is readily seen that in the range of parameters we are dealing with

$$\frac{th^{\alpha-1}}{|x - x_0|^{d+\alpha}} = \frac{O(1)}{(ht)^{d/2}}\left|\frac{x - x_0}{t}\right|^{-d/2} = \frac{O(1)}{(ht)^{d/2}}\left(\det \frac{\partial^2 H}{\partial p^2}(x_0, \hat{p})\right)^{-1/2},$$

and the corresponding term in (2.34) can be estimated by the r.h.s. of (2.33). The second term

$$\frac{1}{(2\pi)^d} \int \exp\{-\sigma|p|^\alpha th^{\alpha-1} + ip(x - x_0)\}\left(\frac{t}{h}\int e^{-ip\xi}(f_1 + f_2)(x, \frac{\xi}{h})\,d\frac{\xi}{h}\right)\,dp$$

is the inverse Fourier transform of the product of two functions, which equals therefore to the convolution of their inverse Fourier transforms

$$\int u_\alpha(t, x - x_0 - \xi)\frac{t}{h}(f_1 + f_2)(x_0, \frac{\xi}{h})h^{-d}\,d\xi. \tag{2.39}$$

Recall that $t \le h = O(|x - x_0|)$. Due to (2.37), Proposition 5.3.5 and Theorem 2.5.3, the first term in (2.39) can be estimated by

$$O(t/h)e^{|\hat{p}|a(x_0)}\int u_\alpha(t, x - x_0 - \xi)h^{-d}f^d_{\alpha,h}(\xi)\,d\xi = O(t/h)e^{|\hat{p}|a(x_0)}h^{-d}f^d_{\alpha,h}$$

$$= \frac{O(|x - x_0|)}{h}(1 + \log^+\frac{|x - x_0|}{t})h^{-d}f^d_{\alpha,h},$$

which when inserted in (2.34) will satisfy the required estimate, because

$$h^{-d}f^d_{\alpha,h} = \frac{O(1)}{(ht)^{d/2}}\left|\frac{x - x_0}{t}\right|^{-d/2} = \frac{O(1)}{(ht)^{d/2}}\left(\det\frac{\partial^2 H}{\partial p^2}(x_0, \hat{p})\right)^{-1/2}.$$

The term with f_2 in (2.39) can be written in the form

$$O(t/h)\int u_\alpha(t, x - x_0 - \xi)\Theta_b\left(\frac{|\xi|}{h}\right)|\hat{p}|e^{|\hat{p}|b}\frac{h^{\alpha-1}\,d\xi}{|\xi|^{d+\alpha-1}}.$$

Since $h = O(|x - x_0|)$, we can choose a small b in such a way that $|\xi| < |x - x_0|/2$ whenever $|\xi| < hb$. For these ξ

$$u_\alpha(t, x - x_0 - \xi) = O(u_\alpha(t, x - x_0)),$$

and therefore the last integral can be estimated by

$$O(t/h)u_\alpha(t, x - x_0)|\hat{p}|e^{|\hat{p}|b}h^{\alpha-1}\int_0^{hb}\frac{d|\xi|}{|\xi|^\alpha} = O(t/h)u_\alpha(t, x - x_0)|\hat{p}|e^{|\hat{p}|b},$$

which is again estimated as the first term in (2.39) and which when inserted in (2.34) gives the term that is readily estimated by the r.h.s. of (2.33).

The whole series obtained in (2.35) can be presented in the form

$$u^{as}_G + u^{as}_G \star F_1 + u^{as}_G \star F_2 + u^{as}_G \star (F_1 \star F_2), \tag{2.40}$$

where

$$F_j = \frac{t}{h}h^{-d}f_{j,h} + \frac{1}{2!}\frac{t}{h}h^{-d}f_{j,h} \star \frac{t}{h}h^{-d}f_{j,h} + ..., \quad j = 1, 2,$$

$\star$ means the standard convolution of the integrable functions (in other words, F_j is the exponent of $f_{j,h}$ in the sense of convolution), and $f_{j,h}(x) = f_j(x/h)$.

Using (5.4.8) one finds that the second term in the expression for $F_1(\xi)$ can be estimated by

$$\frac{1}{2}\left(\frac{t}{h}\right)^2 h^{-2d}\int f^d_{\alpha,h}(\xi-\eta)e^{\hat{p}(\xi-\eta)/h}f^d_{\alpha,h}(\eta)e^{\hat{p}\eta/h}\,d\eta \le \frac{1}{2}\left(\frac{t}{h}\right)^2 h^{-d}f^d_{\alpha,h}(\xi)e^{\hat{p}\xi/h}.$$

By trivial induction one finds further that

$$F_1 \le (e^{\frac{t}{h}}-1)e^{|\hat{p}|a(x_0)}h^{-d}f^d_{\alpha,h} = O(t/h)e^{|\hat{p}|a(x_0)}h^{-d}f^d_{\alpha,h},$$

which gives nothing new as compared with the first term in (2.39). To estimate F_2 we need the following fact, whose elementary proof is omitted.

Lemma 2.2. *Let $\alpha \in (0,1)$, $a > 0$, and let the function $g(x)$ in $\mathcal{R}^d$ be defined by the formula $g_a(x) = \Theta_a(|x|)|x|^{-(d+\alpha-1)}$. Then*

$$(g_a \star g_a) \le C_1 a^{1-\alpha}g_a + C_2 a^{2-2\alpha-d}\Theta_{a,2a}, \qquad (2.41)$$

$$g_a \star f^d_\alpha \le C_3 a^{1-\alpha}(1 + a^{d+\alpha})f^d_\alpha, \qquad (2.42)$$

where positive constants C_1, C_2, C_3 depends only on d and α.

Since

$$(g_{h\delta} \star f^d_{\alpha,h})(x) = (g_\delta \star f^d_\alpha)(x/h)h^{1-\alpha},$$

it follows from the Lemma that uniformly in $\delta \le 1$

$$g_{h\delta} \star g_{h\delta} \le Ch^{1-\alpha}(g_{h\delta} + h^{-d}f^d_{\alpha,h}), \quad g_{h\delta} \star f^d_{\alpha,h} \le Ch^{1-\alpha}f^d_{\alpha,h}. \qquad (2.43)$$

Due to (2.39)

$$h^{-d}f_2(\xi/h) \le h^{\alpha-1}g_{hb}(\xi)e^{|\hat{p}|(b+\delta)}$$

with any $\delta > 0$. Hence, by induction, the result of k convolutions of the function $(t/h)h^{-d}f_2(\xi/h)$ with itself does not exceed the function

$$h^{\alpha-1}C^{k-1}\left(\frac{t}{h}\right)^k e^{|\hat{p}|(b+\delta)}(g_{bh} + (k-1)h^{-d}f^d_{\alpha,h}).$$

Consequently,

$$F_2 \le O(h^{\alpha-1})\frac{t}{h}e^{|\hat{p}|(b+\delta)}(g_{bh} + h^{-d}f^d_{\alpha,h}),$$

which gives nothing new as compared with the second term in (2.39). The contribution of the term $F_1 \star F_2$ from (2.40) is estimated in the same way which completes the consideration of the case $\alpha < 1$.

The case $\alpha \ge 1$ differs by the additional term under the exponent in (2.36) which is increasing in p. To deal with this term we need the following simple statement from the calculus.

Lemma 2.3. *For any $\alpha \in [1, 2)$, $s_0 > 0, v_0 > 0$, there exists a constant K such that if $s \in (0, s_0]$, $w, v \in \mathcal{R}^d$, $|v| > v_0$, then*

$$\left| \int_{\mathcal{R}^d} \exp\{-s|p|^\alpha - i(p, v) + i(p, w)|p|^{\alpha-2}\} \, dp \right| \le K|w|^{d/(2-\alpha)}. \tag{2.45}$$

Proof. Obviously (by changing the variable of integration p to $p|v|$) it is enough to prove (2.53) only for unit vectors v. Consider first a rather trivial one-dimensional situation. In that case one needs to prove the estimate

$$\left| Re \int_0^\infty \exp\{-s|p|^\alpha - ip(1 - wp^{\alpha-2})\} \, dp \right| \le Kw^{1/(2-\alpha)}$$

for any w. Let us consider only positive w (the case of negative w being even simpler). Let a number p_0 be defined by the equation $wp_0^{\alpha-2} = 1$, and let us decompose the domain of integration into the union of two subintervals: $[0, p_0]$ and $[p_0, \infty)$. Obviously the first integral does not exceed in magnitude the $p_0 = w^{1/(2-\alpha)}$. Hence, it is suffice to estimate the integral over $[p_0, \infty)$. Making in this integral the change of the variable of integration $p \to y = p - wp^{\alpha-1}$, one rewrites it in the form

$$\int_0^\infty \exp\{-s(p(y))^\alpha\}p'(y) \cos y \, dy.$$

This integral has the form $\int_0^\infty g(y) \cos y \, dy$ with a positive decreasing function $g(y)$ such that $g(0) = 2 - \alpha$ and $\lim_{y \to \infty} g(y) = 0$. For such a function this integral is bounded by

$$\int_{-\pi/2}^{\pi/2} g(y) \cos y \, dy \le (2 - \alpha) \int_{-\pi/2}^{\pi/2} \cos y \, dy = 2(2 - \alpha),$$

because it is equal to the convergent sum of an alternating series with monotonicaly decreasing terms.

Let us sketch the proof for $d > 1$. For brevity, let $d = 2$. Introducing the circular coordinates r, θ for p in such a way that the vector v has vanishing θ we can write the integral on the l.h.s. of (2.45) in the form

$$I = \int_0^\infty \int_{-\pi}^\pi \exp\{-sr^\alpha - ir \cos \theta + ir^{\alpha-1} \cos \phi w_1 + ir^{\alpha-1} \sin \theta w_2\}r \, dr \, d\theta,$$

where $w = w_1 \cos \phi + w_2 \sin \phi$. Writing

$$(r - r^{\alpha-1} w_1) \cos \theta - r^{\alpha-1} w_2 \sin \theta = R(r) \cos(\theta - \psi)$$

with appropriate R and ψ one can present I in the form

$$I = \int_0^\infty \int_{-\pi}^\pi \exp\{-sr^\alpha\} \cos(R(r) \cos(\theta - \psi))r \, dr \, d\theta,$$

and by periodicity changing θ to $\phi = \theta - \psi$ yields

$$I = \int_0^\infty \int_{-\pi}^\pi \exp\{-sr^\alpha\} \cos(R(r) \cos \phi) r \, dr \, d\theta.$$

Due to the definition of Bessel functions (it is given before the formulation of Proposition 5.2.1), it can be written as

$$I = 2\pi \int_0^\infty \exp\{-sr^\alpha\} r J_0(R(r)) \, dr.$$

As in one-dimensional case it is actually enough to estimate the integral

$$\tilde{I} = 2\pi \int_{r_0}^\infty \exp\{-sr^\alpha\} r J_0(R(r)) \, dr,$$

with $r_0 = r_0^{\alpha-1} w_1$ so that for $r > r_0$ the function $R(r)$ is increasing in r and positive. To estimate this integral one can use the known properties of the Bessel function J_0. The simplest way to do it is using the methods of the proof of Propositions 5.2.2 or 5.2.6. For example, using (5.2.10) one can rewrite $\tilde{I}$ in the form

$$\tilde{I} = 2\sqrt{2\pi} \, Re \int_{r_0}^\infty \exp\{-sr^\alpha\} \frac{r}{\sqrt{R(r)}} \exp\{\frac{\pi i}{4}\} W_{0,0}(2iR(r)) \, dr.$$

Rotating here the contour of integration on any small angle transforms $W_{0,0}$ into an exponentially decreasing function, which gives the required estimate.

End of the proof of Proposition 2.5. Putting $s = t/h$, $v = (x - x_0)/h$, $w = \hat{p}t/h$ in (2.45) yields

$$\frac{1}{(2\pi h)} \int_{\mathcal{R}}^d -\frac{\sigma|p|^\alpha t + ip(x - x_0) - i(\hat{p}, p)\sigma\alpha|p|^{\alpha-2}t}{h} \, dp$$

$$= O(h^{-d})(1 + \log^+ \frac{|x - x_0|}{t})^{d/(\alpha-2)}$$

in the considered range of $|x - x_0|$. Therefore, arguing for the case $\alpha \geq 1$ in the same way as for $\alpha < 1$ one estimates (2.36) by the expression

$$O\left(1 + \log^+ \frac{|x - x_0|}{t}\right)^{d/(\alpha-2)} (1 + 1 \star F_1 + 1 \star F_2 + 1 \star (F_1 \star F_2)),$$

which is estimated in the same way as (2.40). The proof of the Proposition is thus completed.

In order to be able to justify the asymptotics u_{rough} we should be able to estimate the convolutions of the remainder F of form (2.17) with itself and with u_{rough}.

Proposition 2.6. *Let $\tau \in (0,t)$, $h = O(t)$ and $\omega_1, \omega_2 \in [0,2)$. Then*

$$\int \chi(t - \tau, |x - \eta|) u_{rough}^{exp}(t - \tau, x, \eta) \left| \frac{x - \eta}{t - \tau} \right|^{\omega_1}$$

$$\times \chi(\tau, |\eta - x_0|) u_{rough}^{exp}(\tau, \eta, x_0) \left| \frac{\eta - x_0}{\tau} \right|^{\omega_2} d\eta$$

$$= O(1) \left(1 + \log^+ \frac{|x - x_0|}{t} \right)^{(d-1)/2} \left| \frac{x - x_0}{t} \right|^{\omega_1 + \omega_2} u_{rough}^{exp}(t, x, x_0). \qquad (2.46)$$

Proof. It follows from Section 2.5 (see proof of Theorem 2.5.3 and Proposition 2.5.5) that

$$C^{-1} \left(1 + \frac{|x - x_0|}{t} \right)^d \left(1 + \log^+ \left| \frac{x - x_0}{t} \right| \right)^{-(d-1)} \leq \det \frac{\partial^2 H}{\partial p^2}(x_0, \hat{p}(x_0))$$

$$\leq C \left(1 + \left| \frac{x - x_0}{t} \right| \right)^d \left(1 + \log^+ \left| \frac{x - x_0}{t} \right| \right)^{-(d-1)} \qquad (2.47)$$

$$(Ct)^{-1} \left(1 + \left| \frac{x - x_0}{t} \right| \right)^{-1} \leq \frac{\partial^2 S}{\partial x^2}(x_0, \hat{p}(x_0))$$

$$\leq \frac{C}{t} \left(1 + \frac{|x - x_0|}{t} \right)^{-1} \left(1 + \log^+ \left| \frac{x - x_0}{t} \right| \right) \qquad (2.48)$$

with some constant C. The same estimate (2.48) holds for $\frac{\partial^2 S}{\partial x_0^2}$. The integral on the l.h.s. of (2.46) can be estimated then by

$$O(1) \int (2\pi h\tau)^{-d/2} (2\pi h(t - \tau))^{-d/2} \left(1 + \left| \frac{x - \eta}{t - \tau} \right| \right)^{-d/2} \left(1 + \left| \frac{\eta - x_0}{\tau} \right| \right)^{-d/2}$$

$$\times \left(1 + \log^+ \left| \frac{x - \eta}{t - \tau} \right| \right)^{(d-1)/2} \left(1 + \log^+ \left| \frac{\eta - x_0}{\tau} \right| \right)^{(d-1)/2} \left| \frac{x - \eta}{t - \tau} \right|^{\omega_1} \left| \frac{\eta - x_0}{\tau} \right|^{\omega_1}$$

$$\times \exp\left\{ -\frac{S(t - \tau, x, \eta) + S(\tau, \eta, x_0)}{h} \right\} d\eta,$$

where the integral is taken over all η such that $|\eta - x_0| \leq c$, $|x - \eta| \leq c$. To estimate this integral of the Laplace type one can use the Laplace method with estimates given in Propositions B1, B2. To estimate the major term notice that due to the calculus of variations, the minimum of the phase $f(\eta) = S(t - \tau, x, \eta) + S(\tau, \eta, x_0)$ over all possible η is equal to $S(t, x, x_0)$ and is attained at the point $\eta(\tau)$, which lies on the solution of the Hamiltonian system (2.1.2) with Hamiltonian (2.1) joining x_0 and x in time t. Due to the result of Section 2.5, the point $\eta(\tau)$ exists and is unique, and moreover (see (2.5.8) and the proof of Theorem 2.5.1),

$$\eta(\tau) = x_0 + \tau \frac{x - x_0}{t}(1 + O(|x - x_0|)),$$

or (changing the roles of x_0 and x)

$$\eta(\tau) = x - (t - \tau)\frac{x - x_0}{t}(1 + O(|x - x_0|)).$$

It follows that

$$\frac{\eta(\tau) - x_0}{\tau} = \frac{x - x_0}{t}(1 + O(|x - x_0|)), \qquad \frac{x - \eta(\tau)}{t - \tau} = \frac{x - x_0}{t}(1 + O(|x - x_0|)).$$
$$(2.49)$$

From (2.48) it follows that

$$\frac{\partial^2 f}{\partial \eta^2} \geq \frac{C}{t - \tau}\left(1 + \left|\frac{x - \eta}{t - \tau}\right|\right)^{-1} + \frac{C}{\tau}\left(1 + \left|\frac{\eta - x_0}{\tau}\right|\right)^{-1}.$$

Suppose now that $|x-x_0|/t$ does not approach zero (the case of bounded $|x-x_0|/t$ is simpler and we omit it). Therefore, the principle term of the asymptotics of the integral on the l.h.s. of (2.46), due to (2.49) and Proposition B2, can be estimated by

$$(2\pi h)^{-d/2}\exp\{-\frac{S(t, x, x_0)}{h}\}\left(1 + \left|\frac{x - x_0}{t}\right|\right)^{-d}\left(1 + \log^+\left|\frac{x - x_0}{t}\right|\right)^{(d-1)}$$

$$\times \left|\frac{x - x_0}{t}\right|^{\omega_1 + \omega_2}\left[\tau\left(1 + \left|\frac{x - x_0}{t}\right|\right)^{-1} + (t - \tau)\left(1 + \left|\frac{x - x_0}{t}\right|\right)^{-1}\right]^{-d/2},$$

which has the form

$$O(1)(2\pi ht)^{-d/2}\exp\{-\frac{S(t, x, x_0)}{h}\}$$

$$\times \left|\frac{x - x_0}{t}\right|^{\omega_1 + \omega_2 - d/2}\left(1 + \log^+\left|\frac{x - x_0}{t}\right|\right)^{(d-1)},$$

which can be estimated by the r.h.s. of (2.46), due to (2.47). To prove the proposition it remains to estimate the remainder in formula (B5) of Proposition B1. One makes it by taking the constant r of Proposition B1 in form $r = \min(\tau, t - \tau)$. We omit the details.

Proposition 2.7. *Let $h \leq t \leq t_0$ with small enough t_0. Then for all $\tau \in (0, t)$*

$$\int u_{rough}(t - \tau, x, \eta, h)u_{rough}(\tau, \eta, x_0, h)\, d\eta$$

$$= O(1) \left[t^{-\delta} + \left(1 + \log^+ \frac{|x - x_0|}{t} \right)^\kappa \right] u_{rough}^{exp}(t, x, x_0, h) \qquad (2.50)$$

with an arbitrary small δ.

Proof. If both τ and $t - \tau$ are of the order t, the required estimate was proved in Proposition 2.6. Suppose that $\tau \le h \le t/2$. Then $u_{rough}(t - \tau, x, \eta, h)$ can be estimated by u_{rough}^{exp}. Due to Propositions 2.2, 2.5, if $h = O(|\eta - x_0|)$, $u_{rough}(\tau, \eta, x_0, h)$ can be also estimated by the corresponding u_{rough}^{exp} with perhaps an additional multiplier of form $O(1 + \log^+(|\eta - x_0|/\tau)^\kappa$, and for this range of η the integral on the l.h.s. of (2.50) was again estimated as in the Proposition 2.6. Therefore, it remains to estimate this integral over the range of η such that $|x_0 - \eta| \le \epsilon h$ with an arbitrary small ϵ. In that case, u_{rough} is estimated by u_α and thus it remains to prove that

$$\int_{\{|\eta - x_0| < \epsilon h\}} u_{rough}^{exp}(t - \tau, x, \eta, h) u_G^{as}(\tau, \eta, x_0, h) \, d\eta = O(t^{-\delta}) u_{rough}^{exp}(t, x, x_0, h).$$

$$(2.51)$$

The l.h.s. of this equation can be presented in the form

$$\int_{\{|x_0 - \eta| < \epsilon h\}} \frac{O(1)}{(2\pi h(t - \tau))^{d/2}} \left(1 + \left| \frac{x - \eta}{t - \tau} \right| \right)^{-d/2}$$

$$\exp\{-\frac{S(t - \tau, x, \eta)}{h}\} u_\alpha(\tau, \eta, x_0, h) \, d\eta. \qquad (2.52)$$

One has

$$S(t - \tau, x, \eta) = S(t, x, x_0) - \frac{\partial S}{\partial t}(t - \theta_1 \tau, x, x_0) \tau + \frac{\partial S}{\partial x_0}(t, x, x_0 + \theta_2(\eta - x_0))(\eta - x_0)$$

with some $\theta_1, \theta_2 \in (0, 1)$. Since

$$\frac{\partial S}{\partial x_0}(t, x, x_0 + \theta_2(\eta - x_0)) = O(|\log t|),$$

$$\frac{\partial S}{\partial t}(t - \theta_1 \tau, x, x_0) = -H(x_0, p_0(t - \theta_1 \tau, x, x_0)),$$

and H is bounded from below, one finds that

$$S(t - \tau, x, \eta) > S(t, x, x_0) + O(\tau) + O(|\log t|)|\eta - x_0|.$$

Since $|\eta - x_0| \le \epsilon h$ and $\tau < h$, it follows that

$$\exp\{-\frac{S(t - \tau, x, \eta)}{h}\} = O(t^{-\delta}) \exp\{-\frac{S(t, x, x_0)}{h}\}.$$

Furthermore, since

$$|x - x_0| \le |x - \eta| + |\eta - x_0| \le \epsilon h + |x - \eta| \le \epsilon(t - \tau) + |x - \eta|,$$

one has

$$\left(1+\left|\frac{x-\eta}{t-\tau}\right|\right)^{-d/2} = O(1)\left(1+\left|\frac{x-x_0}{t}\right|\right)^{-d/2},$$

and therefore (2.52) can be estimated by

$$O(t^{-\delta})\int_{\{|x-\eta|<\epsilon h\}} u_G^{as}(\tau,\eta,x_0,h)(2\pi ht)^{-d/2}\left(1+\left|\frac{x-x_0}{t}\right|\right)^{-d/2}$$

$$\times \exp\{-\frac{S(t,x,x_0)}{h}\}\, d\eta = O(t^{-\delta})u_{rough}^{exp}(t,x,x_0,h),$$

which completes the proof.

Proposition 2.8. *Let $h \leq t/3$. Then*

$$\int_0^t d\tau \int_{\mathcal{R}^d} d\eta\, u_{rough}(t-\tau,x,\eta,h)F(\tau,\eta,x_0,h)$$

$$= O(u_{rough}(t,x,x_0,h))\left(t^\omega + |x-x_0|\left(\frac{|x-x_0|}{t}\right)^b\right)$$

$$+O(|\log t|)\exp\{-\frac{\Omega}{h}\}\Theta_{c_1,2c_2}(|x-x_0|) \tag{2.53}$$

with some $\Omega > 0$, $\omega \in (0,1)$ and $b > 0$, where b can be chosen arbitrary small by taking a sufficiently small c.

Proof. Let us present the integral on the l.h.s. of (2.53) as the sum of the three integrals over the domains $\tau \leq h$, $h \leq \min(\tau, t-\tau)$, $t-\tau \leq h$. In the second integral both terms can be estimated by the corresponding u_{rough}^{exp} and we get for this integral the required estimate using Propositions 2.3, 2.6. Notice that the last term in (2.53) appears when we estimate u_{rough}^{exp} outside the domain $|x-x_0| \leq c$. Next, the third integral is estimated similarly to the first one, so we need only to consider the first integral over the domain $0 \leq \tau \leq h$. In the range of η, where $h = O(|\eta - x_0|)$, this integral can again be estimated as before due to Proposition 2.5 (and the Remark after it). Therefore, we only need to estimate this integral over the range $\{|\eta - x_0| \leq \epsilon h\}$ with some ϵ. In this domain we estimate $F(\tau,\eta,x_0,h)$ by formulas (2.25), (2.25') and $u_{rough}(t-\tau,x,\eta,h)$ by $u_{rough}^{exp}(t-\tau,x,\eta,h)$. Following the lines of the proof of Proposition 2.7 we can estimate further $u_{rough}^{exp}(t-\tau,x,\eta,h)$ by $O(t^{-\delta})u_{rough}^{exp}(t,x,x_0,h)$. Thus, we need to estimate

$$O(t^{-\delta})u_{rough}^{exp}(t,x,x_0,h)\int_0^t d\tau \int_{\{|\eta-x_0|\leq\epsilon\}} d\eta\, F(\tau,\eta,x_0,h) \tag{2.54}$$

with F given in (2.25), (2.25').

Let $\alpha > 1$. The contributions of the first three terms in (2.25) are estimated using Proposition 2.7. The contribution of the last term is

$$O(t^{-\delta})u_{rough}^{exp}(t,x,x_0,h)\int_0^t d\tau \int_{\{|\eta-x_0|\leq\epsilon\}} d\eta \left(\frac{\tau}{h}\right)^{-1/\alpha}\frac{h^{\alpha-2}|\log\tau|}{|\eta-x_0|^{d+\alpha-2}}$$

$$= O(t^{-\delta})u_{rough}^{exp}(t,x,x_0,h)\int_0^t \left(\frac{\tau}{h}\right)^{-1/\alpha}|\log\tau|\,d\tau$$

$$= O(t^{-\delta})u_{rough}^{exp}(t,x,x_0,h)|\log t|t^{1-1/\alpha}h^{1/\alpha},$$

which obviously contributes to the first term in (2.53).

Now let $\alpha \leq 1$. Estimating the contribution of the third term of (2.25') in (2.54) as above, but taking into account the upper bound $|x_0 - \eta|$ and the assumptions on δ_2, δ_1 from Proposition 2.4, yields for this contribution the estimate

$$O(t^{-\delta})u_{rough}^{exp}(t,x,x_0,h)\int_0^t \tau^{\delta_2(2-\alpha)}h^{-\delta_1(2-\alpha)}h^{\alpha-2+1/\alpha}\tau^{-1/\alpha}|\log\tau|\,d\tau$$

$$= O(t^{-\delta})u_{rough}^{exp}(t,x,x_0,h)\int_0^t \tau^{\delta_2(2-\alpha)-1/\alpha}|\log\tau|\,d\tau = t^{\omega}O(u_{rough}^{exp}(t,x,x_0,h))$$

with some $\omega \in (0,1)$, as required. At last, estimating the contribution of the last term of (2.25') in (2.54) gives for this contribution the estimate

$$O(t^{-\delta})u_{rough}^{exp}(t,x,x_0,h)\int_0^t d\tau \int_{\{|\eta-x_0|\geq t^{\delta_2}h^{-\delta_1}\}}\frac{h^{\alpha-1}\,d\eta}{|x_0-\eta|^{d+\alpha}}$$

$$= O(t^{-\delta})u_{rough}^{exp}(t,x,x_0,h)\int_0^t h^{\alpha-1}(t^{\delta_2}h^{-\delta_1})^{-\alpha}\,d\tau$$

$$= O(t^{-\delta})u_{rough}^{exp}(t,x,x_0,h)\int_0^t h^{(\alpha-1)/(2-\alpha)}\tau^{-\delta_2\alpha}\,d\tau$$

$$= O(u_{rough}^{exp}(t,x,x_0,h))t^{1-\delta\alpha-(1-\alpha)/(2-\alpha)-\delta},$$

which again contributes to the first term in (2.53) due to the assumptions on δ_2 and because δ can be made arbitrary small.

Corollary. *Let $h \leq t/3$. Then for some $\omega \in (0,1)$*

$$\int_0^t d\tau \int_{\mathcal{R}^d} d\eta\, u_{rough}(t-\tau,x,\eta,h)F(\tau,\eta,x_0,h)$$

$$= O(t^{\omega})u_{rough}(t,x,x_0,h) + O(\exp\{-h^{-\omega}\})\Theta_{0,2c}(|x-x_0|) \tag{2.55}$$

Proof. Due to the assumptions, $t^{-1} = O(h^{-1})$. Therefore, for $|x - x_0| \leq t^{\epsilon+b/(1+b)}$, the second term of the r.h.s. of (2.53) can be obviously included in the first term on the r.h.s. of (2.55). The third term of the r.h.s. of (2.53) and for $|x - x_0| \geq t^{\epsilon + b/(1+b)}$ the second term can be included in the last term in (2.55).

Now we can state the main result of this section. We know already from Proposition 2.1 that u_{rough} gives a multiplicative asymptotics for the Green function u_G of equation (1.1), (1.2) for small t and small $x - x_0$. We can prove now that it presents a local asymptotics for small h and small $x - x_0$.

Theorem 2.1. *If t and h/t are small enough, then*

$$u_G(t, x, x_0, h) = u_{rough}(t, x, x_0, h)(1 + O(t^{\omega})) + O(1) \exp\{-h^{-\omega}\}, \qquad (2.56)$$

where $\omega \in (0, 1)$ is a constants and the last term in (2.56) is an integrable function of x.

Proof. We need to estimate the series of type (5.3.26) with u_{rough} instead of u_α and F defined in (2.17) and estimated in (2.18), (2.25), (2.25'). The second term in this series (i.e. the first non-trivial term) is already estimated in (2.55). Other terms are estimated using the same arguments as in the proof of Proposition 2.8, and the general term can be estimated by induction, which proves the theorem (see e.g. the proof of Theorem 4.1 for similar arguments).

The following statement is a direct corollary of the Theorem.

Proposition 2.9. Local principle of large deviations. *For any small enough t and $x - x_0$*

$$\lim_{h \to 0} h \log u_G(t, x, x_0, h) = -S(t, x, x_0),$$

where $S(t, x, x_0)$ is the two-point function corresponding to the Hamiltonian (2.1).

3. Refinement and globalisation

In this section we improve the results of the previous one obtaining first more exact local asymptotics of the Green function u_G and then globalising them.

As in the previous section, let $z(t, v, x_0)$ be the function defined in Theorem 2.5.2 and let $y(t, x, x_0)$ be given by (2.2) in the ball $|x - x_0| \leq 2c$, let $\chi(t, y)$ be a smooth molyfier (as the function of the second variable) of the form $\chi_{c_1}^{c_2}$ from Lemma E1 with $0 < c_1 < c_2 < c$. Define

$$u_G^{as}(t, x, x_0, h) = \frac{\chi(t, |x - x_0|)}{(2\pi h)^d} B(t, x, x_0)$$

$$\times \int_{\mathcal{R}^d} \exp\{-\frac{ipy(t, x, x_0) - H(x_0, ip)t}{h}\} \, dp \qquad (3.1)$$

with

$$B(t, x, x_0) = \left(\det \frac{\partial^2 H}{\partial p^2}(x_0, \hat{p})\right)^{1/2} \phi(t, x, x_0) t^{d/2}, \qquad (3.2)$$

where

$$\phi(t, x, x_0) = J^{-1/2}(t, x, x_0) \exp\{\int_0^t \frac{\partial^2 H}{\partial p \partial x}(X(s), P(s)) \, ds\}, \qquad (3.3)$$

is a solution to the transport equation (3.1.13) corresponding to Hamiltonian (2.1) (see (3.1.18)), and $\hat{p}$ was defined in the formulation of Proposition 2.2. Consequently,

$$u_G^{as}(t, x, x_0, h) = B(t, x, x_0) u_{rough}(t, x, x_0, h), \qquad (3.4)$$

where u_{rough} is defined in (2.3),(2.4). Due to (2.9) and the results of Section 2.5 it follows that

$$J(t, x, x_0) = \det \frac{\partial X}{\partial p_0}(t, x, x_0) = t^d \det \frac{\partial^2 H}{\partial p^2}(x_0, p_0(t, x, x_0))(1 + O(|x - x_0|))$$

$$= t^d \det \frac{\partial^2 H}{\partial p^2}(x_0, \hat{p})(1 + O(|x - x_0|)),$$

and $\frac{\partial^2 H}{\partial p \partial x}(X(s), P(s)) = O(|x - x_0|)/t$. Therefore $B(t, x, x_0) = 1 + O(|x - x_0|)$. It follows that $u_G^{as}(t, x, x_0, h) = (1 + O(|x - x_0|)) u_{rough}(t, x, x_0, h)$ and therefore Proposition 2.1 and Theorem 2.1 are valid also with u_G^{as} instead of u_{rough}. It turns out however that actually u_G^{as} presents a more exact asymptotics of the Green function with respect to $h \to 0$, which is stated in the following main result of this Chapter.

Theorem 3.1. *For small enough c, h_0, and t_0 and any $\delta \in (0, t_0)$*

$$u_G(t, x, x_0, h) = (1 + O(h^\omega)) u_G^{as}(t, x, x_0, h) + O(\exp\{-\frac{\Omega}{h}\}) \qquad (3.5)$$

uniformly in the domain $\delta \le t \le t_0$, $h \le h_0$, where the second term is an integrable function in x.

Proof. It follows the line of the proof of Theorem 2.1 with some technical complications. Multiplying (2.7) by $B(t, x, x_0)$ one obtains for small $\tilde{h}$ (see Proposition 2.2) that

$$u_G^{as}(t, x, x_0, h) = \chi(t, |x - x_0|) u^{exp}(t, x, x_0, h)(1 + O(\tilde{h})) \qquad (3.6)$$

with

$$u^{exp} = B(t, x, x_0) u_{rough}^{exp}(t, x, x_0, h) = (2\pi h)^{-d/2} \phi(t, x, x_0) \exp\{-\frac{S(t, x, x_0)}{h}\}. \qquad (3.7)$$

Consequently, for small $\tilde{h}$ the multiplicative asymptotics of the function u_G^{as} can be given by function (3.7) which is constructed according to the general method of Section 3.1. This fact actually constitutes the reason for definition (3.3). To go further, we need to have an appropriate modification of Propositions 2.3, 2.4, i.e. the estimates for the function

$$F_{as}(t, x, x_0, h) = \left[\frac{\partial}{\partial t} - \frac{1}{h} H(x, -h\nabla) \right] u_G^{as}. \tag{3.8}$$

Using the formula of the commutation of a Lévy-Khintchine PDO with an exponential function from Proposition D1, yields for u_G^{as} the formula of type (2.19), but with χB instead of χ in all expressions. Using instead Proposition D1 a more precise formula from Proposition D2 gives

$$F_{as} = -\frac{1}{(2\pi h)^d h} \int_{\mathcal{R}^d} (\tilde{f}_1 + h\tilde{f}_2 + h\tilde{f}_3 + h\tilde{f}_4) \exp\{-\frac{ipy(t, x, x_0) - H(x_0, ip)t}{h}\} \, dp \tag{3.9}$$

with

$$\tilde{f}_1 = B(t, x, x_0) f_1 - h\chi(t, |x - x_0|)g,$$

$$g(t, x, p) = \frac{\partial B}{\partial t} + \left(\frac{\partial B}{\partial x}, \frac{\partial H}{\partial p}(x, ip\frac{\partial y}{\partial x}) \right) + \frac{1}{2} \, tr \left(\frac{\partial^2 H}{\partial p^2}(x, ip\frac{\partial y}{\partial x})ip_j \frac{\partial^2 y_j}{\partial x^2} \right),$$

$$\tilde{f}_2 = \chi(t, |x - x_0|)\frac{1}{h} \int \exp\{-i(p, \frac{\partial y}{\partial x}\xi)\}\nu(x, d\xi)$$

$$\times \left(\exp\{-hi \int_0^1 (1-s)(p_j \frac{\partial^2 y_j}{\partial x^2}(x + sh\xi)\xi, \xi) \, ds\} - 1 + \frac{h}{2}ip_j(\frac{\partial^2 y_j}{\partial x^2}(x)\xi, \xi) \right),$$

$$\tilde{f}_3 = -\frac{\partial \chi}{\partial t} B - B \left(\frac{\partial \chi}{\partial x}, \frac{\partial H}{\partial p}(x, ip\frac{\partial y}{\partial x}(t, x, x_0)) \right) + \int \left(\frac{\partial(\chi B)}{\partial x}, \xi \right)$$

$$\times \left(\exp\{\frac{h}{i} \int_0^1 (1-s)(p_j \frac{\partial y_j^2}{\partial x^2}(x + sh\xi)\xi, \xi) \, ds\} - 1 \right) \exp\{-i(p, \frac{\partial y}{\partial x}\xi)\}\nu(x, d\xi),$$

$$\tilde{f}_4 = h \int_{\mathcal{R}^d} \left(\int_0^1 (1-s)ds \frac{\partial^2(\chi B)}{\partial x^2}(x + sh\xi)\xi, \xi \right)$$

$$\times \exp\{-hi \int_0^1 (1-s)(p_j \frac{\partial^2 y_j}{\partial x^2}(x + sh\xi)\xi, \xi) \, ds\} \exp\{-i(p, \frac{\partial y}{\partial x}\xi)\}\nu(x, d\xi).$$

Making the same change of the variable in (3.9) as in the proof of Proposition 2.3 will amount to the change of the argument p to $-i\hat{p} + p$ in all expressions and to the change of the phase to $\hat{\Sigma}(p)$ given by (2.13). The first term in $\tilde{f}_1$ is proportional to f_1 from (2.19) and it was proved that f_1 vanishes at the saddle-point. Therefore, the first term of the expansion in small h of integral (3.9) vanishes. The key moment is to prove that unlike (2.19), the second term in this expansion vanishes as well. (except for the boundary layer reflected in $\tilde{f}_3$). Namely, we claim that for small t and $\hbar$

$$F_{as} = O(h)u_G^{as} \left(1 + \left(\frac{|x - x_0|}{t} \right)^{1+b} \right) + O(t^{-1-b})\Theta_{c_1, c_2}(|x - x_0|) \exp\{-\frac{\Omega}{h}\}. \tag{3.10}$$

To prove this, let us simplify the expression for g. Since ϕ satisfies transport equation (3.13), and due to the formula

$$\frac{\partial^2 S}{\partial x^2} = \hat{p}_j \frac{\partial^2 y_j}{\partial x^2} + \frac{\partial \hat{p}}{\partial x}\frac{\partial y}{\partial x}$$

one can write denoting $D = \frac{\partial^2 H}{\partial p^2}(x_0, \hat{p})$:

$$\frac{\partial B}{\partial t} = -\frac{1}{2}B\, tr\left(\frac{\partial^2 H}{\partial p^2}(x, \hat{p}\frac{\partial y}{\partial x})(\hat{p}_j\frac{\partial^2 y_j}{\partial x^2} + \frac{\partial \hat{p}_j}{\partial x}\frac{\partial y_j}{\partial x})\right)$$

$$-t^{d/2}(\det D)^{1/2}\left(\frac{\partial \phi}{\partial x}, \frac{\partial H}{\partial p}(x, \hat{p}\frac{\partial y}{\partial x})\right) + \frac{1}{2}B\, tr\left(\frac{\partial^3 H}{\partial p^3}(x_0, \hat{p})\frac{\partial \hat{p}}{\partial t} \times D^{-1}\right) + \frac{d}{2}t^{-1}B,$$

and

$$\frac{\partial B}{\partial x_j} = t^{d/2}(\det D)^{1/2}\frac{\partial \phi}{\partial x_j} + \frac{1}{2}B\, tr\left(\frac{\partial^3 H}{\partial p^3}(x_0, \hat{p})\frac{\partial \hat{p}}{\partial x_j} \times D^{-1}\right).$$

Consequently,

$$g(-i\hat{p}) = -\frac{1}{2}B\, tr\left(\frac{\partial^2 H}{\partial p^2}(x, \hat{p}\frac{\partial y}{\partial x}) \times \frac{\partial \hat{p}_j}{\partial x}\frac{\partial y_j}{\partial x}\right) + \frac{1}{2}B\, tr\left(\frac{\partial^3 H}{\partial p^3}(x_0, \hat{p})\frac{\partial \hat{p}}{\partial t} \times D^{-1}\right)$$

$$+\frac{1}{2}B\, tr\left(\frac{\partial^3 H}{\partial p^3}(x_0, \hat{p})\frac{\partial \hat{p}}{\partial x_j} \times D^{-1}\right)\frac{\partial H}{\partial p_j}(x, \hat{p}\frac{\partial y}{\partial x}) + \frac{dB}{2t}.$$

Differentiating the defining equation for $\hat{p}$ one obtains

$$\frac{\partial \hat{p}}{\partial t} = D^{-1}(-\frac{1}{t^2}y + \frac{1}{t}\frac{\partial y}{\partial t}) = \frac{1}{tD}\left(\frac{\partial y}{\partial t} - \frac{\partial H}{\partial p}(x_0, \hat{p})\right), \qquad \frac{\partial \hat{p}}{\partial x} = \frac{1}{t}D^{-1}\frac{\partial y}{\partial x},$$

which implies that

$$g(-i\hat{p}) = \frac{B}{2t}\frac{\partial^3 H}{\partial p_i \partial p_j \partial p_k}(x_0, \hat{p})D_{kl}^{-1}\left(\frac{\partial y}{\partial t} + \frac{\partial y}{\partial x}\frac{\partial H}{\partial p}(x, \hat{p}\frac{\partial y}{\partial x}) - \frac{\partial H}{\partial p}(x_0, \hat{p})\right)_l D_{ji}^{-1}$$

$$-\frac{B}{2t}\left(\frac{\partial^2 H}{\partial p^2}(x, \hat{p}\frac{\partial y}{\partial x})\right)_{kl} D_{ij}^{-1}\frac{\partial y_j}{\partial x_k}\frac{\partial y_i}{\partial x_l} + \frac{dB}{2t}.$$

Due to Proposition B4 and to the expression (2.13) for the phase $\tilde{\Sigma}$, in order to prove that the second term in the expansion of (3.9.) in h vanishes, one must show that

$$\frac{1}{2}tr\, \frac{\partial^2 f_1}{\partial p^2}(-i\hat{p})D^{-1} - \frac{1}{2}\frac{\partial f_1}{\partial p_i}(-i\hat{p})\frac{\partial^3 H}{\partial p_j \partial p_k \partial p_l}(x_0, \hat{p})D_{ij}^{-1}D_{kl}^{-1} - g(-i\hat{p}) = 0,$$

which readily follows from the above expression for $g(-i\hat{p})$ and the formulas for the derivatives of f_1 given in the proof of Proposition 2.3. Other arguments for obtaining (3.10) are the same as that used in the proof of Proposition 2.3. Notice only that now, unlike the case of formula (2.18), we must write the contribution of f_3 separately, it has the form $O(u_G^{as})(|x - x_0|/t)^{1+b}$ and does not vanish only for $|x - x_0| \in [c_1, c_2]$, and consequently, it can be written in the form of the last term in (3.10).

Turning to the estimate of F_{as} for small t we claim that the same estimates (2.25), (2.25') of Proposition 2.4 hold for F_{as} as well. To see this, notice that everything is estimated in the same way as in Proposition 2.4 except of the new terms presented by the function g in the expression for $\tilde{f}_1$. We must show that the contribution of g gives nothing new sa compared with other terms. Let us estimate only the contribution of the first term in the expression for g, namely of $\frac{\partial B}{\partial t}$, other terms being estimated similarly.

From the definition of B and using again the notation $D = \frac{\partial^2 H}{\partial p^2}(x_0, \hat{p})$ one gets

$$\frac{1}{B}\frac{\partial B}{\partial t} = \frac{1}{2}tr\left(\frac{\partial D}{\partial t}D^{-1}\right)$$

$$-\frac{1}{2}tr\left(\frac{\partial}{\partial t}\frac{\partial X}{\partial p_0}\left(\frac{\partial X}{\partial p_0}\right)^{-1}\right) + \frac{\partial^2 H}{\partial p \partial x}(x, P(t, x_0, p_0)) + \frac{d}{2t},$$

where $X = X(t, x_0, p_0(t, x, x_0))$ and therefore

$$\frac{\partial}{\partial t}\left(\frac{\partial X}{\partial p_0}\right) = \frac{\partial}{\partial p_0}\dot{X}(t, x_0, p_0) + \frac{\partial^2 X}{\partial p_0^2}\frac{\partial p_0}{\partial t}$$

$$= \frac{\partial^2 H}{\partial p \partial x}(x, P(t, x_0, p_0))\frac{\partial X}{\partial p_0} + \frac{\partial^2 H}{\partial p^2}(x, P(t, x_0, p_0))\frac{\partial P}{\partial p_0}$$

$$-\frac{\partial^2 X}{\partial p_0^2}\left(\frac{\partial X}{\partial p_0}\right)^{-1}\frac{\partial H}{\partial p}(x, P(t, x_0, p_0)).$$

Due to the estimates of the derivatives of H from Theorem 5.3.2, one can write

$$\frac{\partial}{\partial t}\frac{\partial X}{\partial p_0}\left(\frac{\partial X}{\partial p_0}\right)^{-1} = O\left(\frac{|x - x_0|}{t}\right)$$

$$+\frac{\partial^2 H}{\partial p^2}(x, P(t, x_0, p_0))\frac{1 + O(|x - x_0|)}{t}\left(\frac{\partial^2 H}{\partial p^2}(x_0, p_0)\right)^{-1}$$

$$-\frac{\partial^2 X}{\partial p_0^2}\left(\frac{\partial^2 H}{\partial p^2}(x_0, p_0)\right)^{-1}\frac{\partial H}{\partial p}(x, P(t, x_0, p_0))$$

$$\times\frac{1 + O(|x - x_0|)}{t}\left(\frac{\partial^2 H}{\partial p^2}(x_0, p_0)\right)^{-1}.$$

Since

$$\frac{\partial^2 H}{\partial p^2}(x, P(t, x_0, p_0)) = \frac{\partial^2 H}{\partial p^2}(x_0, p_0) + \frac{O(|x - x_0|^2)}{t} = D + \frac{O(|x - x_0|^2)}{t},$$

and similarly estimates hold for $\frac{\partial H}{\partial p}$, and again due to the estimates of Theorem 5.3.2 and the expression for the second derivatives of X from Proposition 2.5.1, it follows that

$$\frac{\partial}{\partial t}\frac{\partial X}{\partial p_0}\left(\frac{\partial X}{\partial p_0}\right)^{-1}$$

$$= 1 - \frac{\partial^3 H}{\partial p^3}(x_0, p_0)D^{-1}\frac{\partial H}{\partial p}(x_0, p_0)D^{-1} + O\left(\frac{|x - x_0|}{t}\right)\left(1 + \log^+\frac{|x - x_0|}{t}\right).$$

Similarly, and due to (2.22),

$$\frac{\partial D}{\partial t}D^{-1} = \frac{\partial^3 H}{\partial p^3}(x_0, \hat{p})\frac{\partial \hat{p}}{\partial t}D^{-1} = \frac{1}{t}\frac{\partial^3 H}{\partial p^3}(x_0, \hat{p})D^{-1}\left(\frac{\partial y}{\partial t} - \frac{\partial H}{\partial p}(x_0, \hat{p})\right)D^{-1}$$

$$= -\frac{1}{t}\frac{\partial^3 H}{\partial p^3}(x_0, p_0)D^{-1}\frac{\partial H}{\partial p}(x_0, p_0)D^{-1} + O\left(\frac{|x - x_0|}{t}\right)\left(1 + \log^+\frac{|x - x_0|}{t}\right)^2.$$

Consequently, one obtains

$$\frac{\partial B}{\partial t} = O\left(\frac{|x - x_0|}{t}\right)\left(1 + \log^+\frac{|x - x_0|}{t}\right)^2 B,$$

This implies that the contribution of $\frac{\partial B}{\partial t}$ does not give anything new as compared with other terms of (2.25), (2.25'). We conclude that the estimates of Proposition 2.4 remain the same for F_{as}.

Following further all the steps of the proof of Theorem 2.1, one obtains instead of (2.53) or (2.55) the estimate

$$\int_0^t d\tau \int_{\mathcal{R}^d} d\eta\, u_{rough}(t - \tau, x, \eta, h)F(\tau, \eta, x_0, h)$$

$$= O(t^\omega)\left[h^\omega u_{rough}(t, x, x_0, h) + \exp\{-\frac{\Omega}{h}\}\Theta_{0,2c}(|x - x_0|)\right] \tag{3.11}$$

with some $\Omega > 0$ and $\omega \in (0, 1)$, because, on the one hand, the estimate t^ω in (2.53) appears from integrating over $\tau \le h$ and can be thus presented in the form $t^\omega h^\omega$ with some (different) $\omega \in (0, 1)$, and on the other hand, we have now additional multiplier h at the second term in (2.53). Therefore, Theorem 3.1 is obtained in the same way as Theorem 2.1.

Clearly, formulas (3.5)-(3.7) implies the statement of Theorem 1.1. Moreover, when the local asymptotics of u_G is obtained, one constructs global small h asymptotics of u_G in the same way as it was done for the standard diffusion in section 3.5.5. Consequently one obtains the following result.

Theorem 3.2. *The formulas of Theorem 3.5.1 and of Propositions 3.5.2-3.5.7 are valid for the equation (1.1), (1.2).*

Let us note for conclusion that the results of this chapter can be generalised to cover some other classes of jump-diffusions, for instance, the localised versions of the stable-like jump-diffusions described in section 5.5.

Chapter 7. COMPLEX STOCHASTIC DIFFUSION OR STOCHASTIC SCHRÖDINGER EQUATION

1. Semiclassical approximation: formal asymptotics

In this chapter we develop the method of semiclassical asymptotics for stochastic complex diffusion equations. Actually we shall consider the equations which appeared recently in the stochastic models of open systems, in the theory of stochastic and quantum stochastic filtering and continuous quantum measurements. Namely, we shall consider the equation

$$ du = \left(\frac{h}{2} tr \left(G \frac{\partial^2 u}{\partial x^2} \right) - \frac{1}{h} V(x) u - \frac{|\alpha|^2}{2} x^2 u \right) dt + \alpha x u \, dW, \qquad (1.1) $$

where $x \in \mathcal{R}^m$, $dW = (dW^1, ..., dW^m)$ is the stochastic differential of the standard Brownian motion in $\mathcal{R}^m$, G is a symmetric complex matrix, α is a complex constants, $|G| > 0$, $Re\, G \geq 0$, and $V(x)$ is an analytic function in the strip $St_b = \{x : |Im\, x| \leq b\}$ with some $b > 0$. Main results presented here can be extended to cover the case of more general equations (1.1) with G and α depending on x. The deduction of this equation in the framework of the theory of continuous quantum measurement is given in Appendix A and the discussion of the simplest examples - in Section 1.4. Clearly, equation (1.1) can be written formally in form (0.9) with the formal random non-homogeneous Hamiltonian

$$ \frac{1}{2} (Gp, p) - V(x) - h \left(\frac{|\alpha|^2}{2} x^2 - \alpha x \dot{B}(t) \right), $$

where $\dot{B}$ is a formal (or generalised) derivative of the Brownian motion. Therefore, one can expect that some generalisation of the procedure of Chapter 3 will lead to the small h or small t asymptotics of the solutions of equation (1.1). This is the subject of the present chapter.

Let us indicate first two simplest particular cases of (1.1), where the semiclassical approximations have been studied in [K1] and [TZ1], [TZ2] (by different approaches). These are the cases when either (i) all coefficients are real (stochastic heat equation), or (ii) α, G, V are purely imaginary (unitary stochastic Schrödinger equation). Both these cases lead to the real Hamilton-Jacobi equations, which can be solved using the theory developed in Section 2.7. At the end of the second section, we shall present for completenes an asymptotic expansion for the Green function of stochastic heat equation, see also [LR] and [So] for small time asymptotics of the heat kernel of stochastic heat equations. Now we shall draw our attention to a more difficult, essentially complex, situation. Namely, we shall suppose that

$$ \alpha_R = Re\, \alpha > 0, \quad G_I = Im\, G > 0, \quad G_R = Re\, G \geq 0. \qquad (1.2) $$

In this section we construct formal asymptotics for the Green function of equation (1.1) (actually, two types of these asymptotics), i.e. to its solution

$u_G(t, x, x_0)$ with the Dirac initial condition $\delta(x - x_0)$. Further on we shall justify these asymptotics. In the last section we consider another approach to the construction of the solutions of equation (1.1), namely, the method of path integration. A new moment in the semiclassical expansion of the path integral formula for the solutions to equation (1.1), (1.2) is the nessecity to move the (infinite dimensional) contour of integration to the complex domain and to use the infinite-dimensional method of saddle-point (and not just Laplace or stationary phase methods needed for this procedure in the case of the (classical or stochastic) heat or unitary Schrödinger equations respectively).

To construct semiclassical asymptotics for equation (1.1) we shall develop first the approach from [K1](applied there for the above indicated real cases) using the results of Section 1.6 on the complex Hamilton-Jacobi equation. To see the main idea, let us look for the solution of equation (1.1) in the form

$$u = \varphi(t, x, [W]) \exp\{-\frac{1}{h} S(t, x, h, [W])\}. \tag{1.3}$$

Note, that this form differs from the standard WKB substitution by more complicated dependence of the phase on h. This dependence will be made more explicit further. By the Ito formula we have

$$du = \left(d\varphi + \varphi(-\frac{1}{h} dS + \frac{1}{2h^2}(dS)^2) - \frac{1}{h} d\varphi dS \right) \exp\{-\frac{1}{h} S\}.$$

Consequently, substituting (1.3) in (1.2) yields

$$d\varphi + \left[-\frac{1}{h} H(x, \frac{\partial S}{\partial x})\varphi + (\frac{\partial H}{\partial p}(x, \frac{\partial S}{\partial x}), \frac{\partial \varphi}{\partial x}) + \frac{1}{2} tr \frac{\partial^2 H}{\partial p^2} \frac{\partial^2 S}{\partial x^2}\varphi + \frac{|\alpha|^2}{2} x^2\varphi \right] dt$$

$$-\frac{h}{2} tr(G\frac{\partial^2 \phi}{\partial x^2}) dt + \varphi(-\frac{1}{h} dS + \frac{1}{2h^2}(dS)^2) - \frac{1}{h} d\varphi dS = \alpha x\varphi \, dW, \tag{1.4}$$

where we denoted by H the classical Hamiltonian

$$H(x, p) = \frac{1}{2}(Gp, p) - V(x) \tag{1.5}$$

of the complex stochastic equation (1.2). The main idea of the approach proposed here is to add additional (linearly dependent on h) terms in the Hamilton - Jacobi equation in such a way that the corresponding transport equation would take the standard deterministic form. To this end, let us write the Hamilton-Jacobi equation in the form:

$$dS + H(x, \frac{\partial S}{\partial x}) dt - \frac{h}{2}(\alpha^2 + |\alpha|^2)x^2 \, dt + h\alpha x \, dW = 0, \tag{1.6}$$

or more explicitly

$$dS + \left(\frac{1}{2}(G\frac{\partial S}{\partial x}, \frac{\partial S}{\partial x}) - V(x) \right) dt - \frac{h}{2}(\alpha^2 + |\alpha|^2)x^2 \, dt + h\alpha x \, dW = 0. \tag{1.7}$$

One sees readily that equation (1.4) is satisfied up to a term of the order $O(h)$, if (1.6) is fulfilled and the following transport equation holds:

$$d\varphi - \alpha x \, d\varphi \, dS + (\frac{\partial H}{\partial p}, \frac{\partial \varphi}{\partial x}) \, dt + \frac{1}{2} \, tr \, \frac{\partial^2 H}{\partial p^2} \frac{\partial^2 S}{\partial x^2} \varphi \, dt = 0.$$

It follows that the differential $d\varphi$ has no stochastic terms, and therefore $d\varphi dS = 0$ and the transport equation takes, in fact, the standard form (see Chapter 3):

$$d\varphi + (\frac{\partial H}{\partial p}, \frac{\partial \varphi}{\partial x}) \, dt + \frac{1}{2} \, tr \left(\frac{\partial^2 H}{\partial p^2} \frac{\partial^2 S}{\partial x^2} \right) \varphi \, dt = 0. \qquad (1.8)$$

To solve equation (1.6) by the method of Sections 1.6, 1.7, we need to consider the corresponding Hamiltonian system

$$\begin{cases} dx = Gp \, dt, \\ dp = (\frac{\partial V}{\partial x} + h(\alpha^2 + |\alpha|^2)x) \, dt - h\alpha \, dW. \end{cases} \qquad (1.9)$$

Along the trajectories of this system equation (1.7) can be written in the form

$$\frac{d\varphi}{dt} + \frac{1}{2} \, tr \, \frac{\partial^2 H}{\partial p^2} \frac{\partial^2 S}{\partial x^2} = 0, \qquad (1.10)$$

which is again the same as in the deterministic case.

Before formulating a general result, let us show how the proposed method works on the simplest example, where the solutions of the Hamilton-Jacobi and transport equations can be written explicitly, and therefore one does not need the general theory. Namely, consider the case of vanishing potential V and the matrix G being a complex constant (i.e. it is proportional to the unit matrix). We obtained the exact Green function for this equation in Section 1.4 using Gaussian solutions. Let us recover this solution using the complex stochastic WKB method described above. The Hamilton-Jacobi equation and the Hamiltonian system in that case have the form (1.7), (1.9) with vanishing V. Obviously, one can write down the solution to (1.9) with vanishing V and G being a constant explicitly:

$$\begin{cases} x = x_0 \cosh \beta t + p_0 G \beta^{-1} \sinh \beta t - h\alpha G \beta^{-1} \int_0^t \sinh \beta(t - \tau) \, dW(\tau), \\ p = x_0 \beta G^{-1} \sinh \beta t + p_0 \cosh \beta t - h\alpha \int_0^t \cosh \beta(t - \tau) \, dW(\tau). \end{cases} \qquad (1.11)$$

where the complex number β is uniquely define by the conditions

$$\beta^2 = hG(\alpha^2 + |\alpha|^2), \quad -\pi/4 < arg \, \beta < \pi/2.$$

Therefore, for all x, x_0 and each $t > 0$ there exists a unique

$$p_0 = \frac{\beta}{G \sinh \beta t} \left(x - x_0 \cosh \beta t + \frac{h\alpha G}{\beta} \int_0^t \sinh \beta(t - \tau) \, dW(\tau) \right) \qquad (1.12)$$

such that the solution (1.11) with initial values x_0, p_0 joins the points x_0 and x in time t. And consequently, the two-point function $S_W(t, x, x_0)$ (see Sections 2.6, 2.7) can be calculated explicitly using the formula

$$S_W(t, x, x_0, h) = \frac{1}{2} \int_0^t (Gp^2(\tau) + h(\alpha^2 + |\alpha|^2)x^2(\tau))\, d\tau - h\alpha \int_0^t x(\tau)\, dW(\tau),$$
(1.13)

where the integral is taken along this solution. Furthermore, it follows from (1.11) that the Jacobian $J = \det(\partial x/\partial p_0)$ is equal to $(G \sinh \beta t/\beta)^m$ and thus does not depend on x. Therefore, the remainder $h\Delta\phi$ in (1.4) vanishes, and the asymptotic Green function of form (1.3) coincides actually with the exact one and is equal to

$$u_G^W(t, x, x_0, h) = (2\pi h)^{-m/2} \left(\frac{\beta}{G \sinh \beta t}\right)^{m/2} \exp\{-\frac{i}{h} S_W(t, x, x_0, h)\}. \quad (1.14)$$

Simple but tedious calculations of S_W from (1.11)-(1.13) show that formula (1.14) coincides with the formula 1.4.13 from Theorem 1.4.1. One can use this example to give a well posedness theorem for the Cauchy problem of equation (1.2) with rather general potential (see [K1]).

Turning to the general case of nonvanishing V suppose that V is analytic and its second derivative is bounded in the strip $St_b = \{x = y + iz \in C^m : |y| \le b\}$ with some $b > 0$. In that case it follows from the theory of Sections 2.6, 2.7 that the boundary value problem for the corresponding Hamiltonian system is solvable for $|x - x_0| \le r$, $t \in (0, t_0]$, $x, x_0 \in St_{b/2}$ with some c and t, the solution giving the saddle-point for the corresponding problem of the complex calculus of variations is unique, the (random) two-point function $S_W(t, x, x_0, h)$ is analytic in x and x_0 under these assumptions and satisfies the Hamilton-Jacobi equation. Furthermore, the function $\phi = J^{-1/2}(t, x, x_0)$, where J is the Jacobian $\det \frac{\partial X}{\partial p_0}$ along this solution is well defined under these assumptions and satisfies the corresponding transport equation. Consequently the function

$$u_{as} = (2\pi h)^{-m/2} \chi(|x - x_0|)\phi_W(t, x, x_0) \exp\{-S_W(t, x, x_0, h)/h\}, \quad (1.15)$$

where χ is a smooth molyfier (which equals one for $|x - x_0| \le r - \epsilon$ with some positive ϵ and vanishes for $|x - x_0 \ge r)$, is smooth, satisfies the Dirac initial conditions (3.1.4), and satisfies equation (1.1) up to the remainder

$$hF(t, x, x_0) = (2\pi h)^{-m/2} h$$

$$\times \left[\frac{1}{2}(G\frac{\partial^2 \phi}{\partial x^2}\chi(|x - x_0|) + O(\phi)(th)^{-1}\Theta_{r-\epsilon, r}(|x - x_0|)\right] \exp\{-\frac{S_W(t, x, x_0, h)}{h}\}.$$
(1.16)

Thus we proved the following result.

Theorem 1.1. *Function (1.15) is well defined and is a formal asymptotic solution to the problem (1.1), (1.2), (3.1.4) in the sense that it satisfies equation*

(1.1) up to a smaller (in h) term of form (1.16) and satisfies the Dirac initial condition (3.1.4).

Next section will be devoted to the justification of this asymptotics. Now we construct another version of semiclassical asymptotics of the Green function of equation (1.1), which was first constructed in [K3], [BK] for a particular case of this equation. For this construction, it is convenient to consider separately two cases.

First let $Re\, G = 0$ and $Re\, V = 0$ on the real plane. Then equation (1.1) can be written in the form

$$du = (\frac{h}{2}i\, tr\,(G\frac{\partial^2 u}{\partial x^2}) - \frac{i}{h}V(x)u - \frac{|\alpha|^2}{2}x^2 u)\, dt + \alpha x u\, dW, \qquad (1.17)$$

with positive constant G (which equals to G_I in previous notations) and a smooth real $V(x)$. One readily sees then that the two-point function S_W can be presented in the form $-iS_1 + hS_2$, where S_1 is real for real x, x_0 and does not depend on W and h. Thus the formal asymptotic solution u_{as} can be rewritten in the form

$$u_{as} = (2\pi h)^{-m/2}\chi(|x - x_0|)\psi_W(t, x, x_0, h) \exp\{iS_1(t, x, x_0)/h\}, \qquad (1.18)$$

with

$$\psi_W(t, x, x_0, h) = \phi_W(t, x, x_0)(1 + S_2(t, x, x_0, h) + O(h)).$$

It turns out moreover that assuming additionally that $V(x)$ is bounded together with its second derivatives in a neightborhood of the real subspace $Im\, x = 0$, one can get get rid of the molyfier χ in the asymptotical formula (1.18). To see this, let us substitute the function of form

$$u_{as} = (2\pi h)^{-m/2}\psi_W(t, x, x_0) \exp\{iS_1(t, x, x_0)\} \qquad (1.19)$$

with a real deterministic S_1 in equation (1.17). Equalising the coefficients at h^{-1} and h^0, one obtains for S_1 the standard (deterministic and real) Hamilton-Jacobi equation of the form

$$\frac{\partial S}{\partial t} + \frac{1}{2}G\left(\frac{\partial S}{\partial x}\right)^2 + V(x) = 0 \qquad (1.20)$$

and for ψ a stochastic version of the transport equation

$$d\psi + \left(G\frac{\partial \psi}{\partial x}\frac{\partial S}{\partial x} + \frac{1}{2}G\Delta S\psi + \frac{|\alpha|^2}{2}x^2\psi\right)\, dt = \alpha x\psi\, dW. \qquad (1.21)$$

As usual in WKB constructions of the Green functions, one takes as the solution of the Hamilton-Jacobi equation (1.20) the two-point function $S_1(t, x, x_0)$ of the corresponding problem of the calculus of variations with the Hamiltonian

$$H(x, p) = \frac{1}{2}(Gp, p) + V(x) \qquad (1.22)$$

and the Lagrangian $L(x, v) = \frac{1}{2}(Gv, v) - V(x)$. Notice the difference of the signs at V in expressions (1.5), (1.22) (which is due to the difference of (1.3) and (1.19)), and also the fact that H of form (1.22) is real. The two-point function S_1 is expressed in terms of the solutions of the real Hamiltonian system $\dot{x} = Gp$, $\dot{p} = -\frac{\partial V}{\partial x}$. Since V is bounded together with its second derivative, the boundary value problem for this system is uniquely solvable for all (real) x, x_0 and $t \leq t_0$ with some $t_0 > 0$ (see Proposition 2.2.9). Therefore, S_1 is well defined and smooth for all such x, x_0, t. Furthermore, to solve (1.21), put $\psi_W = \phi_1 \mu_W$ with the deterministic $\phi_1 = J_1^{-1/2}(t, x, x_0)$, where J_1 is the Jacobian (corresponding to the solutions of the Hamiltonian system with Hamiltonian (1.22)). One obtains then for μ_W the following equation

$$d\mu + \left((G\frac{\partial \mu}{\partial x}, \frac{\partial S}{\partial x}) + \frac{|\alpha|^2}{2} x^2 \mu \right) dt = \alpha x \mu \, dW. \tag{1.23}$$

Since

$$\frac{\partial \mu}{\partial t} + \left(G\frac{\partial \mu}{\partial x}, \frac{\partial S}{\partial x} \right) = \frac{d\mu}{dt}$$

is the full derivative along the solutions of the Hamiltonian system, one can write the solutions to (1.23) similarly to (3.1.18), but using also Ito's formula, to obtain for ψ_W the expression

$$\psi_W(t, x, x_0) = J_1^{-1/2}(t, x, x_0) \exp\{ \int_0^t \left(-|\alpha|^2 x^2(\tau) \, d\tau + \alpha x(\tau) \, dB(\tau) \right). \tag{1.24}$$

Therefore, we have proved the following result.

Proposition 1.2 [K5], [BK]. *If the second derivative of the function V is uniformly bounded, then the function (1.19), where S_1 is the two-point function corresponding to the Hamiltonian (1.22) and ψ_W is given by (1.24), is well defined for all $t \in (0, t_0)$, x, x_0 and is a formal asymptotic solution to the problem (1.17), (3.1.4) in the sense that it satisfies equation (1.17) up to terms of order $O(h)$, and satisfies the initial condition (3.1.4).*

Therefore, in the case $G_R = 0$ in (1.1), (1.2), one can avoid dealing with complex characteristics. It will not be so in the case with $G_R > 0$ in (1.1), which we consider now. Looking for the Green function of (1.1) in the form

$$u_{as} = (2\pi h)^{-m/2} \chi(|x - x_0|) \varphi_1(t, x, x_0, [W]) \exp\{-\frac{1}{h} S_1(t, x, x_0)\}. \tag{1.25}$$

with a deterministic not depending on h phase S_1 (it is not convenient now to write the exponential in form iS), one comes (using (1.4) and the assumption that S_1 is not stochastic) to the deterministic Hamilton-Jacobi equation with the Hamiltonian of form (1.5), namely to the equation

$$\frac{\partial S}{\partial t} + \frac{1}{2} \left(G\frac{\partial S}{\partial x}, \frac{\partial S}{\partial x} \right)^2 - V(x) = 0, \tag{1.26}$$

for the function S_1, and to the stochastic transport equation (1.21). Unlike (1.20), equation (1.26) is still complex (though not stochastic as (1.6)). From the results of Section 2.2 it follows that function (1.25) is well defined. Therefore we obtain the following result.

Proposition 1.2. *If $G_R > 0$, the function (1.25), where S_1 is the two-point function corresponding to the Hamiltonian (1.5) and ψ_W is given by (1.24), with $x(\tau)$ being the characteristic corresponding to Hamiltonian (1.5), is well defined for all $t \in (0, t_0)$, $x, x_0 \in \mathcal{R}^d$ and is a formal asymptotic solution to the problem (3.1), (3.1.4) in the sense that it satisfies equation (1.1) up to terms of order $O(h)$, and satisfies the initial condition (3.1.4).*

Thus, we obtained two asymptotic formulas for the Green function of equation (1.1), given in Theorem 3.1 and in Propositions 1.1, 1.2 respectively, exploiting two approaches to the construction of the WKB type asymptotics of stochastic equations of type (1.1): in the first one, one uses a stochastic Hamilton-Jacobi equation and deterministic transport equation, and in the second one, one uses a deterministic Hamilton-Jacobi equation and the stochastic transport equation. Each of these (in a sense equivalent) approaches has its advantages. Namely, it seems that the formula from Theorem 1.1 gives more adequate asymptotics in a number of situation, for example this formula (and not the one from Propositions 1.1 or 1.2) gives the exact Green function in the case of quadratic potentials. On the other hand, formulas (1.19), (1.25) seem to be easier to justify, as we shall see in the next section.

2. Semiclassical approximation: justification and globalisation

We shall deal with the problem of justification of the asymptotics constructed above following the same line of arguments as for the case of standard diffusions in Chapter 3 paying special attention to the new difficulties which naturally arise in the present context of complex diffusion. Consider first shortly the case of vanishing G_R. The advantage of formula (1.19), as compared with (1.15), is due to the fact that the exponential term $\exp\{iS_1/h\}$ has the same form as for the standard WKB asymptotics of the standard Schrödinger equation, because S_1 is real and deterministic. Stochastic term appears only in the amplitude ψ. Therefore, formal asymptotics (1.19) can be justified in the same way as for the standard Schrödinger equation (see e.g. [M6], [MF1]), which leads directly to the following result.

Theorem 2.1. *The Green function of equation (1.17) exists and has the form $u_{as}(1 + hg)$, with u_{as} from Proposition 1.1, where the function g has a bounded L_2-norm.*

Notice that we obtained here an asymptotic representation for the Green function with the L_2- estimate of the remainder, which is usual in the study of the Schrödinger equation, and not a pointwise estimate, which one obtains usually in the study of the heat kernel. Notice also that the theorem implies automatically also the justification of the asymptotics of Theorem 1.1 for the case of equation (1.17). Moreover, when the asymptotics of the Green function in form (1.18) is justified for $t \leq t_0$, the construction of the global small h asymptotics for all finite times t can be obtained automatically [BK] by taking the convolution of u_{as} with itself $N = t/t_0$ times. The result and calculations are the same as for the deterministic case (see [MF1]), only the amplitude will be now stochastic. In particular, if there exists a finite number of characteristics (of the real Hamiltonian (1.22)), joining x_0 and x in time t, the asymptotics will be equal to the sum of the contributions of each characteristics, and in general, it can be expressed by means of the Maslov canonical operator. Moreover, we wrote the asymptotics only up to the term of the order $O(h)$, but one can easily write the whole expansion in h in the same way as it is done for the deterministic case.

As we already noted, in the case $G_R > 0$ in (1.1), one can not avoid dealing with complex characteristics.

Remark. However, if one is interested only in small time asymptotics (for a fixed h), formula (1.25) can be again simplified in such a way that one can get rid of complex characteristics. Namely, since the parts of S_1 depending on V are of the order $O(t)$, one can move these terms from the phase to the amplitude. In other words, instead of (1.25), one can consider the asymptotic Green function in the form

$$u_{as} = (2\pi ht)^{-m/2}\varphi_2(t, x, x_0, [W]) \exp\{-\frac{(x - x_0)^2}{2thG}\},$$

where φ_2 also depends on h (in a non-regular way), but it is not essential, because we fixed it. Considering here φ to be a positive power series in $t, x - x_0$, and substituting this in equation (1.1), one obtains the recurrent formulas for the coefficients in the same way as one does it for the standard non-degenerate real diffusion equation. The justification presents no special difficulties as compared with the case of standard diffusion, because the phase, though being complex, depends quadratically on $x - x_0$. In particular, one easily obtains in this way two-sided estimates for the Green function for finite times (see next Section), generalising some recent results from [LR] obtained there for the case of (real) stochastic heat equation. We are not going into details of these arguments noting that the global small h asymptotics can not be obtained in this way.

Before discussing the justification of asymptotics (1.25) (or, equivalently, (1.15)), let us give a simple lemma from linear algebra (which must have been known, but the author does not know the reference) that we shall need.

Lemma 2.1. *Let E_m denote the unit matrix in $\mathcal{R}^m$, let $s_0, s_1, ..., s_{n+1}$ be a set of positive real numbers, and let $A_{s_0, s_1, ..., s_{n+1}}$ be the $(n+1)m \times (n+1)m$ block three-diagonal matrix of the form*

$$\begin{pmatrix} c_0 E_m & -a_1 E_m & 0 & \cdots & 0 \\ -a_1 E_m & c_1 E_m & -a_2 E_m & \cdots & 0 \\ 0 & -a_2 E_m & c_2 E_m & \cdots & 0 \\ \cdot & \cdot & \cdot & \cdots & \cdot \\ \cdot & \cdot & \cdot & \cdots & c_n E_m \end{pmatrix}$$

with $c_j = s_j^{-1} + s_{j+1}^{-1}$, $a_j = -s_j^{-1}$. In particular, let A_{n+1}^m denote the matrix $A_{s_0, s_1, \ldots, s_{n+1}}$ with all s_j, $j = 0, \ldots, n+1$, being equal to one, i.e.

$$A_{n+1}^m = \begin{pmatrix} 2E_m & -E_m & 0 & \ldots & 0 \\ -E_m & 2E_m & -E_m & \ldots & 0 \\ 0 & -E_m & 2E_m & \ldots & 0 \\ \cdot & \cdot & \cdot & \ldots & \cdot \\ \cdot & \cdot & \cdot & \ldots & 2E_m \end{pmatrix}.$$

Then

$$\det A_{s_0, s_1, \ldots, s_{n+1}} = \left(\frac{s_0 + s_1 + \ldots + s_{n+1}}{s_0 s_1 \ldots s_{n+1}} \right)^m.$$

Moreover,

$$\min_j s_j^{-1} A_{n+1}^m \le A_{s_0, s_1, \ldots, s_{n+1}} \le \max_j s_j^{-1} A_{n+1}^m. \tag{2.1}$$

Proof. Changing the order of rows and columns one easily reduces the case of an arbitrary m to the case of $m = 1$. In this case one obtains (by decomposing the determinant in the last row) that

$$\det A_{s_0, s_1, \ldots, s_{n+2}} = (s_{n+1}^{-1} + s_{n+1}^{-1}) \det A_{s_0, s_1, \ldots, s_{n+1}} - s_{n+1}^{-2} \det A_{s_0, s_1, \ldots, s_n},$$

and then one finishes the proof by a trivial induction. Inequality (2.1) follows directly from the obvious formula

$$(A_{s_0, s_1, \ldots, s_{n+1}} v, v) = \frac{1}{s_0} v_0^2 + \frac{1}{s_1} (v_1 - v_2)^2 + \ldots + \frac{1}{s_n} (v_n - v_{n-1})^2 + \frac{1}{s_{n+1}} v_n^2,$$

where $v = (v_0, \ldots, v_n)$.

Let us turn to the justification of the semiclassical asymptotics in the case $G_R > 0$. Consider function (1.25) as the first approximation to the exact Green function u_G^W for equation (1.1). Arguing as in Chapter 3 one presents u_G^W in the form of the series

$$u_G^W = u_{as} + h u_{as} \otimes F + h^2 u_{as} \otimes F \otimes F + \ldots, \tag{2.2}$$

where F is given by (1.16) and the convolution-type operation $\otimes$ is defined by the formula

$$(v \otimes w)(t, x, x_0) = \int_0^t \int_{\mathcal{R}^m} v(t - \tau, x, \xi) w(\tau, \xi, x_0) \, d\tau d\xi.$$

For $|x - x_0| \ge r - \epsilon$ (see formula (1.15)) one estimate this series in exactly the same way as in the real situation considered in detail in Chapter 3 Theorem 4.1, if one previously estimate the phase S in all integrals by its real quadratic part. Let us consider the case $|x - x_0| \le r - \epsilon$, which is slightly more subtle. Here the main difference with the real case appear: to get exact asymptotics one can not estimate the terms of series (2.2) recursively using the Laplace method with complex phase (see Appendix B), because each such estimate will destroy analyticity, which is essential for the estimate of the next integral. Therefore, to estimate the k -th term of this series we shall consider it as a Laplace integral over $\mathcal{R}^{mk}$. The phase of this integral is

$$f(\eta; x, x_0, t) = S(t - t_k, x, \eta_k) + \ldots + S(t_2 - t_1, \eta_2, \eta_1) + S(t_1, \eta_1, x_0),$$

which due to the (1.29), (1.30), can be written in the form

$$f(\eta; x, x_0, t) = \frac{(x - \eta_k)^2}{2(t - t_k)G} + \ldots + \frac{(\eta_2 - \eta_1)^2}{2(t_2 - t_1)G} + \frac{(\eta_1 - x_0)^2}{2t_1 G} + O(t).$$

The key moment is to prove that $f_R = Re\, f$ is convex and to estimate its matrix of second derivatives f_R'' from below. Clearly

$$f_R'' \geq \Lambda = (G^{-1})_R A_{s_0, \ldots, s_k} - ct E_{mk} \tag{2.3}$$

with some constant $c > 0$, where $s_j = t_{j+1} - t_j$ with $t_0 = 0$, $t_{k+1} = t$. The spectrum of the matrix A_k^m is well known (and is easy to be calculated explicitly). In particular the lowest eigenvalue of this matrix can be estimated by a/k^2 with a positive constant a. Since G_R is supposed to be positive, the same estmate holds (with possibly different constant a) for the lowest eigenvalue of the matrix $(G^{-1})_R A_k^m$. Let $k \leq t^{-1}\sqrt{2a/c}$. Then, due to (2.1), the lowest eigenvalue of the matrix $(G^{-1})_R A_{s_0, \ldots, s_k}$ is not less than $2ct$. Consequently, if λ_j, $j = 1, \ldots mk$, denote the eigenvalues of the matrix $(G^{-1})_R A_{s_0, \ldots, s_k}$, then the eigenvalues of the matrix Λ on the r.h.s. of (2.3) are not less than $\lambda_j/2$ (this is a very rough estimate but it is sufficient for our purposes). Therefore (also due to Lemma 2.1), inequality (2.3) gives the lower estimate for f_R'' by a convex matrix Λ with

$$\det \Lambda \geq 2^{-km}(\det(G^{-1})_R)^k \left(\frac{t}{t_1(t_2 - t_1)\ldots(t - t_k)} \right)^m. \tag{2.4}$$

Next, by the results of Section 2.6 (and due to the assumption $|x - x_0| < r - \epsilon$), there exists a unique trajectory $x(\tau), p(\tau)$ of the Hamiltonian flow corresponding to Hamiltonian (1.5) and joining x_0 and x in time t. By (1.22), (1.25) the point $\eta(t_1, \ldots, t_k) = (x(t_1), \ldots, x(t_k))$ is a (unique) saddle point to the phase f and

$$f(\eta(t_1, \ldots, t_k); x, x_0, t) = S(t, x, x_0).$$

Moreover, since $Re\, f$ is convex, one can choose the contour of integration in the complex space (using the Cauchy theorem) in such a way that it contains $\eta(t_1, \ldots, t_k)$, and this point is the unique saddle point on this contour and the real part of the phase takes its minimum in this point. Now we can use formula (B3) from Appendix B.

By Proposition 6.3, the amplitude ψ_W of the asymptotic Green function u_{as} is of the order $(2\pi ht)^{-m/2}$. To estimate the amplitude in the expression for F from (3.16), one needs to estimate the second derivative of the Jacobian $J = J(t, x, x_0)$, which is done as in the real case. One has

$$\frac{\partial}{\partial x_i} J = tr \left(\frac{\partial}{\partial x_i} \frac{\partial X}{\partial p_0}(t, x, x_0) \left(\frac{\partial X}{\partial p_0} \right)^{-1} \right) J,$$

and

$$\frac{\partial^2 J}{\partial x_i \partial x_j} = tr \left(\frac{\partial}{\partial x_i} \frac{\partial X}{\partial p_0}(t, x, x_0) \left(\frac{\partial X}{\partial p_0} \right)^{-1} \right) tr \left(\frac{\partial}{\partial x_j} \frac{\partial X}{\partial p_0}(t, x, x_0) \left(\frac{\partial X}{\partial p_0} \right)^{-1} \right) J$$

$$+ tr \left(\frac{\partial^2}{\partial x_i \partial x_j} \frac{\partial X}{\partial p_0}(t, x, x_0) \left(\frac{\partial X}{\partial p_0} \right)^{-1} \right) J$$

$$+ tr \left(\frac{\partial}{\partial x_i} \frac{\partial X}{\partial p_0}(t, x, x_0) \left(\frac{\partial X}{\partial p_0} \right)^{-1} \frac{\partial}{\partial x_j} \frac{\partial X}{\partial p_0}(t, x, x_0) \left(\frac{\partial X}{\partial p_0} \right)^{-1} \right) J.$$

From these formulas and Proposition 6.3 it follows that

$$\frac{\partial J}{\partial x} = O(t)J, \quad \frac{\partial^2 J}{\partial x^2} = O(t^2)J.$$

Therefore, by (2.3), (2.4) and using (B3) (noticing also that though the integral is taken along a complex surface, and not along the plane domain as in (B3), the deformation to the real domain can add an additional multiplier of the order $1 + O(t)$) we obtain for the k-th term of series (2.2), $k \le t^{-1}\sqrt{2a/c}$, the estimate

$$O(th)^k (2\pi h)^{-m/2} \phi_1(t, x, x_0, [W])(\det |G|/ \det Re\, G)^k \exp\{-\frac{Re\, S(t, x, x_0)}{h}\}.$$

And the estimate for the sum of $k \le t^{-1}\sqrt{2a/c}$ terms follows easily. If $k > t^{-1}\sqrt{2a/c}$, we can go on as in the case $|x - x_0| > r - \epsilon$, namely estimating the phase by its quadratic approximation. Because of the coefficient $t^k = O(\exp\{\log t/t\})$, these terms will turn out to be exponentially small as compared with the main term. Consequently, one obtains the following result, which gives the justification to the formal asymptotics obtained in Theorem 1.1 and Proposition 1.2.

Theorem 2.2. *If $G_R > 0$, the Green function u_G^W of problem (1.1), (1.2) exists for small enough t and is given by series (2.2), which converges absolutely. In particular,*

$$u_G^W(t, x, x_0, h) = O(\exp\{-\frac{\Omega}{th}\})$$

$$+ (2\pi h)^{-m/2} \chi(|x - x_0|)\phi_1(t, x, x_0, [W]) \exp\{-\frac{1}{h} S(t, x, x_0)\}(1 + O(th)) \quad (2.5)$$

with some $\Omega > 0$, *where the term* $O(\exp\{-\frac{\Omega}{th}\})$ *is a bounded integrable function in* $x - x_0$.

The Green function of (1.1) for any finite t can be as usually obtained by iteration from the Green function (1.29) for small times (see also the end of the next section). The resulting asymptotic formula will be given by the sum of the contributions of all extremals, which are the saddle points of the action. However, the question of the existence of the complex characteristics joining any two points x, x_0 in time t and giving the saddle-point for the corresponding problem of the calculus of variations, seems to be rather nontrivial in general.

To conclude this section, consider the case of real stochastic heat equation, where everything becomes much simpler. The following result was obtained in [K1] and by different method in [TZ1], [TZ2]. The proof of [K1] is obtained by simplifying the arguments used above for complex situation. Notice only that due to the results of Section 2.7, under the assumptions of Theorem 2.3, the boundary-value problem for corresponding stochastic Hamiltonian system is uniquely globally solvable (for small times), and therefore one do not need to make a cutoff around x_0, which simplifies the situaton essentially.

Theorem 2.3. *Let* $V(x)$ *be a smooth real function with uniformly bounded derivatives of the second and third order, and let* $G = \alpha = 1$. *Then the Green function* u_G *of problem (1.1) exists and for small enough* t *has the form*

$$u_G(t, x, x_0, h) = (2\pi h)^{-m/2} J_W^{-1/2}(t, x, x_0, h) \exp\{-\frac{1}{h} S_W(t, x, x_0, h)\}(1 + O(ht)),$$
(2.6)

where S_W, J_W *are the two-point function and the Jacobian corresponding to the real stochastic Hamiltonian system (1.9) with* $G = \alpha = 1$. *Moreover,* u_G *can be given by the absolutely convergent series of type (4.2).*

Furthermore, in this real situation (unlike the complex case), there appear no additional problems with the globalisation of this result. Namely, on the basis of the results of Section 2.7 and by direct generalisation of the argument of Theorem 3.5.1, one gets the following statement.

Theorem 2.4. *For any* t, x *and* $\tau < t$

$$u_G(t, x, x_0, h) = (2\pi h)^{-m}(1 + O(h)) \int_{\mathcal{R}^m} J_W(t - \tau, x, \eta, h) J_W(\tau, \eta, x_0, h)$$

$$\times \exp\left\{-\frac{S_W(t - \tau, x, \eta, h) + S(\tau, \eta, x_0, h)}{h}\right\} d\eta.$$

In particular, for any $(t, x) \in Reg(x_0)$, *the asymptotics of* u_G *is still given by (2.6).*

3. Applications: two-sided estimates for complex heat kernels, large deviation principle, well-posedness of the Cauchy problem

Here we formulate some direct but important consequences from Theorem 2.2. The following results on the local large deviation principle with respect to small t or small h and on the local two-sided estimates for complex heat kernel follow directly from Theorem 2.2

Proposition 3.1. *Under the assumptions of Theorem 2.2*

$$\lim_{h \to 0} u_G^W(t, x, x_0, h) = -S(t, x, x_0), \quad \lim_{t \to 0} u_G^W(t, x, x_0, h) = (G^{-1})_R(x - x_0)^2/2h$$

for small enough $x - x_0$ and

$$(2\pi h t)^{-m/2} K^{-1} \exp\{-\frac{C_1(x - x_0)^2}{th}\} \le |u_G^W(t, x, x_0, h)|$$

$$\le (2\pi h t)^{-m/2} K \exp\{-\frac{C_2(x - x_0)^2}{th}\}$$

for some constants C_1, C_2, K and for small enough $x - x_0$ and t.

At last, we obtain a result on the well-posedness of the Cauchy problem for equation (1.1).

Proposition 3.2 *Under the assumptions of Theorem 2.2, for any smooth function $u_0 \in L^2(\mathcal{R}^d)$, there exists a unique solution to the Cauchy problem of equation 1.1, which is dissipative in the sense that*

$$\frac{d}{dt} E\|u(t, x)\|^2 \le 0 \tag{3.1}$$

everywhere, where E means the expectation with respect to the Wiener measure of the process W.

Proof. The existence of the solution follows directly from Theorem 2.2. Next, let $u(t, x, h, [W])$ be any solution to (1.1), i.e. u is an adaptive process on the Wiener process, which is almost surely smooth in t and x, and satisfies equation (1.1). Applying vector-valued Ito's formula (see e.g. [Met]) to the square norm (u, u) of the solution yields

$$d(u, u) = -h(G_R \frac{\partial u}{\partial x}, \frac{\partial u}{\partial x}) dt - \frac{2}{h}(V_R u, u) dt + 2\alpha_R(xu, u) dW,$$

which implies the dissipativity condition (3.1). In its turn, condition (3.1) implies the uniqueness of the solution to the Cauchy problem of equation (1.1).

The statement of Proposition 3.2 implies that the Green function u_G^W of equation (1.1) constructed above satisfies the semigroup identity for $t \le t_0$ and therefore it can be extended for all positive t as a smooth function by means of convolutions.

4. Path integration and inifinite-dimensional saddle-point method

After the original papers of Dirac and Feynmann, where it was argued that the solutions to the Shcrödinger equation can be expressed in terms of a heuristically defined path integral, many mathematicians contributed to the development of a well-defined notion of path integration and its connection to the Schrödinger equation. Various approaches were proposed covering different classes of potentials, reviews and references in Appendices G,H. In Chapter 9 we develop an approach to Feyman's integral, which allows to present the solutions to a rather general Schrödinger and stochastic Schrödinger equation (including equation (1.1)) in terms of a genuine integral over a bona fide measure over a path space. Here we only show that various known approaches (which define the integral as a certain generalised functional, and not as a genuine integral in the sense of Riemann or Lebesgue) to the definition of path integral and its applications to Schrödinger equation can be more ot less directly generalised to the case of SSE. An interesting new moment appears in connection with semiclasical asymptotics, which, in the case of SSE, can be obtained (at least formally) from path integral representation by means of an infinite-dimensional version of the complex saddle-point method. We shall use here the theory of normalised Fresnel integrals (NFI) explained in Appendix H. Our integral will be a particular case of the situation described in Proposition H2. Suppose for simplicity that the matrix G in (1.1) is proportional to the unit matrix. As usual for aplications to the Schrödinger equations, we use as a Hilbert space in H7 the Cameron-Martin space $\mathcal{H}_t$ of continuous curves $\gamma : [0, t] \mapsto \mathcal{R}^n$ such that $\gamma(t) = 0$ and the derivative $\dot{\gamma}$ of γ (in the sense of distribution) belongs to $L_2([0, t])$. The scalar product in H_t is defined as

$$(\gamma_1, \gamma_2) = \int_0^t \dot{\gamma}_1(s)\dot{\gamma}_2(s)\, ds.$$

If V, η are Fourier transforms of finite complex measures in $\mathcal{R}^n$, then the function

$$g(\gamma) = \exp\{\int_0^t V(\gamma(s) + x)\, ds\}\eta(\gamma(t) + x)$$

is also the Fourier transform of a finite Borel measure on $\mathcal{H}_t$ (for any x), because the set of such Fourier transforms forms the Banach algebra (see Appendix G). Furthermore, if $B(t)$ denotes the trajectory of the Wiener process, then the curve $l_B(\tau) = \int_\tau^t B(s)\, ds$ belongs to $\mathcal{H}_t$ with probability one, and for $\gamma \in \mathcal{H}_t$

$$(l, \gamma) = \int_0^t B(s)\dot{\gamma}(s)\, ds = \int_0^t \gamma(s)\, dB(s),$$

where the latter integral is understood in the Ito (or Wiener) sense. Finally, if we define the operator L by the formula $(\gamma_1, L\gamma_2) = \int_0^t \gamma_1(s)\gamma_2(s)\, ds$, then L is of trace class in $\mathcal{H}_t$, and under conditions (1.2) the operator $1 + 2|\alpha|^2 hGL$ is invertible with a positive real part. The following fact is a direct generalisation of a theorem from [AKS1], [AKS2], where G was supposed to be purely imaginary and there were no parameter h.

Proposition 4.1. *If V and η are Fourier transforms of finite complex measures on $\mathcal{R}^m$, then there exists a (strong) solution of the Cauchy problem for equation (1.1), (1.2) with initial data $\psi_0 = \eta$, which can be represented in the form*

$$\psi(t,x) = \int_{\mathcal{H}_t}^* \exp\{\int_0^t \left(-\frac{1}{2hG}|\dot{\gamma}(\tau)|^2 - |\alpha|^2(\gamma(\tau) + x)^2 - \frac{1}{h}V(\gamma(\tau) + x)\right) d\tau\}$$

$$\times \exp\{\int_0^t \alpha(\gamma(\tau) + x)\, dW(\tau)\}\, \eta(\gamma(0) + x)\, D\gamma, \tag{4.1}$$

where the integral is well defined in the sense of Proposition 2.1.

Sketch of the proof. The simplest way is to use the Stratonovich form (see e.g. Appendix A)

$$d\psi = (\frac{h}{2}G\Delta\psi - \frac{1}{h}V(x)\psi - |\alpha|^2 x^2\psi)\, dt + \alpha x\psi\, d_S B \tag{4.2}$$

(notice the difference in the coefficient at $x^2\psi\, dt$ in this equation and in (1.1), which is due to the Ito formula) of equation (1.1) and to approximate the trajectories W of the Wiener process by a sequence of smooth curves W_n tending to W as $n \to \infty$. For smooth curves the proof of the theorem can be given in a quite similar way as in the case of the usual Schrödinger equation (see e.g. [AH1], [SS]). We can finish the proof exploiting the fact that the sequence of solutions of the Stratonovoch equation with W_n placed instead of W tends to the solution of (1.1).

As shown in [AKS2], another method of definig the path integral based on the rotation of the potential in the complex space (see Appendix H) leads to a similar formula but with different assumptions on the potential V.

Let us stress now that though the results of this section present only a slight generalisation of that from [AKS2], the introduction of complex coefficients in SSE (1.1) leads to a principally new effect concerning the "Feynman measure" of the corresponding Feynman integral. Namely, though the integral in (4.1) is taken apparently over the path space of real paths, the main contribution in its small h asymptotics comes from a complex path. In other words, the "Feynman measure" is actually concentrated around a complex path, which can be included in the domain of integration by a certain complex shift of the Cameron-Martin space $\mathcal{H}_t$. Speaking more concretely, to get the Green function from (4.1), let us take (at least formally) the Dirac function $\delta(x - x_0)$ instead of η. (Clearly $\delta(x - x_0)$ does not satisfies the assumptions on η considered above. Nevertheless, the corresponding path integral representation can be justified, see e.g. [ABB] for the case of standard Schrödinger equation.) Then the integral in (4.1) will be taken over an appropriate space of curves joining x_0 and x in time

t. For $h \to 0$, the corresponding integral is of Laplace type with the complex analytic phase $\int (|\dot{\gamma}|^2/2G + V(\gamma))\, d\tau$. According to the saddle-point method, the main term of the small h asymptotics of such an integral depends on the behaviour of the amplitude around the saddle-point of this phase. According to the results of Section 2, these saddle-points are the solution to the Hamiltonian system with Hamiltonian (1.5). Therefore, the theory of the previous sections gives a rigorous meaning to the formal application of the infinite-dimensional saddle-point method to the path integral (4.1) with a complex phase. As we already noted, in the case of the oscillatory integrals, the corresponding infinite-dimensional method of stationary phase is discussed in many papers, see e.g. [AH2],[M4], and references therein.

Chapter 8. SOME TOPICS
IN SEMICLASSICAL SPECTRAL ANALYSIS

1. Double-well splitting

In this chapter, we are going to introduce briefly (referring for the proofs and developments to the original papers) three topics in asymptotic spectral analysis closely connected with probability and/or with the methods developed for the corresponding evolutionary equations.

This section is devoted to the asymptotics of the low lying eigenvalues of the Schrödinger equation, whose potential has several symmetric wells. We give here a short introduction to the results of the series of papers [M3], [DK1], [DK2], [DKM1], [DKM2], [K3], where the effective formulas for the calculation of the exponentially small differences between these low lying eigenvalues (splitting) were found and applied to different examples such as the discrete ϕ^4-model on tori or a hydrogen ion in magnetic field. As we shall see in the next section, this theory is closely connected to the problem of the calculation of the low lying eigenvalues of the diffusion operators, and of the life times of the corresponding diffusions. On the other hand, the method of the proof of the main results in [DKM1], [DKM2] is based on the asymptotic theory of the Cauchy problem for the second order parabolic equations, which was the main subject in this monograph.

We recall first two basic facts on the low lying eigenvalues of the Schrödinger equation with a potential having several wells. Consider the Schrödinger operator $H = H(h)$ of the form

$$H = -h^2 \Delta + V(x) + h f(x) \tag{1.1}$$

defined in $L_2(\mathcal{M})$, where $\mathcal{M}$ is an d-dimensional smooth Riemannian manifold, V and f are smooth function on $\mathcal{M}$, V being non-negative, h is a positive small parameter, Δ is the Laplace-Beltrami operator on $\mathcal{M}$. The manifold $\mathcal{M}$ is considered to be either closed or $\mathcal{M} = \mathcal{R}^l$ (in this case we suppose that $\liminf V(x) > 0$ as $x \to \infty$ and $|f(x)| \le aV(x) + b$ for some $a, b > 0$), or $\mathcal{M}$ is a bounded domain in a smooth manifold (in this case H being the Dirichlet realization of the operator (1.1)). Let the function V have only finite number of zeros $\xi_1, ..., \xi_k$ in $\mathcal{M}$ (and all $\xi_j \notin \partial \mathcal{M}$) all of them being non-degenerate, i.e the eigenvalues of the Hesse matrix $\frac{1}{2} V''(\xi_j)$ have the form $(\omega_1^j)^2, ..., (\omega_k^j)^2$, in each ξ_j with some $\omega_l^j > 0$. Let $Sp_{as}(H)$ denote the spectrum of the direct sum of the oscillator approximations of the operator H in the neighbourhoods of all points ξ_j, i.e.

$$Sp_{as}(H) = \{\sum_{i=1}^{d} \omega_i^m (2n_i + 1) + f(\xi_m) : m = 1, ..., k, n_i \in \mathbf{N}\}. \tag{1.2}$$

Let $E_n(h)$ (resp. $e_n(h)$) be the n-th (taking into account the multiplicity) eigenvalue of the operator H (resp. n-th number in $Sp_{as}(H)$).

Proposition 1.1. *For each fixed n and an h small enough, the operator H(h) has at least n eigenvalues and*

$$\lim_{h \to 0} \frac{E_n(h)}{h} = e_n.$$

This fact is known already a long time in the physical literature, and it is intuitively rather clear. The rigorous proof for finite-dimensional case seems to appear first in [Si1], see also [CFKS, HS1].

Now let us recall the notion of the distance $d(E, F)$ for the closed subspaces E, F of the Hilbert space:

$$d(E, F) = \|P_E - P_F P_E\| = \|P_E - P_E P_F\|. \tag{1.3}$$

Here P_E, P_F are the orthonormal projectors on E and F respectively. It is clear that $d(E, F) = 0$ iff $E \subset F$. If E, F are finite dimensional subspaces of equal dimensions, then $d(E, F) = d(F, E)$ and is equal to the sine of the angle between the orthogonal complements of the subspace $E \cap F$ in E and in F respectively.

Proposition 1.2 [HS1,Pa]. *Let A be a self-adjoint operator in the Hilbert space $\mathcal{H}$. Let $[a, b]$ be a compact interval, $\psi_1, ..., \psi_N$ be linear independent elements of $\mathcal{H}$ and let $\mu_1, .., \mu_N \in [a, b]$ such that*

$$A\psi_j = \mu_j \psi_j + r_j, \quad \|r_j\| \le \epsilon.$$

Let for some $\delta > 0$ we have $Sp(A) \cap ([a - \delta, a] \cup ([b, b + \delta]) = \emptyset$. Let E be linear space with the basis ψ_i, $i = 1, ..., N$, and F be the spectral subspace of A associated with $Sp(A) \cap [a, b]$. Then

$$d(E, F) \le \frac{\sqrt{N}\epsilon}{\delta\sqrt{\lambda_{min}}},$$

where λ_{min} is the minimal eigenvalue of the matrix (ψ_i, ψ_j).

The problem of the calculation of the splitting between the low lying energy levels of the Schrödinger operator with symmetric potential wells can be already considered as a classical in quantum mechanics. We shall discuss here only the case of the double-well potential with non-degenerate wells, the corresponding results for more general situations can be found e.g. in [K3], [KM2]. We recall first the exact formulation of the problem. Let

$$H = -\frac{h^2}{2\mu}\Delta + V(x) \tag{1.4}$$

be a Schrödinger operator in $L^2(\mathcal{R}^d)$, where Δ is the Laplace operator, $h > 0$ is a small parameter, and the potential V has the following properties:

(i) V is a smooth nonnegative function,

(ii) there exist two points $\xi_1, \xi_2 \in \mathcal{R}^n$ such that $V(x) = 0$ if and only if x coincides with one of ξ_j,

(iii) V is symmetric, namely, there exists an orthogonal operator R in $\mathcal{R}^n$ with the square R^2 being equal to identity such that $V(Rx) = V(x)$ for all x and $\xi_1 = R\xi_2$,

(iv) V is strictly positive at infinity, namely $\liminf_{x \to \infty} V(x) > 0$,

(v) V has non-degenerate minima, i.e. the matrices $V''(\xi_j), j = 1, 2$, (they are similar, due to the symmetry) have positive eigenvalues, which will be denoted by $\omega_1^2, ..., \omega_d^2$ with all $\omega_j > 0$.

Let $\mathcal{E} = \mu^{-1} \sum_{j=1}^n \omega_j/2$ and F be the set of the fixed points of R.

It is obvious from the physical point of view that when potential barrier between the wells is high, then there should be two quasi-stable states at the bottoms ξ_1, ξ_2 with energy levels E_1, E_2 having equal asymptotic expansions in h:

$$E_j = \mathcal{E}h + e_2 h^2 + e_3 h^3 + ... \tag{1.5}$$

One can easily obtain this expansion using first the oscillator approximation in a neighbourhood of each well (that gives the first term in (1.5)) and then the standard perturbation theory. Rigorously speaking, it follows from Proposition 1.2 that for h small enough the Schrödinger operator (1.4) has exactly two eigenvalues E_1, E_2 with (equal) asymptotic expansion (1.5) and moreover, there exists a constant $C > 0$ such that $E/h - \mathcal{E} > C$ for all other points E in the spectrum of operator (1.4).

Since the asymptotic expansions for E_1, E_2 coincide, the problem of asymptotic calculation of its difference is very subtle. This is the problem we are going to discuss here.

Let us start with some physical motivations. The difference $E_2 - E_1$ is called the splitting of the low lying eigenvalues of the Schrödinger operator (1.1) with a double well potential $V(x)$. This terminology comes from the following "dynamical" interpretation. Imagine for a moment that the potential barrier between the wells is infinity high. Then the low energy level of the Schrödinger operator will be degenerate, the corresponding eigenspace will be two-dimensional and the basis ψ_1, ψ_2 for this space can be chosen in such a way that ψ_1 vanish in a neighbourhood of ξ_2 and conversely. So, each ψ_j stands for the case, when a quantum particle is situated only in a one well. The situation changes crucially when the barrier becomes finite. Although the classical lowest energy level will be still degenerate (the classical particle lying at the bottom of a well can not spring into another one), the quantum mechanical lowest energy level will now split in two non-degenerate ones. Its small difference $E_2 - E_1$ will estimate the inverse time T of the (tunnel) transition of a quantum particle from one well to another. In fact, due to the symmetry, the eigenfunctions corresponding to E_1, E_2 have now the form $\psi_1 + \psi_2$ and $\psi_1 - \psi_2$, where ψ_j, as above, stands for the position of the particle in one well (the asymptotics of this ψ_j in a neighbourhood of a well can be calculated using the oscillator approximation similar to the calculation of series (1.5)). These ψ_j are connected by the symmetry transform: $\psi_1(Rx) = \psi_2(x)$. Thus the time T of transition is the minimal number satisfying the equation $e^{iTH}\psi_1 = \psi_2$. It follows immediately from equations $H(\psi_1 \pm \psi_2) = E_{1,2}(\psi_1 \pm \psi_2)$ that $T = \pi|E_2 - E_1|^{-1}$. Another physical interpretation of the problem is due to the fact that the probability of the spontaneous transition (with photon emission) from the (slightly) excited state E_2 to the real ground state E_1 depends on $E_2 - E_1$. Therefore, this difference stands for the "fate of false vacuum" [CC].

Now we introduce the notations and recall some standard facts concerning the calculation of the difference $E_2 - E_1$. Let

$$S_j(x) = \inf_{t\geq 0} \inf \int_0^t (\frac{\mu}{2}\dot{q}^2(\tau) + V(q(\tau)))\, d\tau,$$

where the second inf is taken over all continuous piecewise smooth curves $q(\tau)$ such that $q(0) = \xi_j$, $q(t) = x$. Let

$$D_1 = \{x \in \mathcal{R}^n : S_1 \leq S_2\}, \quad D_2 = \{x \in \mathcal{R}^n : S_2 \leq S_1\},$$

and D_j^ϵ denote the ϵ-neighbourhoods of these sets. It is not difficult to show (see e.g. [DKM2]) that

(i) for each $x \in D_j^\epsilon$, where ϵ is small enough, there exists a solution (q,p) of the Hamiltonian system

$$\mu\dot{q} = p, \quad \dot{p} = \frac{\partial V}{\partial x}(q) \tag{1.6}$$

such that $q(-\infty) = 0$, $q(0) = x$ and

$$S_j(x) = \int_{-\infty}^0 (\frac{\mu}{2}\dot{q}^2(\tau) + V(q(\tau)))\, d\tau;$$

(ii) for almost all x such a solution is unique, S_j is smooth and satisfies the stationary Hamilton-Jacobi equation

$$\frac{1}{\mu}\left(\frac{\partial S}{\partial x}\right)^2 - V(x) = 0;$$

(iii) the equation $p = \partial S_j/\partial x$ defines locally the unstable manifold W_j^{out} for system (1.6) that corresponds to its hyperbolic singular point $(\xi_j, 0)$ (recall that the unstable manifold W_j^{out} is defined as the set of points (q_0, p_0) in $\mathcal{R}^{2n}$ such that the solution of (1.6) with initial data (q_0, p_0) tends to $(\xi_j, 0)$, as $t \to -\infty$);

(iv) there exists a trajectory (q,p) of system (1.6) such that

$$q(-\infty) = \xi_1, \quad q(+\infty) = \xi_2$$

and

$$S_{12} \equiv S_1(\xi_2) = \int_{-\infty}^{+\infty} (\frac{\mu}{2}\dot{q}^2(\tau) + V(q(\tau)))\, d\tau, \tag{1.7}$$

this trajectory being called an instanton and S_{12} being called the instanton action.

It turns out that

$$S_{12} = -\lim_{h \to 0} h \log(E_2 - E_1), \tag{1.8}$$

i.e. the instanton action describes the logarithmic limit of the splitting value. It follows, in particular, that $E_2 - E_1$ is exponentially small in h. Taking into account formula (1.8), it is natural to seek for the formula for the splitting in the form

$$E_2 - E_1 = A(h) \exp\{-S_{12}/h\}. \tag{1.9}$$

In order to be able to describe the dependence on h of the amplitude $A(h)$ of the splitting, one should make new additional assumptions. For example, let us consider the simplest case, when the instanton $q(t)$ is unique up to a shift in time. It is obvious that $q(t)$ intersects the set F of the fixed points of the symmetry operator R only at one point. We denote this point by ξ_0 and shall fix from now on the parameterisation of the instanton by the condition $q(0) = \xi_0$. It is clear from the symmetry that F belongs to the hyper-surface $\Gamma = D_1 \cap D_2$ and that the instanton q is orthogonal to Γ at ξ_0. Let $M(x), x \in \Gamma$, be the $(n-1) \times (n-1)$ matrix of the second derivatives of the function $S_1(x)$ restricted to Γ. The instanton $q(t)$ is called non-degenerate, if $M(\xi_0)$ is non-degenerate. Now if there exists only one instanton and it is non-degenerate, then it is possible to prove that the amplitude $A(h)$ in (1.9) has the form

$$A(h) = \sqrt{h}a(1 + a_1 h + ... + a_k h^k + O(h^{k+1})). \tag{1.10}$$

The described results were obtained at the beginning of 80 th. Before going further let us give a short historical review. The splitting formula (1.9), (1.10) for one-dimensional case was known in physical literature already for a long time, see for instance [LL]. Heuristically, the logarithmic limit (1.8) of the splitting for finite and simultaneously for infinite dimensions seemed to appear first in [M5] and [Po]. The paper [Po] contained, in particular, a developed program for the application of instantons in gauge theories and thus inspired a lot of new papers on the theory of instantons for quantum field theory as well as for quantum mechanics. Further physical discussions of the corresponding problems in quantum mechanics (including different methods as well as different models) can be found, for instance, in [CC,WH], and many others, see also the book [Gu]. Turning to the mathematical literature we mention first the work of Harrell [Ha], where appeared the first rigorous deduction of the splitting formula (including the calculation of the major term of amplitude (1.10)) for one-dimensional system with the general double well potential. Afterwards, there appeared many works dealing with more general one-dimensional models, (see for instance, [Pa] for one-dimensional systems with several wells) The program of the mathematical study of n-dimensional problem was proposed in the Maslov lecture on the Mathematical Congress in Warsaw [M8]. Then in 1984, three basic papers for the considered subject were published, namely

1) the work of B.Simon [Si2], where the rigorous proof were given of formula (1.7) for the logarithmic limit of the splitting (1.8) in general n-dimensional situation; the proofs in this paper were based on the method of the path integration,

2) the work of B.Helffer and J.Söstrand [HS1], where this result was also obtained together with the theorem of the existence of the asymptotic expansion (1.10) for the amplitude in the case of one non-degenerate instanton (let us stress that the important progress of [HS1] in comparison with [Si2] was the statement that the major term of the asymptotic expansion for the amplitude has the order $\sqrt{h}$); the method of this paper (see also [He] and references therein) was based on the construction of the WKB asymptotics of the low eigenfunctions for the Dirichlet realization of the corresponding Schrödinger operator restricted to a neighbourhood of a well, then these asymptotics were carefully matched together,

3) the work of V.Maslov [M3], where not only the existence of the expansion (1.10) was given, but also the formula for the calculation of the coefficients was presented (see Proposition 1.1 below). The proofs were based on the investigation of the WKB asymptotics of the corresponding heat equation, in order to obtain afterwards the asymptotics of the low lying eigenfunctions and eigenvalues of the Schrödinger operator by a subtle limit procedure with two small parameters h and t^{-1}. One should note however, that the assumptions on the Schrödinger operator in [M3] were more restrictive (namely, the results in [Si2, HS1] were given directly for the Schrödinger operator corresponding to an arbitrary Riemanian manifold and in [M3] only to the plane Euclidian spaces) and that the Maslov arguments were very schematic. Complete version of his proof appeared only in [DKM1, DKM2]. Moreover, the formula for the splitting in [M3] was a geometric one, and was not very convenient for concrete calculations. In [DKM2] an analytic version of this formula was given, which we we are going to present now.

Let us suppose first that the eigenvalues ω_j are non-resonant. Namely, each ω_j can not be represented in the form $\sum_{k \neq j} \nu_k \omega_k$ with some integer nonnegative numbers ν_k such that $\sum \nu_k \geq 2$. Applying the general theory of the normal form of ordinary differential equation in a neighbourhood of a singular point to the restriction of the Hamiltonian flow (1.6) on the unstable manifold W_j^{out} one proves the existence of local coordinate $y(x)$ such that the equation of this restriction in the coordinate y have the form $\dot{y} = \mu^{-1/2} \Omega y$ with the diagonal matrix $\Omega = diag(\omega_1, ..., \omega_n)$. We fix such coordinate system by the additional normalising condition $\det(\partial y / \partial x)(\xi_j) = 1$. These coordinate can be extended to the whole W_j^{out} defining a measure on it such that

$$dy(z_t) = \exp\{2\mu^{-1/2}\mathcal{E}t\}\, dy(z_0),$$

where z_t is the image of a point $z_0 \in W_j^{out}$ under the action of the Hamiltonian flow. Let us define now the Jacobian

$$J_j(x) = \det \frac{\partial x}{\partial y}$$

in non-focal points of W_j^{out} , i.e. in such points that have a neighbourhood in W_j^{out} with a regular projection on the coordinate Eucleadian space $\mathcal{R}^n$.

Theorem 1.1 [M3,DKM1]. *The formula for the amplitude $A(h)$ of the splitting (1.9) has the form*

$$A(h) = 2J_1^{-1}(\xi_0)(\det M)^{-1/2}(\xi_0)|\dot{q}(0)|\sqrt{h/\pi}\sqrt{\omega_1...\omega_n}\mu^{n/4}(1 + O(h)), \quad (1.11)$$

where the matrix $M(\xi_0)$ was introduced above, before formula (1.10).

This formula is not satisfactory from the practical point of view, because to calculate its element one should find special coordinate system and to solve the Hamiltonian system (1.6) in the whole W_j^{out}. In the next statement the calculations are reduced to the solution of the system in variation

$$\mu\ddot{Z} = V''(q(t))Z, \quad Z|_{t=-\infty} = 0, \quad Z|_{t=0} = I \quad (1.12)$$

along the instanton $q(t)$, where I is the unit matrix. Let O be a matrix of the orthogonal transformation that takes the first coordinate vector of $\mathcal{R}^n$ in the unit vector, which is parallel to the velocity $\dot{q}(0)$ of the instanton at zero time.

Theorem 1.2 [DKM2]. *Formula (1.11) holds with*

$$J_1 = \lim_{t \to -\infty} \left(\exp\{2t\mathcal{E}\mu^{-1/2}\}/\det Z(t)\right) \quad (1.13)$$

and $M(\xi_0)$ is the minor of the matrix

$$B = \mu O\dot{Z}(0)O^t \quad (1.14)$$

obtained from it by deleting the first column and the first row.

Remark. This formula holds also without the supposition on non-resonant ω_j. Moreover, instead of the real orthogonal matrix O in (1.14) one can take as well a complex unitary matrix U such that its first vector-column is parallel to $\dot{q}(0)$. Some generalisation of this formula for the case, when the instanton is not unique and the wells are not simply points are given in [K3].

In [DK1], Theorem 1.2 was applied to the calculation of the splitting $E_2 - E_1$ in the case of the potential

$$V = \frac{\alpha}{2}\sum_{k=1}^{n}(x_k - x_{k-1})^2 + \frac{\beta}{4}\sum_{k=1}^{n}(x_k^2 - \xi^2)^2, \quad (1.15)$$

where $\alpha > 0$, $\beta > 0$ are constants and $x_0 = x_n$. This potential stands for a chain of pairwise interacting particles on a circle in the common potential field.

With a special choice of parameters α and β this model tends, as $n \to \infty$, to the field model with the Lagrangian density

$$c^2\left(\frac{\partial\phi}{\partial y}\right)^2 - \gamma^2\left(\phi^2 - \xi^2\right)^2.$$

That is so called ϕ^4-model on the circle . Therefore, the chain with potential (1.15) can be called naturally the discrete ϕ^4-model on a circle. The calculation of the instanton as well as its limit as the number of particle n tends to infinity is quite simple. This problem is reduced to the one-dimensional problem for the potential $\beta(x^2 - \xi^2)^2/4$. But in order to calculate the amplitude A one needs to solve some linear multidimensional system of ordinary differential equations with time depending coefficients. Unexpected fact is the possibility to integrate this system in elementary functions and thus to get the explicit formulas for the determinants $J(\xi_0)$ and $\det M(\xi_0)$. In [DK2], these considerations were generalised to the finite-dimensional case, i.e. to the discrete ϕ^4-model on tori. Let us formulate the corresponding exact result.

Consider the Schrödinger equation

$$\left(-\frac{h^2\Delta}{2\mu} + V(x)\right)\psi = E\psi, \quad \psi \in L^2, \tag{1.16}$$

in Eucleadian space of dimension $|\mathcal{K}| = n_1 \times \ldots \times n_N$ with coordinates x_k, where

$$k = (k_1, \ldots, k_N) \in \mathcal{K} = (\mathcal{Z}/n_1\mathcal{Z}) \times \ldots \times (\mathcal{Z}/n_N\mathcal{Z}),$$

i.e. $k_j \in \mathcal{Z}$ and two indexes k_j are considered to be identical, if their difference is proportional to n_j. Here Δ is the Laplace operator and

$$V(x) = \sum_{k \in \mathcal{K}} \frac{1}{2} \sum_{j=1}^{N} \alpha_j \left(x_k - x_{k-1_j}\right)^2 + \frac{\beta}{4} \sum_{k \in \mathcal{K}} \left(x_k^2 - \xi^2\right)^2, \tag{1.17}$$

where 1_j is the multi-index with elements $(1_j)_m = \delta_j^m$ and α_j, β, μ are positive constants.

For instance, in the one-dimensional case the potential has form (1.15). The potential V is an even function, i.e. it is invariant $V(Rx) = V(x)$ with respect to the reflection $Rx = -x$ (the set of the fixed point for R consists of only one point $\xi_0 = 0$), and has obviously two minimum points $\xi_{1,2}$ with coordinates $x_k = \pm\xi$ for all $k \in \mathcal{K}$. One can prove that these minima are not degenerate and all conditions of Propositions 1 and 2 are satisfied. Therefore, one can calculate the splitting between two low levels E_2 and E_1 by formulas (1.11), (1.13), (1.14). We give now the result of these calculations and discuss the deduction in the next section.

Let us denote

$$b_k = \sqrt{1 + 2(\beta\xi^2)^{-1} \sum_{j=1}^{N} \alpha_j \sin^2\left(\frac{\pi k_j}{n_j}\right)}.$$

Theorem 1.3 [DK2]. *For any fixed $(n_1, \ldots, n_N)$ the following formula holds*

$$E_2 - E_1 = \Delta_0 E(1 + O(h)),$$

$$\Delta_0 E = 4\sqrt{\frac{h n_1 \ldots n_N}{\pi \sqrt{\mu}}} (2\beta)^{3/4} \xi^{5/2} \prod_{k \neq 0} \sqrt{\left(\frac{2b_k + 1}{2b_k - 1}\right)\left(\frac{b_k + 1}{b_k - 1}\right)}$$

$$\times \exp(-2n_1 \ldots n_N \sqrt{2\mu\beta} \xi^3 / 3h). \tag{2.3}$$

Remark. The periodic conditions for the chain is essential. Due to these conditions, the virial theorem (see, [Ra]), which forbids the existence of instantons, is not true.

The quantum field and the thermodynamic limit of this formula are investigated in [DK2]. Other examples of the application of Theorems 1.1, 1.2 can be found in [DKM2, KM2].

We discussed here only the splitting between the low lying levels. The consideration of the splitting between exited levels leads to new effects, (see e.g. [LL], and a modern review in [DS]), because on the classical level it corresponds to the splitting between invariant tori (and not between stable points as in the case of low lying eigenvalues). Another interesting development is the theory of "momentum splitting", where the corresponding invariant classical objects have the same projection on the coordinate space, but different projections on the momentum space. In particular, such situation appears in the case of Laplace-Beltrami operators on Liouville surfaces, see e.g. [KMS] for the main facts of the spectral analysis of these operators, and [K12] for a complete classifications of these operators on two-dimensional compact surfaces. Some results on the splitting in this case can be found in [DS] (see also [Fed3]).

2. Low lying eigenvalues of diffusion operators and the life times of the corresponding diffusions

This section is devoted to the problem of the asymptotic calculations of the low-lying eigenvalues of the diffusion operator $\mathcal{D}$ on a smooth Riemannian manifold $\mathcal{M}$, defined on smooth functions by the formula

$$\mathcal{D} = -h\Delta + (\nabla\varphi, \nabla), \tag{2.1}$$

where Δ is the Laplace-Beltrami operator on $\mathcal{M}$, h is a small positive parameter, the brackets denote the natural inner product on forms defined by the Riemannian metric. The function φ is a Morse function on $\mathcal{M}$, i.e. it is a smooth function with finite number of singular points, all of them being non-degenerate. We will consider two special cases.

A) $\mathcal{M}$ is a compact domain in $\mathcal{R}^d$ with a smooth boundary $\partial\mathcal{M}$, the local minimums of φ are supposed not to belong to the boundary. The operator $\mathcal{D}$ is the Friedrichs closure of the operator defined by (1.1) on smooth functions vanishing in a neighbourhood of $\partial\mathcal{M}$ (the Dirichlet realization of the operator $\mathcal{D}$).

B) The manifold $\mathcal{M}$ is closed or $\mathcal{M} = \mathcal{R}^d$, $\mathcal{D}$ is the closure of the operator defined by (1.1) on smooth functions with compact support. In the case $\mathcal{M} = \mathcal{R}^{\lceil}$, we suppose also the function φ to increase at infinity faster than some positive power of its argument.

Let Ω_i denote the regions of the attraction of the dynamical system

$$\dot{x} = -\nabla\varphi(x). \tag{2.2}$$

We shall call Ω_i the fundamental domains. The geometry of their dispositions proves to play an important role in the study of the diffusion process.

The relation with the theory of the previous section is given by the following (well known) observation. The diffusion operator $\mathcal{D}$ is a non-negative self-adjoint operator in the weighted space $L^2(\mathcal{M}, d\mu)$, where

$$d\mu = exp\{-\frac{\varphi}{\epsilon}\}dx,$$

and dx is the measure on $\mathcal{M}$ associated to its Riemannian metric. Moreover, the operator D is similar to a Schrödinger operator of form (1.1):

$$e^{-\frac{\varphi}{2h}}\epsilon\mathcal{D}e^{\frac{\varphi}{2h}} = H = \left(-h^2\Delta + \left(\frac{1}{4}(\nabla\varphi)^2 - \frac{h}{2}\Delta\varphi\right)\right).$$

Note that each critical point ξ_j of the function φ corresponds to a non-degenerate zero point of the function $V = \frac{1}{4}\nabla\varphi^2$ and $V''(\xi_j) = \frac{1}{2}(\varphi''(\xi_j))^2$. However, only in the minimal points of the function φ the matrix $\varphi''(\xi_j)$ is positive definite and in these points we have $tr[V''(\xi_j)]^{\frac{1}{2}} = \frac{1}{2}\Delta\varphi(\xi_j)$. Therefore, the set (1.2) in this situation has the same number of zeros as the number of minimums of the function φ. Hence the following statement is a direct consequence of Proposition 1.2.

Proposition 2.1. *Let the function φ have N local minimum points on the manifold $\mathcal{M}$. Then the operator $\mathcal{D}$ has a series $\lambda_1(h), \lambda_2(h), ..., \lambda_N(h)$ of N eigenvalues of order $O(h)$ (in case A), $\lambda_1(h) = 0$), and the next eigenvalue is bounded from below by a positive constant (not depending on h).*

Using the variational principle and Proposition 1.2 one readily gets a more precise information about the first N eigenvalues, namely the following result.

Proposition 2.2. *The eigenvalues $\lambda_1(h), \lambda_2(h), ..., \lambda_N(h)$ are exponentially small in h. Moreover, the eigenfunctions corresponding to these eigenvalues are closed to the constant functions in each fundamental region. More precisely, let F be the spectral subspace of the operator $\mathcal{D}$ which stands for the first N eigenfunctions and E be the subspace generated by the indicators χ_i of the domains Ω_i (χ_i equals to one or zero respectively in Ω_i or outside it). Then the distance between the subspaces E and F is exponentially small*

$$dist(F, E) = \mathcal{O}(exp\{-\frac{\alpha}{h}\}), \ \alpha > 0, \tag{2.3}$$

as h tends to zero.

We shall give a proof of this fact after the formulation of Theorem 2.1, noting now that the first rigorous result on the existence of the exponentially small series was obtained by probabilistic methods (and not necessarily for a potential drift) in the works of Wentzell [Wen1],[Wen2], see also [FW]. Now we are going to present a theorem which connects analytical and probabilistic approaches to the problem of the calculation of the eigenvalues $\lambda_1(h), ..., \lambda_N(h)$. This theorem generalises the well known result (see e.g. [FW]) that in case A) the asymptotics of the first eigenvalue of the diffusion operator can be expressed as the inverse value of the mean exit time of the process from the manifold $\mathcal{M}$.

It turns out that in case A) the leading role in the spectral analysis of the low lying eigenvalues of the operator $\mathcal{D}$ is played by the matrix G_{ij} of the mean times that live the process in the fundamental region Ω_i when starting in the region Ω_j. The analogous result holds for case B). This connection allows one to reduce the above mentioned asymptotic problems of the spectral analysis of $\mathcal{D}$ to the study of the asymptotics of the mean life times of the diffusion process and vice versa. From the analytical point of view, this connection reduces the solution of the spectral problem to the solution of some non-homogeneous problems for the operator $\mathcal{D}$.

In case A), let function τ_j on $\mathcal{M}$ be the solution to the problem

$$\mathcal{D}\tau_j = \chi_j, \quad \tau_j|_{\partial\mathcal{M}} = 0. \tag{2.4}$$

In other words, $\tau_j = \mathcal{D}^{-1}\chi_j$, which is well defined, because zero does not belong to the spectrum of D. In case B), let function τ_j be the solution to the problem

$$\mathcal{D}\tau_j = \chi_j - r_j, \quad r_j = \frac{\mu(\Omega_j)}{\mu(\mathcal{M})}. \tag{2.5}$$

The constants r_j here equal to the measures of Ω_j with respect to the normalised stationary distribution of the diffusion process. Let us denote G_{ij} the mean value (with respect to $d\mu$) of the function τ_j in Ω_i:

$$G_{ij} = \frac{\int_{\Omega_i} \tau_j(x)d\mu(x)}{\mu(\Omega_i)}. \tag{2.6}$$

Theorem 2.1 [KMac]. *In the case A) the eigenvalues of the matrix G have the form*

$$\mu_i = \lambda_i^{-1}(h)(1 + \mathcal{O}(h^\infty)), \quad i = 1, 2, ..., N. \tag{2.7}$$

In the case B) the matrix G has one zero eigenvalue $\mu_1 = 0$ and the other eigenvalues have the similar form

$$\mu_i = \lambda_i^{-1}(h)(1 + \mathcal{O}(h^\infty)), \quad i = 2, ..., N. \tag{2.8}$$

Remark 1. One can see from the proof of this theorem that a series of the eigenvectors of G is closed to the series of vectors composed of the projections of the first eigenfunctions on the indicators χ_j.

Remark 2. Due to the Ito formula, the function $\tau_j(x)$ satisfying (2.4) gives the mean life time in Ω_j of the diffusion process with the starting point x.

Therefore G_{ij} is the mean life time in Ω_j of the process starting in the domain Ω_i.

Proof of Proposition 2.2. It would be more convenient to consider instead of χ_j the smooth molyfiers, on which D is well defined. Namely, let U_j be neighbourhoods of the points ξ_j in the fundamental domains Ω_j such that their closures $\overline{U}_j(\eta)$ of its η-neighborhoods are compact and belong to Ω_j for some $\eta > 0$. Let θ_j, $j = 1, ..., N$, be a smooth function $\mathcal{M} \to [0, 1]$ with the support in $\overline{U}_j(\eta)$ that is equal to one in U_j. We claim that the distance $d(E, F)$ between the space E generated by θ_j, $j = 1, ..., N$ and the subspace F of the first N eigenfunctions of the operator $\mathcal{D}$ is exponentially small in h in the norm of $L^2(\mathcal{M}, d\mu)$. To be more precise, for any $\delta > 0$

$$d(E, F) = \mathcal{O}(exp\{-\frac{\Delta - \delta}{h}\}),$$

where

$$\Delta = \min_i \Delta_i, \ \ \Delta_i = \min_{x \in \partial\Omega_i}(\varphi(x) - \varphi(x_i)).$$

To see this, notice first that the functions $\theta_j(x)$ have non intersecting supports and satisfy the equation $\mathcal{D}\theta_j = \alpha_j(x)$, where $supp\,\alpha_j \subset \overline{U}_j(\eta)\backslash U_j$.

On the other hand, by the Laplace method (see e.g. Appendix B) one sees that

$$\|\theta_j\|^2 = (2\pi h)^{-d/2} \exp\{-\phi(x_j)/h\}(\det \phi''(x_j))^{-1/2}(1 + O(h)).$$

Therefore,

$$\frac{\|\mathcal{D}\theta_j\|^2}{\|\theta_j\|^2} \leq C(\eta)exp\{-\frac{\varphi(y_j) - \varphi(x_j) + \delta}{h}\} = O(exp\{-\frac{\Delta_j - \delta}{2h}\}),$$

where

$$\varphi(y_j) = \min_{x \in supp\,\alpha_j} \varphi(x).$$

Thus we display N orthogonal trial functions θ_j for which the variational estimates are exponentially small. Consequently the variational principle (see e.g. [RS]) implies the existence of N exponentially small eigenvalues, and Proposition 1.2 (in its simplest form when all ψ_j are orthogonal) implies that the corresponding subspaces are exponentially close. To finish the proof of the Proposition it remains to notice that since the functions θ_j are closed to the indicators χ_j, the distance between F and the space generated by χ_j is also exponentially small.

Corollary. *Each of the first N eigenfunctions of the operator $\mathcal{D}$ can be represented in the form*

$$\psi_i = \sum_{j=1}^{N} C_{ij}\chi_j + r_i,$$

where all r_j are orthogonal to all χ_i and are exponentially small with respect to the norm of $L_2(\mathcal{M}, d\mu)$. Moreover, the matrix $C = \{C_{ij}\}$ and its inverse matrix C^{-1} are uniformly bounded with respect to small parameter h.

Proof of Theorem 2.1. We start with case A). Let us calculate the bilinear form of the operator $\mathcal{D}^{-1}$ on the normalised indicators $\hat{\chi}_i = \chi_i \|\chi_i\|^{-1}$ of the fundamental domains Ω_i. Due to the definition of the matrix G, we have

$$G_{ij} = \frac{(\chi_i, \mathcal{D}^{-1}\chi_j)}{\|\chi_i\|^2} = (\hat{\chi}_i, \mathcal{D}^{-1}\hat{\chi}_j)\frac{\|\chi_j\|}{\|\chi_i\|},$$

and we conclude that the spectrum of the matrix G coincides with the spectrum of the matrix D^{-1} whose elements are $(\hat{\chi}_i, \mathcal{D}^{-1}\hat{\chi}_j)$.

Let $C = \{C_{ij}\}$ be the matrix of the projections of the function $\hat{\chi}_i$ on the eigenfunction ψ_j of the operator $\mathcal{D}$. Then

$$(\hat{\chi}_i, \mathcal{D}^{-1}\hat{\chi}_j) = (C\Lambda^{-1}C^+)_{ij} + (\hat{\chi}_i^\perp, \mathcal{D}^{-1}\hat{\chi}_j^\perp),$$

where $\hat{\chi}_i^\perp = P^\perp\hat{\chi}_i$ and $\Lambda = diag(\lambda_1, ..., \lambda_N)$ is the diagonal matrix composed of the first eigenvalues of $\mathcal{D}$. Note that $(\hat{\chi}_i, \hat{\chi}_j) = (CC^+)_{ij} + (\hat{\chi}_i^\perp, \hat{\chi}_j^\perp)$. Thus $C^+ = C^{-1} - C^{-1}X^\perp$, where the matrix $X^\perp$ is defined by its matrix elements $X_{ij}^\perp = (\hat{\chi}_i^\perp, \hat{\chi}_j^\perp)$. Consequently,

$$(D^{-1})_{ij} = (C\Lambda^{-1}C^{-1})_{ij} - (C\Lambda^{-1}C^{-1}X^\perp)_{ij} + (\hat{\chi}_i^\perp, \mathcal{D}^{-1}\hat{\chi}_j^\perp).$$

The entries of the matrices $X^\perp$ and $(\chi_i^\perp, \mathcal{D}^{-1}\chi_j^\perp)$ are exponentially small:

$$|X_{ij}^\perp| = |(\hat{\chi}_i^\perp, \hat{\chi}_j^\perp)| = O(exp\{-\frac{\Delta - \delta}{h}\}),$$

$$|\chi_i^\perp, \mathcal{D}^{-1}\chi_j^\perp)| \le \|\chi_i^\perp\| \|\chi_j^\perp\| \|P^\perp\mathcal{D}^{-1}\| = O(exp\{-\frac{\Delta - \delta}{h}\})$$

for any $\delta > 0$. We have used here the fact that the norm of the inverted operator $\mathcal{D}^{-1}$ on the subspace $P^\perp L^2(\mathcal{M}, d\mu)$ is estimated by the inverse to the $(N + 1)$-st eigenvalue of the operator $\mathcal{D}$ which is bounded by some constant, due to Proposition 2.1. Consequently, we have the representation

$$C^{-1}D^{-1}C = \Lambda^{-1} + \Lambda^{-1}A + B = \Lambda^{-1}(I + O(exp\{-\frac{\Delta - \delta}{h}\})),$$

where (due to the Corollary above)

$$A, B = \mathcal{O}(exp\{-\frac{\Delta - \delta}{h}\}).$$

This implies the assertion of Theorem 1.2 in case A).

Let us point out the modifications that one needs for the proof of the case B). Let $Q^\perp$ be the projector in $L^2(\mathcal{M}, d\mu)$ on the subspace of functions orthogonal to the constant functions. (Note that in the case of a closed manifold a constant

function is the eigenvector of the operator $\mathcal{D}$ corresponding to zero eigenvalue.)
We have

$$G_{ij} = \frac{(\chi_i, \mathcal{D}^{-1}Q^{\perp}\chi_j)}{\|\chi_i\|} = (\hat{\chi}_i, \mathcal{D}^{-1}Q^{\perp}\hat{\chi}_j)\frac{\|\chi_j\|^2}{\|\chi_i\|},$$

and thus, as previously, the matrix G and $(\hat{\chi}_i, \mathcal{D}^{-1}Q^{\perp}\hat{\chi}_j)$ have the same spectrum. The existence of zero eigenvalue for the matrix G follows from the degeneracy of the matrix $(\chi_i, \mathcal{D}^{-1}Q^{\perp}\chi_j)$:

$$\sum_i (\chi_i, \mathcal{D}^{-1}Q^{\perp}\chi_j) = (1, \mathcal{D}Q^{\perp}\chi_j) = (Q^{\perp}1, \mathcal{D}^{-1}Q^{\perp}\chi_j) = 0.$$

The remaining part of the proof repeats that of the case A).

A combinatorial method for the calculation of the exponential orders of the exponentially small eigenvalues was proposed in [Wen1], [Wen2].

The main and still open question of the theory is to give a rigorous construction of the precise asymptotics (taking into account the pre-exponential terms) of these first N eigenvalues. Some partial results in the case of different symmetries can be obtained using the method of the previous section, see also [HS2]. A series of papers, see [MS, BM] and references therein, was devoted to the formal calculations of these asymptotics.

Rigorous results in one -dimensional case was obtained in [KMak] on the base of the Theorem 2.1, where we developed the Wentzel technique in a way to allow one to calculate not only the exponential orders of the exponentially small eigenvalues, but also the pre-exponential terms. This method reduces the calculation to a problem of combinatorial analysis that can be easily solved by computer.

3. Quasimodes of diffusion operators around a closed orbit of the corresponding classical system

The correspondence principle of quantum mechanics states that to some "good" sets, invariant with respect to the equations of classical mechanics, one can assign sequences (spectral series) of asymptotic eigenfunctions and eigenvalues (quasimodes) of the corresponding quantum mechanical operators containing a small parameter. The case in which these sets are d-dimensional invariant Lagrangian tori in the $2d$-dimensional is well studied; the answer (the semi-classical quantisation of these tori) is given by the Maslov canonical operator [M1]. Spectral series can be assigned also to such classical objects as critical points, closed invariant curves (see e.g. [BB], [M5], [M9], [Laz]), and k-dimensional isotropic invariant tori (see e.g. [BD]). In probability theory, an analogy of the correspondence principle has been applied for a long time as often and as successfully as in quantum mechanics (see e.g. [M1], [FW], [Var1]). In particular, for the diffusion equation

$$u_t + \mathcal{D}u = 0, \quad \mathcal{D} = V(x)\nabla u - h\Delta u, \quad x \in \mathcal{R}^d, \tag{3.1}$$

this principle states, in a sense, that some asymptotic solutions as $h \to 0$ to this equation can be constructed via the solutions to the dynamical system

$$\dot{x} = V(x), \tag{3.2}$$

which the system of characteristics of the first order partial differential equation obtained from (3.1) by putting $h = 0$. Unlike the case of quantum mechanics, system (3.2) has no non-trivial d-dimensional objects without boundary, but one can expect a relationship between invariant k-dimensional tori of this (3.2) and asymptotic eigenfunctions of the operator D. The simplest case of critical points was discussed in the previous section (in the case of potential field V). We are going to formulae here the results from [DKO1], where it was shown that to each limit cycle Γ of equation (3.2) corresponds complex numbers λ_ν and smooth functions u_ν (quasimodes) "localised " around Γ and such that

$$V(x)\nabla u_\nu - h\delta u_\nu = \lambda_\nu u_\nu + h^{1/2}. \tag{3.3}$$

Here ν is a multi-index that will be defined later.

The difference with the Schrödinger operator consists in the fact that the Hamiltonian H of the classical problem - the symbol of the diffusion operator D - is complex. In fact, multiplying equation (3.1) on ih we get

$$ihu_t = V(x)(-ih\nabla)u - i(-i\epsilon\nabla)^2 u$$

The attempt to present this equation in the pseudo-differential form $ih_t = H(x, -i\frac{\partial}{\partial x})u$ leads inevitably to the complex Hamiltonian $H = V(x)p - ip^2$ (symbol of the operator $\mathcal{D}$). Surely, it can be made real by the complex rotation $p \to -ip$, but this procedure takes us away from the standard approaches to the problems of quantum mechanics. Thus the corresponding Hamiltonian system

$$\dot{x} = V(x) - 2ip, \quad \dot{p} = -(V_x)'p \tag{3.4}$$

is also complex (here $'$ denotes the transpose matrix). That is why the general theory developed in [M5] can not be directly applied in this situation. However it turns out that some modification of this theory gives the solution.

Let the vector field V be smooth, and let $X(t)$ be the solution of (3.2), defining an orbitally asymptotically stable cycle Γ of the vector field V, i.e. $X(t)$ is a solution to (3.2), periodic with period T, the equation of Γ is $x = X(t)$, and all Floquet multipliers of the system in variations

$$\dot{Y} = \frac{\partial V}{\partial x}(X(t))Y \tag{3.5}$$

have the form $e^{-\mu_j T}$, $j = 1, ..., d$, where one of the numbers μ_j (let it be μ_d) is equal to zero and the real parts of the others are positive. function, periodic with period

It turns out that the asymptotic eigenvalues of $\mathcal{D}$ are given by formula

$$\lambda_\nu = \frac{1}{T} \int_0^T divV(\xi)\,d\xi + \frac{2\pi i\nu_n}{T} + \sum_{j=1}^n \mu_j\nu_j + O(\epsilon), \qquad (3.6)$$

where $\nu = (\nu_1, ..., \nu_d)$ is an integer vector ("number of the eigenvalue"), independent of h and $\nu_j \geq 0$ for $j = 1, ..., d - 1$. The formulas for the eigenfunctions contain complex solutions to the system

$$\dot{W} = -\left(\frac{\partial V}{\partial x}\right)' (X(t))W, \quad \dot{Z} = \frac{\partial V}{\partial x}(X(t))Z + 2W. \qquad (3.7)$$

This system is equivalent to the Hamiltonian system with the complex Hamiltonian $(v, p) - ip^2$, linearised on the solution $x = X(t), p = 0$. The fundamental matrix of (3.7) can be written in the form

$$\begin{pmatrix} Z_1...Z_{n-1} & \dot{X} & Y_1...Y_{n-1} & Z'_n \\ W_1...W_{n-1} & 0 & 0...0 & W'_n \end{pmatrix},$$

where Y_j are the Floquet solutions of (3.5), and where the solutions

$$a_j = \begin{pmatrix} Z_j \\ W_j \end{pmatrix}, \quad j = 1, ...d - 1, \quad \begin{pmatrix} Z_n \\ W_n \end{pmatrix} = \begin{pmatrix} \dot{X} \\ 0 \end{pmatrix}$$

satisfy the conditions

$$a_j(t + T) = a_j e^{\mu_j T}.$$

Now let us chose the matrices B and C in the form

$$B = (W_1(\tau), ...W_n(\tau)), \quad C = (Z_1(\tau), ...Z_n(\tau)).$$

The asymptotic eigenfunctions of $\mathcal{D}$ have the form

$$u_\nu = \exp\{T\sum_{j=1}^{d-1}\nu_j\mu_j - 2\pi i\nu_d)\frac{\tau(x)}{T}\}\prod_{j=1}^{d-1}\left(\sqrt{h}\left(Y_j(\tau(x)), \frac{\partial}{\partial x}\right)\right)^{\nu_j} u_0, \qquad (3.8)$$

where

$$u_0 = (\det C(\tau(x)))^{-1/2}\exp\{\frac{1}{2}\int_0^{\tau(x)} divV(x(\xi))\,d\xi - \frac{\tau(x)}{2T}\int_0^T divV(X(\xi))\,d\xi\}$$

$$\times \exp\left\{-\frac{1}{2h}\left(xx(\tau(x)), BC^{-1}(\tau(x))(x - X(\tau(x)))\right)\right\}$$

and $\tau(x)$ is a smooth function satisfying in some (independent of h) neighbourhood of Γ the equation

$$(\dot{X}(\tau), x - X(\tau)) = 0.$$

Outside this neighbourhood $u_\nu = O(h^\infty)$. The function $\tau(x)$ is defined modulo T, but the expression for u_ν does not depend on this choice. The main result is the following.

Theorem 3.1 [DKO1]. *The functions u_ν and the complex numbers λ_ν defined by (3.6), (3.8) satisfy (3.3).*

The proof of this result and its generalisations to k-dimensional invariant tori (see [DKO1,DKO2]) is derived from the asymptotic solutions to the Cauchy problem for the diffusion equation with specially chosen initial data.

Chapter 9. PATH INTEGRATION FOR THE SCHRÖDINGER, HEAT AND COMPLEX DIFFUSION EQUATIONS

1. Introduction.

There exist several approaches to the rigorous mathematical construction of the path integral, the most important of them (as well as an extensive literature on this subject) are reviewed briefly in Appendices G and H. Unfortunately, all these methods cover still only a very restrictive class of potentials, which is clearly not sufficient for physical applications, where path integration is widely used without rigorous justification. On the other hand, most of the known approaches define the path integral not as a genuine integral (in the sense of Lebesgue or Riemann), but as a certain generalised functional. In this chapter we give a rigorous construction of the path integral which, on the one hand, covers a wide class of potentials and can be applied in a uniform way to the Schrödinger, heat and complex diffusion equations, and on the other hand, is defined as a genuine integral over a bona fide σ- additive (or even finite) measure on path space. Moreover, in the original papers of Feynman the path integral was defined (heuristically) in such a way that the solutions to the Schrödinger equation was expressed as the integral of the function $\exp\{iS\}$, where S is the classical action along the paths. It seems that the corresponding measure was not constructed rigorously even for the case of the heat equation with sources (notice that in the famous Feynman-Kac formula that gives a rigorous path integral representation for the solutions to the heat equation, a part of the action is actually "hidden" inside the Wiener measure). Here we construct a measure on a path space (actually on the so called Cameron-Martin space of trajectories with L^2 first derivative) such that the solutions to the Schrödinger, heat and complex diffusion equations can be represented as the integrals of the exponential of the action with respect to this measure, which is essentially the same for all these cases (to within certain bounded densities). However, for the case of the Schrödinger equation the integral is usually not absolutely convergent and needs a certain regularisation. This regularisation is of precisely the same kind as is used to define the finite- dimensional integral

$$(U_0 f)(x) = (2\pi t i)^{-d/2} \int_{\mathcal{R}^d} \exp\{-\frac{|x-\xi|^2}{2ti}\} f(\xi)\, d\xi \qquad (1.1)$$

giving the free propagator $e^{it\Delta/2}f$. Namely, this integral is not well defined for general $f \in L^2(\mathcal{R}^d)$. The most natural way to define it is based on the observation that, according to the spectral theorem, for all $t > 0$

$$e^{it\Delta/2}f = \lim_{\epsilon \to 0_+} e^{it(1-i\epsilon)\Delta/2}f \qquad (1.2)$$

in $L^2(\mathcal{R}^d)$ (the operator $e^{it(1-i\epsilon)\Delta/2}$ defines the free Schrödinger evolution in complex time $t(1 - i\epsilon)$). Since

$$(e^{it(1-i\epsilon)\Delta/2}f)(x) = (2\pi t(i+\epsilon))^{-d/2} \int_{\mathcal{R}^d} \exp\{-\frac{|x-\xi|^2}{2t(i+\epsilon)}\} f(\xi)\, d\xi$$

($\sqrt{i + \epsilon}$ is defined as the one which tends to $e^{\pi i/4}$ as $\epsilon \to 0$) and the integral on the r.h.s. of this equation is already absolutely convergent for all $f \subset L^2(\mathcal{R}^d)$, one can define the integral (1.1) by the formula

$$(U_0 f)(x) = \lim_{\epsilon \to 0_+} (2\pi t(i + \epsilon))^{-d/2} \int_{\mathcal{R}^d} \exp\{-\frac{|x - \xi|^2}{2t(i + \epsilon)}\} f(\xi) \, d\xi. \qquad (1.3)$$

The same regularisation will be used to define the infinite-dimensional integral giving the solutions to the Schrödinger equation with a general potential.

At the end of the Chapter we show that the path integral constructed here has a natural representation in a cetain Fock space, which gives a connection with the Wiener measure and also with non-commutative probability and quantum stochastic calculus.

1.1. The case of potentials which are Fourier transforms of finite measures. The starting point for our construction is a representation of the solutions of the Schrödinger equation whose potential is the Fourier transform of a finite measure, in terms of the expectation of a certain functional over the path space of a certain compound Poisson process. A detailed exposition of this representation, which is due essentially to Chebotarev and Maslov, together with some references on further developments, are given in Appendix G. We begin here with a simple proof of this representation, which clearly indicates the route for the generalisations that are the subject of this chapter.

Let the function $V = V_\mu$ be the Fourier transform

$$V(x) = V_\mu(x) = \int_{\mathcal{R}^d} e^{ipx} \mu(dp) \qquad (1.4)$$

of a finite complex Borel measure μ on $\mathcal{R}^d$. Now (see e.g. Appendix G) for any σ-finite complex Borel measure μ there exists a positive σ-finite measure M and a complex-valued measurable function f on $\mathcal{R}^d$ such that

$$\mu(dy) = f(y) M(dy). \qquad (1.5)$$

If μ is a finite measure, then M can be chosen to be finite as well. In order to represent Feynman's integral probabilisticly, it is convenient to assume that M has no atom at the origin, i.e. $M(\{0\}) = 0$. This assumption is by no means restrictive, because one can ensure its validity by shifting V by an appropriate constant. Under this assumption, if

$$W(x) = \int_{\mathcal{R}^d} e^{ipx} M(dp), \qquad (1.6)$$

then the equation

$$\frac{\partial u}{\partial t} = (W(\frac{1}{i} \frac{\partial}{\partial y}) - \lambda_M) u, \qquad (1.7)$$

where $\lambda_M = M(\mathcal{R}^d)$, or equivalently

$$\frac{\partial u}{\partial t} = \int (u(y + \xi) - u(y)) M(d\xi), \qquad (1.8)$$

defines a Feller semigroup, which is the semigroup associated with the compound Poisson process having Lévy measure M, see e.g. [Br] or [Pr] for the necessary background in the theory of Lévy processes (notice only that the condition $M(\{0\}) = 0$ ensures that M is actually a measure on $\mathcal{R}^d \setminus \{0\}$, i.e. it is a finite Lévy measure). As is well known, such a process has almost surely piecewise constant paths. More precisely, a sample path Y of this process on the time interval $[0, t]$ starting at a point y is defined by a finite number, say n, of jump-times $0 < s_1 < ... < s_n \leq t$, which are distributed according to the Poisson process N with intensity $\lambda_M = M(\mathcal{R}^d)$, and by independent jumps $\delta_1, ..., \delta_n$ at these times, each of which is a random variable with values in $\mathcal{R}^d \setminus \{0\}$ and with distribution defined by the probability measure M/λ_M. This path has the form

$$Y_y(s) = y + Y^{s_1 \ldots s_n}_{\delta_1 \ldots \delta_n}(s) = \begin{cases} Y_0 = y, & s < s_1, \\ Y_1 = y + \delta_1, & s_1 \leq s < s_2, \\ \ldots \\ Y_n = y + \delta_1 + \delta_2 + \ldots + \delta_n, & s_n \leq s \leq t \end{cases} \tag{1.9}$$

We shall denote by $E_y^{[0,t]}$ the expectation with respect to this process.

Consider the Schrödinger equation

$$\frac{\partial \psi}{\partial t} = \frac{i}{2}\Delta\psi - iV(x)\psi, \tag{1.10}$$

where V is a function (possibly complex-valued) of form (1.4). The equation for the inverse Fourier transform

$$u(y) = \tilde{\psi}(y) = (2\pi)^{-d} \int_{\mathcal{R}^d} e^{-iyx}\psi(x)\,dx$$

of ψ (or equation (1.10) in momentum representation) has the form

$$\frac{\partial u}{\partial t} = -\frac{i}{2}y^2 u - iV\left(\frac{1}{i}\frac{\partial}{\partial y}\right)u. \tag{1.11}$$

Proposition 1.1. *Let u_0 be a bounded continuous function. Then the solution to the Cauchy problem of equation (1.11) with initial function u_0 has the form*

$$u(t, y) = \exp\{t\lambda_M\}E_y^{[0,t]}\left[F(Y(.))u_0(Y(t))\right], \tag{1.12}$$

where, if Y is given by (1.9),

$$F(Y(.)) = \exp\{-\frac{i}{2}\sum_{j=0}^{n}(Y_j, Y_j)(s_{j+1} - s_j)\}\prod_{j=1}^{n}(-if(\delta_j)) \tag{1.13}$$

(here $s_{n+1} = t$, $s_0 = 0$, and the function f is as in (1.5)).

In particular, choosing u_0 to be the exponential function e^{iyz_0}, one obtains a path integral representation for the Green function of equation (1.10) in momentum representation.

Remark. In Appendix G some generalisation of (1.10) is dealt with. We restrict ourselves here to (1.10) for simplicity.

We shall now give an equivalent representation for the integral in (1.12), which we shall prove later on.

1.2. Path integral as a sum of finite-dimensional integrals. One way to visualize the integral (1.12) is by rewriting it as a sum of finite dimensional integrals. To this end, let us introduce some notations. Let $PC_p(s,t)$ (abbreviated to $PC_p(t)$, if $s = 0$) denote the set of all right continuous and piecewise-constant paths $[s,t] \mapsto \mathcal{R}^d$ starting from the point p, and let $PC_p^n(s,t)$ denote the subset of paths with exactly n discontinuities. Topologically, PC_p^0 is a point and $PC_p^n = Sim_t^n \times (\mathcal{R}^d \setminus \{0\})^n$, $n = 1, 2, \ldots$, where

$$Sim_t^n = \{s_1, \ldots, s_n : 0 < s_1 < s_2 < \ldots < s_n \le t\} \qquad (1.14)$$

denotes the standard n-dimensional simplex. In fact, the numbers s_j are the jump-times, and the n copies of $\mathcal{R}^d \setminus \{0\}$ represent the magnitudes of these jumps (see (1.9)). To each σ-finite measure M on $\mathcal{R}^d \setminus \{0\}$ (or on $\mathcal{R}^d$, but without an atom at the origin), there corresponds a σ-finite measure $M^{PC} = M^{PC}(t,p)$ on $PC_p(t)$, which is defined as the sum of measures M_n^{PC}, $n = 0, 1, \ldots$, where each M_n^{PC} is the product-measure on $PC_p^n(t)$ of the Lebesgue measure on Sim_t^n and of n copies of the measure M on $\mathcal{R}^d$. Thus if Y is parametrised as in (1.9), then

$$M_n^{PC}(dY(.)) = ds_1 \ldots ds_n M(d\delta_1) \ldots M(d\delta_n).$$

From properties of the Poisson process it follows that (1.12) can be rewritten in the form

$$u(t,y) = \int_{PC_y(t)} M^{PC}(dY(.)) F(Y(.)) u_0(Y(t)), \qquad (1.15)$$

or, equivalently, as the sum

$$u(t,y) = \sum_{n=0}^{\infty} u_n(t,y) = \sum_{n=0}^{\infty} \int_{PC_y^n(t)} M_n^{PC}(dY(.)) F(Y(.)) u_0(Y(t)). \qquad (1.16)$$

The integrals in this series can be written more explicitly (in terms of the parametrisation (1.9) of the paths Y) as

$$u_n(t,y) = \int_{PC_y^n(t)} M_n^{PC}(dY(.)) F(Y(.)) u_0(Y(t))$$

$$= \int_{Sim_t^n} ds_1 \ldots ds_n \int_{\mathcal{R}^d} \ldots \int_{\mathcal{R}^d} M(d\delta_1) \ldots M(d\delta_n) \, F(y + Y_{\delta_1 \ldots \delta_n}^{s_1 \ldots s_n}) u_0(y + \delta_1 + \ldots + \delta_n).$$
$$(1.17)$$

Notice that the multiplier $\exp\{t\lambda_M\}$ in (1.12) arises because the integral in (1.12) is not over the measure M^{PC}, but over a probability measure obtained from M^{PC} by an appropriate normalisation (namely, $M^{PC}(PC_p^1(t)) = t + O(t^2)$

for small t, and the normalised measure of the corresponding Poisson process is such that the probability of $PC_p^1(t)$ is $\lambda_M(t + O(t^2))\exp\{-t\lambda_M\}$ and the jumps are distributed according to the normalised measure $M/\lambda_M)$.

1.3. Connection with perturbation theory and a proof of Proposition 1.1. A simple proof of formula (1.16) can be obtained from non-stationary perturbation theory, which we recall now for use in what follows. First, one can rewrite equation (1.10) in an integral form using the so called interaction representation for the Schrödinger equation, where the evolution of a quantum system is described by the wave function $\phi = e^{-i\Delta t/2}\psi$ rather than the original wave function ψ. From (1.10) one gets that if ψ satisfies the Cauchy problem for equation (1.10) with the initial data ψ_0, then ϕ satisfies the equation

$$\frac{\partial\phi}{\partial t} = -ie^{-i\Delta t/2}Ve^{i\Delta t/2}\phi \qquad (1.18)$$

(here the symbol V is used to denote the operator of multiplication by the function V), with the same initial data $\phi_0 = \psi_0$. Integrating this equation (which is called the Schrödinger equation in the interaction representation) over t and substituting $\phi = e^{-i\Delta t/2}\psi$ one obtains the equation for ψ

$$\psi(t) = e^{i\Delta t/2}\psi_0 - i\int_0^t e^{i\Delta(t-s)/2}V\psi(s)\,ds, \qquad (1.19)$$

which contains not only the information comprised by (1.10), but also the information comprised by the initial data ψ_0. Though, strictly speaking, equation (1.19) is not quite equivalent to (1.10) (because, for example, a solution of (1.19) may not belong to the domain of the operator Δ) under reasonable assumptions on V (for example, if V is bounded, or $V \in L^p + L^\infty$ with $p \ge \max(2, d/2)$, which is quite enough for our purposes) the solutions to (1.19) defines the Schrödinger evolution $e^{it(\Delta/2-V)}$ (see e.g. [Yaj] or the earlier paper [How], where (1.19) is used to prove the existence of the Schrödinger propagator in the more general case of time-dependent potentials).

Substituting expression (1.19) for ψ in the r.h.s. of (1.19) and iterating this procedure one obtains the standard perturbation theory expansion for ψ

$$\psi(t) = \Big[e^{i\Delta t/2} - i\int_0^t e^{i\Delta(t-s)/2}Ve^{i\Delta s/2}\,ds$$

$$+(-i)^2\int_0^t ds\int_0^s d\tau\, e^{i\Delta(t-s)/2}Ve^{i\Delta(s-\tau)/2}Ve^{i\Delta\tau/2} + ...\Big]\psi_0. \qquad (1.20)$$

More precisely, from this procedure one obtains the following: if series (1.20) is convergent, say in L^2-sense, then its sum defines a solution to equation (1.19), and this solution is unique. Clearly, this is the case for bounded functions V, but actually holds also for more general V, see [Yaj].

Clearly in momentum representation (1.19) has the form

$$u(t, y) = e^{-ity^2/2}u_0(y) - i\int_0^t e^{-i(t-s)y^2/2}\left(V(-i\frac{\partial}{\partial y})u_0\right)(y)\,ds. \qquad (1.21)$$

Since the operator $V(-i(\partial/\partial y))$ is that of convolution with the measure μ, in momentum representation the series (1.20) takes the form

$$u(t,y) = \sum_{j=0}^{\infty} I_j(t,y) = I_0(t,y) + (\mathcal{F}I_0)(t,y) + (\mathcal{F}^2 I_0)(t,y) + ..., \qquad (1.22)$$

where $\mathcal{F}$ is the integral operator given by

$$(\mathcal{F}\phi)(t,y) = -i \int_0^t ds \int_{\mathcal{R}^d} M(dv - y)g(t - s, y)f(v - y)\phi(s, v) \qquad (1.23)$$

and

$$g(t,y) = \exp\{-it(y,y)/2\}, \quad I_0 = g(t,y)u(y).$$

It is convenient to consider this series in the Banach space $C_0(\mathcal{R}^d)$ of continuous functions vanishing at infinity. The terms of the series (1.22) can now be obtained from the corresponding terms of the series (1.16), (1.17) by linear change of integration variables. Consequently, if either the series (1.22) or the series (1.16)-(1.17) is absolutely convergent and all its terms are absolutely convergent integrals, as is clearly the case under the assumptions of Proposition 1.1, one obtains the representation (1.15) (and hence also (1.12)) for the solution $u(t,x)$ of the Cauchy problem for equation (1.11).

  1.4. Regularization by introducing complex times or continuous non-demolition measurement. In this chapter we are going to generalise the representations (1.12) or (1.16) to a wide class of potentials. In Section 2 we begin with a class of potentials that have form (1.4) with measure μ having support in a convex cone and being of polynomial growth. In this case (1.15) still holds without any change, even though the measure μ is not finite. However, this case is rather artificial from the physical point of view, because it does not include real potentials. In general, the terms of the series (1.22) would not be absolutely convergent integrals, or, even worse, (1.22) would not be convergent at all. To deal with this situation, one has to use some regularisations of the Schrödinger equation. As we mentioned, this regularisation will be of the same kind as is used to define the standard finite-dimensional (but not absolutely convergent) integral (1.1). Namely, if the operator $-\Delta/2 + V(x)$ is self-adjoint and bounded from below, by the spectral theorem,

$$\exp\{it(\Delta/2 - V(x))\}f = \lim_{\epsilon \to 0+} \exp\{it(1 - i\epsilon)(\Delta/2 - V(x))\}f. \qquad (1.24)$$

In other words, solutions to equation (1.10) can be approximated by the solutions to the equation

$$\frac{\partial \psi}{\partial t} = \frac{1}{2}(i + \epsilon)\Delta\psi - (i + \epsilon)V(x)\psi, \qquad (1.25)$$

which describes the Schrödinger evolution in complex time. The corresponding integral equation (the analogue of (1.19)) can be obtained from (1.19) by replacing i by $i + \epsilon$ everywhere. It has the form

$$\psi(t) = e^{(i+\epsilon)\Delta t/2}\psi_0 - (i + \epsilon) \int_0^t e^{(i+\epsilon)\Delta(t-s)/2}V\psi(s)\, ds. \qquad (1.26)$$

If ψ satisfies (1.25), its Fourier transform u satisfies the equation

$$\frac{\partial u}{\partial t} = -\frac{1}{2}(i + \epsilon)y^2 u - (i + \epsilon)V(\frac{1}{i}\frac{\partial}{\partial y})u. \tag{1.27}$$

We shall define a measure on a path space such that for arbitrary $\epsilon > 0$ and for a rather general class of potentials V, the solution $\exp\{it(1 - i\epsilon)(\Delta/2 - V(x))\}u_0$ to the Cauchy problem of equation (1.25) can be expressed as the Lebesgue (or even Riemann) integral of some functional F_ϵ with respect to this measure, which gives a rigorous definition (analogous to (1.3)) of an improper Riemann integral corresponding to the case $\epsilon = 0$, i.e. to equation (1.10). Thus, unlike the usual method of analytical continuation often used for defining Feynman integrals, where the rigorous integral is defined only for purely imaginary Planck's constant h, and for real h the integral is defined as the analytical continuation by rotating h through a right angle, in our approach, the measure is defined rigorously and is the same for all complex h with non-negative real part. Only on the boundary $Im\, h = 0$ does the corresponding integral usually become an improper Riemann integral.

Of course, the idea of using equation (1.25) as an appropriate regularisation for defining Feynman integrals is not new and goes back at least to the paper [GY]. However, this was not carried out there, because, as was noted in [Ca], there exists no direct generalisation of Wiener measure that could be used to define Feynman integral for equation (1.25) for any real ϵ. Here we shall carry out this regularization using a measure which differs essentially from Wiener measure. The connection with Wiener measure will be discussed in the last section of this chapter.

Equation (1.25) is certainly only one of many different ways to regularise the Feynman integral. However, this method is one of the simplest method, because the limit (1.24) follows directly from the spectral theorem, and other methods may require additional work to obtain the corresponding convergence result. As another regularisation to equation (1.10), one can take, for example, the equation

$$\frac{\partial \psi}{\partial t} = \frac{1}{2}(i + \epsilon)\Delta\psi - iV(x)\psi. \tag{1.28}$$

A more physically motivated regularisation can be obtained from the quantum theory of continuous measurement. Though the work with this regularisation is technically more difficult than with the regularisation based on equation (1.25), we shall describe it, because, firstly, Feynman's integral representation for continuously obserbed quantum system is a matter of independent interest, and secondly, the idea to use the theory of continuous observation for regularisation of Feynman's integral was already discussed in physical literature (see e.g. [Me2]) and it is interesting to give to this idea a rigorous mathematical justification. The idea behind this approach lies in the observation that in the process of continuous non-demolition quantum measurement a spontaneous collapse of quantum states occurs (see e.g. [Di2], [Be2], or Section 1.4), which gives a sort of regularisation for large x (or large momenta p) divergences of Feynman's integral.

As is well known, the standard Schrödinger equation describes an isolated quantum system. In quantum theory of open systems one considers a quantum system under observation in a quantum environment (reservoir). This leads to a generalisation of the Schrödinger equation, which is called stochastic Schrödinger equation (SSE), or quantum state diffusion model, or Belavkin's quantum filtering equation (see Appendix A and Chapter 7). In the case of a non-demolition measurement of diffusion type, the SSE has the form

$$du + (iH + \frac{1}{2}\lambda^2 R^* R)u \, dt = \lambda Ru \, dW, \tag{1.29}$$

where u is the unknown a posterior (non-normalised) wave function of the given continuously observed quantum system in a Hilbert space $\mathcal{H}$, the selfadjoint operator $H = H^*$ in $\mathcal{H}$ is the Hamiltonian of a free (unobserved) quantum system, the vector-valued operator $R = (R^1, ..., R^d)$ in $\mathcal{H}$ represents the observed physical values, W is the standard d-dimensional Brownian motion, and the positive constant λ represents the precision of measurement. The simplest natural examples of (1.29) concern the case when H is the standard quantum mechanical Hamiltonian and the observed physical value R is either the position or momentum of the particle. The first case was considered in detail in Chapter 7. Here we are going to use mostly the second case when R represents the momentum of the particle (and therefore one models a continuous non-demolition observation of the momentum of a quantum particle). In this case the SSE (1.29) takes the form

$$d\psi = \left(\frac{1}{2}(i + \frac{\lambda}{2})\Delta\psi - iV(x)\psi\right) dt + \frac{1}{i}\sqrt{\frac{\lambda}{2}}\frac{\partial}{\partial x}\psi \, dW. \tag{1.30}$$

As $\lambda \to 0$, equation (1.30) approaches the standard Schrödinger equation (1.10). If ψ satisfies the SSE (1.30), the equation on the Fourier transform $u(y)$ of ψ clearly has the form

$$du = \left(-\frac{1}{2}(i + \frac{\lambda}{2})y^2 u - iV(\frac{1}{i}\frac{\partial}{\partial y})u\right) dt + \sqrt{\frac{\lambda}{2}}yu \, dW. \tag{1.31}$$

By Ito's formula, the solution to this equation with initial function u_0 and with vanishing potential V equals $g_\lambda^W(t, y)u_0(t, y)$ with

$$g_\lambda^W(t, y) = g_\lambda^{W(t)}(t, y) = \exp\{-\frac{1}{2}(i + \lambda)y^2 t + \sqrt{\frac{\lambda}{2}}yW(t)\}, \tag{1.32}$$

and therefore the analog of equation (1.21) corresponding to (1.31) has the form

$$u(t, y) = g_\lambda^{W(t)}(t, y)u_0 - i\int_0^t g_\lambda^{W(t)-W(s)}(t - s, .)V(-\frac{1}{i}\frac{\partial}{\partial y})g_\lambda^{W(s)}(t, .)u(s, .) \, ds. \tag{1.33}$$

As we shall see, equation (1.31) can be used to regularise Feynman's integral for equation (1.10) as an alternative to equation (1.25).

For conclusion, let us sketch the content of this chapter. In Section 2 we will obtain the path integral representation for the solutions of equations (1.11), (1.27), (1.31) for rather general scattering potentials V, including the Coulomb potential.

The momentum representation for wave functions is known to be usually convenient for the study of interacting quantum fields (see e.g. [BSch]). In quantum mechanics, however, one usually deals with the Schrödinger equation in x-representation. Therefore, it is desirable to write down Feynman's integral representation directly for equation (1.10). Since in p-representation our measure is concentrated on the space PC of piecewise constant paths, and since classically trajectories $x(t)$ and momenta $p(t)$ are connected by the equation $\dot{x} = p$, one can expect that in x-representation the correspondonding measure is concentrated on the set of continuous piecewise linear paths. In Sections 3 and 4 we shall construct this measure and the corresponding Feynman integral for equation (1.10) with bounded potentials and also for a class of singular potentials. In Section 5 we discuss the connection with the semiclassical asymptotics giving a different path integral representation for the solutions of the Schrödinger or heat equation, which is an integral of the exponential of the classical action. In the last section, we give a repersentation of our measures in Fock space and make some other remarks.

2. Momentum representation and occupation number representation

If a is a unit vector in $\mathcal{R}^d$, $b \in \mathcal{R}$, $\theta \in (0, \pi/2)$, denote

$$Con_a^\theta = \{y : (y, a) \geq |y| \cos \theta\}, \quad \Pi_a(b) = \{y : (y, a) \leq b\}.$$

Let M be a (positive) measure with support in Con_a^θ. Suppose also that M is of polynomial growth, i.e. $M(\Pi_a(b)) \leq Cb^N$ for some positive constants C, N and all positive b. Let V be given by (1.4), (1.5) in the sense of distributions, so that V is the Fourier transform of the measure fM considered as a distribution over the Schwarz space $S(\mathcal{R}^d)$.

Proposition 2.1. *For any continuous function u_0 such that $supp\, u_0 \subset \Pi_a(b)$ for some real b, all terms of the series (1.16) are absolutely convergent integrals representing continuous functions, and this series is absolutely convergent uniformly on compact sets. In particular, there exists a solution to the Cauchy problem of equation (1.11) that has support in $\Pi_a(b)$, has at most polynomial growth, and is represented by means of the Feynman integral (1.15).*

Proof. Clearly, if y does not belong to $\Pi_a(b)$, then all u_n given by (1.17) vanish. Hence, $supp\, u_n \subset \Pi_a(b)$ for all n. If $y \in \Pi_a(b)$, then

$$|u_n(t, y)| \leq \frac{t^n}{n!} \int_{Con_a^\theta} \cdots \int_{Con_a^\theta} M(d\delta_1)...M(d\delta_n)u_0(y + \delta_1 + ... + \delta_n).$$

Since $supp\, u_0 \subset \Pi_a(b)$, the integrand in this formula vanishes whenever $(\delta_1 + ... + \delta_n, a) > b - (y, a)$, and, in particular, if $(\delta_j, a) > b - (y, a)$ for at least one

$j = 1, ..., n$. Hence, denoting $K = \sup\{|u_0(x)|\}$, one has

$$|u_n(t,y)| \leq \frac{t^n}{n!} K \int_{\Pi_a(b-(y,a))} ... \int_{\Pi_a(b-(y,a))} M(d\delta_1)...M(d\delta_n)$$

$$\leq \frac{t^n}{n!} K C^n (b - (y,a))^{Nn},$$

which for fixed b and large y does not exceed $t^n K C^n 2^{Nn}(y,a))^{Nn}/n!$. Consequently, the series (1.16) consists of absolutely convergent integrals, is convergent, and is bounded by $K \exp\{2^N(y,a)^N Ct\}$, which implies the statement of the Proposition.

As we mentioned in the introduction, the potentials of the form occurring in Proposition 2.1 seem to have no physical relevance. As an example of a physically meaningful situation, we consider now the case of scattering potentials using the regularisation (1.27) or (1.31). Let V have form (1.4), (1.5) (again in the sense of distributions) with M being the Lebesgue measure M_{Leb} and $f \in L^1 + L^q$, i.e. $f = f_1 + f_2$ with $f_1 \in L^1$, $f_2 \in L^q$, with q in the interval $(1, d/(d-2))$, $d > 2$. Notice that this class of potentials includes the Coulomb case $V(x) = |x|^{-1}$ in $\mathcal{R}^3$, because for this case $f(y) = |y|^{-2}$.

Proposition 2.2. *Under the given assumptions on V there exists a (strong) solution $u(t, y)$ to the Cauchy problem of equations (1.27) and (1.31) with initial data u_0, which is given in terms of the Feynman integral of type (1.15). More precisely*

$$u(t,y) = \int_{PC_y(t)} M_{Leb}^{PC}(dY(.))F(Y(.))u_0(Y(t)), \qquad (2.1)$$

where, if Y is parametrised as in (1.9),

$$F(Y(.)) = F_\epsilon(Y(.)) = \exp\{-\frac{1}{2}(i+\epsilon)\sum_{j=0}^{n} Y_j^2(s_{j+1} - s_j)\} \prod_{j=1}^{n}(-i(1-i\epsilon)f(\delta_j))$$

$$(2.2)$$

for the case of equation (1.27), and $F(Y(.))$ equals

$$F_\lambda^W(Y(.)) = \prod_{j=1}^{n}(-if(\delta_j))$$

$$\times \exp\{-\sum_{j=0}^{n}\left[\frac{i+\lambda}{2}Y_j^2(s_{j+1} - s_j) - \sqrt{\frac{\lambda}{2}}Y_j(W(s_{j+1}) - W(s_j))\right]\} \qquad (2.3)$$

for the case of equation (1.31).

Proof. Since the proofs for equations (1.27) and (1.31) are quite similar, let us consider only the case of equation (1.31). As is explained in the introduction, it is sufficient to prove that for any bounded continuous function ϕ the integral (1.23), with $g = g_\lambda^W$ as in (1.32), is absolutely convergent (almost surely), and that furthermore, the corresponding series (1.22) is absolutely convergent. To this end, consider the integral

$$J = \int_{\mathcal{R}^d} |f(v - y)| g_\lambda^W(t, y) \, dy.$$

Clearly, the function g_λ^W is bounded (for a.a. W) for times in an arbitrary finite closed subinterval of the positive halfline, and for small t

$$\sup_y\{|g(t, y)|\} = \exp\{W^2(t)/4t\} \leq \exp\{\log|\log t|/2\} = \sqrt{|\log t|}, \qquad (2.4)$$

due to the well known law of the iterated logarithm for the Brownian motion W. Hence, by the assumptions on f and the Hölder inequality

$$J = O(\sqrt{|\log t|}) + O(1)\|g_\lambda^W(t, .)\|_{L^p},$$

where $p^{-1} + q^{-1} = 1$. Since

$$\|g_\lambda^W(t, .)\|_{L^p}^p = \left(\frac{2\pi}{p\lambda t}\right)^{d/2} \exp\{\frac{pW^2(t)}{4t}\},$$

it follows that J is bounded for t in any finite interval of the positive halfline, and $J = O(\lambda t)^{-d/2p}\sqrt{|\log t|}$ for small t. Since the condition $q < d/(d-2)$ is equivalent to the condition $p > d/2$, there exists $\epsilon \in (0, 1)$ such that $J \leq C((\lambda t)^{-(1-\epsilon)})$. Moreover, clearly $I_0(t, y) = g(t, y)u_0(y)$ does not exceed $Kt^{-\epsilon}$ for some constant K. We can now easily estimate the terms of the series (1.22). Namely, we have

$$|\mathcal{F}I_0(t, y)| \leq KC\lambda^{-(1-\epsilon)} \int_0^t (t - s)^{-(1-\epsilon)} s^{-\epsilon} \, ds = KC\lambda^{-(1-\epsilon)} B(\epsilon, 1 - \epsilon),$$

where B denotes the Euler β-function. Similarly,

$$|\mathcal{F}^2 I_0(t, y)| \leq \lambda^{-2(1-\epsilon)} B(\epsilon, 1 - \epsilon) KC^2 \int_0^t (t - s)^{-(1-\epsilon)} \, ds$$

$$= B(\epsilon, 1 - \epsilon) B(\epsilon, 1) KC^2 t^\epsilon.$$

By induction, we obtain the estimate

$$|\mathcal{F}^k I_0(t, y)| \leq KC^k \lambda^{-k(1-\epsilon)} t^{(k-1)\epsilon} B(\epsilon, 1 - \epsilon) B(\epsilon, 1) ... B(\epsilon, 1 + (k-2)\epsilon).$$

Using the representation of the β-function in terms of the Γ-function, one obtainss that the terms of series (1.22) are of order $t^{k\epsilon}/\Gamma(1 + k\epsilon)$, which implies the convergence of this series for all t. Since we have estimated all functions by their magnitude, we have proved also that all terms of series (1.22) are absolutely convergent integrals, and that this series converges absolutely.

The following is a direct consequence of (1.24) and Proposition 2.2.

Proposition 2.3. *Assume the assumptions of Proposition 2.2 hold. If the operator $-\Delta/2 + V(x)$ is self-adjoint and bounded from below, then for any $u_0 \in L^\infty \cap L^2$, the solution to equation (1.10) is given by the improper Feynman integral (1.15), which should be understood as*

$$u(t,y) = \lim_{\epsilon \to 0} \int_{PC_y(t)} M_{Leb}^{PC}(dY(.))F_\epsilon(Y(.))u_0(Y(t)). \qquad (2.5)$$

Since the convergence of solutions of equation (1.31) to solutions of the ordinary Schrödinger equation seems to be unknown, the use of equation (1.31) to obtain a regularisation for the Feynman integral for the Schrödinger equation similar to (2.5) requires some additional work. It seems that this can be done under the assumptions of Proposition 2.2 using the technique from [Yaj]. But we shall restrict ourselves here to the case of a bounded potential, which will be used also in the next Section. Notice that we prove now this result using p-representation, but it automatically implies the same fact for the Schrödinger equation in x-representation.

Proosition 2.4. *Let V be a bounded measurable function. Then for any $u_0 \in L^2(\mathcal{R}^d)$ the solution u_λ^W of equation (1.33) (which exists and is unique, see details in Section 3) tends (almost surely) as $\lambda \to 0$ to the solution of this equation with $\lambda = 0$.*

Proof. Using the boundedness of all operators on the r.h.s. of (1.33), and (2.4) one obtains that

$$\|u_\lambda^W - u\| \le \|g_\lambda^W(t,y)u_0 - g_0^W(t,y)u_0\| + O(t)|\log t|\|u_\lambda^W - u\| + o(\lambda)$$

$$= O(t)|\log t|\|u_\lambda^W - u\| + o(\lambda), \qquad (2.6)$$

where $o(\lambda)$ depends on u_0 but is uniform with respect to finite times t. It follows that $\|u_\lambda^W - u\| = o(\lambda)$ for small t, which proves the Proposition.

Corollary. *If V is a bounded function, and the assumptions of Proposition 2.2 holds, then the solution to (1.11) can be presented in the form*

$$u(t,y) = \lim_{\lambda \to 0} \int_{PC_y(t)} M_{Leb}^{PC}(dY(.))F_\lambda^W(Y(.))u_0(Y(t)). \qquad (2.7)$$

To conclude, let us point out that different kind of interactions can be treated similarly if one changes p- repersentation into occupation number representation. Namely, consider the Schrödinger equation for the anharmonic oscillator

$$i\frac{\partial \psi}{\partial t} = \left(-\frac{\Delta}{2} + \frac{x^2}{2} + V(x)\right)\psi.$$

If V is a smooth function, a path integral representation for its solutions can be obtained as in Section 5. If V is a bounded function, one can also proceed as in Section 3 using the explicit Green function for the quantum harmonic oscillator instead of the "free" Green function. Alternatively, one can use the spectral representation for the harmonic oscillator, i.e. for the operator $\Delta + x^2$. In that approach the corresponding measure on path space will be concentrated on the right continuous piecewise-constant trajectories $q : [0,t] \mapsto \{0,1,...\}$ with values in the countable number of eigenstates ψ_j of the harmonic oscillator, and the analog of formula (1.16) will hold, where the integrand will have the form $\exp\{-i\int_0^t E(q(s)\,ds\}$, where $E(n)$ denotes the energy of the state ψ_n. If the infinite dimensional matrix $(V\psi_j, \psi_k)$ defines finite transition probabilities, we find ourselves in the situation, where the corresponding path integral is expressed as an expectation with respect to a generalised Poisson process in the sense of Combe et al [Com1] (see e.g. [BGR] where this is done in a more general infinite-dimensional model describing a particle interacting with a boson reservoir). If not, we can as usual use the regularisation based on (1.24).

3. Path integral for the Schrödinger equation in x-representation

As we mentioned in introduction, we are going to deal here with measures on paths that are concentrated on the set of continuous piecewise linear paths. Denote this set by CPL. Let $CPL^{x,y}(0,t)$ denote the class of paths $q : [0,t] \mapsto \mathcal{R}^d$ from CPL joining x and y in time t, i.e. such that $q(0) = x$, $q(t) = y$. By $CPL_n^{x,y}(0,t)$ we denote the subclass consisting of all paths from $CPL^{x,y}(0,t)$ that have exactly n jumps of their derivative. Obviously,

$$CPL^{x,y}(0,t) = \cup_{n=0}^{\infty} CPL_n^{x,y}(0,t).$$

Notice also that the set $CPL^{x,y}(0,t)$ belongs to the Cameron-Martin space of curves that have derivatives in $L^2([0,t])$.

To any σ-finite measure M on $\mathcal{R}^d$ there corresponds a unique σ-finite measure M^{CPL} on $CPL^{x,y}(0,t)$, which is the sum of the measures M_n^{CPL} on $CPL_n^{x,y}(0,t)$, where M_0^{CPL} is just the unit measure on the one-point set $CPL_0^{x,y}(0,t)$ and each M_n^{CPL}, $n > 0$, is the direct product of the Lebesgue measure on the simplex (1.14) of the jump times $0 < s_1 < ... < s_n < t$ of the derivatives of the paths $q(.)$ and of n copies of the measure M on the values $q(s_j)$ of the paths at these times. In other words, if

$$q(s) = q_{\eta_1 \ldots \eta_n}^{s_1 \ldots s_n}(s) = \eta_j + (s - s_j)\frac{\eta_{j+1} - \eta_j}{s_{j+1} - s_j}, \quad s \in [s_j, s_{j+1}] \tag{3.1}$$

(where $s_0 = 0, s_{n+1} = t, \eta_0 = x, \eta_{n+1} = y$) is a typical path in $CPL_n^{x,y}(0,t)$ and Φ is a functional on $CPL^{x,y}(0,t)$, then

$$\int_{CPL^{x,y}(0,t)} \Phi(q(.)) M^{CPL}(dq(.)) = \sum_{n=0}^{\infty} \int_{CPL_n^{x,y}(0,t)} \Phi(q(.)) M_n^{CPL}(dq(.))$$

$$= \sum_{n=0}^{\infty} \int_{Sim_t^n} ds_1...ds_n \int_{\mathcal{R}^d} ... \int_{\mathcal{R}^d} M(d\eta_1)...M(d\eta_n)\Phi(q(.)). \tag{3.2}$$

Remark. Nothing is changed if $CPL_n^{x,y}(0,t)$ is defined as the set of paths with at most n jumps in their derivative. In fact, the M_n^{CPL}-measure of the subset $CPL_{n-1}^{x,y}(0,t) \subset CPL_n^{x,y}(0,t)$ vanishes, because if the jump, say, at the time s_j vanishes then $(\eta_j - \eta_{j-1})(s_{j+1} - s_{j-1}) = (\eta_{j+1} - \eta_{j-1})(s_j - s_{j-1})$, therefore s_j can be only one point, and the Lebesgue measure has no atoms.

To express the solutions to the Schrödinger equation in terms of a path integral we shall use the following functionals on $CPL^{x,y}(0,t)$, depending on a given measurable function V on $\mathcal{R}^d$:

$$\Phi_\epsilon(q(.)) = \prod_{j=1}^{n+1} (2\pi(s_j - s_{j-1})(i + \epsilon))^{-d/2}$$

$$\times \exp\{-\sum_{j=1}^{n+1} \frac{|\eta_j - \eta_{j-1}|^2}{2(i + \epsilon)(s_j - s_{j-1})}\} \prod_{j=1}^{n} (-(i + \epsilon)V(\eta_j))$$

$$= \prod_{j=1}^{n+1} (2\pi(s_j - s_{j-1})(i + \epsilon))^{-d/2} \prod_{j=1}^{n} (-(i + \epsilon)V(\eta_j)) \exp\{-\frac{1}{2(i + \epsilon)} \int_0^t \dot{q}^2(s)\,ds\},$$

$$(3.3)$$

and

$$\Phi_\lambda^W(q(.)) = \prod_{j=1}^{n+1} (2\pi(s_j - s_{j-1})(i + \lambda))^{-d/2}$$

$$\times \exp\{-\sum_{j=1}^{n+1} \frac{(\eta_j - \eta_{j-1} - i\sqrt{\frac{\lambda}{2}}(W(s_j) - W(s_{j-1}))^2}{2(i + \lambda)(s_j - s_{j-1})}\} \prod_{j=1}^{n} (-iV(\eta_j)). \quad (3.4)$$

As in Section 2, we shall denote Lebesgue measure on $\mathcal{R}^d$ by M_{Leb}.

Theorem 3.1. *Let V be a bounded measurable function on $\mathcal{R}^d$. Then for arbitrary $\epsilon > 0$ or $\lambda > 0$ and a.a. Wiener trajectories W, there exists a unique solution $G_\epsilon(t,x,x_0)$ or $G_\lambda^W(t,x,x_0)$ to the Cauchy problem of equations (1.25) or (1.30) respectively with Dirac initial data $\delta(x - x_0)$. These solutions (i.e. the Green functions for these equations) are uniformly bounded for all (x,x_0) and t in any compact interval of the open half-line, and they are expressed in terms of path integrals as follows:*

$$G_\epsilon(t,x,x_0) = \int_{CPL^{x,y}(0,t)} \Phi_\epsilon(q(.)) M_{Leb}^{CPL}(dq(.)), \quad (3.5)$$

$$G_\lambda^W(t,x,x_0) = \int_{CPL^{x,y}(0,t)} \Phi_\lambda^W(q(.)) M_{Leb}^{CPL}(dq(.)), \quad (3.6)$$

with Φ_ϵ and Φ_λ^W given by (3.3), (3.4). For arbitrary $\psi_0 \in L^2(\mathcal{R}^d)$ the solution $\psi_0(t,s)$ of the Cauchy problem for equation (1.10) with the initial data ψ_0 has the form of an improper (not absolutely convergent) path integral that can be understood rigorously as either

$$\psi(t,x) = \lim_{\epsilon \to 0+} \int_{CPL^{x,y}(0,t)} \int_{\mathcal{R}^d} \psi_0(y) \Phi_\epsilon(q(.)) M_{Leb}^{CPL}(dq(.)) dy, \qquad (3.7)$$

or (almost surely) as

$$\psi(t,x) = \lim_{\lambda \to 0+} \int_{CPL^{x,y}(0,t)} \int_{\mathcal{R}^d} \psi_0(y) \Phi_\lambda^W(q(.)) M_{Leb}^{CPL}(dq(.)) dy. \qquad (3.8)$$

Proof. Formulas (3.7), (3.8) follow from (3.5), (3.6), (1.24) and Proposition 2.4. The proofs of (3.5), (3.6) are similar and we shall prove only (3.5). To this end, notice that the analogue of series (1.20) for the case of equation (1.25) has form (1.20) with i replaced by $(i + \epsilon)$ everywhere. In particular, for the Green function one has the representation

$$G_\epsilon(t,x,x_0) = G_\epsilon^{free}(t,x,x_0)$$

$$-(i+\epsilon) \int_0^t \int_{\mathcal{R}^d} G_\epsilon^{free}(t-s, x-\eta) V(\eta) G_\epsilon^{free}(s, \eta - x_0) \, d\eta ds + ..., \qquad (3.9)$$

where G_ϵ^{free} is the Green function of the "free" equation (1.25) (i.e. with $V = 0$):

$$G_\epsilon^{free}(t, x-x_0) = (2\pi t(i+\epsilon))^{-d/2} \exp\{-\frac{(x-x_0)^2}{2(i+\epsilon)t}\}.$$

To prove (3.5) one needs to prove that the terms of this series are absolutely convergent integrals and the series is absolutely convergent with a bounded sum. This is more or less straightforward. Namely, to prove that the second integral in this series is absolutely convergent, we must estimate the integral

$$\int_0^t \int_{\mathcal{R}^d} |2\pi(i+\epsilon)|^{-d}((t-s)s)^{-d/2}| \exp\{-\frac{(x-\eta)^2}{2(t-s)(i+\epsilon)} - \frac{(\eta-x_0)^2}{2s(i+\epsilon)}\}| \, dsd\eta$$

$$= \int_0^t \int_{\mathcal{R}^d} (2\pi\sqrt{1+\epsilon^2})^{-d}((t-s)s)^{-d/2}$$

$$\times \exp\{-\frac{\epsilon}{2(1+\epsilon^2)}\left[\frac{(x-\eta)^2}{t-s} + \frac{(\eta-x_0)^2}{s}\right]\} \, dsd\eta.$$

This is a Gaussian integral in η which can be explicitly evaluated using standard integrals to be

$$t(2\pi\epsilon t)^{-d/2} \exp\{-\frac{\epsilon(x-x_0)^2}{2(1+\epsilon^2)t}\}.$$

By induction and similar calculations we obtain the estimate for the n-th term of the series (3.9)

$$C(2\pi t)^{-d/2}(t\epsilon^{-d/2})^k/k!$$

for some real number C. This completes the proof of the Theorem.

The measure M^{CPl} in (3.2) may well be not finite, for example M_{Leb}^{CPL} is not finite. But every Hilbert space can be represented as an L^2 over a probability space. For example, the obvious isomorphism of $L^2(\mathcal{R}, dx)$ with $L^2(\mathcal{R}, e^{-x^2/2}dx)$ is very useful in many situation. In the same way, an integral over M_{Leb}^{CPL} can be rewritten as an integtral over the probability space (up to a normalisation) $(e^{-x^2/2}dx)^{CPL}$. Thus one can always rewrite the integral from (3.2) as an expectation of a ceratin stochastic process, which can be taken to be an integral of the compound Poisson process that stands for the path integral formula for the solutions to the Schrödinger equation in momentum representation. More systematic way of obtaining probabilistic interterpations of path integral constructed is discussed later in Section 6.

4. Singular potentials

There exists an extensive literature devoted to the study of the Schrödinger equations with singular potentials, and more precisely with potentials being Radon measures supported by null sets. As most important examples of such null sets one should mention discrete sets (point interaction), smooth surfaces (surface delta interactions), Brownian paths and more general fractals, see e.g. [BF], [ABD], [AFHL], [AHKS], [AntGS], [DaSh], [Kosh], [Metz], [DeO], [Pav] and references therein for different mathematical techniques used for the study of these models and for physical motivations. We are going to show now that the path integral constructed above can be successfully applied to the construction of solutions to these models.

The one-dimensional situation turns out to be particularly simple in our approach, because in this case no regularisation is needed to express the solutions to the corresponding Schrödinger equation and its propagator in terms of path integral.

Proposition 4.1. *Let V be a bounded (complex) measure on $\mathcal{R}$. Then the solution ψ_G to equation (1.25)), where ϵ is any complex number with $\epsilon \neq i$ and non-negative real part, with initial function $\psi_0(x) = \delta(x - x_0)$ (i.e. the propagator or the Green function of (1.25)) exists and is a continuous function of the form*

$$\psi_G(t, x) = (2\pi(i + \epsilon)t)^{-1/2} \exp\{-\frac{|x - x_0|^2}{2t(i + \epsilon)}\} + O(1) \tag{4.1}$$

uniformly for finite times. Moreover, one has the path integral representation for ψ_G of the form

$$\psi_G(t, x) = \int_{CPL^{x,y}(0,t)} \Phi(q(.))V^{CPL}(dq(.)), \tag{4.2}$$

where V^{CPL} is related to V as M^{CPL} is to M in Section 2, and

$$\Phi(q(.)) = \prod_{j=1}^{n+1} (2\pi(s_j - s_{j-1})(i + \epsilon))^{-1/2} \exp\{-\frac{1}{2(i + \epsilon)} \int_0^t \dot{q}^2(s) \, ds\}.$$

Remark. The cases $\epsilon = 0$, i.e. the Schrödinger equation, and $\epsilon = 1 - i$, i.e. the heat equation, are particular cases of the situation considered in the proposition.

Proof. Since V is a finite measure, in order to prove that the terms of series (3.9) with $\epsilon = 0$ (which expresses the Green function) are absolutely convergent, one needs to estimate the integrals

$$\int_0^t ds_1 \, (2\pi(t - s_1))^{-1/2} \int_0^{s_1} ds_2 \, (2\pi(s_1 - s_2))^{-1/2} \dots \int_0^{s_{n-1}} ds_n (2\pi s_n)^{-1/2},$$

which clearly exist (one-dimensional effect!) and can be expressed explicitly in terms of the Euler β-function. One sees directly that the corresponding series is convergent, which completes the proof.

For the Schrödinger equation in finite-dimensional case one needs a regularisation, say (1.25) with $\epsilon > 0$ or (1.27) with $\lambda > 0$. For simplicity we consider here only the regularisation given by (1.25).

Following essentially [AFHK] (see also [Pus], [KZPS]) we shall say that a number $\alpha \geq 0$ is *admissible* for a finite Borel measure V on $\mathcal{R}^d$, if there exists a constant $C = C(\alpha)$ such that

$$V(B_r(x)) \leq Cr^\alpha \tag{4.3}$$

for all $x \in \mathcal{R}^d$ and all $r > 0$. The least upper bound of all admissible numbers for V is called *dimensionality* of V. It will be denoted by $dim(V)$.

Proposition 4.2. *Let V be a finite Borel measure on $\mathcal{R}^d$ with $dim(V) > d - 2$. Then for any $\epsilon > 0$ and any bounded initial function $\psi_0 \in L^2(\mathcal{R}^d)$ there exists a unique solution $\psi_\epsilon(t, x)$ to the Cauchy problem of equation (1.26) with the initial data $\psi_0(x)$. This solution has the form*

$$\psi_\epsilon(t, x) = \int_{CPL^{x,y}(0,t)} \int_{\mathcal{R}^d} \psi_0(y) \Phi_\epsilon(q(.)) V^{CPL}(dq(.)) dy, \tag{4.4}$$

where

$$\Phi_\epsilon = \prod_{j=1}^{n+1} (2\pi(s_j - s_{j-1})(i + \epsilon))^{-d/2} (-(\epsilon + i))^n \exp\{-\frac{1}{2(i + \epsilon)} \int_0^t \dot{q}^2(s) \, ds\}.$$

Proof. One needs to prove that the terms of the series (1.20), in which i has been replaced by $(i + \epsilon)$, are absolutely convergent integrals and then to estimate the corresponding series. Starting with the first non-trivial term one needs to estimate the integral

$$J = K \int_0^t \int_{\mathcal{R}^{2d}} |2\pi(i + \epsilon)|^{-d} ((t - s)s)^{-d/2}$$

$$\times |\exp\{-\frac{(x - \xi)^2}{2(t - s)(i + \epsilon)} - \frac{(\xi - \eta)^2}{2s(i + \epsilon)}\}| \, ds d\eta |V|(d\xi).$$

$$= K \int_0^t \int_{\mathcal{R}^{2d}} (2\pi\sqrt{1+\epsilon^2})^{-d}((t-s)s)^{-d/2} \exp\{-\epsilon\frac{(x-\xi)^2}{2(t-s)(1+\epsilon)^2}\}$$

$$\times \exp\{-\epsilon\frac{(\xi-\eta)^2}{2s(1+\epsilon)^2}\} \, ds \, |V|(d\xi)d\eta,$$

where $K = \sup\{|\psi_0(\eta)|\}$. Integrating over η yields

$$J \leq K \int_0^t (2\pi\sqrt{1+\epsilon^2})^{-d/2}(t-s)^{-d/2} \exp\{-\epsilon\frac{(x-\xi)^2}{2(t-s)(1+\epsilon)^2}\}\epsilon^{-d/2} \, ds \, |V|(d\xi).$$

Due to the assumptions of the theorem, there exists $\alpha > d - 2$ such that (4.3) holds. Let us decompose this integral into the sum $J_1 + J_2$ of the integrals over the domains D_1 and D_2 with

$$D_1 = \{\xi : |x - \xi| \leq (t-s)^{-\delta+1/2}\}$$

and D_2 its complement. Choosing $\delta > 0$ in such a way that $\alpha(-\delta+1/2) - d/2 > -1$ (which is possible due to the assumption on α) we get from (4.3) that

$$J_1 \leq KC \int_0^t (2\pi\sqrt{1+\epsilon^2})^{-d/2}(t-s)^{\alpha(-\delta+1/2)-d/2} \, ds =$$

$$= KC(1+\alpha(-\delta+1/2)-d/2)^{-1}(2\pi\sqrt{(1+\epsilon^2)})^{-d/2}t^{1+\alpha(-\delta+1/2)-d/2}.$$

In D_2 the integrand is uniformly exponentially small, and therefore using the boundedness of the measure $|V|$ we obtain for J_2 an even better estimate than for J_1. Higher order terms are again estimated by induction giving the required result.

Proposition 4.3. *Assume the assumptions of Proposition 4.2 holds. If the operator $-\Delta/2 + V$ is selfadjoin and bounded from below, then one can take the limit as $\epsilon \to 0$ in (4.4) to obtain the solutions to equation (1.10).*

It was proved in [AFHK] that under the assumptions of Proposition 4.2 there exists a constant ω_0 such that the operator $-\Delta/2 + \omega|V|$ is selfadjoin and bounded below for all $\omega < \omega_0$. Therefore, for these operators the statement of Proposition 4.3 holds. A concrete example of interest is given by measures on $\mathcal{R}^3$ concentrated on a Brownian path, because their dimensionality equals 2 almost surely. As shown e.g. in [AFHK], potentials being the finite sums of the Dirac measures of closed hypersurfaces in $\mathcal{R}^d$ satisfy the assumptions of the above corollary (without an assumption of a small coupling constant), see also [Koch2]. For the particular case of mesaures on spheres of codimension 1, see [AntGS], [DaSh], where one can find the references on physical papers dealing with these models.

Less exotic examples of potentials satisfying the asumptions of Proposition 4.3 are given by measures with a density $V(x)$ having a bounded support and such that $V \in L^p(\mathcal{R}^d)$ with $p > d/2$ (which one checks by the Hölder inequality). Moreover, it is not difficult to check that one can combine the potentials from

Proposition 4.3 and Theorem 3.1, for example, one can take $V \in L^\infty(\mathcal{R}^d) + L^p(\mathcal{R}^d)$ with $p > d/2$, which includes, in particular, the Coulomb potential in $d = 3$.

Notice for conclusion that solutions to the Schrödinger equation with a certain class of singular potentials can be obtained in terms of the Feynman integral defined as a generalised functional in Hida's white noise space [HKPS].

5. Semiclassical asymptotics

In this section we answer the following question: how to define a measure on a path space in such a way that the solutions to the Schrödinger or heat equation (deterministic or stochastic) can be expressed as integrals with respect to this measure of the exponential of the classical action. We start with the heat equation, where no regularisation is needed.

Consider the equation (3.1.1). We are going to interprete the formula for its Green function given in Proposition 3.4.1 of Chapter 3, under the assumptions of regularity of the corresponding diffusion (see Chapter 3), in terms of a path integral of type (1.15). Namely, as in Section 3, let us introduce the set CPC of continuous piecewise classical paths, i.e. continuous paths that are smooth and satisfy the classical equations of motion

$$\dot{x} = p, \quad \dot{p} = \frac{\partial V}{\partial x}, \tag{5.1}$$

where H is the Hamiltonian corresponding to equation (3.1.1) (see Chapter 3), except at a finite number of points, where their derivatives may have discontinuities of the first kind. Let $CPC^{x,y}(0,t)$ denote the class of paths $q : [0,t] \mapsto \mathcal{R}^d$ from CPC joining y and x in time t, i.e. such that $q(0) = y$, $q(t) = x$. We denote by $CPC_n^{x,y}(0,t)$ the subclass consisting of all paths from $CPC^{x,y}(0,t)$ that have exactly n jumps in their derivative. Obviously, to any σ-finite measure M on $\mathcal{R}^d$ there corresponds a unique σ-finite measure M^{CPC} on $CPC^{x,y}(0,t)$, which is the sum of the measures M_n^{CPC} on $CPC_n^{x,y}(0,t)$, where M_0^{CPC} is just the unit measure on the one-point set $CPC_0^{x,y}(0,t)$ and for $n > 0$ M_n^{CPC} is the direct product of the Lebesgue measure on the simplex (1.14) of the jump times $0 < s_1 < ... < s_n < t$ of the derivatives of the paths $q(.)$ and of n copies of the measure M on the values $q(s_j)$ of the paths at these times. In other words, if

$$q_\eta^s = q_{\eta_1...\eta_n}^{s_1...s_n}(s) \in CPC_n^{x,y}(0,t) \tag{5.2}$$

denotes the path that takes values η_j at the times s_j, is smooth and classical between these time-points, and Φ is a functional on $CPL^{x,y}(0,t)$, then

$$\int_{CPC^{x,y}(0,t)} \Phi(q(.)) M^{CPC}(dq(.)) = \sum_{n=0}^{\infty} \int_{CPC_n^{x,y}(0,t)} \Phi(q(.)) M_n^{CPC}(dq(.))$$

$$= \sum_{n=0}^{\infty} \int_{Sim_t^n} ds_1...ds_n \int_{\mathcal{R}^d} ... \int_{\mathcal{R}^d} M(d\eta_1)...M(d\eta_n) \Phi(q(.)). \tag{5.3}$$

We will use this construction only for the case when the measure M is the standard Lebesgue measure M_{Leb}. Clearly, from Proposition 3.4.3 one obtains directly the following result.

Proposition 5.1. *Under the assumptions of Proposition 3.4.3*

$$u_G(t, x, y) = \int_{CPC^{x,y}(0,t)} \exp\{-\frac{1}{h}I(q(.))\}\Phi_{sem}(q(.))M_{Leb}^{CPC}(dq(.)), \qquad (5.4)$$

where $I(y(.)) = \int_0^t L(y(s), \dot{y}(s)) ds$ *denote the classical action on a path* $y(.)$*. The explicit formula for the "semiclassical density"* Φ_{sem} *follows from Proposition 3.4.3. For example, in the simplest case of equation*

$$h\frac{\partial u}{\partial t} = \frac{h^2}{2}\Delta u - V(x)u, \quad x \in \mathcal{R}^d, \quad t > 0, \qquad (5.5)$$

one has

$$\Phi_{sem}(q_\eta^s) = (2\pi h)^{-dn/2} J(t - s_n, x, \eta_n)^{-1/2} \prod_{k=1}^{n} \Delta(J(s_k - s_{k-1}, \eta_k, \eta_{k-1})^{-1/2})h^n \tag{5.6}$$

on a typical path (5.2), where $J(t, x, x_0)$ *is the Jacobian, given by*

$$J(t, x, \xi) = \det \frac{\partial X}{\partial p}(t, \xi, p_0(t, x, \xi)),$$

on the classical path joining x_0 *and* x *in time* t *and where* $s_0 = 0$*,* $\eta_0 = y$*.*

Hence the Green function is expressed as the integral of the exponential of the classical action functional $-\frac{1}{h}I(q(.))\}$ over the measure $\Phi_{sem}M_{Leb}^{CPC}$, which is actually a measure on the Cameron-Martin space of paths that are absolutely continuous and have their derivatives in L_2.

A similar result can be obtained for the stochastic heat equation and for the stochastic Schrödinger equations of Theorems 7.2.2, 7.2.3.

Remark. One sees directly from the path integral (5.4) that the main contribution to the asymptotics as $h \to 0$ of (5.4) comes from a small neighborhood of the classical path joining x and x_0 in time t, since a stationary point of the classical action defined on piecewise classical paths is clearly given by a strictly smooth classical path.

As usual, the case of the Schrödinger equation requires some regularisation. We shall use here the regularisation given by the Schrödinger equation in complex time (1.25), exploiting the semiclassical asymptotics for this equation constructed in Chapter 7. This will give us the required result more or less straightforwardly. However, to use this regularisation one needs rather strong assumption on the potential, namely local analyticity.

Notice that equation (1.25) can be considered as a particular case of equation (7.1.1.) with $\alpha = 0$. Therefore, Theorem 7.2.2 implies straightforwardly the following result.

Proposition 5.2. *Let V be analytic in the strip $St_b = \{x = y + iz : |z| < b\}$ and suppose that all its second and higher order derivatives are uniformly bounded in this strip. Let $S_\epsilon(t, x, x_0)$ be the two-point function for the complex Hamiltoian equation*

$$\frac{\partial S}{\partial t} + (1 - i\epsilon)\left[\frac{1}{2}(\frac{\partial S}{\partial x})^2 + V(x)\right] = 0 \qquad (5.6)$$

with the Hamiltonian $H_\epsilon = (1 - \epsilon i)(p^2/2 + V(x))$. Let χ_R be a smooth bounded function on the positive halfline such that $\chi(s)$ vanishes for $s > R$ and $\chi_R(s) = 1$ for $s \leq R - \delta$ for some $\delta > 0$. Then there exists t_0 such that for any R there exists ϵ_0 and $c > 0$ such that for all $t \leq t_0$, $\epsilon \leq \epsilon_0$ and $x, x_0 \in St_{b/2}$ with $|x - x_0| \leq R$, there exists a unique trajectory of the Hamiltonian system with Hamiltonian H_ϵ which joins x_0 and x in time t, lies completely in St_b, and has the initial momentum $p_0 : |p_0| \leq c/t$. Moreover, the Green function for equation (1.25) exists, and for $t \leq t_0$, $\epsilon \leq \epsilon_0$ it can be represented as the absolutely convergent series

$$u_G^\epsilon = u_{as}^\epsilon + h u_{as}^\epsilon \otimes F + h^2 u_{as}^\epsilon \otimes F \otimes F + ..., \qquad (5.7)$$

with

$$u_{as}^\epsilon = (2\pi hi)^{-m/2} \chi_R(|x - x_0|) J_\epsilon^{-1/2}(t, x, x_0) \exp\{\frac{i}{h} S_\epsilon(t, x, x_0)\}. \qquad (5.8)$$

and $F = \tilde{F} \exp\{\frac{i}{h} S_\epsilon(t, x, x_0)\}$, where $\tilde{F}(t, x, x_0)$ equals

$$[O(t^2) + O((th)^{-1})\Theta_{R-\epsilon,R}(|x - x_0|)](2\pi hi)^{-d/2}\chi_R(|x - x_0|)J_\epsilon^{-1/2}(t, x, x_0)$$

(the exact form of F can be found in Chapter 7).

The series (5.7) can again be interpreted as a path integral. Namely, as in the case of a real diffusion considered above we obtain the following.

Proposition 5.3. *The Green function (5.7) can be written in the form*

$$u_G^\epsilon(t, x, y) = \int_{CPC^{x,y}(0,t)} \exp\{\frac{i}{h} I_\epsilon(q(.))\} \Phi_{sem}^\epsilon(q(.)) M_{Leb}^{CPC}(dq(.)) \qquad (5.9)$$

where the path space CPC is defined by the real Hamiltonian system with the Hamiltonian H_0, but the action and the Jacobian are defined with respect to the complex trajectoties (solutions to the Hamiltonian system with the Hamiltonian H_ϵ) with the same end points as the corresponding real trajectories, and where

$$\Phi_{sem}^\epsilon(q_\eta^s) = (2\pi hi)^{-dn/2} J_\epsilon^{-1/2}(t - s_n, x, \eta_n)$$

$$\times \chi_R(|x - \eta_n|) h^n \prod_{k=1}^n \tilde{F}_\epsilon(s_k - s_{k-1}, \eta_k, \eta_{k-1}). \qquad (5.10)$$

Remark. The path space in (5.9) can also be defined as the set $CPC_\epsilon^{x,y}(0,t)$ of paths that are piecewise classical for the Hamiltonian H_ϵ. This seems to be more appropriate when considering equation (1.25) on its own, but we have defined this measure differently in order to have the same measure for all $\epsilon > 0$.

For $\epsilon = 0$ the integral in (5.9) is no longer absolutely convergent. However, under the assumptions of Proposition 5.2 the operator $-\Delta + V$ is selfadjoint (if V is real for real x). Therefore from (1.24) we obtain the following result.

Proposition 5.4. *Under the assumptions of Proposition 5.2, if V is real for real x, one has the following path integral representation for the solutions $\psi(t,x)$ of the Cauchy problem for equation (1.10) with the initial data $\psi_0 \in L^2(\mathcal{R}^d)$:*

$$\psi(t,x) = \lim_{\epsilon \to 0} \int_{CPC^{x,y}(0,t)} \exp\{\frac{i}{h} I_\epsilon(q(.))\} \Phi_{sem}^\epsilon(q(.)) \psi_0(y) M_{Leb}^{CPC}(dq(.)) dy.$$

6. Fock space representation

The paths of the spaces CRC and CPL used above are parametrised by finite sequences $(s_1, x_1), ..., (s_n, x_n)$ with $s_1 < ... < s_n$ and $x_j \in \mathcal{R}^d$, $j = 1, ..., d$. Denote by $\mathcal{P}^d$ the set of all these sequences and by $\mathcal{P}_n^d$ its subset consisting of sequences of the length n. Thus, functionals on the path spaces CPC or CPL can be considered as functions on $\mathcal{P}^d$. To each measure ν on $\mathcal{R}^d$ there corresponds a measure $\nu_\mathcal{P}$ on $\mathcal{P}^d$ which is the sum of the measures ν^n on $\mathcal{P}_n^d$, where ν^n are the product measures $ds_1...ds_n d\nu(x_1)...d\nu(x_n)$. The Hilbert space $L^2(\mathcal{P}^d, \nu_\mathcal{P})$ is known to be isomorphic to the Fock space Γ_ν^d over the Hilbert space $L^2(\mathcal{R}_+ \times \mathcal{R}^d, dx \times \nu)$ (which is isomorphic to the space of square integrable functions on $\mathcal{R}_+$ with values in $L^2(\mathcal{R}^d, \nu)$). Therefore, square integrable functionals on CPL can be considered as vectors in the Fock space $\Gamma_{V(dx)}^d$. It is well known that the Wiener, Poisson, general Lévy and many other interesting processes can be naturally realised in a Fock space: the corresponding probability space is defined as the spectrum of a commutative von Neumann algebra of bounded linear operators in this space. For example, the isomorphism between $\Gamma^0 = \Gamma(L^2(\mathcal{R}_+))$ and $L^2(W)$, where W is the Wiener space of continuous real functions on halfline is given by the Wiener chaos decomposition, and a construction of a Lévy process with the Lévy measure ν in the Fock space Γ_ν can be found in [Par] or [Mey]. Therefore, using Fock space representation, one can give different stochastic representations for path integrals over CPL or CPC rewriting them as expectations with respect to different stochastic processes.

For example, let us express the solution to the Cauchy problem of equation (1.26) in terms of an expectation with respect to a compound Poisson process. The following statement is a direct consequence of Proposition 4.2 and the standard properties of Poisson processes.

Proposition 6.1 *Suppose a measure V satisfies the assumptions of Proposition 4.2. Let $\lambda_V = V(\mathcal{R}^d)$. Let paths of CPL are parametrised by (3.1) and let E denote the expectation with respect to the process of jumps η_j which are identicaly independently distributed according to the probability measure V/λ_V*

and which occur at times s_j from $[0, t]$ that are distributed according to Poisson process of intensity λ_V. Then the function (4.4) can be written in the form

$$\psi_\epsilon(t, x) = e^{t\lambda_V} \int_{\mathcal{R}^d} \psi_0(y) E(\Phi_\epsilon(q(.))) \, dy. \tag{6.1}$$

As an example of the repersentation of path integral in terms of the Wiener measure, let us consider the Green function (3.5). Let us first rewrite it as the integral of an element of the Fock space $\Gamma^0 = L^2(Sim_t)$ with $Sim_t = \cup_{n=0}^{\infty} Sim_t^n$ (which was denoted $\mathcal{P}^0$ above), where Sim_t^n is as usual the simplex (1.14). Let $g_0^V = G_\epsilon^{free}$ (see (3.9)) and let

$$g_n^V(s_1, ..., s_n) = \int_{\mathcal{R}^{nd}} \Phi_\epsilon(q_{\eta_1...\eta_n}^{s_1...s_n}) \, d\eta_1 ... d\eta_n$$

for $n = 1, 2, ...$, where Φ_ϵ and $q_{\eta_1...\eta_n}^{s_1...s_n}$ are given by (3.3) and (3.1). Considering the series of functions $\{g_n^V\}$ as a single function g^V on Sim_t we shall rewrite the r.h.s. of (3.5) in the following concise notation:

$$\int_{Sim_t} g^V(s) \, ds = \sum_{n=0}^{\infty} \int_{Sim_t^n} g_n^V(s_1, ..., s_n) \, ds_1 ... ds_n. \tag{6.2}$$

Now, the Wiener chaos decomposition theorem states (see e.g. [Mey]) that, if $dW_{s_1}...dW_{s_n}$ denotes the n-dimensional stochastic Wiener differential, then to each $f = \{f_n\} \in L^2(Sim_t)$ there corresponds an element $\phi_f \in L^2(\Omega_t)$, where Ω_t is the Wiener space of continuous real functions on $[0, t]$, given by the formula

$$\phi_f(W) = \sum_{n=0}^{\infty} \int_{Sim_t^n} f_n(s_1, ..., s_n) \, dW_{s_1} ... dW_{s_n}, \tag{6.3}$$

or in concise notations

$$\phi_f(W) = \int_{Sim_t} f(s) \, dW_s.$$

Moreover the mapping $f \mapsto \phi_f$ is an isometric isomorphism, i.e.

$$E_W(\phi_f(W)\bar{\phi}_g(W)) = \int_{Sim_t} f(s)\bar{g}(s) \, ds, \tag{6.4}$$

where E_W denotes the expectation with respect to the standard Wiener process. One easily sees that under the assumptions of Theorem 3.1 the function g^V belongs not only to $L^1(Sim_t)$ (as shown in the proof of Theorem 3.1) but also to $L^2(Sim_t)$. Therefore, the function ϕ_{g^V} is well defined. Since (see e.g. again [Mey])

$$\int_{Sim_t} dW_s = e^{W(t)-t/2},$$

formula (6.4) implies the following result.

Proposition 6.2. *Under the assumptions of Theorem 3.1 the Green function (3.5) can be written in the form*

$$G_\epsilon(t, x, x_0) = E_W(\phi_g v \exp\{W(t) - t/2\}), \tag{6.5}$$

where E_W denotes the expectation with respect to the standard Wiener process.

It is not difficult to see that in order to similarly rewrite formula (4.4) in terms of the Wiener integral one needs a stronger assumption on the measure V than in Proposition 4.2: namely, one needs to assume that $dim(V) > d - 1$. For general V from Propositions 4.1 or 4.2, the corresponding function g^V may well belong to $L^1(Sim_t)$, but not to $L^2(Sim_t)$. In that case, formula (6.5) should be modified. Consider, for example the case of one-dimensional heat or Schrödinger equation with point interactions. Namely, consider the (formal) complex diffusion equation

$$\frac{\partial \psi}{\partial t} = G(\Delta/2 - \sum_{j=1}^{m} a_j \delta_{x_j}(x))\psi, \quad x \in \mathcal{R}, \tag{6.6}$$

where $a_1, ..., a_m$ are positive real numbers, $x_1, ..., x_m$ are some points on the real line, and G is a complex number with a non-negative real part. This is an equation of the type considered in Section 9.4. The path integral representation for its heat kernel from Proposition 4.1 can be written in the form

$$\psi_G^\delta(t, x, x_0) = \sum_{n=0}^{\infty} \int_{Sim_t^n} g_n^\delta(s_1, ..., s_n) \, ds_1...ds_n = \int_{Sim_t} g^\delta(s) \, ds, \tag{6.7}$$

where

$$g_0^\delta = \frac{1}{\sqrt{2\pi t}} \exp\{-\frac{(x - x_0)^2}{2t}\} \tag{6.8}$$

and

$$g_n^\delta(s_1, ..., s_n) = \prod_{k=1}^{n+1} \frac{1}{\sqrt{2\pi(s_k - s_{k-1})}} \sum_{j_1=1}^{m} ... \sum_{j_n=1}^{m}$$

$$\exp\{-\frac{(x - x_{j_n})^2}{2G(t - s_n)} - \frac{(x_{j_n} - x_{j_{n-1}})^2}{2G(s_n - s_{n-1})} - ... - \frac{(x_{j_1} - x_0)^2}{2Gs_1}\} \prod_{k=1}^{n} a_{j_k} \tag{6.9}$$

for $n \geq 1$, where it is assumed that $s_0 = 0$ and $s_{n+1} = t$.

Formula (6.7) has a clear probabilistic interpretation in the spirit of Proposition 6.1: it is an expectation with respect to the measure of the standard Poisson process of jump-times $s_1, ..., s_n$, of the sum of the exponentials of the classical actions $(2G)^{-1} \int \dot{q}^2(s) \, ds$ of all (essential) paths joining x_0 and x in time t each taken with a weight which corresponds to the weights of singularities x_j of the potential. On the other hand, since the function g^δ from (6.7) is not an element of $L^2(Sim_t)$, but only of $L^1(Sim_t)$, formula (6.5) does not hold.

One way to rewrite (6.7) in terms of an integral over the Wiener measure is by factorising g^δ in a product of two functions from $L^2(Sim_t)$. For example, in the case of the heat equation, i.e. when $G = 1$ in (6.6), the function g^δ is positive, which implies the following statement.

Proposition 6.3. *The Green function of equation (6.6) with $G = 1$ can be written in the form*

$$\psi_G^\delta(t, x, x_0) = E_W \left(\int_{Sim_t} \sqrt{g^\delta(s)} \, dW_s \right)^2 .$$

It is also possible to write a regularised version of (6.5). Namely, let $g^{\delta,\alpha}$ with $\alpha < 2$ is defined by (6.8),(6.9) where instead of the multipliers $\sqrt{t}$ and $\sqrt{s_j - s_{j-1}}$ one plugs in the multipliers t^α and $(s_j - s_{j-1})^\alpha$ respectively. If $\alpha \in (0, 1/2)$, the corresponding $g^{\delta,\alpha}$ belongs to $L^2(Sim_t)$ and therefore $\phi_{g^{\delta,\alpha}}$ is well defined and belongs to $L^2(W)$. This implies the following result.

Proposition 6.4. *The Green function ψ^δ of equation (6.6) has the form:*

$$\psi_G^\delta(t, x, x_0) = \lim_{\alpha \to 1/2} E_W \left(\exp\{W(t) - t/2\} \int_{Sim_t} g^{\delta,\alpha}(s) \, dW_s \right) .$$

Similar representations in terms of Poisson or Wiener processes can be given for the path integral over the path space CPL from Section 9.5. Let us notice also that Theorem 3.1 and Propositions 6.1, 6.2 can be easily modified to include the case of the Schrödinger equation for a quantum particle in a magnetic field with a bounded vector-potential.

APPENDICES

A. Main equation of the theory
of continuous quantum measurements

For the author, the main impetus to the development of the complex stochastic method WKB as presented in Chapter 7 was the recent appearance of the complex stochastic versions of the Schrödinger equation in the theory of continuous quantum measurements and, even more generally, in the theory of open quantum systems.. In the general form, the aposterior Schrödinger equation describing the evolution of the state of a quantum system subject to non-direct (but non-demolition) continuous observation of diffusion or counting type was obtained by V.P. Belavkin in [Be1], [Be2]. A simplest particular case corresponding to the continuous observation of a free quantum particle was obtained simultaneously by L. Diosi [Di]. Different approaches and points of view on this equation can be found e.g. in [BHH]. Since this fundamental equation is not till now presented in a monographic literature, we present here a simple self-contained physical deduction of this equation for the case of the continuous observations of diffusion type following [K9], [AKS2]. Mathematical properties of the solutions of this equation are discussed in Sect. 1.4, and Chapter 7.

Consider the continuous measurement of the position of a one-dimensional quantum particle described by the standard Hamiltonian $H = V(x) - \Delta/2$. The Postulates 1,2 given below are largely used in the literature on quantum continuous measurement (see e.g. [Di] and references therein).

Let a measurement at a time t of the position of the particle yields a certain value q.

Postulate 1. *Non-ideal measurement principle.* After such a measurement, the wave function $\psi_t(x)$ transforms into a new function which up to normalisation has the form

$$\psi_t(x) \exp\{-\alpha(x-q)^2\}, \qquad (A1)$$

where the coefficient α characterises the accuracy of the measurement.

This postulate can be deduced from a unitary evolution combined with the standard von Neumann (ideal) measurement postulate. Namely, let us measure the position x of our particle X indirectly by reading out (sharp and direct) the value of "analogous" physical variable y of a measuring apparatus Y (the position of a quantum meter), which is directly connected with (or directly influenced by) the considered quantum system X. More precisely, we consider the state of the quantum meter to be described by vectors in a Hilbert space $L_2(\mathcal{R})$, and the direct influence of X on Y means that the unitary evolution of the compound system in $L_2(\mathcal{R}) \otimes L_2(\mathcal{R}) = L_2(\mathcal{R}^2)$ (the interaction between the quantum particle and the quantum meter in the process of measurement) reduces to the shift of y by the value of x, i.e. it is described by the law $f(x,y) \mapsto f(x, y - x)$. Furthermore, we suppose that we can always prepare a fixed initial state of the quantum meter, say a Gaussian one, $\Phi^\alpha(y) = (2\alpha/\pi)^{1/4} \exp\{-\alpha y^2\}$. Thus the process of measurement is described in the following way. We have a state ϕ of our quantum particle and prepare an initial state Φ^α of the quantum meter Y. Then we switch on the interaction and get as a result the function $\phi(x)\Phi^\alpha(y-x)$ in $L_2(\mathcal{R}^2)$. At last we read out (directly and sharp) the value of the second variable, say $y = q$, which gives us (due to the standard von Neumann reduction postulate) the state (A1) of the considered system X.

Since the square of the magnitude of the state function $\phi(x)\Phi^\alpha(y-x)$ defines the density of the joint distribution of x and $y = q$, the probability density of the measured value q is equal to

$$P_\alpha^q(z) = \sqrt{\frac{2\alpha}{\pi}} \int \phi^2(x) \exp\{-2\alpha(z-x)^2\} \, dx.$$

Now we fix a time t and make non-ideal measurements of the position of the quantum particle at moments $t_k = k\delta$, where $\delta = t/n$, $n \in \mathcal{N}$.

Postulate 2. *Continuous limit of discrete observations.* As $n \to \infty$, the accuracy of each measurement will be proportional to the time between successive measurements, i.e. $\alpha = \delta\lambda$, where the constant λ reflects the properties of the measuring apparatus.

Supposing that between measurements the evolution of the quantum system is described by the law of free Hamiltonian, one concludes that after n measurements the resulting wave function will have the form

$$\psi_t^n = \prod_{j=1}^n \left(\exp\{-\delta\lambda(x-q_j)^2\} \exp\{-i\delta H\}\right) \psi_0,$$

where ψ_0 is the initial state. Applying (at least formally) the Trotter formula $\exp\{A+B\} = \lim_{n\to\infty}(\exp\{A/n\} \exp\{B/n\})^n$ for noncomuting operators A, B (more precisely, its non-homogeneous version) one sees that ψ_t^n, $n \to \infty$, tends to the solution of the equation

$$\dot\psi = -(iH + \lambda(x - q(t))^2)\psi, \qquad (A3)$$

where $q(\tau)$ is the function taking values q_j at times t_j. This is the Schrödinger equation with the complex (and time depending) potential $V(x) - i\lambda(x - q(t))^2$. Therefore, one can write its solution $\psi(t,x)$ in terms of the heuristic path integral

$$\int \psi_0(y)\,dy \int D\xi \exp\left\{i\int_0^t \left(\frac{\dot\xi^2(\tau)}{2} - V(\xi(\tau) + i\lambda|q(\tau) - \xi(\tau)|^2\right) d\tau\right\}, \quad (A4)$$

where the integral being taken over the set of all continuous paths $\xi(\tau)$ joining y and x in time t. This formula was proposed by Menski [Me] in 1979 for the case of quantum oscillator. As we shall see further, the "curve" $q(t)$ is very singular, and thus not only formal measure but already the expression $q(t)^2$ under the integral in (A4) is not well defined.

To make everything correct, let us first modify equation (A3). Observing that the term $\lambda q(t)^2\psi$ on the r.h.s. can be dropped, because the solutions of the equation with and without this term are proportional and this difference is irrelevant for the purposes of quantum mechanics (where we are interested only in normalised states), one gets the equation

$$\dot{\psi} + \left(iH + \lambda x^2\right)\psi = 2\lambda x q(t)\psi. \tag{A5}$$

Introducing the function $Q(t) = \int_0^t q(\tau)\, d\tau$, one can rewrite (A5) in equivalent differential form

$$d\psi + \left(iH + \lambda x^2\right)\psi\, dt = 2\lambda x \psi\, d_S Q(t), \tag{A6}$$

where d_S stands for the Stratonovich differential, which for smooth functions is just standard differential and for stochastic processes that will appear now it gives the limit of the differentials of their smooth approximations. Equation (A6) is formally the Belavkin aposterior linear equation in the Stratonovich form. But Belavkin's result states more, namely, the statistical properties of the trajectories of Q, which imply, in particular, that the function $q(t)$ in (A6) is a rather singular object that almost surely makes sense only as a distribution and not as a continuous curve. In order to see this singularity, notice that due to (A2), if the state function of our quantum particle X was $\psi(\tau, x)$ at the instant τ (in fact, $\tau = t_k$ for some k), then the expectation of the value q at the instant $\tau + \delta$ is given by the formula

$$Eq_{\tau+\delta} = \int z P_{\lambda\delta}^q(z)\, dz = \sqrt{\frac{2\lambda\delta}{\pi}} \int z\psi^2(\tau, x) \exp\{-2\lambda t(x - z)^2\}\, dx\, dz$$

and is equal to the mean value $\langle x \rangle_\tau$ of the position of the particle in the state $\psi(\tau, x)$. By similar calculations, one proves the dispersion of this random variable $E((q_{\tau+\delta} - Eq_{\tau+\delta})^2)$ to be of the form $(4\lambda\delta)^{-1}(1 + O(\delta))$ (at least for wave functions $\phi(\tau, x)$ with a finite second moment with respect to the corresponding probability distribution). Thus the random variable $q_{\tau+\delta} - \langle x \rangle_\tau$ has vanishing expectation and a dispersion (or the second moment) of the order $(4\lambda\delta)^{-1}$. Therefore, one can not consider $q(t)$ as a continuous curve. This object is very singular. The simplest model for the process of errors in the continuous observation (and also the most natural and commonly used in the classical stochastic theory of measurement) is the white noise $\dot{W}$, which can be defined as a formal derivative of the standard Wiener process $W(t)$. Taking into account that $(4\lambda\delta)^{-1}$ is the dispersion of the random value $(2\sqrt{\lambda\delta})^{-1}W(\delta)$, one arrives at the following

Postulate 3. *White noise model of errors.* The difference between $q_{t+\delta}$ and $\langle x \rangle_t$ is approximately equal to $(2\sqrt{\lambda\delta})^{-1}W(\delta)$ for small δ, or more precisely, for the process $Q(t) = \int_0^t q(\tau)\, d\tau$, one has

$$dQ(t) = \langle x \rangle_t\, dt + \frac{1}{2\sqrt{\lambda}}\, dW(t). \tag{A7}$$

Rewriting (A6) in terms of the Ito differential of the process $B(t) = 2\sqrt{\lambda}Q(t)$ (Ito's differential dB is connected with the Stratonovich differential $d_S B$ by the formula $\psi \, d_S B = \psi \, dB + d\psi \, dB/2$) one gets the Belavkin aposterior linear equation [Be2]:

$$d\psi + \left(\frac{1}{2i}\Delta + iV(x) + \frac{\lambda}{2}x^2 \right) \psi \, dt = \sqrt{\lambda}x\psi \, dB(t). \tag{A8}$$

Application of Ito's formula yields for the normalised state $\phi = \psi/\|\psi\|$ a nonlinear aposterior Schrödinger equation, which written in terms of the white noise W has the form

$$d\phi + \left(\frac{1}{2i}\Delta + iV(x) + \frac{\lambda}{2}(x - \langle x \rangle_\phi)^2 \right) \phi \, dt = \sqrt{\lambda}(x - \langle x \rangle_\phi)\phi \, dB(t), \tag{A9}$$

where $\langle x \rangle_\phi$ denotes the mean position in the state ϕ. This equation was written first in 1988 by Belavkin [Be1] in an essentially more general form and by Diosi [Di] for the case of vanishing potential V. The corresponding equations for the mean position and momentum in the case of quantum oscillator had appeared in [Be3].

Notice that if the state vector evolves in time according to equation (A9), then for the expectation (with respect to the Wiener measure of the process W) of the corresponding density matrix $\rho = E(\phi \otimes \bar{\phi})$, one gets directly (using Ito's formula) the equation

$$\dot{\rho} = -i[H, \rho] - \frac{\lambda}{2}[R, [R, \rho]], \tag{A10}$$

which is the famous master equation.

B. Asymptotics of Laplace integrals with complex phase

Here we present the estimates of the remainder in the asymptotic formula for the Laplace integrals with complex phase, i.e. for the integral

$$I(h) = \int_\Omega f(x) \exp\{-S(x)/h\} \, dx, \tag{B1}$$

where Ω is any closed subset of the Euclidean space $\mathcal{R}^d$, the amplitude f and the phase S are continuous complex-valued functions on Ω and $h \in (0, h_0]$ with some positive h_0.

To begin with, let us recall two trivial estimates. First, if $\inf_\Omega \{S(x)\} \geq M$, then obviously

$$I(h) \leq \exp\{-M/h\} \exp\{M/h_0\} \tilde{I}(h_0), \tag{B2}$$

where

$$\tilde{I}(h) = \int_\Omega |f(x)| \exp\{-\operatorname{Re} S(x)/h\} \, dx.$$

Next, let $Re\,S(x)$ be a convex twice differentiable function with $Re\,S''(x) \geq \lambda$ for all x and some positive matrix Λ, and let M be the global minimum of S in $\mathcal{R}^d$, then

$$I(h) \leq \exp\{-M/h\}(2\pi h)^{d/2}(\det \Lambda)^{-1/2}\sup\{|f(x)|\}. \qquad (B3)$$

This formula follows from the estimate $Re\,S(x) \geq M + (\lambda x, x)/2$ and the integration of a Gaussian function.

Let us make now the following assumptions:

(1) integral (B1) is absolutely convergent for $h = h_0$, i.e. $\tilde{I}(h_0) < \infty$;

(2) $S(x)$ is thrice continuously differentiable;

(3) Ω contains a neighbourhood of the origin, $Re\,S(x) > 0$ for $x \neq 0$ and $S(0) = 0$;

(4) $S'(0) = 0$ and $Re\,S''(0)$ is strictly positive;

(5) $\liminf_{x \to \infty,\, x \in \Omega} Re\,S(x) > 0$.

The existence of the asymptotic expansion as $h \to 0$ of integral (B1) for infinitely smooth functions f and S under assumptions (1)-(5) and the (very complicated) recurrent formulas for its coefficients are well known (see e.g. [Fed1]). We are going to present here only the principle term of this expansion but with an explicit estimate for the remainder depending on the finite number of the derivatives of f and S. For real f and S this estimate is an improved and simplified version of the estimate given in [DKM1] and [K10].

Assumptions (3)-(5) imply the existence of a positive r such that

(6) $\inf\{ReS(x) : x \in \Omega \setminus B_r\} = \min\{ReS(x) : x \in \partial B_r\}$,

(7) $ReS''(x) \geq \Lambda$ for all $x \in B_r$ and some positive real Λ.

Let

$$U(h) = \{x : (Re\,S''(0)x, x) \leq h^{2/3}\}.$$

Reducing if necessarily h_0 one can ensure that

(8) $U(h_0) \subset B_r$,

which implies in particular, due to (6),(7), that

$$\inf\{Re\,S(x) : x \in \Omega \setminus U(h)\} = \min\{Re\,S(x) : x \in \partial U(h)\}. \qquad (B4)$$

Remark. Assumption (8) (which is our only assumption of "smallness" of h) will be used further only for brevity. Without this assumption all formulas are essentially the same, which one proves using the neighbourhood $U(h) \cap B_r$ instead of $U(h)$ everywhere in the arguments.

By λ we denote the minimal eigenvalue of the matrix $Re\,S''(0)$. Furthermore, F_j and S_j denote the maximum in B_r of the norms of the j-th order derivatives (whenever they exist) of f and S respectively. At last, let

$$A = \frac{1}{6}S_3\lambda^{-3/2}, \quad D_R = \det Re\,S''(0), \quad D = \det S''(0).$$

Proposition B1. *Under assumptions (1)-(8)*

$$|I(h)| \leq F_0 e^A \left[(2\pi h)^{d/2} D_R^{-1/2} + (2\pi h_0/\Lambda)^{d/2}\exp\{-\frac{1}{2}(h^{-1/3} - h_0^{-1/3})\} \right]$$

$$+\exp\{-\frac{r^2\Lambda}{2h}\}\exp\{\frac{r^2\Lambda}{2h_0}\}\tilde{I}(h_0). \qquad (B5)$$

Proof. From (2)-(4) it follows that in $U(h)$

$$S(x) = \frac{1}{2}(S''(0)x, x) - \sigma(x), \qquad (B6)$$

where

$$|\sigma(x)| \le \frac{1}{6}S_3|x|^3. \qquad (B7)$$

Since obviously the ellipsoid $U(h)$ belongs to the ball of the radius $\lambda^{-1/2}h^{1/3}$ (centred at the origin), (B7) implies

$$|\sigma(x)| \le \frac{1}{6}S_3\lambda^{-3/2}h = Ah. \qquad (B8)$$

Let us now split the integral $I(h)$ in the sum $I(h) = I'(h) + I''(h) + I'''(h)$ of the integrals over the domains $U(h)$, $B_r \setminus U(h)$, and $\Omega \setminus B_r$, respectively. Due to (B4),(B6),(B8), $Re\, S(x) \ge \frac{1}{2}h^{2/3} - Ah$ outside $U(h)$, and using (B2),(B3) one gets the estimate

$$|I''(h)| \le \exp\{-\frac{1}{2}h^{-1/3}\}\exp\{A + \frac{1}{2}h_0^{-1/3}\}F_0(2\pi h_0/\Lambda)^{d/2}. \qquad (B9)$$

Due to (6),(7), $ReS(x) \ge \Lambda r^2/2$ outside B_r, and again using (B2) one gets

$$I'''(h) \le \exp\{-\frac{r^2\Lambda}{2h}\}\exp\{\frac{r^2\Lambda}{2h_0}\}\tilde{I}(h_0). \qquad (B10)$$

On the other hand,

$$|I'(h)| \le F_0 e^A \int_{U(h)} \exp\{-\frac{1}{2h}(Re\, S''(0)x, x)\}\, dx \le F_0 e^A (2\pi h)^{d/2} D_R^{-1/2}.$$

The last three inequalities complete the proof.

Proposition B2. *Let the assumptions of Proposition B1 hold and moreover, the function f (resp. S) is two times (respectively four times) continuously differentiable. Then*

$$I(h) = (2\pi h)^{d/2}(f(0)D^{-1/2} + h(D_R^{-1/2}\delta_1(h) + \Lambda^{-d/2}\delta_2(h)) + \delta_3(h), \qquad (B11)$$

where

$$|\delta_1(h)| \le C_1(d)(F_0 A^2 e^A + F_0 S_4\lambda^{-2} + F_2(1 + A)\lambda^{-1} + F_1 S_3\lambda^{-2}), \qquad (B12)$$

$$|\delta_2(h)| \le C_2(d, h_0)F_0 e^A, \quad |\delta_3(h)| \le \exp\{-\frac{r^2\Lambda}{2h}\}\exp\{\frac{r^2\Lambda}{2h_0}\}\tilde{I}(h_0) \qquad (B13)$$

with $C_1(d), C_2(d, h_0)$ being constants depending continuously on d and on d, h_0 respectively (and which can be written explicitly, see the estimates in the proof below).

Proof. We use the notations introduced in the proof of Proposition B1. Let us present the integral $I'(h)$ as the sum $I_1(h) + I_2(h)$ with

$$I_1(h) = \int_{U(h)} f(x) \exp\{-\frac{1}{2h}(S''(0)x, x)\} \left(e^{\sigma/h} - 1 - \frac{\sigma}{h}\right) dx, \qquad (B14)$$

$$I_2(h) = \int_{U(h)} f(x) \exp\{-\frac{1}{2h}(S''(0)x, x)\} \left(1 + \frac{\sigma}{h}\right) dx. \qquad (B15)$$

Due to (B7),(B8),

$$|e^{\sigma/h} - 1 - \frac{\sigma}{h}| \le \frac{1}{2}(|\sigma|/h)^2 e^{|\sigma|/h} \le \frac{1}{2h^2}(S_3/6)^2|x|^6 e^A \le \frac{e^A S_3^2}{72\,h^2\lambda^3}(Re\,S''(0)x, x)^3$$

in $U(h)$. Hence

$$|I_1(h)| \le \frac{e^A S_3^2}{72\,h^2\lambda^3} F_0 \int_{\mathcal{R}^d} (Re\,S''(0)x, x)^3 \exp\{-\frac{1}{2h}(Re\,S''(0)x, x)\}\, dx$$

$$= \frac{e^A S_3^2}{9\lambda^3} F_0 (2\pi h)^{d/2} \frac{\Gamma(3 + d/2)}{\Gamma(d/2)} h D_R^{-1/2},$$

since

$$\int_{\mathcal{R}^d} |y|^k e^{-|y|^2/2}\, dy = (2\pi)^{d/2} 2^{k/2} \frac{\Gamma((k+d)/2)}{\Gamma(d/2)}.$$

To evaluate $I_2(h)$ we first take the two terms of the Taylor expansion of $\sigma(x)$ presenting $I_2(h)$ as the sum $J_1 + \Delta_1$ with

$$J_1 = \int_{U(h)} f(x) \exp\{-\frac{1}{2h}(S''(0)x, x)\} \left(1 + \frac{1}{6h}(S^{(3)}(0))(x, x, x)\right) dx,$$

$$|\Delta_1| \le \frac{F_0 S_4}{4!h} \int_{U(h)} |x|^4 \exp\{-\frac{1}{2h}(Re\,S''(0)x, x)\}$$

$$\le \frac{F_0 S_4}{6\lambda^2} \frac{\Gamma(2 + d/2)}{\Gamma(d/2)} (2\pi h)^{d/2} D_R^{-1/2} h.$$

Taking three terms of the Taylor expansion of $f(x)$ we write further $J_1 = J_2 + \Delta_2$ with

$$J_2 = \int_{U(h)} (f(0) + f'(0)x) \left(1 + \frac{1}{6h}(S(3)(0))(x, x, x)\right) \exp\{-\frac{1}{2h}(S''(0)x, x)\},$$

$$|\Delta_2| \le \frac{1}{2} F_2(1 + A) \int_{U(h)} |x|^2 \exp\{-\frac{1}{2h}(Re\,S''(0)x, x)\}$$

$$\leq \frac{F_2(1+A)}{\lambda} \frac{\Gamma(1+d/2)}{\Gamma(d/2)} (2\pi h)^{d/2} D_R^{-1/2} h.$$

It remains to calculate the integral J_2. Since

$$\int_{U(h)} g(x) \exp\{-\frac{1}{2h}(S''(0)x, x)\} \, dx = 0$$

for any polylinear form $g(x)$ of an odd order, one can write $J_2 = J_3 + \Delta_3$ with

$$J_3 = \int_{U(h)} f(0) \exp\{-\frac{1}{2h}(S''(0)x, x)\} \, dx,$$

$$|\Delta_3| \leq I_2(h) = \frac{F_1 S_3}{6h} \int_{U(h)} |x|^4 \exp\{-\frac{1}{2h}(Re\, S''(0)x, x)\} \, dx$$

$$\leq \frac{2F_1 S_3}{3\lambda^2} \frac{\Gamma(2+d/2)}{\Gamma(d/2)}.$$

Now present the integral J_3 as the difference $J_4 - \Delta_4$ of the integrals over the whole space $\mathcal{R}^d$ and over $\mathcal{R}^d \setminus U(h)$. The first integral J_4 can be calculated explicitly and is equal to the principle term

$$J_4 = (2\pi h)^{d/2} f(0) D^{-1/2}$$

of the asymptotic expansion (B11). The second integral can be estimated in the same way as $I''(h)$ above:

$$|\Delta_4| \leq |f(0)| \int_{\mathcal{R}^d \setminus U(h)} \exp\{-\frac{1}{2h}(Re\, S''(0)x, x)\} \, dx$$

$$\leq |f(x_0)| \exp\{-\frac{1}{2}h^{-1/3}\} \exp\{\frac{1}{2}h_0^{-1/3}\}(2\pi h_0)^{d/2} D_R^{-1/2}. \qquad (B16)$$

Estimates for $I_1(h), \Delta_1, \Delta_2, \Delta_3$ contribute to $\delta_1(h)$ in (B11); estimates (B9), (B16) contribute to $\delta_2(h)$, where $C_2(d, h)$ is chosen in such a way that

$$2(2\pi h_0)^{d/2} \exp\{-\frac{1}{2}(h^{-1/3} - h_0^{-1/3}) \leq C_2(d, h_0)(2\pi h)^{d/2} h,$$

and estimate (B10) contributes to $\delta_3(h)$. The Proposition is proved.

Let us give also for completeness a more rough estimate of the remainder, which can be useful when the values of F_2 and S_4 are not available.

Proposition B3. *Let the assumptions of Proposition B1 hold and let f be continuously differentiable. Then*

$$I(h) = (2\pi h)^{d/2} \left(f(0) D^{-1/2} + \sqrt{h} D_R^{-1/2} \delta_1(h) + h \Lambda^{-d/2} \delta_2(h) \right) + \delta_3(h),$$

where $\delta_2(h), \delta_3(h)$ satisfy (B13) and

$$|\delta_1(h)| \le C(F_0 S_3 e^A \lambda^{-3/2} + F_1 \lambda^{-1/2}).$$

Proof. It is the same as above but essentially simpler. One also presents $I'(h)$ as the sum $I_1(h) + I_2(h)$ but with

$$I_1(h) = \int_{U(h)} f(x) \exp\{-\frac{1}{2h}(S''(0)x, x)\}(e^{\sigma/h} - 1)\, dx,$$

$$I_2(h) = \int_{U(h)} f(x) \exp\{-\frac{1}{2h}(S''(0)x, x)\}\, dx$$

instead of (B14), (B15), and then makes similar estimates.

As we mentioned, there exist recursive formulas for calculating the coefficients of the whole asymptotic series in powers of h for the integral $I(h)$. However, these formulas are too complicated (especially if the dimension is greater than one) to be of practical use in the most of situations one encounters. Nevertheless, the second term of the expansion begins to be vitally important, if the major term vanishes. We present now a relevant result (with the rough estimates to the remainder, similar to those given in Proposition B3).

Proposition B4. *Let the assumptions of Proposition B2 hold, and moreover, let f be thrice continuously differentiable and $f(0) = 0$. Then*

$$I(h) = (2\pi h)^{d/2}\frac{h}{2}\left[(T_1 - T_2)D^{-1/2} + \sqrt{h}(D_r^{-1/2}\delta_1(h) + \Lambda^{-d/2}\delta_2(h))\right] + \delta_3(h),$$
$$(B17)$$

where $\delta_2(h), \delta_3(h)$ are the same as in Proposition B2 (only with different constants),

$$|\delta_1(h)| \le C_1(d, h_0)\left[e^A A(1 + A)\lambda^{-1/2} F_1\right.$$

$$\left. + S_4 F_1 \lambda^{-5/2} + F_3 \lambda^{-3/2} + (1 + A)F_2 \lambda^{-1}\right],$$
$$(B18)$$

and the coefficients of the main term are

$$T_1 = tr\,(f''(0)(S''(0))^{-1}), \tag{B19}$$

$$T_2 = (f'(0) \otimes S'''(0), (S''(0))^{-1} \otimes (S''(0))^{-1})$$

$$\equiv \frac{\partial f}{\partial x_i}(0)\frac{\partial^3 S}{\partial x_j \partial x_k \partial x_l}(0)(S''(0))_{ij}^{-1}(S''(0))^{-1})_{kl}. \tag{B20}$$

Proof. Following the notations and the lines of the proof of Proposition B2, we are lead to the necessity to estimate the integrals $I_1(h)$ and $I_2(h)$ given by (B14), (B15). Noticing that under the assumptions of the Proposition $|f(x)| \le F_1|x|$ and estimating $I_1(h)$ by the same method as in Proposition B2 one obtains

$$|I_1(h)| \le \frac{e^A S_3^2}{72\,h^2\lambda^{7/2}} F_1 \int_{\mathcal{R}^d} (Re\,S''(0)x, x)^{7/2} \exp\{-\frac{1}{2h}(Re\,S''(0)x, x)\}\, dx$$

$$= \frac{e^A S_3^2}{9\lambda^{7/2}} F_1 (2\pi h)^{d/2} 2^{1/2} \frac{\Gamma((7+d)/2)}{\Gamma(d/2)} h^{3/2} D_R^{-1/2}$$

$$= O(h^{3/2}) e^A A^2 \lambda^{-1/2} F_1 D_R^{-1/2} (2\pi h)^{d/2},$$

which contributes to the first term in the expression for $\delta_1(h)$. Furthermore, similar to the proof of Proposition B2, we write $I_2(h) = J_1 + \Delta_1$ with the same J_1, but for for Δ_1 one obtains now the estimate

$$\Delta_1(h) = O(h^{3/2}) F_1 S_4 \lambda^{-5/2} (2\pi h)^{d/2} D_R^{-1/2},$$

which contributes to the second term in (B18). Next, using for $f(x)$ the Taylor expansion till the third order, one presents J_1 as the sum $J_2 - J_3 + \Delta_2$ with

$$J_2 = \int_{U(h)} \frac{1}{2}(f''(0)x, x) \exp\{-\frac{1}{2h}(S''(0)x, x)\} \, dx,$$

$$J_3 = \int_{U(h)} \frac{1}{6h}(f'(0), x)(S^{(3)}(0))(x, x, x) \exp\{-\frac{1}{2h}(S''(0)x, x)\} \, dx,$$

and the remainder

$$\Delta_2 = \int_{U(h)} \left(\frac{1}{6} F_3 |x|^3 + \frac{1}{12h} F_2 S_3 |x|^5\right) \exp\{-\frac{1}{2h}(\text{Re } S''(0)x, x)\} \, dx,$$

$$= O(h^{3/2})(F_3 \lambda^{-3/2} + F_2 S_3 \lambda^{-5/2})(2\pi h)^{d/2} D_R^{-1/2},$$

which contributes to the third and fourth terms of (B18). Notice now that if in the expressions for J_2 and J_3 one integrates over the whole space instead of over $U(h)$, one will have the differences which are exponentially small in h and proportional to $F_2 \lambda^{-1} D_R^{-1/2}$ and $F_1 \lambda^{-2} S_3 D_R^{-1/2}$ respectively, and which can be therefore included in the first and the last terms in (B18) (with an appropriate $C_1(d, h_0)$). Hence, it remains to estimate the integral of form J_2, J_3 over the whole space, which after the change of the variable x to $y = x/\sqrt{h}$ can be written in the form

$$\tilde{J}_2 = \int_d h^{d/2} h \frac{1}{2}(f''(0)y, y) \exp\{-\frac{1}{2}(S''(0)y, y)\} \, dy,$$

$$\tilde{J}_3 = \int_{\mathcal{R}^d} h^{d/2} h \frac{1}{6}(f'(0), x)(S^{(3)}(0))(y, y, y) \exp\{-\frac{1}{2}(S''(0)y, y)\} \, dy.$$

The explicit calculation of these integrals, which can be carried out by changing the variable y to $z = \sqrt{S''(0)}y$ (in the case of complex $S''(0)$ one can justify this change by rotating the contour of integration in the complex space $\mathcal{C}^d$ and using the Cauchy theorem) yields the main terms in (B17).

Remark. Assuming the existence of five continuous derivatives of f and six continuous derivatives of S in Proposition B4 one can get the remainder of the order $O(h)$ (and not $O(\sqrt{h})$) in (B17), the corresponding calculations however becoming much heavier.

In this book we encounter also the Laplace integrals in the case when the phase has its minimum on the boundary of the domain of integration. As before, the general recursive formulas for the asymptotic expansions of such integrals are well known (see e.g. [Fed1],[Mu]). But we are interested here only in a simple case which is enough for our purposes, namely in the case of linear phase. For this case we are going to present now explicit formulas for the main terms and the estimates of the remainder.

Let us start the discussion with the trivial case of one-dimensional integral. If $f(x)$ is a continuously differentiable function on the interval $[a, b]$, and $c > 0$, then for any $\delta \in (0, b - a)$

$$\int_a^b e^{-cx} f(x)\, dx = \int_a^{a+\delta} e^{-cx}(f(a) + f'(a + \theta(x)\delta)(x - a))\, dx + \int_{a+\delta}^b e^{-cx} f(x)\, dx$$

with some $\theta(x) \in [0, 1]$, and therefore

$$\int_a^b e^{-cx} f(x)\, dx = c^{-1} e^{-ca}(f(a) + c^{-1}\delta_1) + 2\delta_0 c^{-1} e^{-c(a+\delta)}, \qquad (B21)$$

where $|\delta_0|$ does not exceed the maximal magnitude of f on $[a, b]$ and $|\delta_1|$ does not exceed the maximal magnitude of f' on $[a, a + \delta]$.

Now let M be a convex compact set in $\mathcal{R}^d$ with a smooth (thrice continuously differentiable) boundary ∂M having positive Gaussian curvature $\Gamma(x)$ at any point $x \in \partial M$. Notice that since ∂M is compact, it follows that all main curvatures do not approach zero on ∂M. Consider the integral

$$I(\bar{p}) = \int_M \exp\{-\frac{1}{h}(\bar{p}, x)\} f(x)\, dx, \qquad (B22)$$

where $\bar{p}$ is a unit vector, $h > 0$. Let x_0 be a point, where the phase $(\bar{p}, x)$ takes its minimal value $(\bar{p}, x_0)$. Clearly (for instance, from the Lagrange principle) x_0 is uniquely defined and $\bar{p}$ is the unit vector of the inner normal to ∂M at x_0. As before we denote by F_j the maximum of the norms of the derivative $f^{(j)}$ in a neighbourhood of x_0.

Proposition B5. *(i) Generally one has*

$$I(\bar{p}) = (2\pi)^{(d-1)/2} h^{(d+1)/2} \Gamma(x_0)^{-1/2} \exp\{-\frac{1}{h}(\bar{p}, x_0)\}[f(x_0) + O(h)(F_0 + F_1 + F_2)]$$

$$+ O(\max_M |f(x)|) \exp\{-\frac{1}{h}[(\bar{p}, x_0) + \delta]\}, \qquad (B23)$$

where δ is some positive number and $O(h)$ is uniform with respect to $\bar{p}$.
   (ii) If $f(x_0) = f'(x_0) = 0$, then

$$I(\bar{p}) = (2\pi)^{(d-1)/2} h^{(d+1)/2} \Gamma(x_0)^{-1/2} \frac{h}{2} \exp\{-\frac{1}{h}(\bar{p}, x_0)\}[tr\,(\tilde{f}''(x_0)\Phi^{-1}(x_0))$$

$$+O(\sqrt{h})(F_0 + F_1 + F_2 + F_3)] + O(\max_M |f(x)|) \exp\{-\frac{1}{h}[(\bar{p}, x_0) + \delta]\}, \quad (B24)$$

where $\Phi(x_0)$ is the matrix of the second form of ∂M at x_0 and $\tilde{f}$ is the function f restricted to the tangent space to ∂M at x_0 (or to the boundary ∂M itself, which gives the same result, since x_0 is suppose to be a critical point of f).

Proof. Consider an orthonormal system of coordinates $y = (y_1, ..., y_{d-1}, y_d)$ in $\mathcal{R}^d$ such that x_0 is the origin and $\bar{p}$ has coordinates $(0, ..., 0, 1)$. Then ∂M around x_0 can be described by the equation $y_d = \phi(y')$, $y' = (y_1, ..., y_{d-1})$ with some smooth function ϕ. Consequently, for any $\delta > 0$, one has that up to an exponentially small term

$$I(\bar{p}) = \exp\{-\frac{1}{h}(\bar{p}, x_0)\} \int_{\{|y_d|)\leq\delta\}} \exp\{-\frac{y_d}{h}\} f(y) \, dy$$

$$= \exp\{-\frac{1}{h}(\bar{p}, x_0)\} \int_U g(y') \, dy,'$$

where

$$g(y') = \int_{\phi(y')}^{\delta} \exp\{-\frac{y_d}{h}\} f(y', y_d) \, dy_d,$$

and U is a neighbourhood of the origin in $\mathcal{R}^d$ such that $\delta > \phi(y')$ for $y' \in U$. Consequently, due to (B21), up to an exponentially small term

$$I(\bar{p}) = \exp\{-\frac{1}{h}(\bar{p}, x_0)\} \in_U \exp\{-\frac{\phi(y')}{h}\}(f(y', \phi(y')) + hO(F_1)) \, dy'. \quad (B25)$$

From the definition of $\phi(y')$ it follows that the matrix of its second derivatives $\phi''(0)$ at the origin is just the matrix $\Phi(x_0)$ of the second main form of the hypersurface ∂M at x_0 (in coordinate y), the eigenvalues of this matrix are the main curvatures of ∂M at x_0, and $\Gamma(x_0) = \det \phi''(0)$ is the Gaussian curvature. Consequently, applying Proposition B2 to integral (B25) yields (B23). It remains to notice that under condition of statement (ii), one can also write $O(|y'|)F_2$ instead of $O(F_1)$ in (B25), and consequently, applying Proposition B4 to integral (B25) yields (B24).

Sometimes one encounters the integrals depending on a small parameter h in a more complicated way than in (B1). Let us formulate one result on such situation, where the asymptotics and its justification can be obtained by direct generalisation of the arguments of Propositions B1-B4. Let $h \in (0, h_0]$ as usual, and let

$$I(h) = \int_\Omega \exp\{-\frac{1}{h}S(x, h)\} \, dx. \quad (B26)$$

Generalising the asssumptions of Propositions B1,B2 suppose that
(i) $I(h)$ is absolutely convergent for $h = h_0$;
(ii) the function S is four times differential in x and h;

(iii) for any $h \in (0, h_0]$ there exists a unique point $x(h) \in \Omega$ such that $S(x(h), h) = 0$, $S'(x(h), h) = 0$, and $ReS''(x(h), h) > 0$, where by primes we denote the derivatives with respect to x;

(iv) $ReS(x, h) > 0$ whenever $x \neq x(h)$, $\liminf_{x \to \infty} ReS(x, h) > 0$, and the set of the internal points of Ω contains the closure of the set of all $x(h)$, $h \leq h_0$;

(v) there exists r such that $\inf\{ReS(x, h) : x \in \Omega \setminus B_r\} = \min\{ReS(x, h) : x \in \partial B_r\}$;

(vi) there exists $\Lambda > 0$ such that $ReS''(x(h), h) > \Lambda$ for all $x \in B_r$ and all h;

(vii) $U(h_0) \subset B_r$, where

$$U(h) = \{x : (ReS''(x(h), h)(x - x(h)), x - x(h)) \leq h^{2/3}\}.$$

Let $\lambda(h)$ denote the minimal eigenvalue of $ReS''(x(h), h)$, and let S_j denote the maximum in $x \in B_r$, $h \leq h_0$, of the norms of the j-th order derivatives of S with respect to x. Let

$$A(h) = \frac{1}{6}S_3(\lambda(h))^{-3/2}, \quad D_R(h) = \det ReS''(x(h), h), \quad D(h) = \det S''(x(h), h).$$

Proposition B6. *Under these assumptions one has the estimate*

$$|I(h)| \leq \delta_3(h)$$

$$+ e^{A(h)}\left[(2\pi h)^{d/2}D_R(h)^{-1/2} + (2\pi h_0 \Lambda^{-1})^{d/2}\exp\{-\frac{1}{2}(h^{-1/3} - h_0^{-1/3})\}\right],$$
$$(B27)$$

where

$$|\delta_3(h)| \leq \left|\int_{\Omega \setminus B_r} \exp\{-\frac{S(x, h)}{h}\}\,dx\right|$$

$$\leq \exp\{-\frac{r^2 \Lambda}{2h}\}\exp\{\frac{r^2 \Lambda}{2h_0}\}\int_{\Omega}|\exp\{-\frac{S(x, h_0)}{h_0}\}|\,dx,$$

and a more precise formula

$$I(h) = (2\pi h)^{d/2}\left((D(h))^{-1/2} + h((D_R(h))^{-1/2}\delta_1(h) + \Lambda^{-D/2}\delta_2(h))\right) + \delta_3(h),$$
$$(B28)$$

where

$$|\delta_1(h)| \leq C_1(d)(A^2(h)e^{A(h)} + S_4(\lambda(h))^{-2}), \quad |\delta_2(h)| \leq C_2(d, h_0)e^{A(h)}.$$

Notice for conclusion that the phase function S in Propositions B2-B4 was not supposed to be analytic. However, in the case of complex S one usually deals with the Laplace integrals over some contour of integration in the complex space C^d with analytic f and S and the major problem that one encounters is to find a deformation (using the Cauchy theorem) of this contour in such a way that the resulting integral satisfies the conditions of one of the Propositions B1-B6.

C. Characteristic functions of stable laws

This Appendix is devoted mainly to a compact exposition of the standard facts about the characteristic functions of the infinitely divisible distributions and the stable laws. At the end we prove some simple statements on the asymptotic behaviour in the complex domain of the characteristic functions of the localised stable laws disturbed by a compound Poisson process. These results are used in Chapter 6.

A probability distribution in $\mathcal{R}^d$ and its characteristic function $\psi(x)$ are called infinitely divisible, if for any integer n the function $\psi^{1/n}$ is again a characteristic function (of some other distribution). The famous Lévy-Khintchine theorem states (see e.g. [Fel],[GK]) that the logarithm of the characteristic function of an infinitely divisible distribution $\Phi(y) = \log \psi(y)$ can be presented in the following canonical form:

$$\Phi(y) = \log \psi(y) = iAy - \frac{1}{2}(Gy, y) + \int_{\mathcal{R}^d \setminus \{0\}} \left(e^{i(y,\xi)} - 1 - \frac{i(y,\xi)}{1+\xi^2} \right) \nu(d\xi), \quad (C1)$$

where A and G are respectively a real vector and a nonnegative real matrix, and ν is a so called Lévy measure on $\mathcal{R}^d \setminus \{0\}$, which means that

$$\int_{\mathcal{R}^d \setminus \{0\}} \min(1, |\xi|^2)\, \nu(d\xi) < \infty. \quad (C2)$$

An important class of the infinitely divisible distributions is given by the so called compound Poisson distributions. Their characteristic functions are given by (C1) with vanishing G and a finite Lévy measure. In particular, for these distributions

$$\log \psi(y) = iAy + \int_{\mathcal{R}^d \setminus \{0\}} \left(e^{i(y,\xi)} - 1 \right) \nu(d\xi). \quad (C3)$$

A different class of infinitely divisible distributions constitute the so called stable laws. A probability distribution in $\mathcal{R}^d$ and its characteristic function $\psi(x)$ are called stable (resp. strictly stable), if for any integer n there exist a positive constant c_n and a real constant γ_n (resp. if additionally $\gamma_n = 0$) such that

$$\psi(y) = [\psi(y/c_n) \exp\{i\gamma_n y\}]^n.$$

Obviously, it implies that ψ is infinitely divisible and therefore $\log \psi$ can be presented by (C1) with appropriate A, G, ν. It turns out (see e.g. [Fel],[Lu],[ST]) that if ψ is stable, then there exists an $\alpha \in (0, 2]$ which is called the index of stability such that: (i) if $\alpha = 2$, then $\nu = 0$ in the representation (C1) of $\log \psi$, i.e. the distribution is normal; (ii) if $\alpha \in (0, 2)$, then in the representation (C1) the matrix G vanishes and the radial part of the Lévy measure ν has the form $|\xi|^{-(1+\alpha)}$, i.e.

$$\log \psi_\alpha(y) = i(A, y) + \int_0^\infty \int_{S^{d-1}} \left(e^{i(y,\xi)} - 1 - \frac{i(y,\xi)}{1+\xi^2} \right) \frac{d|\xi|}{|\xi|^{1+\alpha}} \mu(ds), \quad (C4)$$

where ξ is presented by its magnitude $|\xi|$ and the unit vector $s = \xi/|\xi| \in S^{d-1}$ in the direction ξ, and μ is some (finite) measure in S^{d-1}.

The integration in $|\xi|$ in (C4) can be carried out explicitly. In order to do it, notice that for $\alpha \in (0,1)$ and $p > 0$

$$\int_0^\infty (e^{irp} - 1) \frac{dr}{r^{1+\alpha}} = -\frac{\Gamma(1-\alpha)}{\alpha} e^{-i\pi\alpha/2} p^\alpha. \quad (C5)$$

In fact, one presents the integral on the r.h.s. of (C5) as the limit as $\epsilon \to 0_+$ of

$$\int_0^\infty (e^{-(\epsilon-ip)r} - 1) \frac{dr}{r^{1+\alpha}} = \frac{\epsilon - ip}{\alpha} \int_0^\infty e^{-(\epsilon-ip)r} r^{-\alpha} \, dr = -\frac{\Gamma(1-\alpha)}{\alpha} (\epsilon - ip)^\alpha,$$

where

$$(\epsilon - ip)^\alpha = (\epsilon^2 + p^2)^{\alpha/2} e^{i\theta\alpha}$$

with $\tan \theta = -p/\epsilon$. Since $\theta \to -\pi/2$ as $\epsilon \to 0_+$, it follows that $(\epsilon - ip)^\alpha$ tends to $p^\alpha e^{-i\alpha\pi/2}$, which gives (C5). Next, for $\alpha \in (1,2)$ and $p > 0$ the integration by parts gives

$$\int_0^\infty \frac{e^{irp} - 1 - irp}{r^{1+\alpha}} \, dr = \frac{ip}{\alpha} \int_0^\infty (e^{ipr} - 1) \frac{dr}{r^\alpha},$$

and therefore, due to (C5), in that case

$$\int_0^\infty \frac{e^{irp} - 1 - irp}{r^{1+\alpha}} \, dr = \frac{\Gamma(\alpha - 1)}{\alpha} e^{-i\pi\alpha/2} p^\alpha. \quad (C6)$$

Note that the real part of both (C5) and (C6) is positive. From (C5), (C6) it follows that for $\alpha \in (0,2)$, $\alpha \neq 1$,

$$\int_0^\infty \left(e^{irp} - 1 - \frac{irp}{1+r^2} \right) \frac{dr}{r^{1+\alpha}} = i a_\alpha p - \sigma_\alpha e^{-i\pi\alpha/2} p^\alpha \quad (C7)$$

with

$$\sigma_\alpha = \alpha^{-1} \Gamma(1-\alpha), \quad a_\alpha = -\int_0^\infty \frac{dr}{(1+r^2)r} \quad (C8)$$

for $\alpha \in (0,1)$ and

$$\sigma_\alpha = -\alpha^{-1} \Gamma(\alpha - 1), \quad a_\alpha = \int_0^\infty \frac{r^{2-\alpha} dr}{1+r^2} \quad (C9)$$

for $\alpha \in (1,2)$. To calculate the l.h.s. of (C7) for $\alpha = 1$ one notes that

$$\int_0^\infty \frac{e^{irp} - 1 - ip \sin r}{r^2} dr$$

$$= -\int_0^\infty \frac{1 - \cos rp}{r^2} dr + i \int_0^\infty \frac{\sin rp - p \sin r}{r^2} dr = -\frac{1}{2}\pi p - ip \log p.$$

In fact, the real part of this integral is evaluated using a standard fact that $f(r) = (1 - \cos r)/(\pi r^2)$ is a probability density (with the characteristic function $\psi(z)$ that equals to $1 - |z|$ for $|z| \leq 1$ and vanishes for $|z| \geq 1$), and the imaginary part can be presented in the form

$$\lim_{\epsilon \to 0} \left[\int_\epsilon^\infty \frac{\sin pr}{r^2} dr - p \int_\epsilon^\infty \frac{\sin r}{r^2} dr \right]$$

$$-p \lim_{\epsilon \to 0} \int_\epsilon^{p\epsilon} \frac{\sin r}{r^2} dr = -p \lim_{\epsilon \to 0} \int_1^p \frac{\sin \epsilon y}{\epsilon y^2} dy = -p \int_1^p \frac{dy}{y},$$

which implies the required formula. Therefore, for $\alpha = 1$

$$\int_0^\infty \left(e^{irp} - 1 - \frac{irp}{1 + r^2} \right) \frac{dr}{r^{1+\alpha}} = ia_1 p - \frac{1}{2}\pi p - ip \log p \qquad (C10)$$

with

$$a_1 = \int_0^\infty \frac{\sin r - r}{(1 + r^2)r^2} dr. \qquad (C11)$$

Using (C7)-(C11) yields for function (C4) the following expression

$$\log \psi_\alpha(y) = i(\tilde{A}, y) - \int_{S^{d-1}} |(y, s)|^\alpha \left(1 - i \, sgn \, ((y, s)) \tan \frac{\pi \alpha}{2} \right) \tilde{\mu}(ds), \quad \alpha \neq 1,$$
$$(C12)$$

$$\log \psi_\alpha(y) = i(\tilde{A}, y) - \int_{S^{d-1}} |(y, s)| \left(1 + i \frac{2}{\pi} sgn \, ((y, s)) \log |(y, s)| \right) \tilde{\mu}(ds), \quad \alpha = 1,$$
$$(C12')$$

where

$$\tilde{A} = A + a_\alpha \int_{S^{d-1}} s\mu(ds)$$

with a_α given in (C8), (C9), (C11) and the measure $\tilde{\mu}$ on S^{d-1} is proportional to μ, more exactly

$$\tilde{\mu} = \begin{cases} \sigma_\alpha \cos(\pi\alpha/2)\mu, & \alpha \neq 1, \\ \pi\mu/2, & \alpha = 1, \end{cases} \qquad (C13)$$

and is called sometimes the spectral measure of a stable law.

For instance, if $d = 1$, S^0 consists of two points. Denoting their $\tilde{\mu}$-measures by μ_1, μ_{-1} one obtains for $\alpha \neq 1$ that

$$\log \psi_\alpha(y) = i\tilde{A}y - |y|^\alpha[(\mu_1 + \mu_{-1}) - i \, sgn \, y(\mu_1 - \mu_{-1}) \tan \frac{\pi \alpha}{2}].$$

This can be written also in the form

$$\log \psi_\alpha(y) = i\tilde{A}y - \sigma|y|^\alpha \exp\{i\tfrac{\pi}{2}\gamma \, sgn \, y\} \tag{C14}$$

with some $\sigma > 0$ and a real γ such that $|\gamma| \le \alpha$, if $\alpha \in (0,1)$, and $|\gamma| \le 2 - \alpha$, if $\alpha \in (1,2)$.

If the spectral measure $\tilde{\mu}$ is symmetric, i.e. $\tilde{\mu}(-\Omega) = \tilde{\mu}(\Omega)$ for any $\Omega \subset S^{d-1}$, then $\tilde{A} = A$ and formulas (C6),(C7) give both the following simple expression:

$$\log \psi_\alpha(y) = i(A, y) - \int_{S^{d-1}} |(y, s)|^\alpha \tilde{\mu}(ds). \tag{C15}$$

In particular, if the measure $\tilde{\mu}$ is uniform, then $\log \psi_\alpha(y)$ is just $i(A, y) - \sigma|y|^\alpha$ with some σ called the scale of a stable distribution.

One sees readily that the characteristic function $\psi_\alpha(y)$ with $\log \psi_\alpha(y)$ from (C12) or (C15) with vanishing A enjoy the property that $\psi_\alpha^n(y) = \psi_\alpha(n^{1/\alpha}y)$, and therefore all stable distributions with the index $\alpha \ne 1$, and for $\alpha = 1$ all symmetric distributions can be made strictly stable, if centred appropriately.

We want to consider now localised versions of the stable laws. They present, on the one hand, a reasonable approximation to the exact stable laws (see e.g [Neg]), and on the other hand, their characteristic functions are analytic, which allows to use powerful analytic tools, when investigating them. In chapter 6, these laws are used as the models for the development of the theory of large deviation. One obtains a localised stable law by cutting off the support of the Lévy measure in the Lévy-Khintchin representation (C4) of the characteristic function of a stable law. More precisely, we shall call a distribution a localised stable distribution of the index of stability $\alpha \in (0,2)$, if for its characteristic function ψ_α^{loc} one has the representation

$$\log \psi_\alpha^{loc}(y) = i(A, y) + \int_0^\infty \int_{S^{d-1}} \left(e^{i(y,\xi)} - 1 - \frac{i(y,\xi)}{1+\xi^2} \right) \frac{\Theta_a(|\xi|)d|\xi|}{|\xi|^{1+\alpha}} \mu(ds). \tag{C16}$$

Remark. We have chosen here the simplest cutoff of the stable measure. Certainly one can choose it in many different ways without changing the results presented further.

Notice that formula (C16) defines an entire analytic function of y. Moreover, the difference between functions (C4) and (C16) for real y is a bounded function (up to an imaginary shift of the form $i(b, y)$). In fact, this difference is given (up to an imaginary shift) by the Lévy-Khintchin formula (C3) for a compound Poisson distribution, which obviously defines a bounded function. It turns out that this property of localised stable laws is preserved after a shift in the complex domain and also after a "small" perturbation in the class of the function of Lévy-Khintchine type, namely for the function

$$\Phi(y) = \log \psi_\alpha^{loc}(y) + \int_{\mathcal{R}^d} (e^{i(y,\xi)} - 1)g(\xi) \, d\xi \tag{C17}$$

with a bounded non-negative g with a support containing in the open ball of the radius a. The corresponding simple results, which we are going to present now, namely formulas (C21)-(C23) below, are used in the proof of the main theorem of Chapter 6.

Further on it will be more convenient to use the "rotated" function

$$H(z) = \Phi(iz) = \log \psi_\alpha^{loc}(iz) + \int_{\mathcal{R}^d} (e^{-(z,\xi)} - 1)g(\xi)\, d\xi \qquad (C18)$$

which is called sometimes the Laplace exponent (or cumulant) of an infinitely divisible process. One has

$$H(z + iy) = \Phi(iz - y) = -(A, z) - i(A, y) + \int (e^{-z\xi - iy\xi} - 1)g(\xi)\, d\xi$$

$$\int \left(e^{-iy\xi - z\xi} - 1 + \frac{z\xi}{1 + \xi^2} + \frac{iy\xi}{1 + \xi^2} \right) \frac{\Theta_a(|\xi|)}{|\xi|^{1+\alpha}}\, d|\xi|\mu(ds)$$

$$= -Az + \log \psi_\alpha(-y) + \int (e^{-z\xi - iy\xi} - 1)g(\xi)\, d\xi + \int d|\xi|\mu(ds)$$

$$\times \left[\left((e^{-z\xi} - 1)e^{-iy\xi} + \frac{z\xi}{1 + \xi^2} \right) \frac{\Theta_a(|\xi|)}{|\xi|^{1+\alpha}} - \left(e^{-iy\xi} - 1 + \frac{iy\xi}{1 + \xi^2} \right) \frac{1 - \Theta_a(|\xi|)}{|\xi|^{1+\alpha}} \right].$$

$$(C19)$$

It follows in particular that

$$H(z) - H(z + iy) + \log \psi_\alpha(-y) = \int e^{-z\xi}(1 - e^{-iy\xi})g(\xi)\, d\xi$$

$$+ \int \frac{d|\xi|\mu(ds)}{|\xi|^{1+\alpha}}$$

$$\times \left[\left(e^{-iy\xi} - 1 + \frac{iy\xi}{1 + \xi^2} \right)(1 - \Theta_a(|\xi|)) + (e^{-z\xi} - 1)(1 - e^{-iy\xi})\Theta_a(|\xi|) \right]. \quad (C20)$$

Notice now that for symmetric μ and $\alpha < 1$ formula (C19) can be rewritten in the form

$$H(z + iy) = \log \psi_\alpha(y) - Az$$

$$+ \int e^{-iy\xi} \left[((e^{-z\xi} - 1)\Theta_a(|\xi|) + 1 - \Theta_a(|\xi|)) \frac{d|\xi|\mu(ds)}{|\xi|^{1+\alpha}} + e^{-z\xi}g(\xi)\, d\xi \right] - K$$

$$(C21)$$

with

$$K = \int \frac{1 - \Theta_a(|\xi|)}{|\xi|^{1+\alpha}} d|\xi|\mu(ds) + \int g(\xi)\, d\xi.$$

On the other hand, for symmetric μ and $\alpha \geq 1$

$$H(z + iy) = \log \psi_\alpha(y) - i\left(z, \frac{\partial}{\partial y} \log \psi_\alpha(y) \right) - Az$$

$$+ \int e^{-iy\xi} \left[((e^{-z\xi} - 1 + z\xi)\Theta_a(|\xi|) + 1 - \Theta_a(|\xi|)) \frac{d|\xi|\mu(ds)}{|\xi|^{1+\alpha}} + e^{-z\xi}g(\xi)\,d\xi \right] - K$$

$$(C22)$$

with the same K sa above.

Proposition C1. *For symmetric μ and H given in (C18), the principle term of the asymptotics of $Re\,(H(z)-H(z+iy))$ as $y \to \infty$ is given by $-\log\psi_\alpha(y)$ with the estimate of the remainder being uniform for z from any compact domain. Moreover,*

$$Re\,(H(z) - H(z + iy)) + \log\psi_\alpha(y) \geq C \qquad (C23)$$

for all y, z and some constant C.

Proof. It follows from (C20) that

$$Re\,(H(z) - H(z + iy)) + \log\psi_\alpha(y) = \int e^{-z\xi}(1 - \cos(y\xi))g(\xi)\,d\xi$$

$$+ \int \Big[(e^{-z\xi} - 1 + z\xi)(1 - \cos(y\xi))\Theta_a(|\xi|)$$

$$- (1 - \cos(y\xi))(1 - \Theta_a(|\xi|)) \Big] \frac{d|\xi|\mu(ds)}{|\xi|^{1+\alpha}}, \qquad (C24)$$

because $\int(z,\xi)(1 - \cos(y,\xi))\nu(d\xi) = 0$ for any (centrally) symmetric measure ν. Formula (C24) implies the statement of the Proposition, because, on the one hand, all unbounded in z terms of the r.h.s. of (C24) are positive, and on the other hand, all terms on the r.h.s. of (C24) are bounded in y.

Remark. Using results from Appendix B, one readily gets the upper bound for the l.h.s. of of (C23), namely that

$$Re\,(H(z) - H(z + iy)) + \log\psi_\alpha(y) \leq C_1 H(z) + C_2 \qquad (C25)$$

with some constants $C_1 > 0$ and C_2.

D. Lévy-Khintchine ΨDO and Feller-Courrège processes

Here we recall the main facts connecting the theory of pseudo-differential operators (ΨDO) and pseudo-differential equation (ΨDE) with the theory of random processes and also give a simple version of the general asymptotic formula of the commutation of a ΨDO with an exponential function for the class of ΨDO arising in the analytical description of random processes. This formula has two special features as compared with the general one. On the one hand, the symbols of ΨDO appearing in the theory of stochastic processes may not belong to the standard classes of symbols, for example, they may not be smooth (see a detailed discussion in [Ja]); on the other hand, they have a special form which allows to write down an explicit expression for the remainders in the standard asymptotic expansions.

Let us recall first the main notations of the theory of ΨDO. For an appropriate function $\Psi(x,p)$ (a symbol), $x, p \in \mathcal{R}^d$, the action of the ΨDO $\Psi(x, -i\nabla)$

on a function f is defined by the integral (which may exists, perhaps, in some generalised sense)

$$[\Psi(x, -i\nabla)f](x) = (2\pi)^{-d/2} \int_{\mathcal{R}^d} e^{ipx} \Psi(x, p) \hat{f}(p) \, dp$$

with $\hat{f}$ being the Fourier transform of f, or equivalently

$$[\Psi(x, -i\nabla)f](x) = (2\pi)^{-d} \int_{\mathcal{R}^{2d}} e^{ip(x-\xi)} \Psi(x, p) f(\xi) \, d\xi dp.$$

With each ΨDO one can associate the evolutionary equation

$$\frac{\partial u}{\partial t} = \Psi(x, -i\Delta)u \qquad (D1)$$

The resolving operator of the Cauchy problem corresponding to this equation is given by the semigroup of operators $\exp\{t\Psi(x, -i\Delta)\}$ (whenever it is well-defined).

In asymptotic theory of ΨDE one usually considers the asymptotic solutions with respect to a small positive parameter h being the "weight" of the derivative operators ∇ and $\partial/\partial t$. More precisely, one associates with a symbol $\Psi(x, p)$ the so called h-ΨDO (see e.g [MF1]) defined by the formula

$$[\Psi(x, -ih\nabla)f](x) = (2\pi h)^{-d/2} \int_{\mathcal{R}^d} e^{ipx/h} \Psi(x, p) \hat{f}_h(p) \, dp$$

with $\hat{f}_h$ being the h-Fourier transform of f:

$$\hat{f}_h(p) = (2\pi h)^{-d/2} \int_{\mathcal{R}^d} e^{-ipx/h} f(x) \, dx.$$

The resolving operator to the Cauchy problem for the corresponding evolutionary equation

$$h\frac{\partial u}{\partial t} = \Psi(x, -ih\nabla)u \qquad (D2)$$

can be written formally as $\exp\{\frac{t}{h}\Psi(x, -ih\Delta)\}$. Since in quantum mechanics the limit of the solution of the Schrödinger equation (which is of type (D2) with h being the so called Planck constant) as $h \to 0$ describes the classical limit, in general theory of ΨDe the asymptotics of the solutions of equation (D2) as $h \to 0$ are called semi-classical or quasi-classical.

Turning now to the connection of the theory of ΨDO with probability, let us recall first the following famous characterisation of the class of the Lévy-Khintchine functions (C1): it coincides with the set of the generators of the translation invariant and positivity preserving semigroups. More precisely, if Φ is a complex valued function on $\mathcal{R}^d$ with a bounded from below real part, then the resolving operator $\exp\{t\Phi(-i\nabla)\}$ of the Cauchy problem of the ΨDE

$\partial u/\partial t = \Phi(-i\nabla)u$ preserves positivity, if and only if $\Phi(y)$ has the form (C1) up to a real additive constant. A purely analytic proof of this fact can be found e.g. in [RS]. From the probabilistic point of view, this fact is surely not surprising, because due to the Lévy-Khintchine theorem the semigroups of operators $\exp\{t\Phi(-i\nabla)\}$ with Φ of form (C1) correspond to general random processes with independent increments. The important generalisation of this fact is given by the fundamental theorem of Courége [Cou], [BCP]. To formulate it, let us recall first that a Feller semigroup is by definition a strongly continuous semigroup T_t, $t \geq 0$, of linear contractions on the Banach space of continuous functions on $\mathcal{R}^d$ vanishing at infinity such that $0 \leq u(x) \leq 1$ for all x implies that $0 \leq T_t u(x) \leq 1$ for all t and x. In particular, each operator T_t preserves positivity. The Courrège theorem states that if the generator of essentially any Feller semigroup is a PDO with symbols of form (C1) "with varying coefficients", i.e. these semigroups are defined by the equations of the form (D1) with

$$\Psi(x,p) = i(A(x),p) - \frac{1}{2}(G(x)p,p) + \int_{\mathcal{R}^d\setminus\{0\}} \left(e^{i(p,\xi)} - 1 - \frac{i(p,\xi)}{1+\xi^2} \right) \nu(x,d\xi),$$

$$(D3)$$

where $\nu(x,d\xi)$ and $G(x)$ are respectively a Lévy measure and a nonnegative matrix for all x. Notice however that Courrège theorem gives only a necessary condition on the generator and does not state that any operator of form (D3) defines a Feller semigroup. It is proven in the probability theory that to each Feller semigroup corresponds a Markov stochastic process, which is called in that case a Feller process. In particular, the transition probability densities of this Markov process (whenever they exist) satisfy the corresponding equation (D2),(D3). The ΨDO with symbols of form (D3) can be naturally called the Lévy-Khintchine ΨDO and the corresponding semigroups (and stochastic processes) can be called the Courrège-Feller semigroups. If for all x function (D3) corresponds to a stable process, we shall say that the corresponding process is a stable Courrège-Feller process or a stable diffusion (usual diffusions obviously correspond to stable generators of the index $\alpha = 2$). If $G(x)$, $A(x)$, $\nu(x,d\xi)$ do not depend actually on x, the corresponding Courrège-Feller process is a process with independent equally distributed increments, called Lévy process. In stable case such process is sometimes called Lévy stable motion. There exists enormous literature on Lévy processes (see [ST], [Ber], and references there).

Since $e^{a\Delta}f(x) = f(x+a)$, for symbols of form (D3) the action of the corresponding ΨDO can be given by the formula

$$[\Psi(x,-ih\nabla)f](x) = h\left(A(x), \frac{\partial f}{\partial x}\right) + \frac{h^2}{2}\, tr\left(G(x)\frac{\partial^2 f}{\partial x^2}\right) + (L_{int}^h f)(x) \quad (D4)$$

with

$$(L_{int}^h f)(x) = \int_{\mathcal{R}^d\setminus\{0\}} \left(f(x+h\xi) - f(x) - \frac{h(\frac{\partial f}{\partial x},\xi)}{1+\xi^2} \right) \nu(x,d\xi). \qquad (D5)$$

In the probabilistic framework, the parameter h has a clear meaning: it controls the mean amplitude of jumps in the corresponding random process. In the theory of diffusion processes, the asymptotics corresponding to $h \to 0$ are called the small diffusion approximation.

Notice now that introducing a function

$$H(x, p) = \Psi(x, ip) = \frac{1}{2}(G(x)p, p) - (A(x), p)$$

$$+ \int_{\mathcal{R}^d \setminus \{0\}} \left(e^{-(p, \xi)} - 1 + \frac{(p, \xi)}{1 + \xi^2} \right) \nu(x, d\xi), \qquad (D6)$$

one can rewrite equation (D2) in the "real" form

$$h \frac{\partial u}{\partial t} = H(x, -h\nabla)u. \qquad (D7)$$

The use of the function H instead of the symbol Ψ turns out to be more convenient for the construction of semiclasssical approximation for Courrège-Feller processes, because the function H, and not Ψ appears in the corresponding Hamilton-Jacobi equation that plays a central role in WKB-type asymptotics (see Chapter 6).

When solving ΨDO an important tool is the formula for the commutation of a ΨDO with an exponential function (see, e.g. [M4], [MF1]). We present now a version of this formula for ΨDO of type (D4),(D5).

Proposition D1. *Let*

$$u(x) = \phi(x) \exp\{-\frac{S(x)}{h}\}. \qquad (D8)$$

with some complex-valued smooth functions ϕ and S, and let Ψ be a symbol of type (D3) of a Lévy-Khintchine ΨDO. Then

$$\exp\{\frac{S(x)}{h}\}[\Psi(x, -ih\nabla)u](x) = \phi(x)H(x, \frac{\partial S}{\partial x}) - h \left(\frac{\partial \phi}{\partial x}, \frac{\partial H}{\partial p}(x, \frac{\partial S}{\partial x}) \right)$$

$$- \frac{h}{2}\phi(x)\, tr\, (G(x)\frac{\partial^2 S}{\partial x^2}) + \frac{h^2}{2}\, tr\, (G(x)\frac{\partial^2 \phi}{\partial x^2}) + R_\nu(x), \qquad (D9)$$

where

$$R_\nu(x) = \int_{\mathcal{R}^d \setminus \{0\}} \exp\{-(\frac{\partial S}{\partial x}, \xi)\} \nu(x, d\xi)$$

$$\times \left[\left(\phi(x) + h(\frac{\partial \phi}{\partial x}, \xi) \right) \left(\exp\{-h \int_0^1 (1 - \theta_1) \left(\frac{\partial^2 S}{\partial x^2}(x + \theta_1 h\xi)\xi, \xi \right) d\theta_1\} - 1 \right) \right.$$

$$\left. + h^2 \int_0^1 (1 - \theta_2) \left(\frac{\partial^2 \phi}{\partial x^2}(x + \theta_2 h\xi)\xi, \xi \right) d\theta_2 \right]$$

$$\times \exp\{-h \int_0^1 (1-\theta_1) \left(\frac{\partial^2 S}{\partial x^2}(x+\theta_1 h\xi)\xi, \xi \right) d\theta_1 \}], \qquad (D10)$$

if all terms on the r.h.s. of (D8) are well defined. For instance, they are well defined if the function ϕ and the Lévy measure ν have both bounded supports.

Proof. It is straightforward, because from (D4),(D5) it follows that

$$\exp\{\frac{S(x)}{h}\}[\Psi(x,-ih\nabla)u](x)$$

$$= \frac{1}{2} \left(G(x)\frac{\partial S}{\partial x}, \frac{\partial S}{\partial x} \right) - (A(x), \frac{\partial S}{\partial x}) - h \left(G(x)\frac{\partial S}{\partial x}, \frac{\partial \phi}{\partial x} \right)$$

$$+h(A(x), \frac{\partial \phi}{\partial x}) - \frac{h}{2}\phi(x)\, tr\, (G(x)\frac{\partial^2 S}{\partial x^2}) + \frac{h^2}{2}\, tr\, (G(x)\frac{\partial^2 \phi}{\partial x^2}) + \exp\{\frac{S(x)}{h}\}(L_{int}^h u)(x)$$

with

$$\exp\{\frac{S(x)}{h}\}(L_{int}^h u)(x) = \int_{\mathcal{R}^d\backslash\{0\}} [\phi(x+h\xi)\exp\{-\frac{1}{h}(S(x+h\xi)-S(x))\}$$

$$-\phi(x) + \frac{\phi(x)(\frac{\partial S}{\partial x},\xi) - h(\frac{\partial \phi}{\partial x},\xi)}{1+\xi^2}] \nu(x,d\xi)$$

$$= \int_{\mathcal{R}^d} \nu(x,d\xi)[\left(\phi(x) + h(\frac{\partial \phi}{\partial x},\xi) + h^2 \int_0^1 (1-\theta_2) \left(\frac{\partial^2 \phi}{\partial x^2}(x+\theta_2 h\xi)\xi, \xi \right) d\theta \right)$$

$$\exp\{-(\frac{\partial S}{\partial x},\xi)\} \exp\{-h \int_0^1 (1-\theta_1) \left(\frac{\partial^2 S}{\partial x^2}(x+\theta_1 h\xi)\xi, \xi \right) d\theta_1\} - \phi(x)$$

$$+\frac{\phi(x)(\frac{\partial S}{\partial x},\xi) - h(\frac{\partial \phi}{\partial x},\xi)}{1+\xi^2}].$$

Sometimes it is useful to rewrite (D9), (D10) in a slightly different form.

Proposition D2. *Under the assumptions of Proposition D1*

$$\exp\{\frac{S(x)}{h}\}[\Psi(x,-ih\nabla)u](x) = \phi(x)H(x,\frac{\partial S}{\partial x})$$

$$-h\left[\left(\frac{\partial \phi}{\partial x}, \frac{\partial H}{\partial p}(x,\frac{\partial S}{\partial x}) \right) + \frac{1}{2}\phi(x)\, tr\, \left(\frac{\partial^2 H}{\partial p^2}(x,\frac{\partial S}{\partial x})\frac{\partial^2 S}{\partial x^2} \right) \right]$$

$$+\frac{h^2}{2}\, tr\, (G(x)\frac{\partial^2 \phi}{\partial x^2}) + \tilde{R}_\nu(x) \qquad (D11)$$

with

$$\tilde{R}_\nu(x) = \int_{\mathcal{R}^d\backslash\{0\}} \exp\{-(\frac{\partial S}{\partial x},\xi)\} \nu(x,d\xi)$$

$$\times [\phi(x)(\exp\{-h \int_0^1 (1-\theta_1) \left(\frac{\partial^2 S}{\partial x^2}(x+\theta_1 h\xi)\xi, \xi \right) d\theta_1\} - 1 + \frac{h}{2} \left(\frac{\partial^2 S}{\partial x^2}(x)\xi, \xi \right))$$

$$+h(\frac{\partial \phi}{\partial x},\xi)\left(\exp\{-h\int_0^1(1-\theta_1)\left(\frac{\partial^2 S}{\partial x^2}(x+\theta_1 h\xi)\xi,\xi\right)d\theta_1\}-1\right)$$

$$+h^2\int_0^1(1-\theta_2)\left(\frac{\partial^2 \phi}{\partial x^2}(x+\theta_2 h\xi)\xi,\xi\right)d\theta_2$$

$$\times \exp\{-h\int_0^1(1-\theta_1)\left(\frac{\partial^2 S}{\partial x^2}(x+\theta_1 h\xi)\xi,\xi\right)d\theta_1\}]. \tag{D12}$$

The proof is straightforward. The advantage of (D11) as compared with (D9) consists in the fact that for small h the remainder $\tilde{R}_\nu$ is of the order $O(h^2)$, and therefore formulas (D11),(D12) give the asymptotic representation of the result of the commutation of a Lévy-Khintchine ΨDO with an exponential function up to a remainder of the order $O(h^2)$.

E. Equivalence of convex functions

It is proved here that any two smooth convex functions on Eucleadian space, each having a non-degenerate minimum are smoothly equivalent. This result is simple and natural but I did not find it in the literature. It is used in the construction of the uniform small time and small diffusion asymptotics for Feller semigroups given in Chapter 6.

Proposition E1. *Let f be an infinitely smooth (resp. of the class C^k with $k \geq 2$) convex function on $\mathcal{R}^d$ such that $f(0) = 0$ and $f(x) > 0$ for all $x \neq 0$. Suppose also that the matrix of the second derivatives $f''(0)$ of f at the origin is not degenerate (and therefore it is positive). Then there exists an infinitely smooth (resp. of the class C^{k-2}) diffeomorphism $D : \mathcal{R}^d \mapsto \mathcal{R}^d$ such that $f(D^{-1}y) = (y,y)/2$. Moreover, D can be presented as the composition $D = D_3 D_2 D_1$, where D_1 is a linear operator in $\mathcal{R}^d$, D_2 differs from the identity only in a neighbourhood of the origin and D_3 is a dilation*

$$D_3(x) = (1+\omega(x))x \tag{E1}$$

with some scalar function ω vanishing in a neighbourhood of the origin.

We begin with two lemmas. The first is rather standard. We present here the formula and the estimates for a smooth molyfier in a form convenient for our purposes.

Lemma E1. *There exists a constant $C > 0$ such that for any $a,b: 0 < b < a$ there exists an infinitely smooth non-increasing function χ_b^a on $\mathcal{R}$ such that χ_b^a vanishes for $x \geq a$, is equal to one for $x \leq b$ and $\chi_b^a(x) \in (0,1)$ for $x \in (b,a)$. Moreover, this function depends smoothly on a,b, and for all a,b,x*

$$|(\chi_b^a)'(x)| \leq \frac{C}{a-b}, \quad |(\chi_b^a)''(x)| \leq \frac{C}{(a-b)^2}, \quad |\frac{\partial \chi_b^a}{\partial a}(x)| \leq \frac{C}{a-b}. \tag{E2}$$

Proof. Let

$$g(y) = \begin{cases} K\exp\{-\frac{1}{y(y-1)}\}, & y \in (0,1), \\ 0, & y \in (-\infty,0]\cup[1,\infty), \end{cases}$$

where the constant K is chosen in such a way that $\int_0^1 g(y)\,dy = 1$. One readily sees that g is an infinitely differentiable function on $\mathcal{R}$. It follows that the function

$$\chi_b^a(x) = \frac{1}{a-b} \int_x^\infty g\left(\frac{z-b}{a-b}\right) dz \equiv \int_{(x-b)/(a-b)}^\infty g(y)\,dy \qquad (E3)$$

satisfies the requirements of the Lemma with

$$C = \max_{y \in [0,1]} \max(g(y), g'(y)).$$

The following lemma is crucial.

Lemma E2. *Let a function f satisfies the assumptions of Theorem E1 and moreover, the matrix $f''(0)$ is the unit matrix. Then there exists a diffeomorphism $D : \mathcal{R}^d \mapsto \mathcal{R}^d$ such that D differs from the identity only in a neighbourhood of the origin and $f(D^{-1}y)$ is a convex function on $\mathcal{R}^d$ which equals $(y,y)/2$ in a neighbourhood of the origin.*

Proof. The idea is to sew the local diffeomorphisms used in the standard proof of the Morse lemma with the identical diffeomorphism in such a way that the resulting function $f(D^{-1}y)$ will be again convex. For brevity, let us give the proof for $d = 2$. The general case is obtained by the similar modification of the proof of the Morse lemma. If $d = 2$, it follows from the assumptions of the Lemma that $f(x) = f(x_1, x_2)$ can be presented in the form

$$f(x_1, x_2) = \frac{1}{2} A(x) x_1^2 + B(x) x_1 x_2 + \frac{1}{2} C(x) x_2^2 \qquad (E4)$$

with some smooth functions A, B, C such that

$$A = 1 + O(|x|), \quad B = O(|x|), \quad C = 1 + O(|x|).$$

Let $r > 0$ be chosen in such a way that $A(x) > 0$ and $(AC - B^2)(x) > 0$ for $|x| \le r$. Clearly for $|x| \le r$ the function f can be presented in the equivalent form

$$f(x_1, x_2) = \frac{1}{2} A(x) \left(x_1 + \frac{B(x)}{A(x)} x_2 \right)^2 + \frac{1}{2} \left(C(x) - \frac{B^2(x)}{A(x)} \right) x_2^2.$$

For any positive $\epsilon < r/2$ one can now define the mapping $D_\epsilon : x \mapsto y$ in $\mathcal{R}^2$ by the formula

$$y_1 = \chi_\epsilon^{2\epsilon}(|x|) \sqrt{A(x)} \left(x_1 + \frac{B(x)}{A(x)} x_2 \right) + (1 - \chi_\epsilon^{2\epsilon}(|x|)) x_1,$$

$$y_2 = \left[\chi_\epsilon^{2\epsilon}(|x|) \sqrt{C(x) - \frac{B^2(x)}{A(x)}} + (1 - \chi_\epsilon^{2\epsilon}(|x|)) \right] x_2.$$

Obviously D_ϵ is a smooth (infinitely smooth, if f is infinitely smooth, or of the class C^{k-2}, if f is of the class C^k) mapping $\mathcal{R}^2 \mapsto \mathcal{R}^2$ that differs from the identity only inside the ball $B_{2\epsilon}$, and $f(D^{-1}y)$ equals $(y,y)/2$ in $D(B_\epsilon)$ and equals $f(y)$ outside $D(B_{2\epsilon})$. Let us estimate the derivatives of D_ϵ. First of all,

$$y_1 = (1 + O(|x|))x_1 + O(|x|)x_2, \quad y_2 = (1 + O(|x|))x_2.$$

Furthermore, due to (E2),

$$\frac{\partial y}{\partial x} = E + O(|x|) + O(\epsilon^{-1}|x|^2), \quad \frac{\partial^2 y}{\partial x^2} = O(1 + \epsilon^{-1}|x| + \epsilon^{-2}|x|^2).$$

Consequently, in $B_{2\epsilon}$ one has

$$\frac{\partial y}{\partial x} = E + O(\epsilon), \quad \frac{\partial^2 y}{\partial x^2} = O(1), \quad \frac{\partial^2 x}{\partial y^2} = O(1)$$

uniformly for $\epsilon \to 0$. Hence, D_ϵ is a global diffeomorphism for small enough ϵ. At last, since

$$\frac{\partial^2 f}{\partial y^2} = \left(\frac{\partial x}{\partial y}\right)^t \frac{\partial^2 f}{\partial x^2} \frac{\partial x}{\partial y} + \frac{\partial f}{\partial x} \frac{\partial^2 x}{\partial y^2},$$

it follows that $\frac{\partial^2 f}{\partial y^2} = E + O(\epsilon)$ and therefore $f(D^{-1}y)$ is a convex function, if ϵ is small enough.

Proof of Proposition E1. It is now almost straightforward. One takes first a linear mapping D_1 in $\mathcal{R}^d$ such that the matrix of the second derivatives of $f(D_1^{-1}y)$ at the origin is the unit matrix. Then one uses Lemma E2 to find a diffeomorphism D_2 such that $f(D_1^{-1}D_2^{-1}y)$ is a convex function coinciding with $(y,y)/2$ in a neighbourhood of the origin. At last, one easily verifies that any two convex functions coinciding in a neighbourhood of their minimum points can be transformed one to another by a diffeomorphism of form (E1).

Proposition E2. *For any two functions f_1, f_2 satisfying the assumptions of Proposition E1 there exists a diffeomorphism $D : \mathcal{R}^d \mapsto \mathcal{R}^d$ such that $f_1(y) = f_2(D^{-1}y)$ and moreover, outside a neighbourhood of the origin, D can be presented as the composition of a linear transform of $\mathcal{R}^d$ and of a dilation of form (E1).*

Proof. It is a direct consequence of Proposition E1.

F. Unimodality of symmetric stable laws

Here we discuss the property of unimodality of stable laws, which is used in Chapter 5. We present a short but essentially selfcontained exposition of the main facts of the theory of unimodality of finite dimensional distributions, which is used in Chapter 5. A full account on the main results discussed here can be found in [DJ1]. Roughly speaking, the significance of the property of unimodality for the study of stable distributions and more generally stable diffusions consists in the fact that when one gets the asymptotic expansions for the behaviour of stable densities for small and large distances (see Sections 5.1, 5.2) one needs this property to fill the gap, namely to describe the behaviour of stable densities for the distances that lie between the regions of "large" and "small" distances.

To begin with let us recall that a probability law (or a finite measure) on the real line with the distribution function F is called unimodal with the mode (or vertex) $a \in \mathcal{R}$, if $F(x)$ is convex (possibly not strictly) on $(-\infty, a)$ and concave on (a, ∞). Clearly, if F has a continuous density function f, the unimodality means that f is non-decreasing on $(-\infty, a)$ and non-increasing on (a, ∞). It was proved in [Wi] that all symmetric one-dimensional stable laws are unimodal (with the mode at the origin); there are now many proofs of this well-known result (see e.g. [Lu], [Zo]). The case of non-symmetric stable laws turned out to be essentially more difficult, it was proved only in [Yam] (see also [Zo]) that all stable distributions are unimodal, some important preliminary results being obtained in [IC].

One can imagine several extensions of the notion of unimodality from one dimensional case to several dimensions. We shall mention here the two definitions which are mostly relevant to the study of symmetric stable distributions. We shall denote by $V_k(M)$ the Lebesgue volume of a measurable set $M \subset \mathcal{R}^k$, or just $V(M)$ if the value of K is clear from the context. For any convex set A in $\mathcal{R}^d$ let $d(A)$ be its dimension (which is the dimension of the minimal subspace containing A, $d(A) \leq d$), and let μ_A be the measure in $\mathcal{R}^d$ which is uniformly distributed in A, i.e. $\mu_A(B) = V_{d(A)}(A \cap B)$ for any measurable $B \subset \mathcal{R}^d$.

Definition F1 [And]. *A measure with a density f is called convex unimodal, if the function f has convex sets of upper values, i.e. the sets $\{x : f(x) \geq c\}$ are convex for all c.*

One of the disadvantages of this definition is the fact that the class of convex unimodal measures is not closed under convex linear combinations. The following more general concept improves the situation (at least for the symmetric case with which shall deal here).

Definition F2 [She]. *An elementary unimodal symmetric measure in $\mathcal{R}^d$ is a measure μ_A with some compact convex $A \subset \mathcal{R}^d$. A centrally symmetric finite measure on $\mathcal{R}^d$ is called central convex unimodal (CCU), if it is a weak limit of a sequence of the finite linear combinations of elementary symmetric unimodal measures.*

The class CCU measures is by definition closed with respect to linear combinations (with positive coefficients) and the pass to the weak limit. Moreover, if a convex unimodal measure (as defined in Definition F1) is centrally symmetric, then it is CCU. Notice also that since any convex set can be approximated by a sequence of convex sets with nonempty interiors, the above definition will not change, if one would consider there only the compact convex sets with nonempty interiors. Such convex sets will be called here convex bodies. The main nontrivial fact about CCU measures is that the class of such measures is closed under convolution. The proof of this fact is based on the Brunn-Minkowski theory of mixed volume, which we shall recall now.

The famous Brunn-Minkowski inequality states that for any non-empty compact sets $A, B \subset \mathcal{R}^d$

$$V^{1/d}(A + B) \geq V^{1/d}(A) + V^{1/d}(B), \qquad (F1)$$

where V denotes the standard volume in $\mathcal{R}^d$ and the sum of two sets is defined as usual by $A + B = \{a + b : a \in A, b \in B\}$. This classical result has a long history (it was first proven by Brunn for convex sets, then Minkowski gave necessary and sufficient conditions when the equality sign holds in (F1), and then it was generalised to all compact sets by Lusternik) and can be proved by different methods, see e.g. [Sch]. We sketch here for completeness a beautiful elementary proof taken from [BZ]. Namely, let us say that a compact set A in $\mathcal{R}^d$ is elementary, if it is the union of the finite number $l(A)$ of non-degenerate cuboids with sides parallel to the coordinate axes and such that their interiors do not intersect. Each compact set A can be approximated by a sequence of elementary sets A_i so that $V(A_i) \to V(A)$ (see [BZ]; at least, it is clear for convex compact sets). Therefore , it is suffice to prove (F1) for elementary sets only. Consider first the case, when each of A, B consists of only one cuboid with edges $a_i > 0$ and $b_i > 0$ respectively. Then (F1) takes the form

$$\prod_{i=1}^{d} (a_i + b_i)^{1/d} \geq \prod_{i=1}^{d} a_i^{1/d} + \prod_{i=1}^{d} b_i^{1/d},$$

which follows from the inequality

$$\left(\prod_{i=1}^{d} \frac{a_i}{a_i + b_i}\right)^{1/d} + \left(\prod_{i=1}^{d} \frac{b_i}{a_i + b_i}\right)^{1/d} \leq \frac{1}{d} \sum_{i=1}^{d} \frac{a_i}{a_i + b_i} + \frac{1}{d} \sum_{i=1}^{d} \frac{b_i}{a_i + b_i} = 1.$$

For general non-empty elementary sets A, B the proof can be carried out by induction over $l(A) + l(B)$. Assume that (F1) is true when $l(A) + l(B) \leq k - 1$. Suppose that $l(A) \geq 2$. Clearly there exists a hyperplane P which is orthogonal to one of the coordinate axes and which splits A into elementary sets A', A'' such that $l(A') < l(A)$ and $l(A'') < l(A)$. Then $V(A') = \lambda V(A)$ with some $\lambda \in (0, 1)$. Since parallel translations do not change the volumes, one can choose the origin of coordinates on the plane P and then shift the set B so that the same hyperplane P splits B into sets B', B'' with $V(B') = \lambda V(B)$. Plainly $l(B') \leq l(B), l(B'') \leq L(B)$. The pairs of sets A', B', and A'', B'' each lies in its own half-space with respect to P and in each pair there are no more than $k - 1$ cuboids. Hence

$$V(A + B) \geq V(A' + B') + V(A'' + B'')$$

$$\geq [V^{1/d}(A') + V^{1/d}(B')]^d + [V^{1/d}(A'') + V^{1/d}(B'')]^d$$

$$= \lambda[V^{1/d}(A) + V^{1/d}(B)]^d + (1 - \lambda)[V^{1/d}(A) + V^{1/d}(B)]^d = [V^{1/d}(A) + V^{1/d}(B)]^d,$$

which completes the proof of (F1). From (F1) one easily obtains a more general form of Brunn-Minkowski inequality, which states that for any compact non-empty A, B and any non-negative t_1, t_2

$$V^{1/d}(t_1 A + t_2 B) \geq t_1 V^{1/d}(A) + t_2 V^{1/d}(B). \qquad (F2)$$

Consider now a convex body A in $\mathcal{R}^d$. Recall the following definition (see e.g. [Gar]). Let S be a k-dimensional subspace, $k < d$. The k-dimensional X-ray of A (or, in other terminology, the section function, or the k-plane Radon transform of A) parallel to S is the function of $x \in S^\perp$ defined by the formula $X_S A(x) = V_k(A \cap (S + x))$, where V_k denotes the k-dimensional volume This function can be defined also for any compact set A, but then, generally speaking, it will be defined only for almost all x. We shall need the following well known corollary (see, e.g. [Gar]) of the Brunn-Minkowski inequality: for any convex body A and any k-dimensional subspace S the function $(X_S A)^{1/k}$ is concave on its support. It follows directly from (F2) and a simple observation that if A_0, A_1 are convex k-dimensional bodies sitting in the parallel k-dimensional hyperplanes $x_1 = 0$ and $x_1 = 1$, respectively, in $\mathcal{R}^{k+1}$, then

$$(1 - t)A_0 + tA_1 = conv\,(A_0 \cup A_1) \cap \{x : x_1 = t\},$$

where $conv$ denotes the convex hull of a set.

After this short introduction to the Brunn-Minkowski theory (a complete survey see in [Sch]), let us return to the unimodal measures.

Proposition F1. *The class CCU is closed with respect to the operation of convolution $\star$.*

Proof. It is more or less straightforward corollary of the Brunn- Minkowski inequality. In fact, obviously, it is enough to prove that if f_1 and f_2 are the characteristic functions of the centrally symmetric convex bodies B_1 and B_2, then $f_1 \star f_2$ is a density of a CCU measures. Notice that

$$(f_1 \star f_2)(x) = V_d((x - B_1) \cap B_2) = V_d((x + B_1) \cap B_2). \qquad (F3)$$

It turns out that this function has convex sets of upper values, i.e. it is a density of a convex unimodal measure in the sense of Definition F1, and consequently of a CCU mesaure. In fact, one obtains from the Brunn-Minkowski inequality even a more stronger result, which is called the Fary-Redei Lemma (obtained in [FR]), namely that the $(f_1 \star f_2)^{1/d}$ is concave on its support. To see this, let us consider two d-dimensional planes L_1 and L_2 in $\mathcal{R}^{2d}$ intersecting only at the origin and having the angle $\phi < \pi/2$ between them. Let M_1 and M_2 denote the bodies which are equal to B_1 and B_2 respectively, but lie on the planes L_1 and L_2 respectively. Then the measure with the density $f_1 \star f_2$ can be considered as the limit as $\phi \to 0$ of the measures with the densities f_ϕ being the convolutions of the characteristic functions of M_1 and M_2 in $\mathcal{R}^{2d}$. One sees that f_ϕ is equal to the characteristic function of the set

$$M_\phi = \{x + y : x \in M_1, y \in M_2\}$$

multiplied by $(\sin\phi)^{-d}$ (in fact, the linear transformation of $\mathcal{R}^{2d}$ which is identical on L_1 and which makes L_2 perpendicular to L_1, has the determinant $(\sin\phi)^d$ and leads to the situation, where M_1 and M_2 lie in perpendicular planes, and where the corresponding statement is therefore obvious). Hence

$$(f_1 \star f_2)(x)1/d = \lim_{\phi\to 0}(\sin\phi)^{-1}V_d^{1/d}((x+L_1^\perp)\cap M_\phi)$$

$$= \lim_{\phi\to 0}(\sin\phi)^{-1}(X_{L_i^\perp}M_\phi(x))^{1/d},$$

and the concavity of $(f_1 \star f_2)^{1/d}$ follows from the property of the X-ray stated above.

Definition F3. *A measure μ on $\mathcal{R}^d$ is called monotone unimodal, if for any $y \in \mathcal{R}^d$ and any centrally symmetric convex body $M \subset \mathcal{R}^d$, the function $\mu(M+ty)$ is nonincreasing for $t > 0$.*

The following important fact was proved in [And] for convex unimodal measures (in the sense of Definition F1), and in [She] for general case.

Proposition F2. *All CCU measures are monotone unimodal.*

Proof. Let us prove it here only for CCU measures with densities. Notice first that if $\mu = \mu_A$ with some compact convex A, and if M is compact, the required statement about the function $\mu(M+ty)$ follows directly from the Fary-Redei Lemma (see the proof of Proposition F1). For a non-compact set M, the statement is obtained by a trivial limiting procedure. For a general absolutely continuous μ it is again obtained by a limiting procedure, due to the well known fact that if a sequence μ_n of measures on $\mathcal{R}^d$ converges weakly to a measure μ, then $\mu_n(K)$ converges to $\mu(K)$ for any compact set K such that $\mu(\partial K) = 0$.

Proposition F3. *Let a CCU measure $\mu \in SU$ has a continuous density f. Then for any unit vector v the function $f(tv)$ is non-increasing on $\{t \geq 0\}$, and moreover, for any $m < d$ and any m-dimensional subspace S the integral of f over the plane $tv + S$ is a non-increasing function on $\{t \geq 0\}$. (In other words, the Radon transform of f is non-increasing as well as the Radon transforms of the restrictions of f on any subspace.)*

Proof. It is a direct consequence of the previous Proposition. For instance, to prove that $f(tv)$ is non-increasing one supposes that $f(t_1v > f(t_2v)$ with some $t_1 > t_2$ and then uses the statement of Proposition F2 with a set M being the ball B_ϵ of sufficiently small radius ϵ to come to a contradiction. The result of Proposition F2 arises a natural question, does all symmetric and monotone unimodal measures are CCU. A positive answer to this question was conjectured in [She]. On the level of convex bodies this conjecture holds, as shows the following simple result.

Proposition F4. *A centrally symmetric compact set M in $\mathcal{R}^d$ is convex if and only if it is a starlike set (in the sense that together with any point x it contains the whole closed interval $[0, x]$) and its one-dimensional X-ray function is non-increasing when moving away from the origin, i.e. for any unit vector v and any straight line l from $v^\perp$, the function $V_1((tv+l)\cap M)$ is a non-increasing function on $\{t \geq 0\}$.*

Remark. The characterisation of convex bodies given in Proposition F4 can be essentially improved, at least if one supposes some regularity property of the boundary. Namely, it is possible to show that a symmetric compact set with nonempty interior and a piecewise-smooth boundary is convex if and only if its X-ray is non-increasing when moving away from the origin (i.e. being starlike is in fact redundant in the characterisation of convex bodies given in Proposition F4). We shall omit here a simple geometric proof of Proposition F4 (which we shall not use further on). Notice only that this Proposition is apparently close (but is far from being identical) to the well known theorem on the characterisation of convex bodies given in [Fal], which states that a compact set M with a nonempty interior is convex if and only if for any hyperspace P in $\mathcal{R}^d$ its one-dimensional X-ray function $V_1((x + P^{\perp}) \cap M)$ is concave when restricted to the set of $x \in P$ such that $(x + P^{\perp}) \cap M$ is not empty. However, this result of Falconer is of no use for the study of distribution, because as one easily sees the property of concavity of the X-ray is destroyed when considering the linear combinations of the convex bodies.

Surprisingly enough, the conjecture of Sherman on general CCU measures was disapproved in [Wel] following the previous indications from [DJ2].

Now we are going to obtain the main result of this Appendix, which was proved first in [Kan].

Proposition F5. *All symmetric stable laws are unimodal.*

Proof. It will be given in three steps.

Step 1. Reduction to the case of finite Lévy measure. Recall that the density of a general symmetric stable law with the index of stability $\alpha \in (0, 2)$ (we shall not consider the case of $\alpha = 2$ which is the well known Gaussian distribution) is given by the Fourier transform

$$S(x, \alpha, \mu) = \frac{1}{(2\pi)^d} \int \psi_\alpha(x) e^{ipx} \, dp \qquad (F4)$$

of the characteristic function ψ_α, which can be given either by formula (C4) or by formula (C14) with symmetric measures μ and $\tilde{\mu}$ on S^{d-1} connected by formula (C13) and with vanishing drift A. For any $\epsilon > 0$ consider the finite Lévy measure

$$\nu_\epsilon(d|\xi|, ds) = \begin{cases} |\xi|^{-1-\alpha} d|\xi| \mu(ds), & |\xi| \geq \epsilon \\ \epsilon^{-1-\alpha} d|\xi| \mu(ds), & |\xi| \leq \epsilon \end{cases} \qquad (F5)$$

and the corresponding infinite divisible distribution with the characteristic function ψ_α^ϵ defined by the formula

$$\log \psi_\alpha^\epsilon(y) = \int_0^\infty \int_{S^{d-1}} \left(e^{i(y,\xi)} - 1 - \frac{i(y, \xi)}{1 + \xi^2} \right) \nu_\epsilon(d|\xi|, ds). \qquad (F6)$$

Let P_ϵ denote the corresponding probability distribution. One sees that $\psi_\alpha^\epsilon \to \psi_\alpha$ as $\epsilon \to 0$ uniformly for y from any compact set, because

$$|\log \psi_\alpha^\epsilon(y) - \log \psi_\alpha(y)| \leq \int_0^\epsilon \int_{S^{d-1}} \left| e^{i(y,\xi)} - 1 - \frac{i(y, \xi)}{1 + \xi^2} \right| |\xi|^{-1-\alpha} d|\xi| \mu(ds)$$

$$= O(1)|y|^2 \int_0^\epsilon |\xi|^{1-\alpha} d|\xi| = O(1)|y|^2 |\epsilon|^{2-\alpha}.$$

The convergence of characteristic functions (uniform on compacts) implies the week convergence of the corresponding distributions. Therefore, it is enough to prove the unimodality of the distribution P_ϵ for any ϵ.

Step 2. Reduction to the unimodality property of the Lévy measure. We claim now that in order to prove the unimodality of P_ϵ it is suffice to prove the unimodality of the Lévy measure (F5). In fact, since this measure is finite, formula (F6) can be rewritten in the form

$$\log \psi_\alpha^\epsilon(y) = \int_0^\infty \int_{S^{d-1}} e^{i(y,\xi)} \nu_\epsilon(d|\xi|, ds) - C_\epsilon$$

with some constant C_ϵ. Therefore $\psi_\alpha^\epsilon(y)$ is the exponent of the Fourier transform of measure (F5) (up to a multiplier) and hence it is a limit (uniform on compacts) of the finite linear combinations of ν_ϵ and its convolutions with itself. Therefore our assertion follows from Proposition F1.

Step 3. It remains to prove that the Lévy measure (F5) is unimodal. To this end, notice that any measure μ on S^{d-1} can be approximated weakly by a sequences of discrete measures (concentrated on a counted number of points). Hence, by linearity, it is enough to prove the unimodality of measure (F5) in the case of $\mu(ds)$ concentrated in one point only. But in this case measure (F5) is one-dimensional and the statement is obvious, which completes the proof of Proposition F5.

The same arguments prove the following fact.

Proposition F6. *If the Lévy measure ν of an infinitely divisible distribution F in $\mathcal{R}^d$ (with polar coordinates $|\xi|, s = \xi/|\xi|$) has the form*

$$\nu(d\xi) = f(|\xi|) d|\xi| \mu(ds) \qquad (F7)$$

with any finite (centrally) symmetric measure μ on S^{d-1} and any non-increasing function f, then F is symmetric unimodal.

The result of Proposition F5 was generalised in [Wol] to a more general class of infinitely divisible distributions, namely to the distributions of class L (see [Wol] or [DJ1]).

The following statement is a direct consequence of Propositions F5, F3.

Proposition F7. *If the drift A in (C4) or (C14) vanishes, then the corresponding density (F4) enjoys the property described in Proposition F3, in particular, when restricted to any straight line going through the origin, it is non-increasing when moving away from the origin.* For conclusion, let us notice that no general results seem to be known now on the unimodality of nonsymmetric stable laws in dimension more than one. Due to the following statement, it is difficult to expect that the monotone unimodality will be proved in general case.

Proposition F8. *A compact convex body is (centrally) symmetric if and only if its one-dimensional X-ray function is non-increasing when moving away from the origin.*

A simple proof of this fact is based on the observations that, on the one hand, it is enough to prove this fact for two-dimensional convex bodies, and on the other hand, symmetricity for two-dimensional bodies means that the tangents to the opposite points of the boundary are parallel which is an obvious consequence of the non-increasing of the X-ray when moving away from the origin.

G. Infinitely divisible complex distributions and complex Markov processes

We present here a general approach to the construction of the measures on the path space that can be used for the path integral representation of evolutionary equations. In particular, we give an exposition of the important results of Maslov and Chebotarev (see [M7], [MC2], [Che], [CheKM], [HuM]) on the representation of the solution to the Schrödinger equation as an expectation of a certain functional on the trajectories of a ceratain Poisson prosess. An interpretation of this result from the point of view of non-commutative probability is given in [Par2]. Our exposition will be given in terms of more or less standard probabilistic concepts generalised to the complex case. Various generalisations of the representation from [MC2] to other classes of equations can be found e.g. in [BGR],[Com1]-[Com3], [Ich], [Gav], [PQ]. For example, in [Com1] one can find generalisations to some quantum field models and to a case of the Schrödinger equation with a potential depending on momentum. For these cases Feynman's integral is presented as an expectation with respect to certain generalisations of Poisson processes, which were called generalised Poisson in [Com1] and which can be met in literature on probability theory under different names, see e.g. [Meti], where these processes are called pure jump Markov processes.

The following general construction of the mesaures on path space is especially close in spirit to the construction from [Ich], which, in turn, adapts Nelson's approach (see [Nel2]) to the construction of the Wiener measure to the case of the measures corresponding to the hyperbolic systems of the first order.

Let $\mathcal{B}(\Omega)$ denote the class of all Borel sets of a topological space (i.e. it is the σ-algebra of sets generated by all open sets). If Ω is locally compact we denote (as usual) by $C_0(\Omega)$ the space of all continuous complex-valued functions on Ω vanishing at infinity. Equipped with the uniform norm $\|f\| = \sup_x |f(x)|$ this space is known to be a Banach space. It is also well known (Riesz-Markov theorem) that if Ω is a locally compact space, then the set $\mathcal{M}(\Omega)$ of all finite complex regular Borel measures on Ω equipped with the norm $\|\mu\| = \sup |\int_\Omega f(x)\mu(dx)|$, where sup is taken over all functions $f \in C_0(\Omega)$ with $\|f(x)\| \leq 1$, is a Banach space, which coincides with the set of all continuous linear functionals on $C_0(\Omega)$. Clearly, any complex σ-additive measure μ on $\mathcal{R}^d$ has the form

$$\mu(dy) = f(y)M(dy) \qquad (G1)$$

with some positive measure M (which can be chosen to be finite whenever μ is finite) and some bounded complex-valued function f (in fact, a possible choice of M is $|Re\,\mu| + |Im\,\mu|$, where $|\nu|$ for a real signed measure ν denotes, as usual, its total variation measure, i.e. $|\nu| = \nu_+ + \nu_-$, where $\nu = \nu_+ - \nu_-$ is the Hahn decomposition of ν on its positive and negative parts). Representation (G1) is surely not unique; however, the measure M in (G1) is uniquely defined under additional assumption that $|f(y)| = 1$ for all y. If this condition is fulfilled, the positive measure M is called the total variation measure of the complex measure μ and is denoted by $|\mu|$. Clearly, if a complex measure μ is presented in form (G1) with some positive measure M, then $\|\mu\| = \int |f(y)| M(dy)$.

We say that a map ν from $\mathcal{R}^d \times \mathcal{B}(\mathcal{R}^d)$ into $\mathcal{C}$ is a *complex transition kernel*, if for every x, the map $A \mapsto \nu(x, A)$ is a (finite complex) measure on $\mathcal{R}^d$, and for every $A \in \mathcal{B}(\mathcal{R}^d)$, the map $x \mapsto \nu(x, A)$ is $\mathcal{B}$-measurable. A (time homogeneous) *complex transition function* (abbreviated CTF) on $\mathcal{R}^d$ is a family ν_t, $t \geq 0$, of complex transition kernels such that $\nu_0(x, dy) = \delta(y - x)$ for all x, where $\delta_x(y) = \delta(y - x)$ is the Dirac measure in x, and such that for every non-negative s, t, the Chapman-Kolmogorov equation

$$\int \nu_s(x, dy)\nu_t(y, A) = \nu_{s+t}(x, A)$$

is satisfied. (We consider only time homogeneous CTF for simplicity, the generalisation to non-homogeneous case is straightforward).

A CTF is said to be (spatially) homogeneous, if $\nu_t(x, A)$ depends on x, A only through the difference $A - x$. If a CTF is homogeneous it is natural to denote $\nu_t(0, A)$ by $\nu_t(A)$) and to write the Chapman-Kolmogorov equation in the form

$$\int \nu_t(dy)\nu_s(A - y) = \nu_{t+s}(A).$$

A CTF will be called *regular*, if there exists a positive constant K such that for all x and $t > 0$, the norm $\|\nu_t(x, .)\|$ of the measure $A \mapsto \nu_t(x, A)$ does not exceed $\exp\{Kt\}$.

CTFs appear naturally in the theory of evolutionary equations: if T_t is a strongly continuous semigroup of bounded linear operators in $C_0(\mathcal{R}^d)$, then there exists a time-homogeneous CTF ν such that

$$T_t f(x) = \int \nu_t(x, dy) f(y). \qquad (G2)$$

In fact, the existence of a measure $\nu_t(x, .)$ such that (G2) is satisfied follows from the Riesz-Markov theorem, and the semigroup identity $T_s T_t = T_{s+t}$ is equivalent to the Chapman-Kolmogorov equation. Since $\int \nu_t(x, dy) f(y)$ is continuous for all $f \in C_0(\mathcal{R}^d)$, it follows by the monotone convergence theorem (and the fact that each complex measure is a linear combination of four positive measures) that $\nu_t(x, A)$ is a Borel function of x.

We say that the semigroup T_t is *regular*, if the corresponding CTF is regular. Clearly, this is equivalent to the assumption that $\|T_t\| \leq e^{Kt}$ for all $t > 0$ and some constant K.

Now we construct a measure on the path space corresponding to each regular CTF, introducing first some (rather standard) notations. Let $\mathcal{R}_d$ denote the one point compactification of the Euclidean space $\mathcal{R}^d$ (i.e. $\dot{\mathcal{R}}_d = \mathcal{R}^d \cup \{\infty\}$ and is homeomorphic to the sphere S^d). Let $\dot{\mathcal{R}}_d^{[s,t]}$ denote the infinite product of $[s,t]$ copies of $\dot{\mathcal{R}}_d$, i.e. it is the set of all functions from $[s,t]$ to $\dot{\mathcal{R}}_d$, the path space. As usual, we equip this set with the product topology, in which it is a compact space (Tikhonov's theorem). Let $Cyl_{[s,t]}^k$ denote the set of functions on $\dot{\mathcal{R}}_d^{[s,t]}$ having the form

$$\phi_{t_0,t_1,\ldots t_{k+1}}^f(y(.)) = f(y(t_0), \ldots, y(t_{k+1}))$$

for some bounded complex Borel function f on $(\dot{\mathcal{R}}^d)^{k+2}$ and some points t_j, $j = 0, \ldots, k+1$, such that $s = t_0 < t_1 < t_2 < \ldots < t_k < t_{k+1} = t$. The union $Cyl_{[s,t]} = \cup_{k \in \mathcal{N}} Cyl_{[s,t]}^k$ is called the set of cylindrical functions (or functionals) on $\dot{\mathcal{R}}_d^{[s,t]}$. It follows from the Stone-Weierstrasse theorem that the linear span of all continuous cylindrical functions is dense in the space $C(\dot{\mathcal{R}}_d^{[s,t]})$ of all complex continuous functions on $\dot{\mathcal{R}}_d^{[s,t]}$. Any CTF ν defines a family of linear functionals $\nu_{s,t}^x$, $x \in \mathcal{R}^d$, on $Cyl_{[s,t]}$ by the formula

$$\nu_{s,t}^x(\phi_{t_0 \ldots t_{k+1}}^f)$$

$$= \int f(x, y_1, \ldots, y_{k+1}) \nu_{t_1-t_0}(x, dy_1) \nu_{t_2-t_1}(y_1, dy_2) \ldots \nu_{t_{k+1}-t_k}(y_k, dy_{k+1}). \quad (G3)$$

Due to the Chapman-Kolmogorov equation, this definition is correct, i.e. if one considers an element from $Cyl_{[s,t]}^k$ as an element from $Cyl_{[s,t]}^{k+1}$ (any function of l variables $y_1, \ldots, y_l$ can be considered as a function of $l+1$ variables $y_1, \ldots, y_{l+1}$, which does not depend on y_{l+1}), then the two corresponding formulae (G3) will be consistent.

Proposition G1. *If the semigroup T_t in $C_0(\mathcal{R}^d)$ is regular and ν is its corresponding CTF, then the functional (G3) is bounded. Hence, it can be extended by continuity to a unique bounded linear functional ν^x on $C(\dot{\mathcal{R}}_d^{[s,t]})$, and consequently there exists a (regular) complex Borel measure $D_x^{s,t}$ on the path space $\dot{\mathcal{R}}_d^{[s,t]}$ such that*

$$\nu_{s,t}^x(F) = \int F(y(.)) D_x^{s,t} y(.) \quad (G4)$$

for all $F \in C(\dot{\mathcal{R}}_d^{[s,t]})$. In particular,

$$(T_t f)(x) = \int f(y(t)) D_x^{s,t} y(.).$$

Proof. It is a direct consequence of the Riesz-Markov theorem, because the regularity of CTF implies that the norm of the functional $\nu_{s,t}^x$ does not exceed $\exp\{K(t-s)\}$.

If E is a measurable subset of $\dot{\mathcal{R}}_d^{[s,t]}$, we shall say (using probabilistic language) that E is an event on $\dot{\mathcal{R}}_d^{[s,t]}$, and we shall denote by $\nu_{s,t}^x(E)$ the value of the functional $\nu_{s,t}^x$ on the indicator χ_E of E, i.e.

$$\nu_{s,t}^x(E) = \int \chi_E(y(.))D_x^{s,t}y(.). \tag{G4'}$$

Formula (G3) defines the family of finite complex distributions on the path space , which gives rise to a finite complex measure on this path space (under the regularity assumptions). Therefore, this family of measures can be called a complex Markov process. Unlike the case of the standard Markov processes, the generator, say A, of the corresponding semigroup T_t is not self-adjoint, and the corresponding bilinear "Dirichlet form" (Av, v) is complex. Such forms present a natural generalisation of the real Dirichlet forms that constitute an important tool in modern probability theory, see e.g. [Fu], [MR] and references therein. In the complex situation, only some particular special cases have so far been investigated, see [AU].

The following simple fact can be used in proving the regularity of a semigroup.

Proposition G2. *Let B and A be linear operators in $C_0(\mathcal{R}^d)$ such that A is bounded and B is the generator of a strongly continuous regular semigroup T_t. Then $A + B$ is also the generator of a regular semigroup, which we denote by $\tilde{T}_t$.*

Proof. Follows directly from the fact that $\tilde{T}_t$ can be presented as the convergent (in the sense of the norm) series of standard perturbation theory

$$\tilde{T}_t = T_t + \int_0^t T_{t-s}AT_s\,ds + \int_0^t ds \int_0^s d\tau T_{t-s}AT_{s-\tau}AT_\tau + ... \tag{G5}$$

Of major importance for our purposes are the spatially homogeneous CTFs. Let us discuss them in greater detail, in particular, their connection with infinitely divisible characteristic functions. Let $\mathcal{F}(\mathcal{R}^d)$ denote the Banach space of Fourier transforms of elements of $\mathcal{M}(\mathcal{R}^d)$, i.e. the space of (automatically continuous) functions on $\mathcal{R}^d$ of form

$$V(x) = V_\mu(x) = \int_{\mathcal{R}^d} e^{ipx}\,\mu(dp) \tag{G6}$$

for some $\mu \in \mathcal{M}(\mathcal{R}^d)$, with the induced norm $\|V_\mu\| = \|\mu\|$. Since $\mathcal{M}(\mathcal{R}^d)$ is a Banach algebra with convolution as the multiplication, it follows that $\mathcal{F}(\mathcal{R}^d)$ is also a Banach algebra with respect to the standard (pointwise) multiplication. We say that an element $f \in \mathcal{F}(\mathcal{R}^d)$ is *infinitely divisible* if there exists a family $(f_t, t \geq 0,)$ of elements of $\mathcal{F}(\mathcal{R}^d)$ such that $f_0 = 1$, $f_1 = f$, and $f_{t+s} = f_t f_s$ for all positive s, t. Clearly if f is infinitely divisible, then it has no zeros and a continuous function $g = \log f$ is well defined (and is unique up to an imaginary shift). Moreover, the family f_t has the form $f_t = \exp\{tg\}$ and is defined uniquely up to a multiplier of the form $e^{2\pi ikt}$, $k \in \mathcal{N}$. Let us say that a continuous function g on $\mathcal{R}^d$ is a *complex characteristic exponent* (abbreviated CCE), if e^g is an infinitely divisible element of $\mathcal{F}(\mathcal{R}^d)$, or equivalently, if e^{tg} belongs to $\mathcal{F}(\mathcal{R}^d)$ for all $t > 0$.

Remark. The problem of the explicit characterisation of the whole class of infinite divisible functions (or of the corresponding complex CCEs) seems to be quite nontrivial. When dealing with this problem, it is reasonable to describe first some natural subclasses. For example, it is easy to show that if $f_1 \in \mathcal{F}(\mathcal{R})$ is infinite divisible and such that the measures corresponding to all functions $f_t, t > 0$, are concentrated on the half line $\mathcal{R}_+$ (complex generalisation of subordinators) and have densities from $L_2(\mathcal{R}_+)$, then f_1 belongs to the Hardy space H_2 of analytic functions on the upper half plane (see e.g. [Koo]), which have no Blaschke product in its canonical decomposition.

It follows from the definitions that the set of spatially homogeneous CTFs $\nu_t(dx)$ is in one-to-one correspondence with CCE g, in such a way that for any positive t the function e^{tg} is the Fourier transform of the transition measure $\nu_t(dx)$.

Proposition G3. *If V is a CCE, then the solution to the Cauchy problem*

$$\frac{\partial u}{\partial t} = V(\frac{1}{i}\frac{\partial}{\partial y})u \qquad (G7)$$

defines a strongly continuous and spatially homogeneous semigroup T_t of bounded linear operators in $C_0(\mathcal{R}^d)$ (i.e. $(T_t u_0)(y)$ is the solution to equation (G7) with the initial function u_0). Conversely, each such semigroup is the solution to the Cauchy problem of an equation of type (G7) with some CCE g.

Proof. This is straightforward. Since (G7) is a pseudo-differential equation, it follows that the Fourier transform $\tilde{u}(t, x)$ of the function $u(t, y)$ satisfies the ordinary differential equation

$$\frac{\partial \tilde{u}}{\partial t}(t, x) = V(x)\tilde{u}(t, x),$$

whose solution is $\tilde{u}_0(x)\exp\{tV(x)\}$. Since e^{tV} is the Fourier transform of the complex transition measure $\nu_t(dy)$, it follows that the solution to the Cauchy problem of equation (G7) is given by the formula $(T_t u_0)(y) = \int u_0(z)\nu_t(dz - y)$, which is as required.

We say that a CCE is *regular*, if equation (G7) defines a regular semigroup.

It would be very interesting to describe explicitly all regular CCE. We only give here two classes of examples. First of all, if a CCE is given by the Lévy-Khintchine formula (i.e. it defines a transition function consisting of probability measures), then this CCE is regular, because all CTF consisting of probability measures are regular. Another class is given by the following result.

Proposition G4. *Let $V \in \mathcal{F}(\mathcal{R}^d)$, i.e. it is given by (G6) with $\mu \in \mathcal{M}(\mathcal{R}^d)$. Then V is a regular CCE. Moreover, if the positive measure M in the representation (G1) for μ has no atom at the origin, i.e. $M(\{0\}) = 0$, then the corresponding measure $D_x^{0,t}$ on the path space from Proposition G1 is* concentrated on the set of piecewise-constant paths in $\dot{\mathcal{R}}_d^{[0,t]}$ with a finite number of jumps. In other words, $D_x^{0,t}$ is the measure of a jump-process.

Proof. Let $W = W_M$ be defined by the formula

$$W(x) = \int_{\mathcal{R}^d} e^{ipx} M(dp). \tag{G8}$$

The function $\exp\{tV\}$ is the Fourier transform of the measure $\delta_0 + t\mu + \frac{t^2}{2}\mu \star \mu + ...$ which can be denoted by $\exp^\star(t\mu)$ (it is equal to the sum of the standard exponential series, but with the convolution of measures instead of the standard multiplication). Clearly $\| \exp^\star(t\mu)\| \le \| \exp^\star(t\bar{f}M)\|$, where we denoted by $\bar{f}$ the supremum of the function f, and both these series are convergent series in the Banach algebra $\mathcal{M}(\mathcal{R}^d)$. Therefore $\|e^{Vt}\| \le \|e^{Wt}\| \le \exp\{t\bar{f}\|\mu\|\}$, and consequently V is a regular CCE. Moreover, the same estimate shows that the measure on the path space corresponding to the CCE V is absolutely continuous with respect to the measure on the path space corresponding to the CCE W. But the latter coincides up to a positive constant multiplier with the probability measure of the compound Poisson process with the Lévy measure M defined by the equation

$$\frac{\partial u}{\partial t} = (W(\frac{1}{i}\frac{\partial}{\partial y}) - \lambda_M)u, \tag{G9}$$

where $\lambda_M = M(\mathcal{R}^d)$, or equivalently

$$\frac{\partial u}{\partial t} = \int (u(y+\xi) - u(y)) M(d\xi), \tag{G10}$$

(see e.g. [Br], [Meti] or [Pr] for the necessary background in compound Poisson processes), because the condition $M(\{0\}) = 0$ ensures that M is actually a measure on $\mathcal{R}^d \setminus \{0\}$, i.e. it is a finite Lévy measure. It remains to note that as is well known in the theory of stochastic processes (see e.g. [Pr],[Fel]) the measures of compound Poisson processes are concentrated on piecewise-constant paths.

Therefore, we have two different classes (essentially different, because they obviously are not disjoint) of regular CCE: those given by the Lévy-Khintchine formula, and those given by Proposition G4. It is easy to prove that one can combine these regular CCEs, more precisely that the class of regular CCE is a convex cone, see [K13].

Let us apply the simple results obtained so far to the case of the pseudo-differential equation of the Schrödinger type

$$\frac{\partial \tilde{u}}{\partial t} = -G(-\Delta)^\alpha \tilde{u} + (A, \frac{\partial}{\partial x})\tilde{u} + V(x)\tilde{u}, \tag{G11}$$

where G is a complex constant with a non-negative real part, α is any positive constant, A is a real-valued vector (if $\operatorname{Re} G > 0$, then A can be also complex-valued), and V is a complex-valued function of form (G6). The standard Schrödinger equation corresponds to the case $\alpha = 1$, $G = i$, $A = 0$ and

V being purely imaginary. We consider a more general equation to include the Schrödinger equation, the heat equation with drifts and sources, and also their stable (when $\alpha \in (0,1)$) and complex generalisations in one formula. This general consideration also shows directly how the functional integral corresponding to the Schrödinger equation can be obtained by the analytic continuation from the functional integral corresponding to the heat equation, which gives a connection with other approaches to the path integration (see Appendix H). The equation on the inverse Fourier transform

$$u(y) = (2\pi)^{-d} \int_{\mathcal{R}^d} e^{-iyx} \tilde{u}(x)\, dx$$

of $\tilde{u}$ (or equation (G11) in momentum representation) clearly has the form

$$\frac{\partial u}{\partial t} = -G(y^2)^{\alpha} u + i(A,y)u + V\left(\frac{1}{i}\frac{\partial}{\partial y}\right)u. \qquad (G12)$$

One easily sees that already in the trivial case $V = 0$, $A = 0$, $\alpha = 1$, equation (G11) defines a regular semigroup only in the case of real positive G, i.e. only in the case of the heat equation. It turns out however that for equation (G12) the situation is completely different. The next result generalises the corresponding result from [M7], [MC2] on the standard Schrödinger equation to equation (G11).

Proposition G5. *The solution to the Cauchy problem of equation (G12) can be written in the form of a complex Feynman-Kac formula*

$$u(t,y) = \int \exp\{-\int_0^t [G(q(\tau)^2)^{\alpha} - (A,q(\tau))]\,d\tau\} u_0(q(t)) D_y^{0,t} q(.), \qquad (G13)$$

where D_y is the measure of the jump process corresponding to equation (G7).

Proof. Let T_t be the regular semigroup corresponding to equation (G7). By the Trotter formula, the solution to the Cauchy problem of equation (G12) can be written in the form

$$u(t,y) = \lim_{n\to\infty} \left(\left(\exp\{-\frac{t}{n}(G(y^2)^{\alpha} - (A,y))\} T_{t/n} \right)^n u_0 \right)(y)$$

$$= \lim_{n\to\infty} \int \exp\{-G\frac{t}{n}[(q_1^2)^{\alpha} + \dots + (q_n^2)^{\alpha}] + \frac{t}{n}(A, q_1 + \dots + q_n)\} u_0(q_n)$$

$$\times \nu_{t/n}(y, dq_1)\nu_{t/n}(q_1, dq_2)\dots\nu_{t/n}(q_{n-1}, dq_n).$$

Using (G3), we can rewrite this as

$$u(t,y) = \lim_{n\to\infty} \nu_{0,t}^y(F_n) = \lim_{n\to\infty} \int F_n(q(.)) D_y^{0,t} q(.),$$

where F_n is the cylindrical function

$$F_n(q(.)) = \exp\{-G\frac{t}{n}[(q(t/n)^2)^{\alpha} + (q(2t/n)^2)^{\alpha} + \dots + (q(t)^2)^{\alpha}]\}$$

$$\times \exp\{\frac{t}{n}(A, q(t/n) + q(2t/n) + ... + q(t))\}u_0(q(t)).$$

By the dominated convergence theorem this implies (G13).

The statement of the Proposition can be generalised easily to the following situation, which includes all Schrödinger equations, namely to the case of the equation

$$\frac{\partial \phi}{\partial t} = i(A - B)\phi,$$

where A is selfadjoint operator, for which therefore exists (according to spectral theory) a unitary transformation U such that UAU^{-1} is the multiplication operator in some $L^2(X, d\mu)$, where X is locally compact, and B is such that UBU^{-1} is a bounded operator in $C_0(X)$.

As another example, let us consider the case of complex anharmonic oscillator. i.e. the equation

$$\frac{\partial \tilde{\psi}}{\partial t} = \frac{1}{2}\left(G\Delta - x^2 - iV(x)\right)\tilde{\psi}, \tag{G15}$$

where $V = V_\mu$ is an element of $\mathcal{F}(\mathcal{R}^d)$. The Fourier transform of this equation (or, equation (G15) in the p-representation) has the form

$$\frac{\partial \psi}{\partial t} = \frac{1}{2}\left(\Delta - Gp^2 - iV(\frac{1}{i}\frac{\partial}{\partial p})\right)\psi. \tag{G16}$$

Proposition G6. *If $Re\,G \geq 0$, the Cauchy problem of equation (G16) defines a regular semigroup of operators in $C_0(\mathcal{R}^d)$, and thus can be presented as the path integral from Proposition G1.*

Proof. If $V = 0$, the Green function for equation (G17) can be calculated explicitly (see e.g. Section 1.4), and from this formula one easily deduced that in case $V = 0$ the semigroup defined by equation (G17) is regular. For general V the statement follows from Proposition G2.

For numerical calculations of path integral (see e.g. [CheQ]), it is convenient to write a path integral as an integral over a positive, and not a complex measure. This surely can be done, because any complex measure has a density with respect to its total variation measure. To be more concrete, suppose that V is given by (G6), (G1), where the positive measure M has no atom at the origin. Then it follows from Proposition G4 that the complex measure on the path space defined (according to Proposition G1) by equation (G7) has a density with respect to the measure of the Poisson process described by equation (G9). To conclude this Appendix we shall calculate this density. This will imply an alternative form of the path integral in (G13), which has a more clear probabilistic interpretation. We give here a probabilistic proof of this result, which includes a study of the main properties of the complex measure $D_y^{s,t}$ and the corresponding functional $\nu_{s,t}^y$, which are defined by equation (G7) (according to Proposition G1) and which are similar to the properties of the underlying Poisson process defined by equation G9. Another, more direct and more simple proof is given in the first section of Chapter 9.

Since the trajectories of a compound Poisson process are piecewise constant (and finite) almost surely, a typical random path Y of such a process on the interval of time $[0, t]$ starting at a point y is defined by a finite, say n, number of the moments of jumps $0 < s_1 < ... < s_n \leq t$, which are distributed according to the Poisson process N with the intensity $\lambda_M = M(\mathcal{R}^d)$, and by the independent jumps $\delta_1, ..., \delta_n$ at these moments, each of which is a random variable with values in $\mathcal{R}^d \setminus \{0\}$ and with the distribution defined by the probability measure M/λ_M. This path has the form

$$Y_y(s) = y + Y^{s_1...s_n}_{\delta_1...\delta_n}(s) = \begin{cases} Y_0 = y, & s < s_1, \\ Y_1 = y + \delta_1, & s_1 \leq s < s_2, \\ ... & \\ Y_n = y + \delta_1 + \delta_2 + ... + \delta_n, & s_n \leq s \leq t \end{cases} \qquad (G17)$$

We shall denote by $E^{[0,t]}_y$ the expectation with respect to this process.

Let us now obtain some properties of the complex measure $D^{s,t}_y$ and the corresponding functional $\nu^y_{s,t}$, which are defined by equation (G7) (according to Proposition G1) and which are similar to the corresponding properties of positive measures of compound Poisson processes defined by equation (G9).

As a first consequence from Proposition G4 we conclude that typical paths for the measure $D^{s,t}_y$ do not go to infinity in finite times almost surely (from now on, almost surely will be always understood in the sense of the Poisson process described by equation (G9)) and can be parametrised by (G17). In particular, we can consider only the restricted path space $\mathcal{R}^{[s,t]}_d$ (instead of $\dot{\mathcal{R}}^{[s,t]}_d$), which is more convenient, because of its linear structure. Let us say that a cylindrical functional F on $\mathcal{R}^{[s,t]}_d$ (respectively an event E on $\mathcal{R}^{[s,t]}_d$) is translation invariant, iff $F(y + Y(.)) = F(Y(.))$ for any path Y and any vector y (respectively if $Y(.) \in E$ implies $y + Y(.) \in E$). It follows from the definition of ν that if a cylindrical functional F or an event E are translation invariant, then $\nu^y_{s,t}(F)$ or $\nu^y_{s,t}(E)$ respectively does not depend on the initial point y, and will be denoted simply by $\nu_{s,t}(F)$ or $\nu_{s,t}(E)$ respectively. From the Chapman-Kolmogorov equation we obtain now directly the following result, which is the analogy of the probabilistic concept of the independence of increments: if E_1 and E_2 are some events on $\mathcal{R}^{[s,\tau]}_d$ and $\mathcal{R}^{[\tau,t]}_d$ respectively, and if E_2 is translation invariant, then

$$\nu^y_{s,t}(E_1 \cap E_2) = \nu^y_{s,\tau}(E_1)\nu_{\tau,t}(E_2). \qquad (G18)$$

for an arbitrary y.

Let $E_j(t)$ denote the event that there are exactly j jumps of the process on the interval of the length t, and the trajectories are constant between the jumps. As we noted already, the whole measure is concentrated on the union $\cup_{j=0}^{\infty} E_j$.

Proposition G7. Let $\phi^j_t = \nu_{0,t}(E_j(t))$ and $\lambda_\mu = \int \mu(dy) = \mu(\mathcal{R}^d)$. Then

$$\phi^j_t = (\lambda_\mu t)^j / j!. \qquad (G19)$$

Proof. Notice first that $\phi_t^j = O(t^j)$, $t \to 0$, for all j, because of Proposition G4 and by the well known properties of compound Poisson processes. Let $\tilde{E}_0(t)$ denote the event that the end-points of trajectories coincide, i.e. $Y(0) = Y(t)$. Clearly, $E_0(t) \subset \tilde{E}_0(t)$ and the intersection $E_1(t) \cap \tilde{E}_0(t)$ is empty. Since $\nu_{0,t}(\cup_{j>1} E_j) = O(t^2)$, it follows that

$$\phi_t^0 = \nu_{0,t}(\tilde{E}_0(t)) + O(t^2)$$

for small t. But

$$\nu_{0,t}(\tilde{E}_0(t)) = \int \nu_t(dy)\chi_{\{0\}}(y) = 1 + O(t^2),$$

where $\chi_{\{0\}}$ denotes the indicator of the one-point set $\{0\}$, because $\nu_t = \exp^*(t\mu)$ (see the proof of Proposition G4) and $\mu(\{0\}) = 0$ by our assumptions. Therefore, $\phi_t^0 = 1 + O(t^2)$ for small t. Noticing that $\phi_t^0 = (\phi_{t/n}^0)^n$ for all n by (G18) we conclude that $\phi_t^0 = \lim_{n\to\infty}(1 + O(t/n)^2)^n = 1$, which proves the statement of the proposition for $j = 0$.

Next, since

$$\nu_{0,t}(\mathcal{R}_d^{[0,t]}) = \nu_{0,t}(\cup_{j=0}^\infty E_j) = \nu_t(\mathcal{R}^d) = \exp^*(t\mu)(\mathcal{R}^d) = e^{t\lambda_\mu},$$

it follows that

$$\phi_t^1 = e^{t\lambda_\mu} - 1 - O(t^2) = t\lambda_\mu + O(t^2). \tag{G20}$$

The proof can be completed now as in the case of the standard Poisson process (see e.g. [Pr]). Namely, let N_t denote the number of jumps of the process on the interval $[0, t]$. Consider the random functional α^{N_t}, where α is a parameter from $(0, 1)$, and its average

$$\omega(t) = \omega_\alpha(t) = \nu_{0,t}(\alpha^{N_t}).$$

By (G18), $\omega(t + s) = \omega(t)\omega(s)$. Moreover, the function ω is clearly measurable and bounded. This implies (see e.g. [Br]) that $\omega(t) = e^{t\psi(\alpha)}$ with some function ψ. In particular, $\omega'(0) = \psi(\alpha)$. But clearly

$$\omega(t) = \sum_{n=0}^\infty \alpha^n \phi_t^n,$$

and consequently, due to (G20), $\omega'(0) = \alpha(\phi_t^1)' = \alpha\lambda_\mu$, and therefore $\omega(t) = \exp\{t\alpha\lambda_\mu\}$, which evidently implies (G19).

Proposition G8. *Suppose U is a Borel subset of $\mathcal{R}^d \setminus \{0\}$ and $I = [t_1, t_2] \subset [0, t]$ is an interval of the length $t_2 - t_1 = \tau$. Let E_I^U denotes the event that in the interval $[0, t]$ there is only one jump, and moreover, this jump occurs on a moment of time from I, and the size of this jump belongs to U. Then, for small τ,*

$$\nu_{0,t}(E_\tau^U) = \mu(U)\tau + O(\tau^2). \tag{G21}$$

Proof. Due to the equation $\nu_{0,t_1}(E_0(t_1 - t)) = \nu_{t_2,t}(E_0(t - t_2)) = 1$, which follows from Proposition G7, and using formula (G18), one concludes that $\nu_{0,t}(E_I^U) = \nu_{t_1,t_2}(E_I^U)$. Furthermore, since the measure of more than one jump on any interval of the length τ is of order $O(\tau^2)$, one has

$$\nu_{0,t}(E_I^U) = \int \chi_U(y)\nu_\tau(dy) + O(\tau^2) = \nu_\tau(U) + O(\tau^2) = \tau\mu(U) + O(\tau^2).$$

Let us note by passing that the weak topology on $\dot{\mathcal{R}}_d^{[s,t]}$ restricted to the set $E_j(t - s)$ generates on E_j a natural topology, in which two trajectories from $E_j(t - s)$ are close, if the times and the sizes of all their jumps are close.

Now we can prove the central property of the measure $\nu_{s,t}^x$.

Proposition G9. *Let μ and M are connected by (G1), M is a positive finite measure without an atom at the origin and f is a bounded Borel-measurable complex-valued function. Then the measure $\nu_{s,t}^y$ constructed from μ has the density with respect to $\tilde{\nu}_{s,t}^y$ constructed from M and this density is the function ϕ such that $\phi(Y(.)) = \prod_{j=1}^n f(\delta_j)$ on the path $Y = Y_y(s)$ of the form (G17).*

Proof. One can prove the statement separately on each event E_n. The case $n = 0$ follows from Proposition G7 with $j = 0$. Consider the case $n = 1$. Then, if I is an interval of the length τ containing the moment of time s_1, and if $U = U_\epsilon(\delta_1)$ is the ball of the radius ϵ with the centre δ_1, then by Proposition G8

$$\frac{\nu_{0,t}(E_I^U)}{\tilde{\nu}_{0,t}(E_I^U)} = \frac{\mu(U)}{M(U)} + O(\tau).$$

It follows from (0.2) that the r.h.s. of this expression tends to $f(\delta_1)$ as $\tau \to 0$ and $\epsilon \to 0$, which proves the required result for $n = 1$. The case of an arbitrary n is considered similarly.

Since obviously the measure $\tilde{\nu}_{s,t}^y$ has the (constant) density $\exp\{(t - s)\lambda_M\}$ with respect to the probability measure of the Poisson process defined by equation (G9), the following statement follows straightforwardly from Propositions G5 and G9.

Proposition G10. *Let u_0 be a bounded continuous function. Then the solution to the Cauchy problem of equation (G12) with the initial function u_0 has the form*

$$u(t,y) = \exp\{t\lambda_M\} E_y^{[0,t]} [F(Y(.))u_0(Y(t))], \qquad (G22)$$

where if Y has form (G17),

$$F(Y(.)) = \exp\{-\sum_{j=0}^n [G(Y_j, Y_j)^\alpha - i(A, Y_j)](s_{j+1} - s_j)\} \prod_{j=1}^n (f(\delta_j))$$

(s_{n+1} is assumed to be equal to t in this formula, and the function f is defined in (G1)).

In particular, choosing u_0 to be the exponential function e^{iyx_0} one obtains a path integral representation for the Green function of equation (G11) in momentum representation.

H. A review of main approaches
to the rigourous construction of path integral

We give here a short review of main approaches to the definition of path integrals, where the integral is defined as a sort of generalised functional in some functional space, and not as a genuine integral over a bona fide measure on a path space, (as in the approach described in Appendix G or in chapter 9). As shown in Chapter 7, these approaches can also be used to construct path integral representation to complex stochastic Schrödinger equations.

1. *Analytic continuation, or complex rotation.*

This is one of the earliest approaches to the mathematical theory of Feynman integrals (see e.g. [Ca], [Nel2], [Joh1], [CaS], [Kal2], [KKK], [Chu] and referencess therein). In particular, in [KKK], [Chu], this integral is used for the repesentation of the fundamental solution of the Schrödinger equation. In this approach one considers first one of the main parameters, say the mass m, in the standard Schrödinger equation

$$\frac{\partial \psi}{\partial t} = (\frac{i}{2m}\Delta - iV(x))\psi \qquad (H1)$$

to be imaginary, i.e. of the form $m = i\tilde{m}$, with $\tilde{m} > 0$. In this case, equation (H1) is a diffusion equation (with a complex source), whose solution can be therefore written in terms of an integral over the standard Wiener measure (using the Feynman-Kac formula). This define a function of m for imaginary m. The analytic continuation of this function (if it exists) can be considered as a definition of the path integral for complex (in particular, real) m. This continuation is often called the analytic Feynman integral. Equivalently one can carry out the analytical continuation in time.

A similar, but slightly different approach is obtained by the idea of rotation in configuration space, see e.g. [HuM]. Namely, changing the variables x to $y = \sqrt{i}x$ in equation (H1) leads to the equation

$$\frac{\partial \psi}{\partial t} = (-\frac{1}{2m}\Delta - iV(-\sqrt{i}y))\psi,$$

which is again of diffusion type and can be thus treated by means of the Feynman-Kac formula and the Wiener measure. To use this approach one needs certain analytic assumptions on V.

Interesting applications of the analytic Feynman integral to the mathematical theory of Feynman's non-commutative operational calculus can be found in the series of papers [JoL1], [JoL2], [DeFJL], see also references therein.

2. *Parceval equality.*

This approach was first systematically developed in [AH1].

Let h be a complex constant with a non-negative real part and let L be a complex matrix such that $1 + L$ is non-degenerate with a positive real part. To see the motivation for the main definition given below, suppose first that a function g has the form

$$g(x) = g_\mu(x) = \int_{\mathcal{R}^d} e^{-ipx}\hat{g}(p)\,dp = \int_{\mathcal{R}^d} e^{-ipx}\mu(dp)$$

with some $\hat{g}$ from the Schwarz space S, where we denoted by μ the finite measure on $\mathcal{R}^d$ with the density $\hat{g}(p)$. Then, g also belongs to the Schwarz space and moreover, due to the Parceval equality,

$$\int \exp\{-\frac{1}{2h}((1+L)x,x)\}g(x)\,dx$$

$$= (2\pi h)^{d/2}(\det(1+L))^{-1/2}\int \exp\{-\frac{h}{2}((1+L)^{-1}p,p)\}\mu(dp), \qquad (H2)$$

where both sides of this equation are well defined as Riemann integrals. Suppose more generally, that

$$g(x) = g_\mu(x) = \int_{\mathcal{R}^d} e^{-ipx}\mu(dp)$$

for a finite Borel complex measure μ (not necessarily with a density), i.e. g belongs to the space $\mathcal{F}(\mathcal{R}^d)$ of the Fourier transforms of finite Borel measures on $\mathcal{R}^d$. In this case, though the l.h.s. of (H2) may not be well defined in the sense of Riemann or Lebesgue, the r.h.s. is still well defined and can be therefore considered as some sort of the regularisation of the (possibly divergent) integral on the l.h.s. of (H2). In other words, in this case, the integral on the l.h.s. of (H2) can be naturally defined by the r.h.s. expression of this equation. In order to get in (H2) an expression not depending on the dimension (which one needs to pass succesfully to the infinite dimensional limit), one needs to normalise (or, in physical language, renormalise) this integral by the multiplier $(2\pi h)^{-d/2}$. This leads to the following definition [AH1], which can be now given directly in the infinite dimensional setting. Let H be a real separable Hilbert space, let

$$g(x) = g_\mu(x) = \int_H e^{-ipx}\mu(dp) \qquad (H3)$$

be a Fourier transform of a finite complex Borel measure μ in H and let L be a selfadjoint trace class operator in H such that $1 + L$ is an isomorphism of H with a non-negative real part. Define the (normalised) Fresnel integral

$$\int_H^* \exp\{-\frac{1}{2h}((1+L)x,x)\}g(x)\,Dx$$

$$= (\det(1+L))^{-1/2}\int_H \exp\{-\frac{h}{2}((1+L)^{-1}p,p)\}\mu(dp). \qquad (H4)$$

For application to the Schrödinger equation one takes as the Hilbert space H in (H4) the space H_t (sometimes called the Cameron-Martin space) of continuous curves $\gamma : [0,t] \mapsto \mathcal{R}^d$ such that $\gamma(t) = 0$ and the derivative $\dot{\gamma}$ of γ (in the sense of distributions) belongs to $L_2([0,t])$, the scalar product in H_t being defined as

$$(\gamma_1, \gamma_2) = \int_0^t \dot{\gamma}_1(s)\dot{\gamma}_2(s)\,ds.$$

It is not very difficult to prove the following result:

Proposition H1 [AH1]. *If $\psi_0 \in \mathcal{F}(\mathcal{R}^d) \cap L^2(\mathcal{R}^d)$ and $V \in L^2(\mathcal{R}^d)$, then the (obviously unique) solution to the Cauchy problem of equation (H2) with the initial data ψ_0 has the form*

$$\psi(t,x) = \int_{H_t}^{*} \exp\{\frac{i}{2m}\|\gamma\|^2\} \exp\{-i \int_0^t V(\gamma(s) + x) \, ds\} \, D\gamma,$$

where this Fresnel integral is well defined in the sense of formula (H4) (with $h = im$, $L = 0$).

The definition of the normalised (infinite dimensional) Fresnel integral (H4) can be generalised in various ways. The most advanced definition in this direction was given in [CW1], where the (infinite dimensional)differential $D_{\Theta,Z}$ was defined (in a sense, axiomatically) by the formula

$$\int_{\Phi} \Theta(\phi, J) D_{\Theta,Z}\phi = Z(J),$$

where Φ and Φ' are two Banach spaces and $\Theta : \Phi \times \Phi' \mapsto \mathcal{C}$, $Z : \Phi \mapsto \mathcal{C}$ are two given maps.

3. Discrete approximations (see e.g. [Tr1],[Tr2],[ET]).

One says that a Borel measurable complex valued function f on $\mathcal{R}^d$ is $\mathcal{F}_h$-*integrable*, if the limit

$$\lim_{\epsilon \to 0} (2\pi h)^{-d/2} \int_{\mathcal{R}^d} \exp\{-\frac{1}{2h}|x|^2\} f(x)\psi(\epsilon x) \, dx \qquad (H5)$$

exists for any $\psi \in S(\mathcal{R}^d)$: $\psi(0) = 1$, and is ψ-independent. The limit (H5) is then called the *normalised Fresnel integral* (abbreviated NFI) of f (with parameter h) and will be denoted by

$$\int_{\mathcal{R}^d}^{*} \exp\{-\frac{1}{2h}x^2\} f(x) \, Dx.$$

Let H be a separable real Hilbert space. A Borel measurable complex-valued function f on H is called $\mathcal{F}_h$-*integrable* iff for any increasing sequence $\{P_n\}$ of finite dimensional orthogonal projections in H, which is strongly convergent to the identity operator, the limit

$$\lim_{n \to \infty} \int_{P_n H}^{*} \exp\{-\frac{1}{2h}x^2\} f(x) \, dx$$

exists and its value is independent of the choice of the sequence $\{P_n\}$. In such a case, their common value denoted by

$$\int_H^{*} \exp\{-\frac{1}{2h}|x|^2\} f(x) \, Dx \qquad (H6)$$

is called the NFI of f (with parameter h). In the case of positive h (resp. purely imaginary h), the NFI (H6) is called the normalised Gaussian (resp. oscillatory) integral. Not surprisingly, it turns out that this definition leads to the same formula (H4), which was the starting point for the definition of [AH1].

We formulate here a result from [AKS2] on the existence of NFI in a slightly more general context than in (H4), which is neccessary for applications to stochastic Schrödinger equations. The prooof of this generalisation does not differ from the proof in the case (H4) (see e.g. [ET], [AlB]), and we shall not give it here.

Proposition H2 [ET],[AlB],[AKS2]. *Let a function g be the Fourier transform of a complex Borel finite measure on a Hilbert space H, i.e. it is given by (H3). Let $l \in H$, $\omega \in C$, and let L be a selfadjoint trace class operator in H such that $1 + 2\omega hL$ is an isomorphism of H with a non-negative real part. Then the function $\exp\{(l, x) - \omega(Lx, x)\}g(x)$ is $\mathcal{F}_h$-integrable, and*

$$\int_H^* \exp\{-\frac{1}{2h}|x|^2 - \omega(Lx, x) + (l, x)\}g(x)\, Dx$$

$$= \det(1 + 2\alpha hL)^{-1/2} \int_H \exp\{-\frac{h}{2}((1 + 2\alpha hL)^{-1}(p - il), p - il)\}\mu(dp). \quad (H7)$$

An application of this fact to the theory of stochastic Schrödinger equation is given in Chapter 7.

NFI (H6) was defined above as a limit of descrete approximations, which is not dependent on the choice of approximation. The most natural concrete type of discrete approximations for heuristic Feynman's integral giving the solutions to the Schrödinger equation can be obtained from the Trotter product formula. A detailed account of the approach to Feynman's integral based on this particular approximation, can be found e.g. in [Bere], see also [Joh2] and references therein.

4. Path integral as a symbol for perturbation theory. This approach was systematically developed in [SF]. Here one considers the path integral simply as a convenient concise symbol, which encodes the rules of perturbation theory in a compact form. From this point of view, one can develope rigorously (at least for Gaussian type integrals, which are important for quantum field theory) a technique of calculations and transformations, which contains all combinatorial aspects of the method of Feynman's diagramms.

Notice that this approach considered from the point of view of complex Markov processes from Appendix G leads to the theory of Feynman's integral for Schrödinger equations developed in Chaprter 9.

5. Path interal from the point of view of white noise calculus. In this approach, see [HKPS], [SH] and references therein, path integral is considered as a distribution in Hida's infinite dimensional calculus, called also the white noise analysis, which is a calculus on the dual S' to the Schwarz space $S(\mathcal{R}^d)$.

Various extensions of the approaches described above and their applications are developed in many papers, see e.g. [AlB],[ABB],[ACH],[AH2], [Bere], [CW2], [DMN], [El], [SS], and references therein. In particular, one can find applications to semiclassical asymptotics in [ABB], [AH2], [M1], to differential equations on manifolds in [El],[CW2], to stochastic and infinite dimensional generalisations of Schrödinger's equation in [AKS1], [AKS2], [K1], [TZ1], to rigorous calculations of important quantities of quantum mechanics and quantum field theory in [Bere], [DMN], and the definition of the Feynman integral over the phase space in [DNM], [SS], [SF], [Bere]. A detailed discussion of the connections between different approaches is given in [SS]. Let us mention also the paper [Joh2], where one can find a nice elementary discussion of general mathematical and physical ideas leading to the notion of Feynman's integral, in particular, of the background of the original papers of R. Feynman.

I. Perspectives and problems

We discuss here some problems which arise from the theory developed in this book, and indicate some possible generalisations.

1) One of the central questions arising from the developed theory is to find the general class of Feller-Courrège processes, for which one can construct semiclassical asymptotics by the methods of Chapter 3 or 6. The question concerns the characterisation of the necessary properties of the Lévy measure ν and (possibly degenerate) matrix G of diffusion coefficients. Already for $\nu = 0$ the question is not trivial, see example from Section 3.6 showing that for a non-regular degenerate diffusions, the small time asymptotics of its Green function are not regular, but the semiclassical small h asymptotics can be still constructed by the standard scheme. More general, it is important to try to justify this construction for general Maslov's tunnel equation (as defined in Chapter 6) and the corresponding tunnel systems, which comprise important physical examples going far beyond the range of the problems of the probability theory.

2) An important question is whether it is possible to prove the unimodality of general non-symmetric stable laws. Probably, some ideas from [IC] or [Yam] corresponding to the one-dimensional case can be used here for such a generalisation, which would allow the whole theory of Chapters 5,6 to be automatically generalised to the case of general (non-symmetric) stable jump- diffusions. Next, apparently there exists a deeper connection between the theory of stable Lévy motions and the Brunn-Minkowski theory. Namely, due to a famous result of Minkowski, to each symmetric measure μ on a sphere (with some additional weak non-degeneracy assumption) corresponds a unique convex body whose surface area measure coincides with μ (see e.g. [Sch] for the proof of this result and the definition of the surface area measure). It seems that the properties of local times and excursions of finite dimensional stable motions with the spectral measure μ should be governed by the surface structure of the corresponding convex body.

3) It was demonstrated in Section 5.6 how the analytic results of Chapter 5 can be used to obtain the generalisation of the lim sup law of stable motions to the case of corresponding diffusions. It seems that in the same way lots of other results of stable motions, for instance on the behaviour of lim inf (see

[Ber], [We] and references therein), can be generalised to stable and stable-like jump-diffusions. Furthermore, in our exposition, we avoided the study of completely skew stable laws, especially those corresponding to the subordinators, which are stable motions with $\alpha < 1$, $d = 1$, and a Lévy measure concentrated on the positive half line only. These processes play an important role in the general theory of stochastic processes. It seems possible to develop the theory of the corresponding stable diffusions (subordinators non-homogeneous in space) using the approaches from the present paper. Lots of the beautiful properties of subordinators (see e.g. [Ber]) can then be generalised to the corresponding stable diffusions.

4) The two-sided estimate for the (generalised) heat kernels we proved, say in Theorem 3.2 of Chapter 5 or in Theorem 2.3 from Chapter 7, are restricted to finite times, i.e. they are not uniform with respect to $t \to \infty$, which means that they can not be used for the study of the behaviour of the corresponding processes for large times. It would be interesting to generalise these estimates for all times using perhaps some ideas from the well studied case of the standard diffusion (see e.g. [Da1],[Da2]). If such results would be obtained, one can use them, for example, to obtain the information on the sample path properties of stable jump-diffusions, as time tends to infinity, in the same way, as the estimates for finite times are used in section 5.6 to the study of the sample path properties for small times.

Two-sided estimates of the heat kernel of diffusion equations can be used to deduce the Harnack inequalities for positive solutions of these equations (see references in the introduction). Using the two-sided estimates from Chapter 5 one can hope to obtain some analogous inequalities for stable jump-diffusions.

5) In order to globalise the asymptotics for complex stochastic equations of Chapter 7, one needs the results on the global existence of the solutions to the boundary-value problem of complex Hamiltonians.

6) An interesting problem is to develop the scattering theory for the stochastic Schrödinger equation, whose "free" evolution is given by Theorem 4.3 of hapter 1.

7) An alternative way of study the stochastic Schrödinger equations of Chapter 7 can be based on the "elementary formula" of Elworthy-Truman, see e.g. [TZ1], [TZ2] for this approach in the case of the unitary evolution or of the stochastic heat equation. It is natural to try to develop this approach in the general complex stochastic situation. This development may help to answer many questions arising around the representation of the solutions of stochastic and quantum stochastic equations in terms of the Feynmann path integral.

8) Stochastic Schrödinger equations considered in Chapter 7 do not exaust the class of SSE which appear in physics. One of natural generalisations convern the equations where the white noise is substituted by a general stable (or even Lévy) noise, or by a general semimartingale. The theory developed in Chapter 7 should work for this case, if one developes the corresponding method of stable stochastic characteristics (or genreally stochastic characteristics of Lévy type) similar to the case of white noise considered in Section 2.7.

REFERENCES.

[AcW] L.Accardi, W. von Waldenfels (Eds.). Quantum Probability II. Proceedings, Heidelberg 1984. Springer Lecture Notes Math. **1136**, Springer 1985.

[Ak] N. Akhiezer. The calculus of variations. Blaisdell Publishing Company. N.Y., London, 1962 (translated from Russian).

[AlB] S. Albeverio, Zd. Brzezniak. Finite dimensional approximations approach to oscillatory integrals in finite dimensions. J. Funct. Anal. **113** (1993), 177-244.

[ABB] S. Albeverio, A. Boutet de Monvel-Bertier, Zd. Brzezniak. Stationary Phase Method in Infinite Dimensions by Finite Approximations: Applications to the Schrödinger Equation. Poten. Anal. **4:5** (1995), 469-502.

[ABD] S. Albeverio, Z. Brzezniak, L. Dabrowski. Fundamental solution of the heat and Schrödinger equations with point interaction. J. Funct. Anal. **130** (1995), 220-254.

[ACH] S. Albeverio, Ph. Combe, R. Hoegh-Krohn, G. Rideau, R.Stora (Eds.). Feynman path integrals. LNP 106, Springer 1979.

[AFHK] S. Albeverio, J.E. Fenstad, R. Hoegh-Krohn, W. Karwowski, T. Lindstrom. Schrödinger operators with potentials supported by null sets. In: Ideas and Methods in Quantum and Statistical Physics. Vol. 2 in Memory of R. Hoegh-Krohn, ed. S Albeverio et al, Cambridge Univ. Press, 1992, 63-95.

[AFHL] S. Albeverio, J.E. Fenstad, R. Hoegh-Krohn, T. Lindstrom. Nonstandard methods in stochastic analysis and mathematical physics. Academic Press, New York, 1986.

[AHK1] S.Albeverio, A. Hilbert, V.N. Kolokoltsov. Transience for stochastically perturbed Newton systems. Stochastics and Stochastics Reports **60** (1997), 41-55.

[AHK2] S. Albeverio, A.Hilbert, V.N. Kolokoltsov. Sur le comportement asymptotique du noyau associé à une diffusion dégénérée. Inst. Math. Ruhr Universität Bochum. Preprint **320** (1996).

[AHK3] S. Albeverio, A. Hilbert, V.N. Kolokoltsov. Estimates uniform in time for the transition probability of diffusions with small drift and for stochastically perturbed Newton equations. J. Theor. Prob. **12:3** (1999), 293-300.

[AHZ] S. Albeverio, A. Hilbert, E. Zehnder. Hamiltonian systems with a stochastic force: nonlinear versus linear, and a Girsanov formula. Stochastics and Stochastics Reports **39** (1992), 159-188.

[AH1] S. Albeverio, R.J. Hoegh-Krohn. Mathematical Theory of Feynman Path Integrals. LNM **523**, Springer Verlag, Berlin, 1976.

[AH2] S. Albeverio, R.J. Hoegh-Krohn. Oscillatory integrals and the method of stationary phase in infinitely many dimensions. Inventions Math. **40** (1977), 59-106.

[AH3] S. Albeverio, R.J. Hoegh-Krohn. Dirichlet forms and Markov semigroups on C^*-algebras. Comm. Math. Phys. **56** (1977), 173-187.

[AHS] S. Albeverio, R. Hoegh-Krohn, L. Streit. Energy forms, Hamiltonians and distorted Brownian paths. J. Math. Phys. **18** (1977), 907-917.

[AHO] S. Albeverio, R. Hoegh-Krohn, G. Olsen. Dynamical semigroups and Markov processes on C^*-algebras. J. Reine Ang. Math. **319** (1980), 29-37.

[AJPS] S. Albeverio, J. Jost, S. Paycha, S. Scarletti. A mathematical introduction to string theory. London Math. Soc. Lect. Notes Math., 1997.

[AK] S.Albeverio, V.N. Kolokoltsov. The rate of escape for some Gaussian Processes and the scattering theory for their small perturbations. Stochastic Analysis and Appl. **67** (1997), 139-159.

[AKS1] S. Albeverio, V.N. Kolokoltsov, O.G. Smolyanov. Représentation des solutions de l'équation de Belavkin pour la mesure quantique par une version rigoureuse de la formule d'intégration fonctionnelle de Menski. C.R.Acad. Sci.Paris, Sér. 1, **323** (1996), 661-664.

[AKS2] S. Albeverio, V.N. Kolokoltsov, O.G. Smolyanov. Continuous Quantum Measurement: Local and Global Approaches. Rev. Math. Phys. **9:8** (1997), 907-920.

[AU] S. Albeverio, S. Ugolini. Complex Dirichlet Forms: Non-Symmetric Diffusion Processes and Schrödinger Operators. Preprint 1997.

[ATF] V.M. Alexeev, V.M. Tikhomirov, S.V. Fomin. Optimal Control. Nauka, Moscow, 1979 (in Russian).

[AntGS] J.P. Antoine, F. Gesztesy, J. Shabani. Exactly solvable models of sphere interactions in quantum mechanics. J. Phys. A **20:12** (1987), 3687-3712.

[And] T.W. Anderson. The integral of a symmetric unimodal function over a symmetric convex set and some probability inequalities. Proc. Amer. Math. Soc. **6** (1955), 170-176.

[ApB] D. Applebaum, M. Brooks. Infinite series of quantum spectral stochastic integrals. J.Operator Theory **36** (1996), 295-316.

[Ar] L. Arnold. The log log law for multidimensional stochastic integrals and diffusion processes. Bull. Austral. Math. Soc. **5** (1971), 351-356.

[Aro1] D.G. Aronson. Bounds for the fundamental solution of a parabolic equation. Bull. Amer. Math. Soc. **73:6** (1967), 890-896.

[Aro2] D.G. Aronson. Non-negative solutions of linear parabolic equations. Ann. Scuola Norm. Sup. Pisa **22** (1968), 607-694.

[Aru] A.V. Arutyunov. Conditions of extremum. Anonrmal and degenerate problems. Moscow, "Faktorial", 1997 (in Russian).

[At] M.F. Atiyah. Resolution of singularities and division of distribusions. Comm. Pure and Appl. Math. **23:2** (1970), 145-150.

[Az1] R. Azencott. Grandes déviations et applications. Cours de la probabilite de Saint-Flour, LNM 774, 1978.

[Az2] R. Azencott. Densité des diffusions en temps petit: développement asymptotique. Seminaire de Probabilités XVIII, LNM **1059** (1984), 402-494.

[BB] V.M. Babich, V.S. Buldyrev. Asymptotic methods in short wave diffraction problems. Moscow, Nauka, 1972 (in Russian).

[Bas] R.F. Bass. Uniqueness in law for pure jump type Markov processes. Prob. Rel. Fields **79** (1988), 271-287.

[Be1] V.P. Belavkin. Nondemolition measurement, nonlinear filtering and dynamic programming of quantum stochastic processes. In: Modelling and Control of Systems. Proc. Bellman Continuous Workshop, Sophia-Antipolis, 1988. LNCIS **121** (1988), 245-265.

[Be2] V.P. Belavkin. A new wave equation for a continuous nondemolition measurement. Phys. Lett. A **140** (1989), 355-358.

[Be3] V.P. Belavkin. Optimal filtering of marcovian signals in white noise. Radio Eng. Electr. Physics **18** (1980), 1445-1453.

[Be4] V.P. Belavkin. Chaotic states and stochastic integration in quantum systems. Russian Math. Surveys **47:1** (1992), 47-106.

[Be5] V.P. Belavkin. Quantum stochastic positive evolutions. Preprint **96-16** Univ. of Nottingham Math. Preprint Series, 1996. To appear in CMP.

[BHH] V.P. Belavkin, R.Hudson, R.Hirota (Eds). Quantum Communications and Measurements. Proc. Intern. Workshop on Quantum Communications and Measurement, held July 11-16, 1994, Nottingham, England. N.Y., Plenum Press, 1995.

[BK] V.P. Belavkin, V.N. Kolokoltsov. Quasi-classical asymptotics of quantum stochastic equations. Teor. i mat. Fiz. **89** (1991), 163-178. Engl. transl. in Theor Math. Phys.

[Bel] D.R. Bell. The Malliavin Calculus. Pitman Monograph and Surveys in Pure and Applied Math **34**, Longman and Wiley, 1987.

[BD] V.Belov, S. Dobrockotov. Quasiclassical Maslov's asymptotics with complex phases Teor.Matem. Fizika **92:2** (1992), 215-254. Engl.transl. in Theor. Math. Phys.

[BA1] G. Ben Arous. Développement asymptotique du noyau de la chaleur hypoelliptique hors du cut-locus. Annales Sci de l'Ecole Norm Sup. **21** (1988), 307-331.

[BA2] G. Ben Arous. Développement asymptotique du noyau de la chaleur hypoelliptique sur la diagonale. Ann. Inst. Fourier (Grenoble) **39:1** (1989), 73-99.

[BAG] G. Ben Arous, M. Gradinary. Singularities of Hypoelliptic Green Functions. POten. Anal. **8** (1998), 217-258.

[BAL] G. Ben Arous, R. Léandre. Décroissance exponentielle du noyau de la chaleur sur la diagonale I, II. Prob. Theory Rel. Fields **90** (1991), 175-202, 377-402.

[Ben1] A. Bendikov. Asymptotic formulas for symmetric stable semigroups. Expo. Math. **12** (1994), 381-384.

332

[Ben2] A. Bendikov. On the Lévy measures for Symmetric Stable Semi-groups on the Torus. Potential Analysis **8** (1998), 399-407.

[Bere] F.A. Beresin. Path integral in phase space. Uspehki Fis. Nauk **132:3** (1980), 497-548. Engl. Transl. in Russian Phys. Surveys.

[BG] I.N. Bernstein, S.I. Gelfand. Meromorphity of the function p^λ. Funk. Anal. i ego Pril. **3:1** (1969), 84-86. Engl.transl. in Funct. Anal and Appl.

[Berr] M. V. Berry. Scaling and nongaussian fluctuations in the catastrophe theory of waves. Preprint 1984. Published in "Prometeus" (UNESCO publication) 1984.

[BD] M.V. Berry, D.H.J. O'Dell. Diffraction by volume gratings with imaginary potentials. J. Phys. A, Math. Gen. bf 31:8 (1998), 2093-2101.

[Ber] J. Bertoin. Lévy processes. Cambridge Univ. press, 1996.

[BGR] J. Bertrand, B. Gaveau, G. Rideau. Poisson processes and quantum field theory: a model. In: [AcW], 74-80.

[BF] F. Berezin, L. Faddeev. A remark on Schrödinger equation with a singular potential. Sov. Math. Dokl. **2** (1961), 372-375.

[Bi] N.H. Bingham. Maxima of Sums of Random Variables and Suprema of Stable Processes. Z. Warscheinlichkeitstheorie verw. Geb. **26** (1973), 273-296.

[Bis] J.M. Bismut. Large deviations and the Malliavin Calculus. Progress in Math. v. 45. Birkhäuser, Boston, 1984.

[BG] R.M. Blumenthal, R.K. Getoor. Some Theorems on Stable Processes. Trans. Amer. Math. Soc. **95** (1960), 263-273.

[BSch] N.N. Bogolyubov, D.V. Shirkov. Introduction to the Theory of Quantized Fields. Interscience, N.Y. 1959.

[BCP] J.-M. Bony, Ph. Courrège, P. Priouret. Semi-groupes de Feller sur une variété a bord compacte et problèmes aux limites intégro- différentiels du second ordre donnant lieu au principe du maximum. Ann. Inst. Fourier, Grenoble **18:2** (1968), 369-521.

[Bo] A.A. Borovkov. Boundary problem for random walks and large deviation in functional spaces. Theor. Prob. Appl. **12:4** (1967), 635-654.

[Br] L. Breiman. Probability. SIAM Philadelphia, 1992.

[BM] V.A. Buslov, K.A. Makarov. The Life-times and low lying eigenvalues of the small diffusion operator. Matem. Zametki **51** (1992), 20-31. Engl. transl. in Math. Notes.

[BZ] Yu. D. Burago, V.A. Zalgaller. Geometric Inequalities. Springer, New York, 1988.

[Ca] R.H. Cameron. A family of integrals serving to connect the Wiener and Feynman integrals. J. Math. and Phys. **39:2** (1960), 126-140.

[CaS] R.H. Cameron, D.A. Strovick. Feynman integral of variations of functionals. In: [ItH], 144-157.

[CW1] P. Cartier, C. DeWitt-Morette. Integration fonctionnelle; éléments d'axiomatique. CRAS, Ser.2, **316** (1993), 733-738.

[CW2] P. Cartier, C. DeWitt-Morette. A new perspective on functional integration. J. Math. Phys. **36** (1995), 2237-2312.

[CC] J. Callan, S. Coleman. Fate of the false vacuum. Phys. Rev. D **16:6** (1977), 1762-1768.

[CME] M. Chaleyat-Maurel, L. Elie. Diffusions Gaussiennes. Astérisque **84-85** (1981), 255-279.

[CMG] M. Chaleyat-Maurel, J.-F. Le Gall. Green functions, capacity and sample path properties for a class of hypoelliptique diffusions processes. Prob. Theory Rel. Fields **83** (1989), 219-264.

[Cha] S. Chandrasekhar. Stochastic problems in physics and astronomy. Rev. Modern Phys. **15:1** (1943), 1-89.

[Che] A.M. Chebotarev. On a stochastic representation of solutions for the Scgrödinger equation. Mathematical Notes **24:5** (1978), 873-877.

[CheKM] A.M. Chebotarev, A.A. Konstantinov, V.P. Maslov. Probabilistic representation of solutions of the Cauchy problem for Schrödinger, Pauli and Dirac equations. Russian Math. Surv. **45 :6** (1990), 3-24.

[CheQ] A.M. Chebotarev, R.B. Quezada. Stochastic approach to time-dependent quantum tunneling. Russian J. of Math. Phys. **4:3** (1998), 275-286.

[Chu] D.M. Chung. Conditional Feynman integrals for the Fresnel class of functions on abstract Wiener space. In: [ItH], 172–186.

[Com1] Ph. Combe et al. Generalised Poisson processes in quantum mechanics and field theory. Phys. Rep. **77** (1981), 221-233.

[Com2] Ph. Combe et al. Poisson processes on groups and Feynman path integrals. Comm. Math. Phys. bf 77 (1980), 269-298.

[Com3] Ph. Combe et al. Feynman path integral and Poisson processes with classical paths. J. Math. Phys. **23** (1980), 405-411.

[Com4] Ph. Combe et al. Quantum dynamic time evolution as stochastic flows on phase space. Physica A **124** (1984), 561-574.

[Coo] J. Cook. Convergence of the Moeler wave matrix. J. Math. and Phys. **36** (1957), 82-87.

[Cou] Ph. Courrège. Sur la forme integro-différentiélle du générateur infinitésimal d'un semi-groupe de Feller sur une varieté. Sém. Théorie du Potentiel, 1965-66. Exposé 3.

[CFKS] H.L. Cycon, R.F. Froese, W. Kirsh, B.Simon. Schrödinger Operators with Applications to Quantum mechanics and Global Geometry. Springer-Verlag, 1987.

[DaSh] L. Dabrowski, J. Shabani. Finitely many sphere interactions in quantum mechanics - non separated boundary conditions. J. Math. Phys. **29:10** (1988), 2241-2244.

[Dan] V.G. Danilov. A representation of the delta function via creation operators and Gaussian exponentials and multiplicative fundamental solution asymptotics for some parabolic pseudodifferential equations. Russian J. Math. Phys. **3:1** (1995), 25-40.

[DF1] V.G. Danilov, S.M. Frolovicvev. Reflection method for constructing of the Green function asymptotics for parabolic equations with variable coefficients. Mat. Zametki **57:3** (1995), 467-470. Engl. transl. in Math. Notes.

[DF2] V.G. Danilov, S.M. Frolovichev. Reflection method for constructing multiplicative asymptotic solutions of the boundary value problems for linear parabolic equations with variable coefficients. Uspehki Mat. Nauk **50:4** (1995), 98.

[Da1] E.B. Davies. Heat Kernels and Spectral Theory. Cambridge Univ. Press., 1992.

[Da2] E.B. Davies. Gaussian upper bounds for the heat kernel of some second order operators on Riemannian manifolds. J. Func. Anal. **80** (1988), 16-32.

[Da3] E.B. Davies. Quantum Theory of Open Systems.. Academic Press, N.Y., 1972.

[DeFJL] B. DeFacio, G.W. Johnson, M.L. Lapidus. Feynman's operational calculus and evolution equations. Acta Appl. Math. **47:2** (1997), 155-211.

[DeO] Y.N. Demkov, V.N. Ostrowski. Zero-Range Potentials and Their Applications in Atomic Physics. Plenum Press, New York, 1988.

[DMN] C. DeWitt-Morette, A. Maheshwari, B. Nelson. Path integration in non-relativistic quantum mechanics. Phys. Rep. **50** (1979), 256-372.

[Di1] L. Diosi. Continuous quantum measurement and Ito formalism. Phys. Let. A **129** (1988), 419-423.

[Di2] L. Diosi. Localised solution of a simple nonlinear quantum Langevin equation. Phys. Let. A **132** (1988), 233-236.

[DJ1] S. Dharmadhikari, K. Joag-Dev. Unimodality, Convexity, and Applications. Academic Press 1988.

[DJ2] S. Dharmadhikari, K. Joag-Dev. Multivariate Unimodality. Ann. Statist. **4**, 607-613.

[DK1] S.Yu. Dobrokhotov, V.N. Kolokoltsov. Splitting amplitude of the lowest energy levels of the Schrödinger operator with double-well potential. Teor. i Matem. Fiz., **94:3** (1993), 426-434. Engl. transl. in Theor. Math. Phys.

[DK2] S.Yu. Dobrokhotov, V.N. Kolokoltsov. The double-well splitting of the low energy levels for the Schrödinger operator of discrete ϕ^4- models on tori. J. Math. Phys., **36** (1994), 1038-1053.

[DKM1] S.Yu. Dobrokhotov, V.N. Kolokoltsov, V.P. Maslov. The splitting of the low lying energy levels of the Schrödinger operator and the asymptotics of the fundamental solution of the equation $hu_t = (h^2\Delta/2 - V(x))u$. Teoret. Mat. Fizika **87:3** (1991), 323-375. Engl. transl. in Theor. Math. Phys.

[DKM2] S.Yu. Dobrokhotov, V.N. Kolokoltsov, V.P. Maslov. Quantisation of the Bellman equation, exponential asymptotics, and tunneling. In: V. Maslov, S. Samborski (eds.). Idempotent Analysis, Adv. Sov. Math. **13**, 1-47.

[DKO1] S.Yu. Dobrokhotov, V.N. Kolokoltsov, V.M. Olivé. Quasimodes of the diffusion operator $-\epsilon\delta + v(x)\nabla$, corresponding to asymptotically stable limit cycles of the field $v(x)$. Sobreito de Aportaciones Matematicas III Simposio de Probabilidad y Processes Estocasticos **11**. Sociedad Matematica Mexicana (1994), 81-89.

[DKO2] S.Yu. Dobrokhotov, V.N. Kolokoltsov, V.M. Olivé. Asymptotically stable invariant tori of a vector field $V(x)$ and the quasimodes of the operator $V(x)\nabla - \epsilon\delta$. Matem Zametki **58:2** (1995), 301-306. Engl. transl. in Math. Notes.

[DS] S.Yu. Dobrokhotov, A.I. Shafarevich. Momentum tunneling between tori and the splitting of eigenvalues of the Beltrami-Laplace operator on Liuoville surfaces. Preprint 599, Inst. Prob. Mech. Russian Acad. Sci., 1997.

[DNF] B.A. Dubrovin, S.P. Novikov, A.T. Fomenko. Modern Geometry. Nauka, Moscow, 1979.

[El] K.D. Elworthy. Stochastic Differential Equations on Manifolds. CUP, Cambridge, 1982.

[ET] K.D. Elworthy, A. Truman. Feynman maps, Cameron-Martin formulae and anharmonic oscillators. Ann. Inst. Henri Poincaré, **41:2** (1984), 115-142.

[Fal] K.J. Falconer. A result on the Steiner symmetral of a compact set. J.London Math. Soc. **14** (1976), 385-386.

[FR] I.Fary, L.Redei. Der zentralsymmetrische Kern und die Zentralsymmetrische Hülle von konvexen Körpern. Math. Ann. **122** (1950), 205-220.

[Fed1] M.V. Fedorjuk. Asymptotics, integrals and series. Moscow, Nauka, 1987 (in Russian).

[Fed2] M.V. Fedorjuk. Asymptotics of the Green function a parabolic pseudo-differential equation. Differenz. Uravneniya **14:7** (1978), 1296-1301. Engl Transl. in Diff. Equation **14:7** (1978), 923-927.

[Fed3] M.V. Fedorjuk. Asymptotic Analysis: Linear Ordinary Differential Equations, Springer Verlag, Berlin - N.Y., 1993.

[Fel] W. Feller. An Introduction to Probability. Theory and Applications. Sec. Edition, v. 2., John Wiley and Sons, 1971.

[FW] M.I. Freidlin, A.D. Wentzel. Random Perturbations of Dynamical Systems. Springer Verlag, N.Y., Berlin, Heidelberg, 1984.

[Fu] M. Fukushima. Dirichlet forms and Markov processes. North Holland Publisher, 1980.

[Gar] R. Gardner. Geometric Tomography. Cambridge Univ. Press, 1995.

[Gav] B. Gaveau. Representation formulas of the Cauchy problem for hyperbolic systems generalising Dirac system. J. Func. Anal. **58** (1984), 310-319.

336

[GH] M. Giaquinta, S. Hildebrandt. Calculus of Variations, vol 1,2. Grundlehren der mathematischen Wissenschaften, v. 310,311. Springer-Verlag, Berlin-Heidelberg, 1996.

[GY] I.M. Gelfand, A.M. Yaglom. Integration in functional spaces and applications in quantum physics. Uspehki Mat. Nauk **11:1** (1956), 77-114. Engl. transl. in J.Math. Phys. **1:1** (1960), 48-69.

[GK] B.V. Gnedenko, A.N. Kolmogorov. Limit Distributions for Sums of Independent Random Variables. Moscow, 1949 (in Russian). Engl. transl. Addison-Wesley, 1954.

[Gr] P. Greiner. An asymptotic expansion for the heat equation. Arch. Ret. Mec. Anal. **41** (1971), 163-218.

[Gui] A. Guichardet. Symmetric Hilbert Spaces and Related Topics. Springer Lecture Notes Math. **261**, Springer-Verlag 1972.

[Gu] M.G. Gutzwiller. Chaos in Classical and Quantum Mechanics. Springer-Verlag, N.Y., 1982.

[Ha] E.M. Harrel. Double wells. Comm. Math. Phys. **75** (1964), 239-261.

[He] B. Helffer. Introduction to the semiclassical analysis for the Schrödinger operator and applications.

[HS1] B. Helffer, J. Sjöstrand. Multiple wells in the semiclassical limit, 1. Comm. PDE **9(4)** (1984), 337-408.

[HS2] B. Helffer, J. Sjöstrand. Multiple wells in the semiclassical limit, 4. Comm. PDE **10:3** (1985), 245-340.

[HKPS] T. Hida, H.-H. Kuo, J. Potthoff and L. Streit. White noise, An infinite dimensional calculus. Kluwer Acad. Publishers, 1993.

[Ho] W. Hoh. Pseudo-differential operators with negative definite symbols and the martingale problem. Stochastics and Stochastics Rep. **55** (1995), 225-252.

[How] J. Howland. Stationary scattering theory for the time dependent Hamiltonians. Math. Ann. **207** (1974), 315-335.

[Hor] L. Hörmander. Hypoelliptic second order differential equations. Acta Math. **119** (1967), 147-171.

[HuM] Y.Z. Hu, P.A. Meyer. Chaos de Wiener et intégrale de Feynman. In: Séminaire de Probabilités XXII (Ed. J. Azéma, P.A. Meyer, M. Yor). Springer Lecture Notes Math. **1321** (1988), 51-71.

[HP] R. Hudson, K.R. Parthasarathy. Quantum Ito's formula and stochastic evolution. Comm. Math. Phys. **93** (1984), 301-323.

[IC] I.A. Ibragimov, K.E. Chernin. On the unimodality of stable laws. Teor. Veroyatn. i Primeneniya 1 (1956), 283-288. Engl. transl. in Theor. Prob. Appl.

[IK] K. Ichihara, H. Kunita. A classification of the second order degenerate elliptic operators and its probabilistic characterization. Z. Wahrsch. Verw. Gebiete **30** (1974), 235-254. Corrections in **30** (1977), 81-84.

[Ich] T. Ichinose. Path integral for a hyperbolic system of the first order. Duke Math. Journ. **51:1** (1984), 1-36.

[IKO] A.M. Il'in, A.C. Kalashnikov, O.A. Olejnik. Linear second-order equations of parabolic type. Uspekhi Mat. Nauk **17:3** (1962), 3-141. Engl. transl. in Russian Math. Surveys.

[Is] Y. Ishikawa. On the upper bound of the density for truncated stable processes in small time. Potential Analysis **6** (1997), 1-37.

[ItH] K. Ito, T. Hida (Eds.). Gaussian Random Fields. The Third Nagoya Lévy Seminar, Nagoya, Japan, 15020 Aug. 1990. World Scientific 1991.

[Ja] N. Jacob. Pseudo-Differential Operators and Markov Processes. Berlin, Akademie Verlag, Mathematical Research, v. **94**, 1996.

[JS] N. Jacod, A.N. Shiryajev. Limit Theorems for Stochastic Processes. Berlin-Heidelberg, Springer-Verlag, 1987.

[JS1] D. Jerison, A. Sanchez. Subellitic second order differential operator in complex analysis III, LNM 1277, Ed. E. Berenstein (1987), 46-78.

[Joe] E. Joergensen. Construction of the Brownian motion and the Orstein-Uhlenbeck Process in a Riemannian manifold. Z. Wahrscheinlichkeitstheorie Verw. Gebiete **44** (1978), 71-87.

[Joh1] G.W. Johnson. The equivalence of two approaches to the Feynman integral. J. Math. Phys. **23:11** (1982), 2090-2096.

[Joh2] G.W. Johnson. Feynman's paper revisited. In: Functional integration with emphasis on the Feynman integral. Proc. Workshop held at the Univ. Sherbrooke, Quebec Canada July 1986. Supplemento ai Rendiconti del Circolo Matematico di Palermo, Ser II, **17** (1987), 249-270.

[JoL1] G.W. Johnson, M.L. Lapidus. Feynman's operation calculus, generalised Dyson series and the Feynman integral. Contemporary Math. **62** (1987), 437-445.

[JoL2] G.W. Johnson, M.L. Lapidus. Noncommutative Operations on Wiener Functionals and Feynman's Operational Calculus. J. Func. Anal. **81:1** (1988), 74-99.

[Ju] D. Juriev. Belavkin-Kolokoltsov watch-dog effect in interectively controlled stochastic computer-graphic dynamical systems. A mathematical study. E-print (LANL Electronic Archive on Chaos Dyn.): chao-dyn/9406013+9504008 (1994,1995).

[Kal] G. Kallianpur. Stochastic Filtering Theory. Springer-Verlag. N.Y., Heidelberg, Berlin, 1980.

[Kal2] G. Kallianpur. Traces, natural extensions and Feynman distributions. In: [ItH], 14-27.

[KKK] G. Kallianpur, D. Kannan, R.L. Karandikar. Analytic and sequential Feynman integrals on abstract Wiener and Hilbert Spaces and a Cameron-Martin formula. Ann. Inst. H. Poincaré **21** (1985), 323-361.

[Kana] K. Kanai. Short time asymptotics for fundamental solutions of partial differential equations. Comm. Part. Diff. Eq. **2:8** (1977), 781-830.

[Kan] M. Kanter. Unimodality and dominance for symmetric random vectors. Trans. Amer. Math. Soc. **229**, 65-85.

[Kh] A. Khintchin. Sur la crosissance locale des prosessus stochastiques homogènes à acroissements indépendants. Isvestia Akad. Nauk SSSR, Ser Math. (1939), 487-508.

[Ki] Y.I. Kifer. Transition density of diffusion processes with small diffusion. Theor. Prob. Appl. **21** (1976), 513-522.

[KN] K. Kikuchi, A.Negoro. On Markov processes generated by pseudodifferential operator of variable order. Osaka J. Math. **34** (1997), 319-335.

[KSZ] J. Klafter, M. Schlesinger, G. Zumoven. Beyond Brownian Motion. Physics Today, February 1996, 33-39.

[Koch] A.N. Kochubei. Parabolic pseudodifferential equations, supersingular integrals and Markov processes. Izvestia Akad. Nauk, Ser. Matem. **52:5** (1988), 909-934. Engl. Transl. in Math. USSR Izvestija **33** (1989), 233-259.

[Koch2] A.N. Kochubei. Elliptic operators with boundary conditions on a subset of measure zero. Func. Anal. Appl. **16** (1982), 137-139.

[K1] V.N. Kolokoltsov. Stochastic Hamilton-Jacobi-Bellman equation and stochastic WKB method. Proc. Intern. Workshop "Idempotency", held in Bristol, October 1994 (Ed. J.Gunawardena), Cambridge Univ. Press, 1997, 285-302.

[K2] V.N. Kolokoltsov. Stochastic Hamilton-Jacobi-Bellman equation and stochastic Hamiltonian systems. J. Dynamic Contr. Syst. **2:3** (1996), 299-319.

[K3] V.N. Kolokoltsov. On the asymptotics of low eigenvalues and eigenfunctions of the Schrödinger operator. Dokl. Akad. Nauk **328:6** (1993). Engl. transl. in Russian Acad. Sci. Dokl. Math. **47:1** (1993), 139-145.

[K4] V.N. Kolokoltsov. Scattering theory for the Belavkin equation describing a quantum particle with continuously observed coordinate. J. Math. Phys. **36:6** (1995) 2741-2760.

[K5] V.N. Kolokoltsov. Application of quasi-classical method to the investigation of the Belavkin quantum filtering equation. Mat. Zametki **50** (1993), 153-156. Engl. transl. in Math. Notes.

[K6] V.N. Kolokoltsov. A note on the long time asymptotics of the Brownian motion with application to the theory of quantum measurement. Potential Analysis **7** (1997), 759-764.

[K7] V.N. Kolokoltsov. Localization and analytic properties of the simplest quantum filtering equation. Rev Math.Phys. **10:6** (1998), 801-828.

[K8] V.N. Kolokoltsov. Long time behavior of the solutions of the Belavkin quantum filtering equation. In: [BHH], 429-439.

[K9] V.N. Kolokoltsov. Short deduction and mathematical properties of the main equation of the theory of continuous quantum measurements. In:

GROUP21 (Eds. H.-DdDoebner, P.Nattermann, W.Scherer): Proc. XXI Intern. Colloq. on Group Theoret. Methods in Physics July 1996, World Scientific 1997, v.1, 326-330.

[K10] V.N. Kolokoltsov. Estimates of accuracy for the asymptotics of the Laplace integrals. Trudy Mat. Inst. Steklova **203** (1994), 113-115. Engl. transl. in Proc. Steklov Inst. Math. **3** (1995), 101-103.

[K11] V.N. Kolokoltsov. Symmetric stable laws and stable-like jump-diffusions. Research Rep 6/98, Dep. Math. Stat. and O.R., Nottingham Trent University, 1998. To appear in PLMS.

[K12] V.N. Kolokoltsov. Geodesic flows on two-dimensional manifolds with an additional integral quadratic in momentum. Isvestia. Acad. Nauk USSR **46:5** (1982), 994-1010.

[K13] V.N. Kolokoltsov. Complex measures on path space: an introduction to the Feynman integral applied to the Schrödinger equation. To appear in MCAP.

[K14] V.N. Kolokoltsov. Long time behavior of continuously observed and controlled quantum systems (a study of the Belavkin quantum filtering equation). In: Quantum Probability Communications, QP-PQ, V. 10, Ed. R.L. Hudson, J.M. Lindsay, World Scientific, Singapore (1998), 229-243.

[KMak] V.N. Kolokoltsov, K. Makarov. Asymptotic spectral analysis of the small diffusion operator and life times of the diffusion process. Russian Journ. Math. Phys. **4:3** (1996), 341-360.

[KM1] V.N. Kolokoltsov, V.P. Maslov. Idempotent Analysis and its Application to Optimal Control. Moscow, Nauka, 1994 (in Russian).

[KM2] V.N. Kolokoltsov, V.P. Maslov. Idempotent Analysis and its Applications. Kluwer Publishing House, 1997.

[Kom1] T. Komatsu. On the martingale problem for generators of stable processes with perturbations. Osaka J. Math. **21** (1984), 113-132.

[Kom2] T. Komatsu. Pseudo-differential operators and Markov processes. J. Math. Soc. Japan **36** (1984), 387-418.

[Koo] P. Koosis. Introduction to H_p Spaces. London Math. Soc. Lecture Note Series **40**, Cambridge University Press, 1980.

[Kosh] V. Koshmanenko. Singular Quadratic Forms in Perturbation Theory. Kluwer Academic 1999.

[KMS] D.V. Kosygin, A.A. Minasov, Ya. G. Sinai. Statistical properties of the spectra of the Laplace-Beltrami operators on Liouville surfaces. Russ. Math. Surv. **48:4** (1993), 3-130.

[KZPS] M.A. Krasnoselskii, P.P. Zabreiko, E.I. Pustylnik, P.E. Sobolewskii. Integral operators in spaces of summable functions. Noordhoff, Leiden, 1976.

[KS] S. Kusuoka, D. Stroock. Application of the Malliavin calculus, part III. J. Fac. Sci. Univ. Tokyo, Sect. Math. **34** (1987), 391-442.

[LSU] O.A. Ladishenskaya, V.A. Solonnikov, N.N. Ural'tseva. Linear and quasilinear equations of parabolic type. Nauka, Moscow, 1967 (in Russian).

[LL] L.D. Landau, E.M. Lifshits. Quantum mechanics. Moscow, 1963. Engl. transl. Addison-Wesley, Reading, 1965.

[Las] P.I. Lasorkin. Description of spaces $L_p^r(\mathcal{R}^n)$ in terms of difference singular integrals. Matem. Sbornik **81:1** (1970), 79-91.

[Laz] V.F. Lazutkin. KAM-theory and semiclassical approximations to eigenfunctions. Springer-Verlag, Berlin 1993.

[Le1] R. Léandre. Estimation en temps petit de la densité d'une diffusion hypoelliptique. CRAS Paris **301:17**, Ser 1 (1985), 801-804.

[Le2] R. Léandre. Minoration en temps petit de la densité d'une diffusion dégénérée. J. Funct. Anal. **74** (1987), 399-414.

[Le3] R. Léandre. Volume de boules sous-riemannienes et explosion du noyau de la chaleur au sens de Stein. Séminaire de Probabilités XXIII, LNM 1372 (1990), 426-448.

[Le4] R. Léandre. Dévéloppement asymptotique de la densité d'une diffusion dégénéré. Forum Math. **4** (1992), 45-75.

[Le5] R. Léandre. Uniform upper bounds for hypoelliptic kernels with drift. J. Math. Kyoto Univ. **34:2** (1994), 263-271.

[LR] R. Léandre, F. Russo. Estimation de la densité de la solution de l'équation de Zakai robuste. Potential Analysis 4 (1995), 521-545.

[Li]G. Lindblad. On the geneators of quantum dynamical semigroups. Comm. Math. Phys. **48** (1976), 119-130.

[LP] J.M. Lindsay, K.R. Parthasarathy. On the Generators of Quantum Stochastic Flows. J. Funk. Anal. **158:2** (1998), 521-549.

[Lis] V.S. Lisitsa. Stark broadening of hydrogen lines in plasmas. Usp. Fiz. Nauk **122** (1977), 449-495. Engl. transl. in Sov. Phys. Usp. **20:7** (1977), 603-630.

[Lu] E. Lukacs. Chartacteristic functions. Second edition. Griffin, London, 1971.

[MR] Z.M. Ma, M. Röckner. Introduction to the Theory of Non-Symmetric Dirichlet Forms. Springer-Verlag, 1992.

[Mal] P. Malliavin. Stochastic calculus of variation and hypoelliptic operators. In: Proc. Intern. Symp. on Stochastic Differential Equations, ed. by K. Ito. Kinokuniya, Tokyo and Wiley, New York, 1978.

[M1] V.P. Maslov. Perturbation theory and asymptotical methods. MGU, Moscow, 1965 (in Russian). French Transl. Dunod, Paris, 1972.

[M2] V.P. Maslov. Global exponential asymptotic behavior of the solution of equations of tunnel type. Dokl. Akad. Nauk **261:5** (1981), 1058-1062. Engl. transl. in Soviet Math. Dokl.

[M3] V.P. Maslov. Global exponential asymptotics of the solutions of the tunnel equations and the large deviation problems. Trudi Mat. Inst. Steklov **163** (1984), 150-180. Engl. transl. in Proc. Steklov Inst. Math. **4** (1985).

[M4] V.P. Maslov. Méthodes opératorielles. Editions MIR, Moscow, 1987.

[M5] V.P. Maslov. The Complex WKB Method in Nonlinear Equations. Nauka, Moscaw (in Russian), 1977.

[M6] V.P. Maslov. Quasiclassical asymptotics of the solutions of some problems of mathematical physics. Jurn. Vychisl. matem. i matem. Fisiki **1** (1961), 113-128, 638-663 (in Russian).

[M7] V.P. Maslov. Complex Markov Chains and Functional Feynman Integral. Moscow, Nauka, 1976 (in Russian).

[M8] V.P. Maslov. Nonstandard characteristics in asymptotic problems. In Proc. Intern. Congr. Math., Warsawa 1983, vol 1, 139-184.

[M9] V.P. Maslov. The Complex WKB method for Nonlinear Equations. I. Lenear Theory. Birkhäuser, 1994.

[MC1] V.P. Maslov, A.M. Chebotarev. On the second term of the logarithmic asymptotics of functional integrals. Itogi Nauki i Tehkniki, Teoriya Veroyatnosti **19** (1982). VINITI, Moscow, 127-154 (in Russian).

[MC2] V.P. Maslov, A.M. Chebotarev. Processus à sauts et leur application dans la mécanique quantique, in [ACH], p. 58-72.

[MF1] V.P. Maslov, M.V. Fedoryuk. Semi-classical approximation in quantum mechanics. Nauka, Moscow, 1976. Engl. transl. Reidel, Dordrecht, 1981.

[MF2] V.P. Maslov, M.V. Fedoryuk. Logarithmic asymptotics of the Laplace integrals. Mat. Zametki **30:5** (1981), 763-768. Engl.transl. in Math. Notes.

[MS] B.J. Matkowsky, Z. Schuss. Eigenvalues of the Fokker-Planck operator and the approach to eqilibrium for diffusion in potential field. SIAM J. Appl. Math. **40** (1981), 242-254.

[Me1] M.B. Menski. Quantum restriction on the measurement of the parameters of motion of a macroscopic oscillator. Sov. Phys. JETP **50** (1979), 667-674.

[Me2] M.B. Menski. The difficulties in the mathematical definition of path integrals are overcome in the theory of continuous quantum measurements. Teoret. i Matem. Fizika **93:2** (1992), 264-272. Engl. transl. in Theor. Math. Phys.

[Meti] M. Métivier. Semimartingales: a course on stochastic processes. de Gruyter, Berlin, New York 1982.

[Metz] V. Metz. Potentialtheorie auf dem Sierpinski gasket. Math. Ann. **289** (1991), 207-237.

[Mey] P.A. Meyer. Quantum Probability for Probabilists. Springer Lecture Notes Math. **1538**, Springer-Verlag 1991.

[Mol] S.A. Molchanov. Diffusion processes and riemannien geometry. Russian Math. Surveys **30:1** (1975), 1-63.

[Mos1] J. Moser. A Harnack inequality for parabolic differential equations. Comm. Pure Appl. Math. **17:1** (1964), 101-134; and corrections in: Comm. Pure Appl. Math. **20:1** (1967), 231-236.

[Mos2] J. Moser. On a pointwise estimate for parabolic differential equations. Comm. Pure Appl. Math. **24:5** (1971), 727-740.

[Mu] J.D. Murray. Asymptotic Analysis. Clarendon Press, 1974.

[NSW] A. Nagel, E.M. Stein, S. Wainger. Balls and Metrics defined by vector fields, I. Basic Properties. Acta Math. **155** (1985), 103-147.

[Nas] J. Nash. Continuity of solutions of parabolic and elliptic equations. Amer. J. Math. **80:4** (1958), 931-954.

[Neg] A. Negoro. Stable-like processes: consruction of the transition density and the behavior of sample paths near $t = 0$. Osaka J.Math. **31** (1994), 189-214.

[Nel1] E. Nelson. Dynamical theory of Brownian motion. Princton Univ. Press, 1967.

[Nel2] E. Nelson. Feynman integrals and the Schrödinger equation. J. Math. Phys. **5:3** (1964), 332-343.

[NS] J.R. Norris, D.W. Stroock. Estimates on the fundamental solution to heat flows with uniformly elliptic coefficients. Proc. London Math. Soc. **62** (1991), 373-402.

[Nu] D. Nualart. The Malliavin Calculus and Related Topics. Springer-Verlag, 1995.

[Pa] T.F. Pankratova. Quasimodes and the splitting of eigenvalues. Dokl. Akad. Nauk SSSR **276** (1984), 795-798. Engl. transl. in Sov. Math. Dokl. **29** (1984).

[Par] K.R. Parthasarathy. An Introduction to Quantum Stochatic Calculus. Birkhäuser Verlag, Basel, 1992.

[Par2] K.R. Parthasarathy. Some remarks on the integration of Schrödinger equation using the quantum stochastic calculus. In: [AcW], 409-419.

[Pav] B.S. Pavlov. The theory of extensions and explicitly solvable models. Russian Math. Surveys **42** (1987), 127-168.

[PQ] P. Pereyra, R. Quezada. Probabilistic representation formulae for the time evolution of quantum systems. J. Math. Phys. **34** (1993), 59-68.

[Pol] G. Polya. On the zeros of an integral function represented by Fourier's integral. Messenger of Math. **52** (1933), 185-188.

[Po] A.M. Polyakov. Quark confinement and topology of gauge theories. Nucl. Phys. B, **120** (1977), 429-456.

[PE] F.O. Porper, S.D. Eidelman. Two-sided estimates of fundamental solutions of second-order parabolic equations, and some applications. Uspehki Mat. Nauk **39:3** (1984), 107-156. Engl. transl. in Russian Math. Surv. **39:3** (1984), 119-178.

[Por] N.I. Portenko. Some perturbations of drift-type for symmetric stable processes with locally unbounded drift. Random Oper. and Stoch. Eq. **2:3** (1994), 211-224.

[PP] N.I. Portenko, S.I. Podolynny. On multidimensional stable processes with locally unbounded drift. Random Oper. and Stoch. Eq. **3:2** (1995), 113-124.

[Pr] P. Protter. Stochastic Integration and Differential Equations. Applications of Mathematics 21, Springer-Verlag, 1990.

[Pu] A. Puhalskii. Large deviations of semimartingales: a maxingale problem approach 1. Limits as solutions to a maxingale problem. Stochastics and Stochastics Reports **61** (1997), 141-243.

[Pus] E.I. Pustylnik. Operators of potential type for the case of sets with abstract measures. Dokl. Akad. Nauk SSSR **156** (1964), 519-520.

[QO] Quantum and Semiclassical Optics **8:1** (1996). Special Issue on Stochastic Quantum Optics.

[Ra] R. Rajaraman. Solitons and Instantons in Quantum Physics. North-Holland, Amsterdam-N.Y.-Oxford, 1982.

[RS] M.Reed, B.Simon. Methods of Modern Mathematical Physics, v.4, Analysis of Operators. Academic Press, N.Y. 1978.

[RY] D. Revuz, M. Yor. Continuous Martingales and Brownian motion. Springer Verlag, 1991.

[Rob] D. Robert. Autour de l'approximation semi-classique. Birkhäuser, Bochum, 1987.

[Roc] R.T. Rockafellar. Convex Analysis. Princeton Univ. Press, Princeton, 1970.

[Roe] J. Roe. Elliptic operators, topology and asymptotic methods. Pitman Research Notes in Math. Series **179**. Longman Scientific, 1988.

[SV] Yu. Safarov, D. Vassiliev. The Asymptotic Distribution of Eigenvalues of Partial Diffeential Operators. Translations of Math. Monographs **155**, AMS, 1997.

[Sam1] S.G. Samko. Hypersingular Intergals and Applications. Rostov-na-Donu Univ. Press, 1984 (in Russian).

[Sam2] S.G. Samko. Generalised Riesz potentials and hypersingular integrals with homogeneous characteristics, their symbols and inversion formulas. Trudi Steklov Math. Inst. **156** (1980), 157-222.

[ST] G. Samorodnitski, M.S. Taqqu. Stable non-Gaussian Random Processes, Stochastic Models with Infinite Variance. Chapman and Hall, N.Y., 1994.

[SC] A. Sanchez-Callé. Fundamental solutions and the sums of squares of vector fields. Inv. Math. **78** (1984), 143-160.

[Schi] R.L. Schilling. On Feller processes with sample paths in Besov spaces. Math. Ann. **309** (1997), 663-675.

[Sch] R. Schneider. Convex Bodies: The Brunn-Minkowski Theory. Cambridge Univ. Press, 1993

[She] S. Sherman. A theorem on convex sets with applications. Ann. Math. Statist. **26** (1955), 763-766.

[Si1] B. Simon. Semiclassical analysis of low lying eigenvalues, 1. Ann. Inst. H. Poincaré **38** (1983), 295-307.

[Si2] B. Simon. Semiclassical analysis of low lying eigenvalues, 2. Ann. Math. **120** (1984), 89-118.

[SF] A.A. Slavnov, L.D. Faddeev. Gauge Fields, Introduction to Quantum Theory. Moscow, Nauka, 1988 (in Russian). Engl. transl. by Benjamin/Cumminge, Reading Mass,. 1980

[SS] O.G. Smolyanov, E.T. Shavgulidze. Kontinualniye Integrali. Moscow Univ. Press, Moscow, 1990 (in Russian).

[So] R.B. Sowers. Short-time Geometry of Random Heat Kernels. Memoirs of AMS, **132**, 1998.

[Sp] F. Spitzer. Principles of Random Walks. Van Nostrand, Princton, 1964.

[SH] L. Streit, T. Hida. Generalised Brownian functionals and the Feynman integral. Stochastic Processes and their Applications **16** (1983), 55-69.

[SV] D.W. Stroock, S.R. Varadhan. On the support of diffusion processes with applications to the strong maximum principle. Proc. Sixth Berkley Symp., v. **3** (1972), 333-359.

[Su] H. Sussman. On the gap between detrministic and stochastic ordinary differential equations. Ann. Prob. **6:1** (1978), 19-41.

[To] L. Tonelli. Fondamenti del calcolo delle variazioni. Zanichelli, Bologna, 1921-1923, 2 vols.

[Tr1] A. Truman. The Feynman maps and the Wiener integral. J. Math. Phys. **19** (1978), 1742.

[Tr2] A. Truman. The Polygonal Path Formulation of the Feynman Path Integral. In [ACH], p. 73-102.

[TZ1] A. Truman, Z. Zhao. The stochastic H-J equations, stochastic heat equations and Schrödinger equations. Univ. of Wales, Preprint 1994. Published in: A.Truman, I.M.Davies, K.D. Elworthy (Eds), Stochastic Analysis and Application. World Scientific Press (1996), 441-464.

[TZ2] A. Truman, Z. Zhao. Stochastic Hamilton-Jacobi equation and related topics. In: A.M. Etheridge (Ed.). Stochastic Partial Differential Equations. London Math. Soc. Lecture Notes Series **276**, Cambridge Univ. Press, 1995, 287-303.

[Ts] M. Tsuchiya. Lévy measure with generalised polar decomposition and associate SDE with jumps. Stochastics and stochastics Rep. **38** (1992), 95-117.

[Ue] H. Uemura. On a short time expansion of the fundamental solution of heat equations by the method of Wiener functionals. J. Math. Kyoto Univ. **27:3** (1987), 417-431.

[Var1] S.R. Varadhan. Asymptotic probabilities and differential equations. Comm. Pure Appl. Math. **19:3** (1966), 231-286.

[Var2] S.R. Varadhan. On the behavior of the fundamental solution of the heat equation with variable coefficients. Comm. Pure Appl. Math. **20** (1967), 431-455.

[Var3] S.R. Varadhan. Diffusion Processes in a Small Time Interval. Comm. Pure Appl. Math. **20** (1967), 659-685.

[Var4] S.R. Varadhan. Large Deviations and Applications. SIAM 1984.

[We] I. Wee. Lower functions for processes with stationary independent increments. Prob. Th. Rel. Fields **77** (1988), 551-566.

[Wel] D.R. Wells. A result in multivariate unimodality. Ann. Statist. **6** (1978), 926-931.

[Wen1] A.D. Wentzel. On the asymptotics of eigenvalues of matrices with entries of order $\exp\{-V_{ij}/(2\epsilon^2)\}$. Dokl. Akad. Nauk SSSR **202** (1972), 263-266. Engl. transl. in Sov. Math. Dokl. **13** (1972), 65-68.

[Wen2] A.D. Wentzel. On the asymptotic behavior of the greatest eigenvalue of a second order elliptic differential operator with a small parameter at the higher derivatives. Dokl. Akad. Nauk SSSR **202** (1972), 19-21. Engl. transl. in Sov. Math. Dokl. **13** (1972), 13-17.

[WH] M. Wilkinson, J.H. Hannay. Multidimensionaol tunneling between excited states. Phys. D **27** (1987), 201-212.

[Wi] A. Wintner. On the stable distribution laws. Amer. J. Math. **55** (1933), 335-339.

[WW] E.T. Whittaker, G.N. Watson. Modern Analysis. Third edition. Cambridge Univ. Press 1920.

[Wol] S.J. Wolfe. On the unimodality of multivariate symmetric distribution functions of class L. J. Multivar. Anal **8** (1978), 141-145.

[WZ] E. Wong, H. Zakai. Riemann-Stiltjes approximation of stochastic integrals. Z. Wahrscheinlichkeitstheorie Verw. Gebiete **12** (1969), 87-97.

[Yaj] K. Yajima. Existence of solutions for Schrödinger evolution equations. Comm. Math. Phys. **110** (1987), 415-426.

[Yam] M. Yamazato. Unimodality of infinitely divisible distribution functions of class L. Ann. Prob. **6** (1978), 523-531.

[Za] M. Zakai. On the optimal filtering of diffusion processes. Z. Wahrscheinlichkeitstheorie Verw. Gebiete **11** (1969), 230-243.

[Zo] V.M. Zolotarev. One-dimensional Stable Distributions. Moscow, Nauka, 1983 (in Russian). Engl. transl. in vol. 65 of Translations of Mathematical Monographs AMS, Providence, Rhode Island, 1986.

MAIN NOTATIONS

$Re\, a, Im\, a$ -real and imaginary part of a complex number a

$B_r(x)$ (resp. B_r)- the ball of radius r centred at x (resp. at the origin)

$\mathcal{N}, \mathcal{Z}$- natural and integer numbers respectively

$C^d, \mathcal{R}^d$ - complex and real d-dimensional spaces

S^d - d-dimensional unit sphere in $\mathcal{R}^{d+1}$

(x, y) or xy - the scalar product of the vectors x, y

1_m or E_m - the $m \times m$-unit matrix

if A is a $d \times d$ matrix, then A' denotes the transpose to A,

if A is a $d \times d$ selfadjoint matrix and δ is a real number,

then $A \geq \delta$ means $A \geq \delta 1_d$

$Sp(A)$ - spectrum of the operator A

$Ker\, A$ - kernel of the matrix (or operator) A

$tr\, A$ - the trace of the operator A

$\Theta_{a,b}$ (resp. Θ_a) - the indicator of the closed interval $[a, b]$ (resp. $[0, a]$), i.e.
$\Theta_{a,b}(x)$ equals one or zero according to whether $x \in [a, b]$ or otherwise

$f = O(g)$ means $|f| \leq Cg$ for some constant C

C_k^n - binomial coefficients

$C(X)$ - the Banach space of bounded continuous functions on
a topological space X equipped with the uniform norm;
if X is locally compact, then $C_0(X)$ denotes the subspace
of $C(X)$ of functions vanishing at infinity

Summation over repeating indices will be always assumed. The numeration of formulas and theorems is carried out independently in each chapter. A reference to, say, formula (2.4) in chapter 3, when referred to from another chapter, will be given as to formula (3.2.4).

SUBJECT INDEX

Printing: Weihert-Druck GmbH, Darmstadt
Binding: Buchbinderei Schäffer, Grünstadt

Lecture Notes in Mathematics

For information about Vols. 1–1530
please contact your bookseller or Springer-Verlag

Vol. 1574: V. P. Havin, N. K. Nikolski (Eds.), Linear and Complex Analysis – Problem Book 3 – Part II. XXII, 507 pages. 1994.

Vol. 1575: M. Mitrea, Clifford Wavelets, Singular Integrals, and Hardy Spaces. XI, 116 pages. 1994.

Vol. 1576: K. Kitahara, Spaces of Approximating Functions with Haar-Like Conditions. X, 110 pages. 1994.

Vol. 1577: N. Obata, White Noise Calculus and Fock Space. X, 183 pages. 1994.

Vol. 1578: J. Bernstein, V. Lunts, Equivariant Sheaves and Functors. V, 139 pages. 1994.

Vol. 1579: N. Kazamaki, Continuous Exponential Martingales and *BMO*. VII, 91 pages. 1994.

Vol. 1580: M. Milman, Extrapolation and Optimal Decompositions with Applications to Analysis. XI, 161 pages. 1994.

Vol. 1581: D. Bakry, R. D. Gill, S. A. Molchanov, Lectures on Probability Theory. Editor: P. Bernard. VIII, 420 pages. 1994.

Vol. 1582: W. Balser, From Divergent Power Series to Analytic Functions. X, 108 pages. 1994.

Vol. 1583: J. Azéma, P. A. Meyer, M. Yor (Eds.), Séminaire de Probabilités XXVIII. VI, 334 pages. 1994.

Vol. 1584: M. Brokate, N. Kenmochi, I. Müller, J. F. Rodriguez, C. Verdi, Phase Transitions and Hysteresis. Montecatini Terme, 1993. Editor: A. Visintin. VII. 291 pages. 1994.

Vol. 1585: G. Frey (Ed.), On Artin's Conjecture for Odd 2-dimensional Representations. VIII, 148 pages. 1994.

Vol. 1586: R. Nillsen, Difference Spaces and Invariant Linear Forms. XII, 186 pages. 1994.

Vol. 1587: N. Xi, Representations of Affine Hecke Algebras. VIII, 137 pages. 1994.

Vol. 1588: C. Scheiderer, Real and Étale Cohomology. XXIV, 273 pages. 1994.

Vol. 1589: J. Bellissard, M. Degli Esposti, G. Forni, S. Graffi, S. Isola, J. N. Mather, Transition to Chaos in Classical and Quantum Mechanics. Montecatini Terme, 1991. Editor: 2S. Graffi. VII, 192 pages. 1994.

Vol. 1590: P. M. Soardi, Potential Theory on Infinite Networks. VIII, 187 pages. 1994.

Vol. 1591: M. Abate, G. Patrizio, Finsler Metrics – A Global Approach. IX, 180 pages. 1994.

Vol. 1592: K. W. Breitung, Asymptotic Approximations for Probability Integrals. IX, 146 pages. 1994.

Vol. 1593: J. Jorgenson & S. Lang, D. Goldfeld, Explicit Formulas for Regularized Products and Series. VIII, 154 pages. 1994.

Vol. 1594: M. Green, J. Murre, C. Voisin, Algebraic Cycles and Hodge Theory. Torino, 1993. Editors: A. Albano, F. Bardelli. VII, 275 pages. 1994.

Vol. 1595: R.D.M. Accola, Topics in the Theory of Riemann Surfaces. IX, 105 pages. 1994.

Vol. 1596: L. Heindorf, L. B. Shapiro, Nearly Projective Boolean Algebras. X, 202 pages. 1994.

Vol. 1597: B. Herzog, Kodaira-Spencer Maps in Local Algebra. XVII, 176 pages. 1994.

Vol. 1598: J. Berndt, F. Tricerri, L. Vanhecke, Generalized Heisenberg Groups and Damek-Ricci Harmonic Spaces. VIII, 125 pages. 1995.

Vol. 1599: K. Johannson, Topology and Combinatorics of 3-Manifolds. XVIII, 446 pages. 1995.

Vol. 1600: W. Narkiewicz, Polynomial Mappings. VII, 130 pages. 1995.

Vol. 1601: A. Pott, Finite Geometry and Character Theory. VII, 181 pages. 1995.

Vol. 1602: J. Winkelmann, The Classification of Three-dimensional Homogeneous Complex Manifolds. XI, 230 pages. 1995.

Vol. 1603: V. Ene, Real Functions – Current Topics. XIII, 310 pages. 1995.

Vol. 1604: A. Huber, Mixed Motives and their Realization in Derived Categories. XV, 207 pages. 1995.

Vol. 1605: L. B. Wahlbin, Superconvergence in Galerkin Finite Element Methods. XI, 166 pages. 1995.

Vol. 1606: P.-D. Liu, M. Qian, Smooth Ergodic Theory of Random Dynamical Systems. XI, 221 pages. 1995.

Vol. 1607: G. Schwarz, Hodge Decomposition – A Method for Solving Boundary Value Problems. VII, 155 pages. 1995.

Vol. 1608: P. Biane, R. Durrett, Lectures on Probability Theory. Editor: P. Bernard. VII, 210 pages. 1995.

Vol. 1609: L. Arnold, C. Jones, K. Mischaikow, G. Raugel, Dynamical Systems. Montecatini Terme, 1994. Editor: R. Johnson. VIII, 329 pages. 1995.

Vol. 1610: A. S. Üstünel, An Introduction to Analysis on Wiener Space. X, 95 pages. 1995.

Vol. 1611: N. Knarr, Translation Planes. VI, 112 pages. 1995.

Vol. 1612: W. Kühnel, Tight Polyhedral Submanifolds and Tight Triangulations. VII, 122 pages. 1995.

Vol. 1613: J. Azéma, M. Emery, P. A. Meyer, M. Yor (Eds.), Séminaire de Probabilités XXIX. VI, 326 pages. 1995.

Vol. 1614: A. Koshelev, Regularity Problem for Quasilinear Elliptic and Parabolic Systems. XXI, 255 pages. 1995.

Vol. 1615: D. B. Massey, Le Cycles and Hypersurface Singularities. XI, 131 pages. 1995.

Vol. 1616: I. Moerdijk, Classifying Spaces and Classifying Topoi. VII, 94 pages. 1995.

Vol. 1617: V. Yurinsky, Sums and Gaussian Vectors. XI, 305 pages. 1995.

Vol. 1618: G. Pisier, Similarity Problems and Completely Bounded Maps. VII, 156 pages. 1996.

Vol. 1619: E. Landvogt, A Compactification of the Bruhat-Tits Building. VII, 152 pages. 1996.

Vol. 1620: R. Donagi, B. Dubrovin, E. Frenkel, E. Previato, Integrable Systems and Quantum Groups. Montecatini Terme, 1993. Editors:M. Francaviglia, S. Greco. VIII, 488 pages. 1996.

Vol. 1621: H. Bass, M. V. Otero-Espinar, D. N. Rockmore, C. P. L. Tresser, Cyclic Renormalization and Auto-morphism Groups of Rooted Trees. XXI, 136 pages. 1996.

Vol. 1622: E. D. Farjoun, Cellular Spaces, Null Spaces and Homotopy Localization. XIV, 199 pages. 1996.

Vol. 1623: H.P. Yap, Total Colourings of Graphs. VIII, 131 pages. 1996.

Vol. 1624: V. Brînzanescu, Holomorphic Vector Bundles over Compact Complex Surfaces. X, 170 pages. 1996.

Vol.1625: S. Lang, Topics in Cohomology of Groups. VII, 226 pages. 1996.